D0153924

Nonlinear waves,
solitons and chaos

Nonlinear waves, solitons and chaos

E. INFELD

Institute for Nuclear Studies, Warsaw

G. ROWLANDS

Department of Physics, University of Warwick

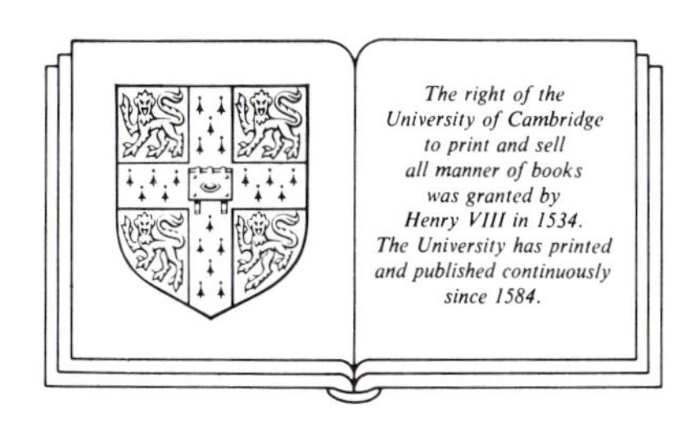

CAMBRIDGE UNIVERSITY PRESS

Cambridge

New York   Port Chester

Melbourne   Sydney

Published by the Press Syndicate of the University of Cambridge
The Pitt Building, Trumpington Street, Cambridge CB2 1RP
40 West 20th Street, New York NY 10011, USA
10 Stamford Road, Oakleigh, Melbourne 3166, Australia

First published 1990

British Library cataloguing in publication data

Infeld, Eryk
Nonlinear waves, solitons and chaos.
1. Nonlinear waves
I. Title   II. Rowlands, George
531′.1133

Library of Congress cataloguing in publication data

Infeld, E. (Eryk)
Nonlinear waves, solitons, and chaos/Eryk Infeld, George Rowlands.
p. cm.
Bibliography: p.
Includes indexes.
ISBN 0-521-32111-5. ISBN 0-521-37937-7 (pbk.)
1. Solitons. 2. Chaotic behavior in systems. 3. Nonlinear waves.
I. Rowlands, George. II. Title.
QC174.26.W28I55 1990
530.1′4–dc20 89-34313 CIP

ISBN 0 521 32111 5 hard covers
ISBN 0 521 37937 7 paperback

VN

Contents

Foreword

The last few decades have seen three important developments in nonlinear classical physics, all of which extend across the board of physical disciplines. They have, however, received uneven coverage in the literature.

Perhaps the best known outburst of activity is associated with the soliton, and the most famous development here is the inverse scattering method which has been with us now for over twenty years. There are, however, several other, less known methods for treating solitons. Indeed these compact, single hump wave entities have been known to scientists for over a century and a half (it might be interesting to look through some old ships' log books!). Nevertheless, books on the subject tend to concentrate on the inverse scattering method.

The second much publicized development is a new understanding of some deterministic aspects of chaos as well as the various roads a physical system can take to reach a chaotic state. Established views are being revised and new concepts and indeed even universal constants are being found. These important new developments derive from a realization that complex chaotic behaviour can be described by simple equations. The field has now reached the stage where a summary of basic theory can be given, though applications to specific physical problems are largely in the research stage.

The third development is somewhat less well publicized. Over the last three decades or so, scientists working on fluid dynamics and plasma theory have developed a multitude of new methods to deal with nonlinear waves. Some of these people were aware of the shortcomings of our linear physics education even before the above-mentioned two developments brought them to the attention of the scientific community.

We believe that, although there are now a number of books on all three topics, the time has come to try to bring them together in one volume. Thus the present book documents the three important developments in classical physics jointly, and, when possible, points out the similarities of approach

The authors' research interest over the past twenty years has been in fluid dynamics and plasma theory and this is reflected in the book. However, the main aim is to cover a wider range of nonlinear wave phenomena than hitherto. A few examples of what is done are: treatment of both surface and volume wave phenomena, including recent results (e.g. instabilities and their pictorial representation, period doubling, wave dynamics in three dimensions, splitting of signals observed experimentally, the universal theory of wave envelope dynamics); new developments in soliton studies (e.g. many soliton experiments in rectangular, cylindrical and spherical devices and their theory); and a bringing together of theoretical and numerical results on various scenarios for reaching chaos. An example of what is not attempted is a coverage (or even mention) of the 100 or so instabilities found in plasmas and fluids. Instead we present the basic physics of a few of them, each representing a whole category in some sense. Thus, all in all, the ambition of the book could be summarised by the adage 'not many but much'. Some unsolved problems are indicated. References are extensive and exercises are given at the end of each chapter. Thus the more ambitious reader should be able to get into the field. On the other hand, little knowledge is assumed, thus also giving the general science graduate (or senior undergraduate), who would like to learn what these new developments are about, a chance to do so.

As one of us is based in Warsaw, an attempt has been made to do some justice to research performed in Poland and the Soviet Union.

The authors would like to thank Drs P. Frycz, P. Goldstein, T. Lenkowska-Czerwińska, and K. Murawski for critical reading of the manuscript, and Professors P.N. Butcher, J.P. Dougherty, A. Kuszell, E. Kuznetsov, R. Raczka, A.A. Skorupski, K.N. Stepanov and R. Zelazny for remarks on parts of the text. Additional thanks are due to Dr Frycz for preparing the material used in Sections 7.10 and A.1 and Figs. 9.2 and 9.3. We would also like to acknowledge a huge debt to Ms H. Gilder for typing the manuscript several times over. Finally, Cambridge University Press in the person of Dr S. Capelin has been both helpful and patient.

E. Infeld and G. Rowlands
Warwick

1

Introduction

1.1 Occurrence of nonlinear waves and instabilities in Nature

This book is concerned with the propagation of waves and instabilities both linear and nonlinear, but concentrating on the latter.

The main advances in this subject have quite naturally come from studies involving fluids and more recently plasmas. The latter primarily because of the possibility of 'cheap, unlimited' (and hopefully safe) power from thermonuclear reactions. Everyone is of course familiar with waves on water if only being aware of the many instances where they provide examples of natural beauty. It is not so obvious that very similar waves can exist in a plasma which, to a good approximation, is usually a very dilute assembly of ions and electrons. We shall see later in this Chapter that this is indeed so and fluids and plasmas have much in common. However, plasmas also show a much wider range of phenomena basically because they are two or more component and also can be made strongly anisotropic by the introduction of magnetic fields, something that is not possible for simple fluids. This richer variety of phenomena has also been a reason why plasmas have had more than their share of attention.

The above remarks notwithstanding, there are numerous media other than plasmas and fluids which can support waves and/or propagate instabilities. As we will see, some of these are more intriguing than others.

1.1.1 Nonlinear phenomena in our everyday experience

As most of us are aware, waves generated by the wind can propagate across a field of corn. In this case the microscopic model is that of the ears moving, due to the stalk bending, in an harmonic manner, and ear interacting with adjacent ear only when in contact (a hard sphere potential). On a macroscopic level the corn can be considered as a dense fluid and now with the moving air flow over it one has the classic situation of a Kelvin–Helmholtz instability (Chapter 4).

The wind drives the instability and the stalks of corn bend in an analogous manner to how water waves are formed on lakes by the wind. The nonlinear requirement is different, in that if the bending of the stalk is too great it will break and produce a permanent record of the wave. Kelvin–Helmholtz type instabilities occur in plasma physics and they have been controlled to a certain extent by introducing perpendicular periodicities. The intriguing question naturally arises if similar conditions would stabilise the motion of corn heads and stop the breaking of the stalks. By analogy this could be done by planting trees periodically spaced in a line (or lines) perpendicular to the usual direction of the wind. Another example where Nature leaves a permanent record of a surface wave instability (at least until diffusion processes slowly remove it) is the so-called herringbone cloud formation. Here the sky is broken up into alternate bands of cloud (high density of moisture regions) and apparently clear regions (low density regions). The moisture in this case is responsible for the permanent record, two layers of air moving relative to each other giving rise to the Kelvin–Helmholtz instability (Chapter 4) and subsequent nonlinear effects (Fig. 1.1).

Television coverage of soccer matches usually shows, incidentally, the swaying of the crowd. This is seen as a wave moving through the stadium. Unfortunately, in some circumstances this wave can build up and those near the barriers can get crushed. Lighthill and Whitham (1955) and Richards (1956) have considered the flow of cars (discrete objects) in a fluid context and explained phenomena such as the effects of traffic lights in terms of the propagation of waves and, in particular shocks.

The conclusion to be reached from these quite disparate examples is that they can be effectively studied in terms of the propagation of waves and instabilities, leading to nonlinear effects, in continuous media. If the natural wavelength is large compared to the underlying microscopic length, the above picture should be applicable.

1.1.2 Nonlinear phenomena in the laboratory

Waves in solids have received considerable attention both at the microscopic (atomic) and the macroscopic (continuous) level. The most interesting phenomenon concerning the propagation of disturbances in solids at a microscopic level is the effect of anisotropies and non-homogeneities in the media. Until recently, nonlinear effects in solids have received little attention as the energy needed to produce them is very large. However, with the advent of intense power sources such as lasers it has been possible to show that the flow of heat in solids is closely related to the flow of solitons. The basic relationships between soliton amplitude, width and

velocity have been verified in this context (Section 1.2 and, in some detail, Chapters 5 and 7).

At the microscopic level it is usually necessary to quantize the system and instead of talking about sound waves one talks about phonons. However, because of the lattice periodicity the phonons have a dispersive nature. Most interesting phenomena, such as thermal conductivity, depend on phonon–phonon interaction. Until recently, such interaction has been studied in what could be called weakly nonlinear theories. However, a major breakthrough was made when it was realized that a number of phenomena could best be explained by introducing solitons as elementary excitations. Then it was found that a statistical mechanics based on weakly interacting solitons and phonons gave better results than previous theories which attempted to consider phonon–phonon interaction outside the control of a weak interaction.

Many years ago, Fermi, Ulam and Pasta (1965), though the actual work was performed much earlier and published in a Los Alamos report in 1955, studied numerically the problem of the strong interaction of phonons. They found, somewhat perplexingly at that time, that the phonons did not come to thermal equilibrium, but rather they underwent nearly periodic variations.

Much later Zabusky and Kruskal (1965) showed that this was the correct behaviour and *could* be explained in terms of solitons, Fig. 1.2. (A name given to reflect their quantal nature.) Nowadays people realize that solitons are not *necessary* to explain this phenomenon (Thyagaraja (1983)) Chapter

Fig. 1.2. Evolution of an initially periodic profile, $\cos \pi x$, as given by the Korteweg–de Vries equation (1.2.8). The breaking time for the wave profile (when the third term is neglected) is t_B. From Zabusky and Kruskal (1965).

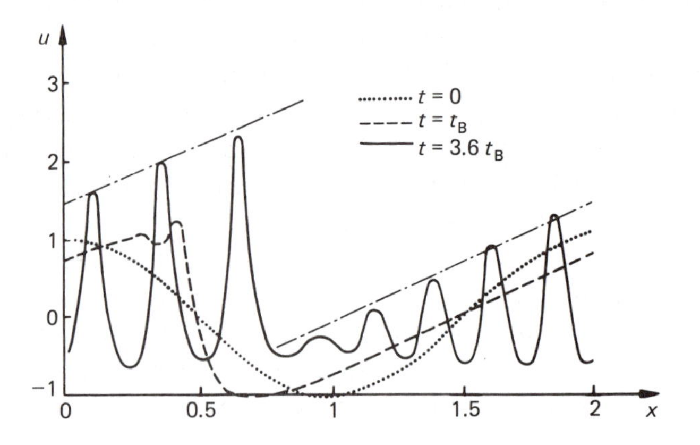

5. A theory that relies on the interaction of a small number of periodic modes also exists, see Infeld (1981c). Solitons are special wave pulses which interact with one another so as to keep their basic identity and so that they act as particles. Now the soliton has come of age in its mother subject, though in the meantime promising to be a useful concept in many other branches of physics, most of them being well out of the range of quantum effects. One of the first detailed experimental verifications of soliton type behaviour was in the study of nonlinear ion acoustic waves.

Davydov (1978), (1985) has applied some of the rules of solid state physics to the transport of energy down protein chains. He has shown that the idea of soliton propagation is essential to a study of chemical changes taking place in long protein molecules. This leads to the transfer of ATP (adenosine triphosphate) and could be the basis for an understanding of muscle contraction. Though much may still remain to be done in molecular biological problems it seems likely that the concept of a soliton will remain the cornerstone of any future theory.

The motion of electrons in solids is surprisingly well understood in terms of a simple Drude picture where the current J is linearly related to the electric field E by $J=(ne^2\tau/m)E$. Here n is the electron density, e its charge, and τ the mean free time. The effective mass 'm' takes into account the presence of the periodic nature of atomic structure. Normally all the quantities are constants and we simply have Ohm's law. However, it can happen, particularly in semiconductors, that if E is large enough the electron can be excited to a higher band which can have a different value of m. Thus the conductivity $\sigma(=ne^2\tau/m)$ is dependent on E and Ohm's law is no longer linear. If the mass is larger in the higher band, then an increase in E causes a decrease in J and thus we have a negative differential resistance. Obviously this is an instability mechanism and such a mechanism is observed in GaAs (gallium arsenide). The instability itself leads to the propagation of nonlinear stable pulses called Gunn domains, analogues of a soliton. These have been observed and are in fact the bases of many modern day oscillators. See Butcher (1967) for a review of Gunn domains; and Butcher and Rowlands (1968) for a study of the stability of the domain.

Another instability, leading to the propagation of nonlinear pulses that are also kinsmen of solitons and have been studied in the general area of solid state physics, is that associated with the acousto-electric effect. Here the instability mechanism arises because of a piezo-electric coupling between the propagation of sound waves and the flow of electrons. A nonlinear pulse can propagate down a crystal, for example of CdS or ZnO, but reach such an amplitude as to cause permanent distortion to the crystal. For a brief account of the general theory and for a discussion of the instability and nonlinear effects see Pawlik and Rowlands (1975).

It took a long time for it to be accepted that a homogeneous mixture of chemicals could lead to a periodic time variation in the concentration of a particular chemical or to an inhomogeneous spatial separation of the chemicals. Turing demonstrated that nonlinear chemical reactions together with diffusion could lead to a spatial separation of the chemicals and explored the idea in connection with the theory of morphogenesis, the formation of life. Zhabotinsky and later Zhabotinsky and Belousov found experimentally that a homogeneous mixture of certain chemicals could lead to a time periodic variation of the colour of the mixture. Later experiments showed that the same mechanism could lead to spatial colour patterns. All these phenomena can be explained in terms of nonlinear waves in time and space that have the inherent stability of a soliton.

1.2 Universal wave equations

It has now been realized that the study of many different types of waves in numerous media can often be based on a few universal nonlinear equations. These replace the usual linear wave equations, such as

$$\left(\frac{\partial^2}{\partial t^2} - \nabla^2\right)\Phi = 0.$$

1.2.1 The Korteweg–de Vries and Kadomtsev–Petviashvili equations and a first look at solitons

Most classical media propagate longitudinal, plane waves at or near a characteristic velocity c (acoustic-type waves). These waves are, for very small amplitude, given by

$$u = a\cos(\mathbf{k}\cdot\mathbf{x} - \omega t), \tag{1.2.1}$$

where ω is taken to be a function of $\mathbf{k}$, the form of which is dictated by the medium. When this medium is isotropic, we expect ω to depend on the modulus k only. All acoustic waves are such that, for small k,

$$\omega^2 = c^2k^2 + \ldots, \tag{1.2.2}$$

where c is the velocity of sound or another velocity specific to the medium. Thus long-wave acoustic modes propagate with little or no dispersion and the signal (group) velocity $\partial\omega/\partial\mathbf{k}$ is almost equal to the phase velocity c:

$$\frac{\partial\omega}{\partial\mathbf{k}} \simeq \omega k^{-2}\mathbf{k} = ck^{-1}\mathbf{k}. \tag{1.2.3}$$

However, dispersion will come in for k other than very small and it is perhaps more natural for the signal to lag behind the phase. This stipulation and symmetry suggest that the simplest possible correction to (1.2.2) is a negative quartic term in k:

$$\omega^2 = c^2k^2 - \beta^2k^4. \tag{1.2.4}$$

We have arrived at a very general small **k** dispersion relation. It covers all isotropic media that propagate accoustic modes such that the signal lags behind the phase. For the moment we will consider one space dimension, returning to general **k** later on. If we follow a wave propagating from left to right (admittedly thus losing some generality) we have

$$\omega = ck - (\beta^2/2c)k^3 + \ldots. \tag{1.2.5}$$

We can look at the wave behaviour in a coordinate system moving with velocity c and renormalize lengths so as to get rid of the $\beta^2/2c$ coefficient. Thus, in the new system,

$$\omega^* = -k^3, \tag{1.2.6}$$

corresponding to the following differential equation for u:

$$\frac{\partial u}{\partial t} + \frac{\partial^3 u}{\partial x^3} = 0. \tag{1.2.7}$$

This equation has several drawbacks. For one, it is not Galilean-invariant (this is partly due to our choice of coordinate system). It also leads to spreading of all finite extent initial profiles $u(x,0)$ (dispersion). However, if we replace the first term by the more general convective derivative, we remove both these shortcomings. Thus we suggest

$$\frac{\partial u}{\partial t} + u\frac{\partial u}{\partial t} + \frac{\partial^3 u}{\partial x^3} = 0 \tag{1.2.8}$$

as a more adequate equation. This is known as the Korteweg–de Vries (KdV) equation (1895) and will be derived rigorously in two physical contexts in Chapter 5 (water surface gravity waves and ion acoustic waves in a plasma). Some subsequent developments are reviewed by Miura (1976) and Miles (1981). However, it is already seen here to be the simplest possible *unidirectional* wave equation including dispersion and non-linearity, but not dissipation. All this allows us to hope for a stationary, pulse-like solution to exist if the nonlinearity, leading to wave steepening, can just counteract the dispersion (this wave steepening can easily be seen graphically by noticing that a solution to (1.2.8) without the third term is $u(x,t) = u(x - ut, 0)$, taking any pulse shaped $u(x,0)$, and drawing u for later times, Fig. 8.1.d; but the initial stages can be seen from the first two profiles of Fig. 1.2). Indeed,

$$u = 3\eta \operatorname{sech}^2 \tfrac{1}{2}\eta^{\frac{1}{2}}(x - \eta t), \tag{1.2.9}$$

where η is a constant, can be seen to be a solution just by inspection. This entity is called a soliton and evidently propagates at a uniform speed proportional to the amplitude. It transpires that many soliton solutions exist and n solitons with different amplitudes η_i will sooner or later line up with the tallest in the front and the shortest at the rear, rather as one might imagine a jungle queue! These solitons are very durable. See Fig. 1.2 again.

A generalization of (1.2.8) to two space dimensions is very simple on intuitive grounds if we assume a potential flow, as we indeed will do in much of this book. Thus

$$\mathbf{u} = \nabla \phi \tag{1.2.10}$$

and we wish to treat x and y dynamics on an unequal footing, just adding a y dependent term to (1.2.8), which has been seen to be a good model so far and thus worthy of some protection. Thus we go back to (1.2.4) in two dimensions, taking $c = 1$, $\beta^2 = 2$ for convenience,

$$\omega^2 = k_x^2 2 + k_y^2 2 - 2(k_x^2 2 + k_y^2 2)^2, \tag{1.2.11}$$

extract the positive root and assume k_x small but $k_y/k_x \ll 1$ (y variation weaker than x variation):

$$\omega = k_x - k_x^3 + \tfrac{1}{2}k_y^2 k_x^{-1} + \dots . \tag{1.2.12}$$

When multiplied through by k_x this suggests, as a generalization of (1.2.8); (1.2.10) and

$$\frac{\partial^2 \phi}{\partial x \partial t} + \frac{\partial \phi}{\partial x}\frac{\partial^2 \phi}{\partial x^2} + \frac{\partial^4 \phi}{\partial x^4} + \frac{1}{2}\frac{\partial^2 \phi}{\partial y^2} = 0 \tag{1.2.13}$$

known as the Kadomtsev–Petviashvili equation (1970) (KP). Soliton solutions to (1.2.13) are all of infinite extent in one direction and never propagate along the y axis. When they intersect at an angle they either just go through one another, or there is a region of interaction at the intersection known as a virtual soliton (Figs. 1.4(a), 7.1 and 7.5). At two critical angles the solitons merge to produce one. This phenomenon is known as soliton resonance. The solitons as defined here are stable (Chapter 8). They are by now familiar not only on water surfaces, but also in plasmas, see Tran (1979).

A variant of (1.2.13) with a minus sign in front of the last term also appears in classical physics (hydrodynamics, solid state, Chapters 5, 7 and

8). The same type of soliton is found, but soliton interaction is now simpler, there being no merging (Fig. 7.3). However, the solitons are unstable (Chapter 8).

We have seen how the equations governing wave behaviour in many classical media are the simplest possible undirectional, nonlinear, dispersive equations. To 'derive' them we neglected dissipative effects and, in the two dimensional case, assumed y variation to be weaker than x variation. Solitons emerged as the simplest stationary solutions to these equations.

Having established two equations that claim to describe wave motions in a reasonably complete manner, we now move on to an equation that concentrates on the wave amplitude. This is the ubiquitous nonlinear Schrödinger equation (NLS). We will in fact see in Chapter 5 how it can be derived for the amplitude of a wave satisfying the KdV or KP equation, among other possible provenances.

1.2.2 The Nonlinear Schrödinger equation

To study linear waves and instabilities in homogeneous media one expresses a disturbance $\Phi(x,t)$ in Fourier modes and for one of these modes writes

$$\Phi(x,t)=A\mathrm{e}^{\mathrm{i}(kx-\omega t)}, \tag{1.2.14}$$

where the amplitude A is a constant. The requirement that solutions of this form exist is found to be given in the form of the dispersion relation $D(\omega,k,\mu)=0$. A simple example of such a relation is given by (1.2.2). Here we have introduced the idea of a control parameter μ which 'controls' the system externally. For example, when one studies the propagation of waves in a fluid the pressure or density could be considered the control parameter. Usually the dispersion relation has more than one branch and we can choose a particular branch and then the condition $D(\omega,k,\mu)=0$ reduces to $\omega=\omega(k,\mu)$ where ω is now to be considered as a known single valued function. For weakly nonlinear systems one can imagine that the effect of the nonlinearity, apart from generating harmonics, is simply to affect the dispersion relation in a parametric manner. This we formally do by assuming that the dispersion relation is of the form $\omega=\omega(k,\mu,|\Phi|^2)$.

It is sometimes convenient to express the disturbance in terms of a fundamental linear frequency ω_o and wave number k_o and relegate the remaining variation into the amplitude factor:

$$\Phi(x,t)=a(x,t)\mathrm{e}^{\mathrm{i}(k_ox-\omega_ot)}. \tag{1.2.15}$$

For example ω_o, k_o may specify a marginally stable mode. (Throughout the book, lower case a will accompany ω_o and k_o when considering weakly nonlinear waves.)

Comparing (1.2.14) and (1.2.15) we obtain

$$a=a_o e^{i[(k-k_o)x-(\omega[k,\mu,|a|^2]-\omega_o)t]}. \tag{1.2.16}$$

For ks near k_o we may expand ω in a Taylor series and for weakly nonlinear situations

$$\begin{aligned}\omega(k,\mu,|a|^2)&=\omega(k_o,\mu_o,0)+\omega_k(k-k_o)\\&+\tfrac{1}{2}\omega_{kk}(k-k_o)^2+\omega_\mu(\mu-\mu_o)+\omega_{|a|^2}|a|^2+\ldots,\end{aligned} \tag{1.2.17}$$

where the suffix denotes partial differentiation with respect to the particular variable. We now choose ω_o such that $\omega_o=\omega(k_o,\mu_o,0)$ and then from the above we find from a naive reasoning analogous to that of Subsection 1.2.1:

$$i\left(\frac{\partial a}{\partial t}+vg\frac{\partial a}{\partial x}\right)+\frac{1}{2}\omega_{kk}\frac{\partial^2 a}{\partial x^2}-[(\mu-\mu_o)\omega_\mu+\omega_{|a|^2}|a|^2]a=0, \tag{1.2.18}$$

where we have identified ω_k with the group velocity v_g.

The term proportional to 'a' is identified in the above with a change of μ. This procedure is relevant for example when a change in μ causes the onset of instability and the subsequent build up of the amplitude (for a detailed analysis of such a problem see for example Pawlik and Rowlands (1975)). However, such a term can arise in a different context. If a wave is impressed on the system in some external manner such that the amplitude 'a' takes the constant value a_o, say, as $x\to\pm\infty$, then the nonlinear correction is more appropriately proportional to $|a|^2-|a_o|^2$. Then, even in the case of $\Delta\mu=0$, the above equation has a term linear in 'a', not obtainable from the above reasoning. Specific examples are discussed in detail in Chapter 5. The above (1.2.18) is the nonlinear Schrödinger equation (NLS). We see that it is an equation for the slow time and space modulation of the amplitude of a basic linear wave brought about by variations in the background medium, (that is variations in μ), and by weakly nonlinear effects. The above is of course not a derivation, but it serves to illustrate the universality of this equation so that one expects it to appear in most studies of the modulation of plane waves. Explicit and rigorous derivations are given in Chapter 5. It should be noted that V_g, ω_{kk} and ω are all quantities that can be obtained from linear theory (but not $\omega_{|a|^2}$).

In the above we chose to specify the mode in terms of the frequency, but we could also specify it in terms of a wave number and write $k=k(\omega,\mu,|a|^2)$ as the solution of the dispersion relation. Proceeding as above we again get a NLS equation, but in different variables such that ω and k are interchanged. This form of the equation could be used when discussing an instability that causes growth in space but such that time dependence is controlled.

Equation (1.2.18) has soliton solutions. This means that the *envelope* of a nonlinear wave can take the shape of a simple pulse. These solitons have some of the same properties as the Korteweg–de Vries solitons met socially in the previous subsection, such as clean collisions and overtaking of two solitons, stability in x, t space etc. These and other solutions will be investigated in Chapters 6, 7 and 8.

1.3 What is a plasma?

Many of the wave modes considered in this book propagate in a plasma medium. A plasma is often called the fourth state of matter, though unfortunately this label is sometimes used for different media, such as superfluid helium. However, as a new state of matter is usually obtained by heating a given state, and a plasma can be so obtained from a gas (undoubtedly the third state), a plasma's claim to this name would seem to be strongest. A plasma is a fully (or almost fully) ionized gas in quantities larger than one so-called Debye sphere. This sphere corresponds to the extent of an electric charge's influence on other charged particles. Outside its 'sphere of influence' an ion's potential field is screened by the surrounding cloud of charged particles. We will assume the plasma to be non-relativistic and classical (non-degenerate).

The above paragraph, laying down the law on what we will dignify with the name of a plasma in this book, can also be treated as the demarcation of the relevant region in, say, electron density, temperature parameter space (Fig. 1.3). Here electron, ion and atomic temperatures are the same, $T_i = T_e = T$, these values being given in degrees Kelvin, and n_e is the total number of electrons.

By rereading the first paragraph sentence by sentence and finding a specific condition from each one, we can gain a feeling for what is implied in terms of parameters. Actual values in Fig. 1.3 are taken for a deuterium plasma.

The left hand boundary of the plasma region derives from the condition that α, the ionization degree, exceed $\frac{1}{2}$, admittedly a rather arbitrary criterion. Values of n_e, the total number of electrons both free and bound, as a function of T for $\alpha = \frac{1}{2}$:

$$n_e(T, \alpha = \tfrac{1}{2}),$$

are found from Saha's formula. This formula describes the balance of various ionization and recombination processes in equilibrium. At large temperatures it gives $n_e \sim T^{3/2}$ for fixed α, and this asymptotic behaviour can be seen in Figure 1.3 at the top of the drawing (see Bates and Esterman (1970) for more theory).

We would like to be able to fit more than one Debye sphere into the body of the plasma, taken to cover one cubic cm, a reasonable volume for many current plasma experiments. Thus

$$\lambda_D < 1\,\text{cm} \tag{1.3.1}$$

$$\lambda_D = (K_B T / 4\pi n_e e^2)^{\frac{1}{2}} = V_T / (4\pi n_e m_e^{-1} e^2)^{\frac{1}{2}} \sim (T/n_e)^{\frac{1}{2}}$$

(for a derivation of the Debye length λ_D, see Ichimaru (1973)). We thus just draw the line $\lambda_D = 1$ cm in Fig. 1.3, obtaining unit slope ($n_e \sim T$).

For the plasma to be non-relativistic, we must have, say, $V_{Te} < 0.9c$. This condition, which is only temperature dependent, gives the cutoff on the right.

Finally, since the electrons are to be non-degenerate if the classical methods of this book are to model the plasma, we must have roughly $n_e < 10^{25}\,\text{cm}^{-3}$, one million times the number density of atoms in air at room temperature.

Fig. 1.3. Existence regions for a non-relativistic, classical deuterium plasma in parameter space. The region denoted by MHD is maximal for applicability of a magnetohydrodynamic model, by V for the Vlasov model. Some regions in parameter space are indicated: SC = Solar corona; SA = Solar atmosphere; MT = magnetic traps; TP = Thermonuclear plasma; S = Centre of the Sun; L = Laser produced plasma. Where electron temperature differs from ion temperature, the former is taken. A typical length scale is taken to be 1 cm, thus excluding dilute cosmic plasmas.

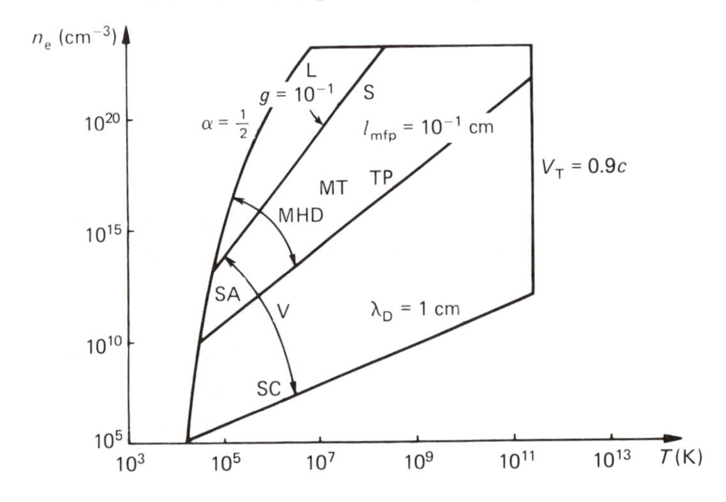

All in all, we have obtained the bread-loaf shaped region of Fig. 1.3. For other light atom plasmas of interest (hydrogen, deuterium–tritium, helium) diagrams would be very similar. Thus, our very first meeting with plasma physics, in the course of which we merely found its proper place in parameter space, nevertheless involved diverse fields of contemporary physics.

Several simplified models will be used to describe plasma waves and solitons in this book. One of them is the Vlasov model, (Sections 2.5 and 6.3) obtained from Liouville's equation when the number of electrons (and ions) in a Debye sphere is large (at least 50 in practical terms, see Montgomery and Tidman (1964)). Thus, for the Vlasov model to apply,

$$g = n_e^{-1}\lambda_D^{-3} < 10^{-1} \tag{1.3.2}$$

(g is known as the plasma parameter and we have taken $4\pi/3 \simeq 5$ in this rough calculation). We obtain the upper boundary of the Vlasov region from (1.3.2). The cube power dependence for this line, $n \sim T^3$, is easily deduced from equations (1.3.1) and (1.3.2).

Another plasma model that will put in an appearance in this book is that of a magnetofluid, applicable when a strong magnetic field is present (the model is called magnetohydrodynamics or MHD). This model assumes, among several other restrictions, that our plasma is collison-dominated and this alleviates some of the complications introduced by the magnetic field. We will take this to mean that the mean free path, or distance between two collisions, is smaller than 10^{-1} times our agreed value of 1 cm. This is of course very rough, as the magnitude of the magnetic field will enter any detailed considerations. There will also be characteristic lengths other than the mean free path and the extension of the plasma as a whole, see Sutton and Sherman (1965). The electron mean free path is (again see Ichimaru (1973) for details)

$$\ell_{\text{mfp}} \sim n\lambda^4{}_D/\ln\Lambda \sim T^2/n\ln\Lambda$$

$$\ell_{\text{mfp}} < 10^{-1}\,\text{cm}, \; \Lambda = 12\pi/g \tag{1.3.3}$$

This gives the bound of the magnetohydrodynamic region shown in Fig. 1.3. The $n \sim T^2$ dependence is almost exact, as the Coulomb logarithm is of order 10 for almost all parameters involved (Montgomery and Tidman (1964), p. 31; Krall and Trivelpiece (1973), p. 306).

Fortunately for plasma physicists, Nature was kind enough to place thermonuclear plasmas, which are important as a possible future energy source, in the region of overlap of the two basic theoretical models (TP in Fig. 1.3).

Our considerations so far are too restricted to include hybrid models for which $T_e \neq T_i$, and which appear in Chapters 5, 8 and 9. Usually when this happens $T_e > T_i$.

The regions shown in Fig. 1.3 are just the *maximal* regions for applicability of the two general models considered. Nevertheless, the above, admittedly simplified analysis may give the reader unfamiliar with plasma physics an idea of the complexity of the field. A plasma is indeed a rather tricky state of matter, differing from water, say, in that, even when considering only one kind of plasma (deuterium) and classical, non-relativistic physics, a uniform approach to the medium in which waves propagate is no longer possible. This is perhaps not really surprising, as water can only exist in a tiny region of our n, T plot. The story of our physical universe is therefore the story of a plasma to a much greater extent than that of bodies of water or indeed any fluids!

1.4 Wave modes on a water surface

As opposed to a plasma, people rarely argue about what water is and do not speak of something being 'weak water' or 'almost water' (except perhaps for bathers in one of our more polluted rivers). As mentioned above, water can only exist in a tiny region of n, T parameter space. Nevertheless, wave phenomena can be very complicated (though indeed, fewer types of wave propagate as compared with a plasma).

Although many interesting water depth wave modes can lead to phenomena analogous to those described in this book, such as coupled solitons, we will concentrate here on surface waves with one brief exception, Subsection 6.2.2 (for in-depth solitons, see Liu (1984) and his references). This is just due to the fact that surface water waves furnish excellent illustrations of the kind of nonlinear phenomena we will be looking at. In most of this book, viscous and turbulent effects will be neglected, though Chapter 10 will be more ambitious. However, models introduced there will have to be simplified in other ways.

A water surface is difficult to keep track of precisely because it describes the co-existence of two media. In the simplest, though still very complicated form, these media are: inviscid, vorticity-free water, and . . . nothing. The atmosphere which covers all bodies of water encountered in our everyday experience, plays no role in some of the phenomena considered here. This is because we will not consider how our initial conditions were set up, and in the course of the motion we will bar strong winds, tornados etc. The physical ingredients we will retain are the inertial forces, a constant gravitational field, the water pressure and, sometimes, water surface tension and finite depth. Gone (with the wind) are viscosity, the heat generated by water molecules scraping the bottom, compressibility, water evaporation, and its opposite, the entrapping of air by water (foam), to name a few effects. Not all are of equal importance and they can be divided into two main categories; those of absolute significance (dynamic and pressure effects) and those that can play either a significant or insignificant role *depending on the flow* (the wind, compressibility, viscosity, surface tension, and even gravity in the sense that increasing g to $2g$ would not influence capillary waves significantly).

1.4.1 Mathematical theory

Having removed as much physics as we dare, we are left with just four conditions, all expressible as equations for the velocity and the water pressure (density variations have gone and $\rho = 1\ \mathrm{g/cm^3}$). These are:

I What is left of the equation of continuity when ρ = constant, i.e.

$$\nabla \cdot \mathbf{v} = 0 \tag{1.4.1}$$

where $\mathbf{v}$ is the fluid velocity at a point.

II The Newtonian force on a moving fluid element (hence the nonlinear convective term) derives from a pressure gradient and gravity:

$$\frac{\partial \mathbf{v}}{\partial t} + (\mathbf{v} \cdot \nabla)\mathbf{v} = -\frac{1}{\rho} \nabla p - g. \tag{1.4.2}$$

Viscosity, which we have neglected, would add a third force, $\eta \rho^{-1} \nabla^2 \mathbf{v}$, where η/ρ, known as kinematic viscosity is $10^{-2}\ \mathrm{cm^2/sec}$ for water at 20 °C. A model derived from (1.4.2) plus this viscous term, known as the Navier–Stokes equation, will be described in Chapter 10. This is the Lorenz model. The minus sign in front of the gravitational term assumes the vertical coordinate z to increase with height. Surface tension, neglected in (1.4.2) will be reinstated in some considerations (Chapter 5). Having simplified so much already we are certainly not in a mood to complicate life by introducing sinks or the mouths of underground rivers, so

III The normal component of the fluid velocity must satisfy

$$\mathbf{v} \cdot \mathbf{n} = 0 \tag{1.4.3}$$

at the bottom, assumed flat ($z = -h_0$). Thus we have innocuously passed on to surface conditions. The fourth condition is that water elements on the surface remain parts of the main body of water. As already mentioned, sprays, evaporation and the generation of foam are not for us. Thus, if the equation of the surface is

$$f(\mathbf{x}, t) = 0 \tag{1.4.4}$$

then

IV $$\frac{\partial f}{\partial t} + (\mathbf{v} \cdot \nabla) f = 0, \tag{1.4.5}$$

where $\mathbf{v}$ is the value at the surface.

The above equations, though free from quite a lot of the physics, are still difficult to solve. The reader may have noticed that the mysterious f function, describing the shape of the water surface at time t, was introduced *very* discretely. There was a reason for this. Not only do we not know f, but it is also just about the most important quantity we would like to know! Thus, all in all, equation (1.4.5) is a boundary condition to be satisfied on an unknown surface.

Considerable interest attaches to the irrotational case

$$\nabla \wedge \mathbf{v} = 0, \tag{1.4.6}$$

in which case we can introduce a velocity potential $\mathbf{v} = \nabla\phi$ and (1.4.1) reduces to

$$\nabla^2 \phi = 0 \qquad \eta > z > -h_o \tag{1.4.7}$$

We have written the equation of the surface (1.4.4) in the form

$$z = \eta(x, y, t),$$

also

$$\frac{\partial \phi}{\partial z} = 0 \qquad \text{on} \qquad z = -h_o \tag{1.4.8}$$

at the bottom as follows from (1.4.3). Two very nasty surface conditions arise from (1.4.2) integrated over z and (1.4.5):

$$\frac{\partial \phi}{\partial t} + \tfrac{1}{2}(\nabla\phi)^2 + g\eta = 0 \tag{1.4.9}$$

and

$$\eta_t + \phi_x \eta_x + \phi_y \eta_y - \phi_z = 0 \tag{1.4.10}$$

both for $z = \eta$.

All model equations introduced in Section 1.3 can be derived from (1.4.7)–(1.4.10), as we will see in Chapter 5. Inclusion of surface tension in the above scheme is relatively simple and this agent can take over from gravity as a restoring force for very short wave motions.

1.4.2 Comments

Very little will be said by way of apology for the ruin of so much physics presented here. However, experiments do show that wave and soliton dynamics as following from (1.4.1)–(1.4.5) or even (1.4.7)–(1.4.10) *can* be extremely good fits to what is observed in experimental pools, canals, rivers and on ocean surfaces (Fig. 1.4).

Single humped solitons such as are seen in Fig. 1.4(a) are obtainable from the above model. These oddities will make more frequent appearances on the pages of this book than in the everyday lives of most of its readers. It is true that theoretical physicists tend to get carried away by the tremendous progress recently made in treating solitons mathematically. However, it should be remembered that the compact solitons treated here are akin to more prolific wave phenomena, such as mentioned in Section 1.1 or, to take an example from some of our rivers, as bores (hydraulic jumps) propagating upstream at high tide. They progress with the speed of a trotting horse when the conditions are right (strong tide, gradual slope of the river bottom near the river mouth, proper shape of estuary). These at least *can* influence our everyday lives, occasionally even terminating them (Victor Hugo's daughter Leopoldine was drowned by the tidal bore on the Seine in 1843).

Finally, lest he or she get carried away by the success of the idealized

Fig. 1.4. Some wave phenomena that can be described by simple model equations: (a) Oblique interaction of two solitons in shallow water, observed off the coast of Oregon (Photograph by T. Toedtemeier).

theory presented here, we owe the reader at least one example of a flow that cannot even roughly be described by our model. This is the flow of a fluid around a solid cylinder. When the fluid is water at a fixed temperature (and hence constant viscosity) and the Mach number is small, the flow pattern is uniquely determined by the product of the upstream velocity and the cylindrical diameter (yes, if you shrink the cylinder you must *increase* the

Fig. 1.4(b). Disintegration of a train of Stokes waves. The upper photograph shows a regular pattern of plane waves. In the lower photograph, some 60 m (28 wavelengths) further along the tank, the same wavetrain has suffered distortion due to the Benjamin–Feir instability which will be treated in detail in Chapter 5. Photograph taken at the Ship Division of the National Physical Laboratory (Feltham, Greater London). From Benjamin (1967).

velocity to stay in the same regime). The dimensionless parameter used is the Reynolds number, $Re = V_{US} D_{CYL}/\eta$. Flow patterns for five values of Re are shown in Fig. 1.5.

Our theory as presented above (zero vorticity and only normal component of $\mathbf{v}$ vanishing on cylindrical surface) could only be used to describe the laminar region of the large Re flows. However, both the small Re, and the vortex dominated and, subsequently turbulent regions of the large Re flows are extremely important to theorist and experimentalist alike. All we can conclude is that the simplified theory of this Section is in general not suited to the steady flow of water around submerged, solid objects, a discipline that has developed largely in the Twentieth Century due to Ludwig Prandtl and his school. A possible reference for these flows is Walshaw and Jobson (1972). See also van Dyke (1982) for more photographs.

Figures 1.4 and 1.5 contrast two spectacular successes with an 80% failure of one and the same theoretical model. This duality is often observed in theoretical physics and will keep us company throughout this book.

1.5 Linear stability analysis and its limitations

In this book by 'instability' we will generally mean instability with respect to small, or linear, perturbations of the shape, say, of a wave. We will not worry about just how conclusive a linear stability analysis is. However, finite amplitude perturbations can lead to distinct effects. When and where is then the question.

Galdi and Straughan (1985) addressed this very question for Bénard convection of a fluid layer of thickness d heated from below and rotating around the vertical. *Without* the rotation what happens is well known. Below a critical value for the temperature gradient one has straight heat convection and no flow. When this critical temperature gradient is reached, however, the configuration destabilizes and then changes to a flow pattern (rolls or hexagonal cells: Fig. 1.6). Galdi and Straughan considered the stabilizing effect of a rotation of the apparatus around the vertical on this phenomenon. Their results for water are shown on Fig. 1.7, which compares unstable regions in which microscopic motion is observed as following from a linear perturbation analysis, from Rossby's experiments, and also from Galdi and Straughan's finite amplitude stability analysis (continuous line on Fig. 1.7 roughly represents both).

Here Ra, the Rayleigh number, is proportional to the square root of the temperature gradient and T, the Taylor number, to the velocity of rotation (to be more precise, $Ra^2 \sim (T_o - T_1)d^3$ and $T \sim \Omega d^2$). A linear perturbation analysis is seen to give good results for the critical value $Ra_c(T)$ for the onset of Bénard convection when $T \leqslant 5 \times 10^4$.

Fig. 1.5. Flow of water past a circular cylinder at Reynolds numbers $Re = 0.16$; 9.6; 26; 2×10^3; 10^4. Photographs (*a*), (*b*) and (*c*) by S. Taneda, *Rep. Res. Inst. Appl. Mech.*, Kyushu Univ. **4** 29–40 (1955); (*d*) by H. Werlé and M. Gallon, *Aeronaut. Astronaut* **34** 21–33 (1972); (*e*) by T. Corke and Hassan Nagib.

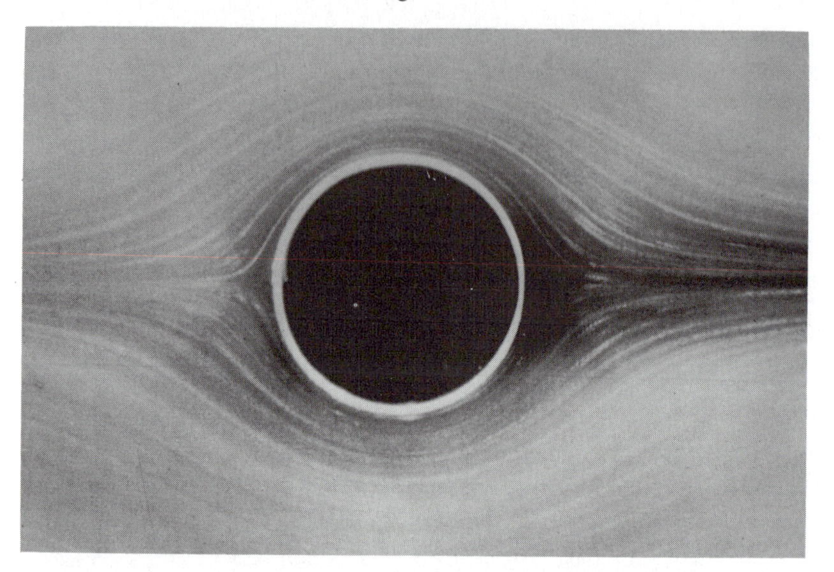

(*a*)

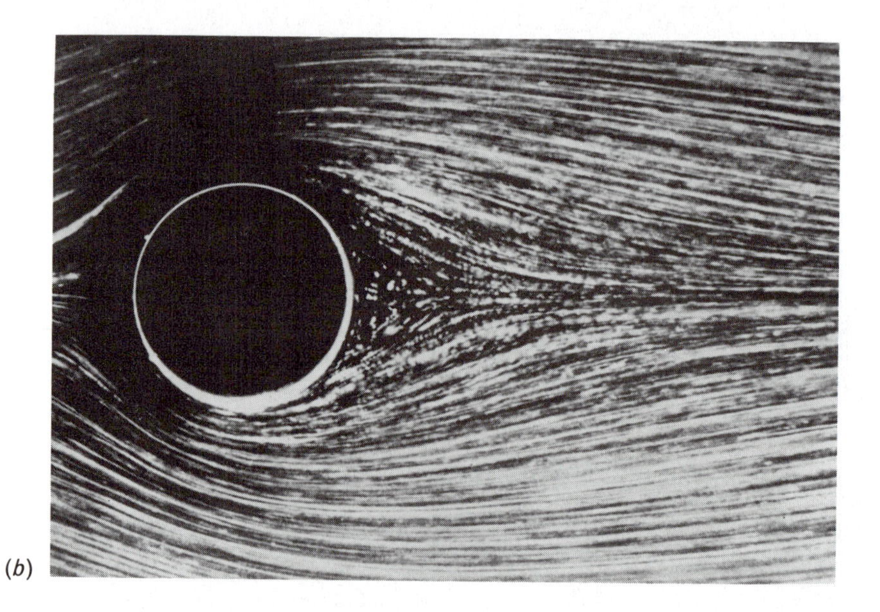

(*b*)

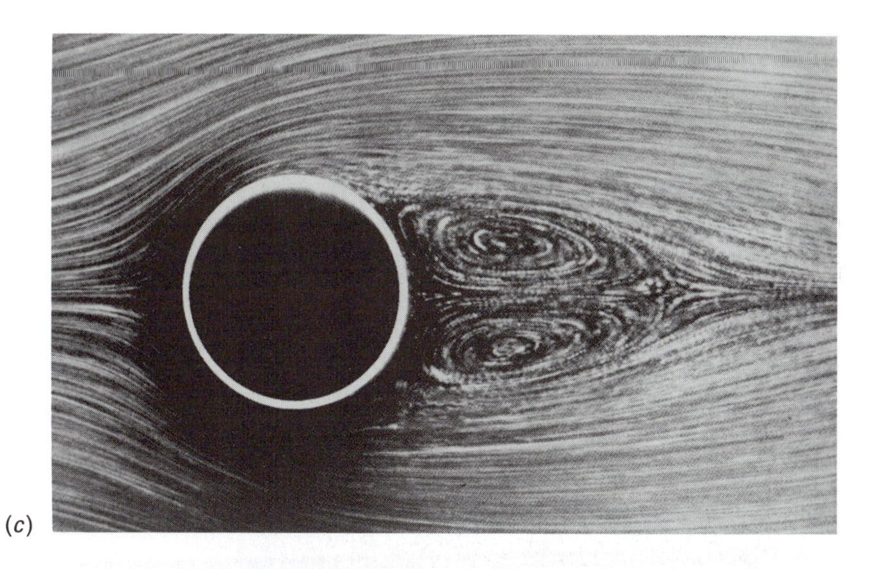

(c)

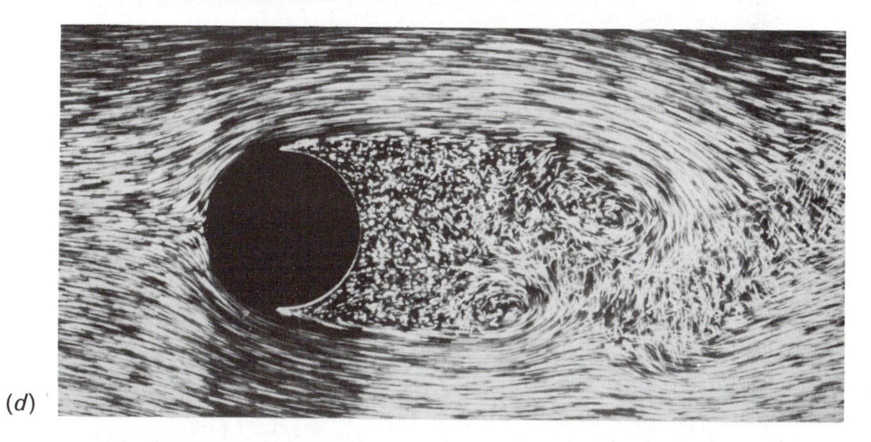

(d)

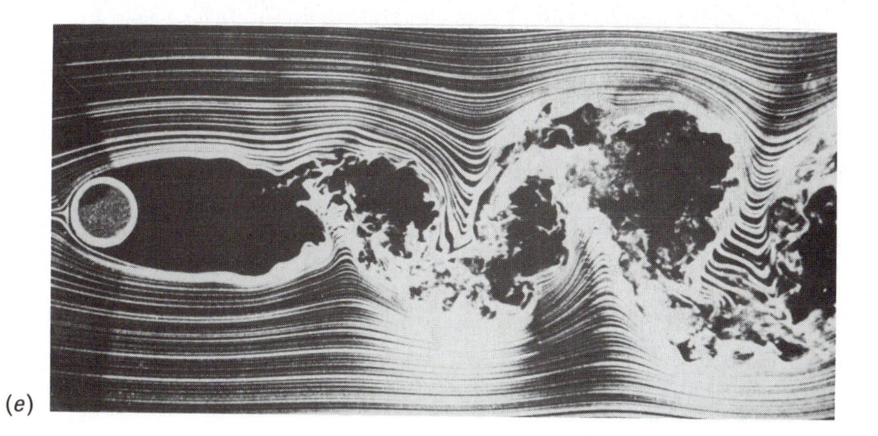

(e)

Fig. 1.6. Bénard convection cells: (a) Rolls at Rayleigh number 2.9 times critical value. (b) Hexagonal convection pattern. Photographs by E.L. Koschmieder, *Adv. Chem. Phys.* **26** 177–212 (1974).

(*a*)

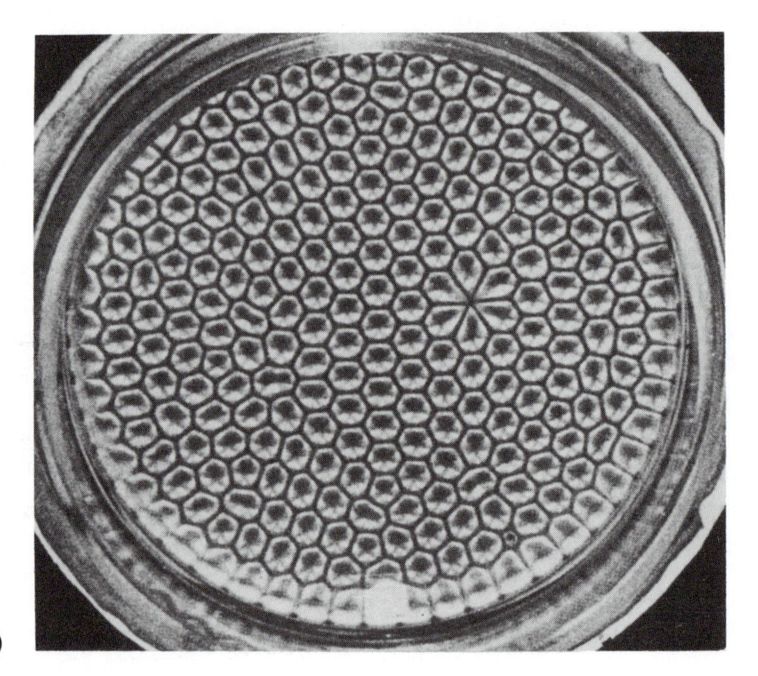

(*b*)

Fig. 1.7. Critical Rayleigh number against Taylor number for Bénard convection of a fluid rotating around the vertical. Dashed line represents results of a linear analysis, continuous line denotes both experiental results due to Rossby and theoretical analysis based on an arbitrary amplitude, nonlinear stability analysis. From Galdi and Straughan (1985).

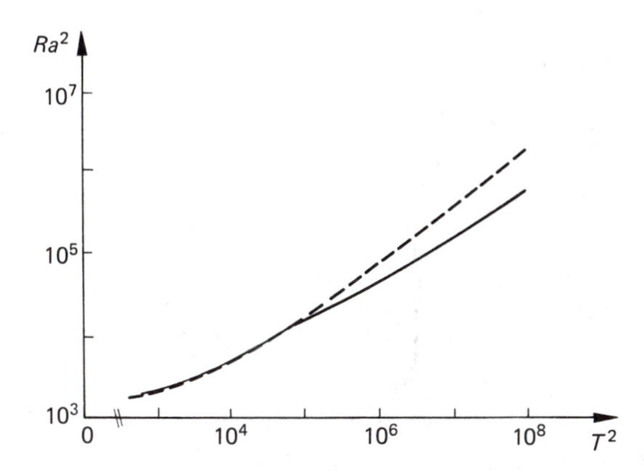

In general we should not be too surprised if, in a given physical situation, instabilities set in before our linear perturbation analysis predicts them. We should then ask ourselves how large the perturbations can be and whether their being of finite amplitude could be responsible for the discrepancy.

1.6 Chaos, turbulence and strange attractors

A substantial part of this book is concerned with a study of the coherent structures that are formed when dispersive effects are counter-balanced by nonlinear effects. The KdV equation is an expression of this balancing. Certain of these coherent states are stable and have been found experimentally. An example is furnished by ion accoustic solitons.

Somewhat similar coherent states can also be formed by the counter-balancing of an instability mechanism by nonlinear effects. A classical example is furnished by the beautiful rolls generated in a liquid container between two concentric cylinders which rotate relative to one another (Fig. 1.8). The rolls are first observed when ω is increased above a critical value ω_c (just as a critical temperature difference was needed for Bénard convection). Mathematically this behaviour can be understood in terms of what is called the Landau equation for the amplitude A of any disturbance

$$\frac{\partial A}{\partial t} = (\omega - \omega_c)A - A^3$$

For $\omega < \omega_c$, $A \to 0$ for large t and this corresponds to the uniform state. For

Fig. 1.8. Taylor vortices. Only inner cylinder rotates. Ratio of rotation speed to critical value ω/ω_c (ω_c for roll structure) is: (*a*) 1.6; (*b*) 8.5; (*c*) 1625. Photographs by E.L. Koschmieder, *J. Fluid Mech.* **93**, 515–27 (1979).

(*a*)

(*b*)

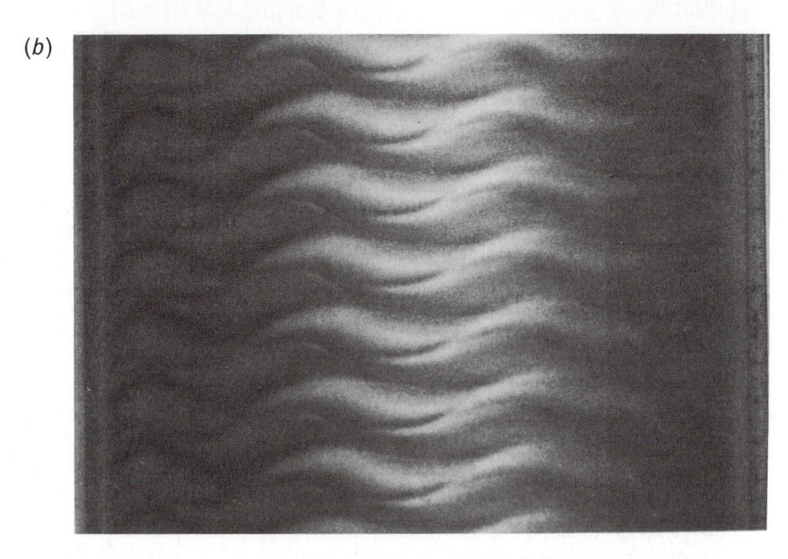

$\omega > \omega_c$, $A \rightarrow \pm(\omega - \omega_c)^{\frac{1}{2}}$ for large t and either of these solutions is to be associated with the simple roll structure. One says that as ω becomes greater than ω_c, the structure bifurcates to one of the two solutions $\pm(\omega - \omega_c)^{\frac{1}{2}}$. Though the equation for A is simple, the associated physical state, namely the roll structure, is spatially complicated. It is not a simple linear mode but still comprises a well defined disturbance and may be called a quasi-mode.

With further increase in the value of ω it is found experimentally that the roll structure is replaced by a more complicated but still coherent structure (a wavy structure superimposed on the rolls).

One can imagine this as simply a gradual change in the form of the quasi-mode and possibly the excitation of further quasi-modes.

Experimentally it is found that the increase of ω above another critical value leads to a structure which has random behaviour both in space and time. The system is now in a turbulent state. Until fairly recently it has been implicitly assumed that such a state can only be described by the presence of a large number of non-interacting quasi-modes, each with their basic structures, both temporal as well as spatial. This is essentially the Landau–Hopf theory of turbulence. However, it is now realized that if the quasi-modes interact strongly, even a small number (greater than two) of them can give rise to a pseudo-random behaviour. To all intents and purposes the behaviour is random but arises from a small set of coupled deterministic equations. Such a phenomenon is called CHAOTIC. In

(*c*)

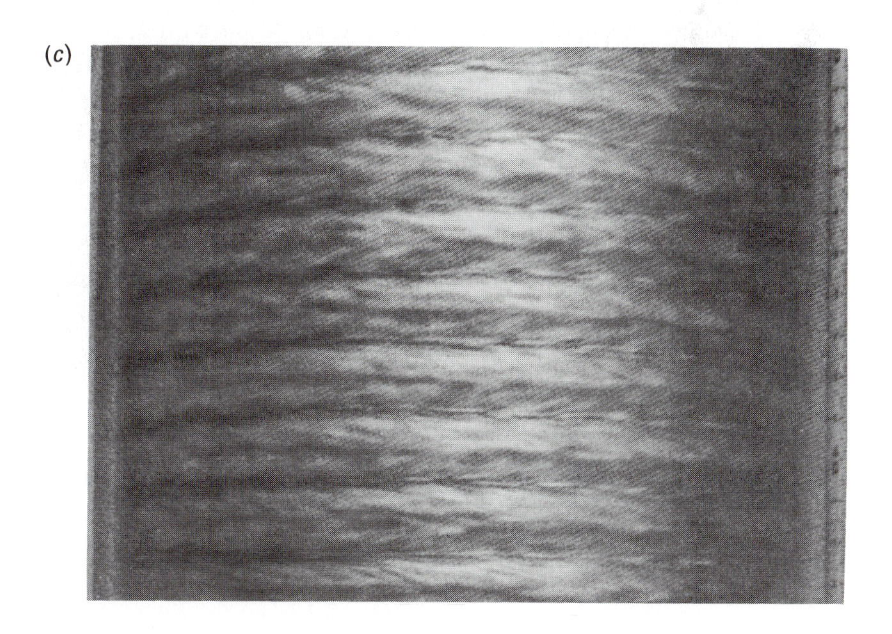

Chapter 10 various simple model sets of equations exhibiting chaos are described. One may consider a system which evolves to a simple equilibrium state to be describable by a simple attractor. By this one means that all initial states of the system evolve with time such that they are 'attracted' to the equilibrium state. A more complicated attractor is a limit cycle where a dynamical state evolves to a final periodic state. In some abstract phase plane a simple attractor is represented by a point and a limit cycle by a closed curve. Higher dimensional analogues such as toroidal surfaces, as the attractor, are easily envisaged. However, Ruelle and Takens (1971) pointed out that such generalizations are not the ones that usually occur, that is they are not generic. The attractors to be found in Nature have a complicated structure in a direction transverse to the toroidal surfaces. Such attractors are called STRANGE because of this complexity. They can in fact only exist in three or more dimensions. Simple examples are discussed in Chapter 10.

1.7 Contents of Chapters 2–10

Although this book concentrates on nonlinear wave phenomena, an in-depth treatment will only be possible if some linear wave theory is introduced first. Thus Chapters 2 and 3 deal with linear waves and instabilities, largely in fluids and plasmas.

Chapter 2 reviews the concepts of phase and group velocity, energy propagation and ways of representing wave behaviour in various media. Presentation is geared to future, nonlinear applications to be encountered later in the book. Thus the polar phase diagrams of ω/k introduced by Clemmow, Mullaly and Allis, but not so far widely known outside plasma physics, are reviewed (sure enough, they will be generalized to nonlinear waves in Chapter 8). Some characteristic instabilities found both in fluids and obtainable from a fluid model of a plasma (MHD) will be considered. Other plasma instabilities, treated here as illustrations of the importance of the kinetic theory, wind up the Chapter.

Chapter 3 reviews the theory of convective and non-convective (absolute) instabilities. This classification has important experimental implications. Group velocity is generalized to unstable media.

Chapter 4 on surface waves first treats these phenomena linearly and then opens the field of nonlinear waves and instabilities. In a way we are taking quite a plunge, as the theory of large amplitude water surface waves has progressed tremendously during the last twenty years. Instabilities on interfaces such as the Rayleigh–Taylor, Kelvin–Helmholtz and gravity wave instabilities are treated in one chapter. We also anticipate finite amplitude effects and wavelength doubling. All this is something of an experiment. Surface waves will reappear in later chapters.

Chapter 5 on weakly nonlinear theory and small amplitude waves and solitons covers two related, but definitely distinct areas. One is the rigorous derivation of the universal model equations, three of which were introduced via general and rather naive arguments in Section 1.2. Both standard coordinate stretching and a somewhat more original, Lagrangian, method are used. The second main theme is the behaviour of small amplitude, almost cosine waves as follows from a^2 theory. Many phenomena, characteristic of fully nonlinear waves, such as the splitting of characteristics and the sad fate of the concept of energy propagation can be quite effectively introduced through the back door of (small) a^2 theory. Long time behaviour and the Fermi–Ulam–Pasta recurrence are also considered.

Chapters 6 and 7 give a wide review of exact methods for finding both nonlinear wave and soliton structures. These include phase plane analysis, the Bernstein–Greene–Kruskal class of nonlinear plasma waves, the Lagrangian formulation for nonlinear plasma oscillations, and various methods for solitons. Among these the inverse scattering method will be covered briefly. However, readers primarily interested in this all too fashionable approach will be best advised to refer to other books (some of which are listed in Section 8.1). Some attention will be paid to the powerful Zakharov–Shabat operator dressing method for starting from a given equation known to be solvable and finding new equations and *their* solutions at one blow.

On the level of soliton kinetics obtainable from the above methods, we will see diverse examples of soliton behaviour, including soliton fusion, their weak and strong interaction etc. Chapter 8 treats the one and higher dimensional stability of one dimensional nonlinear waves and solitons. Several methods are described in some detail: those of Whitham, Hayes and the present authors are all covered. The results will help us assess the model equations introduced in earlier chapters.

Chapter 9 covers cylindrical and spherical solitons in plasmas and other media. This may be the first treatment in a more or less homogeneous book. There is more emphasis on numerics and experimental results than in other chapters, reflecting the developing state of the field.

Chapter 10 introduces non-coherent phenomena. There is some logic in this, as instabilities often help build up a nonlinear wave structure which in turn can go unstable, and these phenomena are the themes of much of this book. When a nonlinear structure does go unstable there are many possible fates it can meet with (such as the abovementioned Fermi–Ulam–Pasta recurrence) but the most probable class is that of turbulence. To date the only way to investigate this state relatively painlessly is to simplify the models so only the bare essentials are left. However, most people believe the salient features of turbulent behaviour can be studied by looking at these very simplified models. This Chapter was written for them (and of course, for the reader who approaches the subject with an open mind).

2

Linear waves and instabilities in infinite media

2.1 Introduction

It has been stressed in the last Chapter that waves can propagate in many distinct types of media. However, they can all be studied in the linear (vanishing amplitude) limit using the same mathematical methods. In this Chapter we consider the propagation of 'bulk' waves, that is waves in media such that surface effects may be neglected, whilst in Chapter 4 we consider the propagation of waves along surfaces, where interface effects are paramount. The basic techniques and analyses in the present Chapter are illustrated by considering relatively simple examples from plasma physics and fluid dynamics. However, it must be stressed that the techniques are universal and can be applied to waves and instabilities in a wide range of media.

2.2 Plasma waves

For the sake of simplicity, the plasma is treated by the simplest of models, namely an electron gas described in terms of a mass density ρ (identical to mn), a velocity u, an isothermal pressure p and a static uniform background of cold ions of density ρ_0. In one spatial dimension, mass conservation gives

$$\frac{\partial \rho}{\partial t}+\frac{\partial}{\partial x}(\rho u)=0, \tag{2.2.1}$$

whilst momentum conservation implies

$$\frac{\partial u}{\partial t}+u\frac{\partial u}{\partial x}=-\frac{1}{\rho}\frac{\partial p}{\partial x}+\frac{e}{m}E. \tag{2.2.2}$$

Here E is the electric field which must satisfy Poisson's equation

$$\frac{\partial E}{\partial x}=\frac{4\pi e}{m}(\rho-\rho_o). \tag{2.2.3}$$

To complete these equations an equation of state, relating the pressure and the density, must be specified. We take the simplest case and write $p=\alpha^2\rho$, α^2 constant, corresponding to an isothermal plasma medium.

The above constitute a set of coupled nonlinear equations which, even for this simplest of cases, are not in general tractable analytically. (The cold plasma model where $p=0$ is solvable, but this is the exception rather than the rule. For non-zero p only special solutions are known, see Section 6.4.) In common with the more general problem of waves in other media, the above system of equations has a simple static solution $\rho=\rho_o$, $p=p_o$ and both E and u zero. This state is usually called the equilibrium state (such equilibrium states do not necessarily correspond to *thermodynamic* equilibria. In fact many of the equilibrium states of most interest in this book are unstable and hence are not thermodynamic equilibria, though this one in fact is).

The mathematical analysis of sets of equations such as the above is immensely simplified if one restricts attention to solutions which are in some sense close to the stationary equilibrium state. To do this we linearize, that is write $\rho(x,t)=\rho_o+\delta\rho(x,t)$, $p=p_o+\delta p(x,t)$, $u(x,t)=\delta u(x,t)$, $E(x,t)=\delta E(x,t)$, substitute these expressions into the above equations and neglect all products such as $\delta\rho\delta u$.

Such a procedure gives rise to the set of linear equations:

$$\frac{\partial\delta\rho}{\partial t}+\rho_o\frac{\partial\delta u}{\partial x}=0, \tag{2.2.4}$$

$$\frac{\partial\delta u}{\partial t}=-\frac{1}{\rho_o}\frac{\partial}{\partial x}\delta p+\frac{e\delta E}{m}, \tag{2.2.5}$$

and

$$\frac{\partial\delta E}{\partial x}=\frac{4\pi e}{m}\delta\rho. \tag{2.2.6}$$

The inclusion of products such as $\delta\rho\delta u$ would lead to a *nonlinear* theory and various aspects of such theories are the subject matter of the last seven chapters of this book. The important simplifying property of these equations, besides being linear, is that the coefficients of the various terms are all constant. This suggests that one looks for solutions where all the perturbed values, $\delta\rho$, δp, δu and δE are proportional to $e^{(ikx-i\omega t}$. If this procedure is carried out the set of partial differential equations reduces to algebra:

$$\omega\overline{\delta\rho} - k\rho_o\overline{\delta u} = 0, \tag{2.2.7}$$

$$-\omega\overline{\delta u} = -\alpha^2 k\overline{\delta\rho}/\rho_o - \mathrm{i}e\overline{\delta E}, \tag{2.2.8}$$

$$\mathrm{i}k\overline{\delta E} = 4\pi e\overline{\delta\rho}/m, \tag{2.2.9}$$

where $\overline{\delta\rho}$, for example, is the constant amplitude of the density perturbation $(\delta\rho(x,t) = \overline{\delta\rho}\exp(\mathrm{i}kx - \mathrm{i}\omega t))$. These equations may be solved to give

$$(\omega^2 - \omega_p^2 - \alpha^2 k^2)\overline{\delta\rho} = 0, \tag{2.2.10}$$

where ω_p is the plasma frequency $(\omega_p^2 = 4\pi\rho_o e^2/m^2)$. Ignoring the trivial solution $\overline{\delta\rho} = 0$ (implying $\overline{\delta E} = 0$ and $\overline{\delta u} = 0$) we see that a solution to the equations only exists if the coefficient of $\overline{\delta\rho}$ is zero, that is

$$D(\omega, k) = \omega^2 - \omega_p^2 - \alpha^2 k^2 = 0. \tag{2.2.11}$$

This condition is called a dispersion relation and gives a condition on the allowed values of ω and k.

Generally one restricts the discussion to waves in infinite media, in which case x takes all values from $-\infty$ to $+\infty$, and if $\delta\rho(x,t)$ is to remain bounded in space, k must be real. Therefore, we take this to be the case. For each real k, the allowed values of ω are obtained by solving the dispersion relation $D(\omega, k) = 0$. Each solution is called a normal mode. Since the equations are linear we may construct a more general solution by simply adding the normal modes. Thus we write

$$\delta\rho(x,t) = \sum_n \int_{-\infty}^{+\infty} \mathrm{e}^{\mathrm{i}kx}\overline{\delta\rho}(k)\mathrm{e}^{-\mathrm{i}\omega_n(k)t}\mathrm{d}k$$

where the summation is over the roots of $D(\omega_n(k), k) = 0$. In the present case we simply have two roots $\pm\sqrt{(\omega_p^2 + \alpha^2 k^2)}$ which are evidently real for all k.

Implicit in the above is the idea that the normal modes form a complete set and so by integrating (and summing) over them one can get the most general solution of the set of partial differential equations. This is in fact usually, though not always the case. The notable exception, which will be discussed later, is connected with the phenomenon of Landau damping.

The proof of completeness is fraught with subtle difficulties. They can be avoided by considering the initial value problem. Here one assumes that the system is in its stationary state and some small perturbation is introduced (switched on) at $t = 0$. One then studies the subsequent time evolution of the system. To carry out this procedure one introduces the Laplace–Fourier transform of the various perturbed quantities. For example, one writes

$$\delta\hat{\rho}(k,p)=\frac{1}{2\pi}\int_{o}^{+\infty} \mathrm{e}^{-pt}\,\mathrm{d}t \int_{-\infty}^{+\infty} \mathrm{e}^{-ikx}\,\delta\rho(x,t)\mathrm{d}x, \tag{2.2.12}$$

and for the associated inverse transform

$$\delta\rho(x,t)=\int_{\gamma}\mathrm{e}^{pt}\mathrm{d}p \int_{-\infty}^{\infty} \mathrm{e}^{ikx}\delta\hat{\rho}(k,p)\mathrm{d}k. \tag{2.2.13}$$

The k integral is taken along the real axis as we demand that $\delta\rho(x,t)$ be bounded in all space. However, the p integral, as is always the case with Laplace transforms, is such that the real part of p is always greater than the real part of any pole of the integrand $\delta\rho(k,p)$. This condition has to be imposed to allow the time integral defining p to exist. The formal treatment of the time dependence is in terms of the Laplace transform as in (2.2.13) where it is seen that the time dependence is proportional to e^{pt}. However, later on when we consider a particular solution, it will prove more convenient to consider a time dependence of $\mathrm{e}^{-i\omega t}$. Thus throughout this Chapter we sometimes consider dispersion relations as functions of p and sometimes as functions of ω. The simple replacement $p\to -\mathrm{i}\omega$ connects these formulae. Unfortunately it is not always expedient to stick to one fixed convention!

If we multiply (2.2.4) by $\exp(-pt+ikx)$ and integrate over all allowed values of x and t, we obtain

$$\mathrm{i}p\delta\hat{\rho}+k\rho_o\delta\hat{u}=\delta\hat{\rho}(k,t=0) \tag{2.2.14}$$

where

$$\delta\hat{\rho}(k,t=0)=\frac{\mathrm{i}}{2\pi}\int_{-\infty}^{+\infty} \delta\rho(x,t=0)\mathrm{e}^{ikx}\mathrm{d}x.$$

The source term $\delta\rho(k,t=0)$ arises from a necessary integration by parts in obtaining the transform of $\partial\rho/\partial t$. Equation (2.2.14) should be compared to (2.2.7) with $p\to -\mathrm{i}\omega$. The only difference is the presence of the source term. The transform method now applied to the set of equations (2.2.4)–(2.2.6) gives rise to a set of algebraic equations of the form of (2.2.7)–(2.2.9) with ω replaced by $\mathrm{i}p$ together with source terms on the right hand side. These are readily solved and the solution (2.2.10) is replaced by

$$D(\mathrm{i}p,k)\delta\hat{\rho}(k,p)=k\rho_o\{\mathrm{i}p\delta\hat{\rho}(k,t=0)+\delta\hat{u}(k,t=0)\}=S(k), \tag{2.2.15}$$

where D is still defined by (2.2.11). Substitution of this expression for $\delta\rho$ into

(2.2.13) gives the complete solution

$$\delta\rho(x,t)=\int_{\gamma} e^{pt}\mathrm{d}p \int_{-\infty}^{+\infty} \frac{S(k)e^{ikx}\mathrm{d}k}{D(ip,k)}.$$

The p integration is carried out using the method of residues. There will be contributions at the poles, namely the zeros of $D(ip,k)$ so that

$$\delta\rho(x,t)=\sum_{n}\int_{-\infty}^{+\infty} 2\pi\frac{S(k)e^{ikx-i\omega_n(k)t}\mathrm{d}k}{D'(\omega_n,k)}, \qquad (2.2.16)$$

where $D'(\omega,k)=\partial D/\partial\omega$. This we see is exactly the same form as obtained above using the normal mode analysis so the transform method gives a justification to this approach.

In performing the p integration to obtain the above form for $\delta\rho(x,t)$, we implicitly continued the dispersion function $D(ip,k)$ analytically into the whole of the p plane. This is necessary when one uses the method of residues. However, it must be remembered that in obtaining (2.2.15) the real part of p has to be sufficiently positive for the time integral in (2.2.12) to exist. This is why the process of analytic continuation is necessary. For dispersion relations such as (2.2.11) this is simple but the procedure can lead to very unexpected results. This point will be resumed in Section 2.6.

Perturbed quantities such as $\delta\rho$ represent physical entities, and so must be real. Imposition of this condition leads to a constraint on $\omega(k)$. Taking the complex conjugate of $\delta\rho$ as given by (2.2.16) and replacing k by minus k shows that $\delta\rho$ is real if

$$\omega(k)=-\omega^*(-k)$$

where the asterisk denotes the complex conjugate. If ω is a real function of k then this implies that ω is an odd function of k as for ion acoustic waves (Section 1.2). This can also be achieved by ω^2 being a function of k^2.

In practice one usually discusses wave phenomena in terms of a single normal mode. That is to assume that all perturbed quantities are proportional to $\exp(ikx)$. This is equivalent to taking $S(k')=S_o\delta(k'-k)$ and for a single mode (2.2.16) reduces to

$$\delta\rho(x,t)=Ae^{ikx-i\omega(k)t}, \qquad (2.2.17)$$

where A, (the amplitude) is equal to $S_o/D'(\omega,(k))$ and $D(\omega(k),k)=0$. For a warm plasma we have $\omega^2(k)=\omega_p^2+\alpha^2k^2$.

In many situations, such as for sound waves in various simple media, gravity water waves, or simple ion acoustic waves in a plasma, the solution of the dispersion relation for very long waves is of the form $\omega=ck$ where c is constant and referred to as the 'velocity of sound'. Such waves are said to be non-dispersive, that is an initial disturbance of arbitrary shape propagates without distortion at velocity c. This is readily seen from (2.2.16) by considering just one term in the summation and taking $\omega(k)=ck$ for it. Then $\delta\rho(x,t)$ will simply be a function of $x-ct$ only, corresponding to the uniform translation of the initial disturbance with velocity c. In general, however, $\omega(k)/k$ is not constant and there is no unique velocity of translation. Such waves are called dispersive. Warm waves are of course dispersive ($\omega/k=\sqrt{(\alpha^2+\omega_p^2/k^2)}$). For such waves, each k value has a different phase velocity, still defined by $v_{ph}(k)=\omega(k)/k$. The major effect of a pulse propagating in a medium, where v_{ph} is a function of k, is that any initial pulse changes shape as it propagates and in general 'disperses' so that a compact initial pulse eventually gives rise to a general background disturbance. The essential nature of this can be seen by taking $2\pi S(k)/D'(\omega,k)=S_o$, a constant, and $\omega(k)=\alpha k+\beta k^3$ in (2.2.16) as is the case for gravity waves on a water surface or ion acoustic plasma waves (Sections 1.2 and 5.2). Then for this particular dispersion relation

$$\delta\rho(x,t)=S_o\int_{-\infty}^{+\infty} e^{ikx-i\beta k^3t-i\alpha kt}dk \qquad (2.2.18)$$

For $t=0$, $\delta\rho(x,0)=(S_o/2\pi)\delta(x)$, that is a Dirac delta function type of initial disturbance. The integral can be expressed in terms of an Airy function of the first kind (Abramowitz and Stegun (1965) – Chapter 10) to give

$$\delta\rho(x,t)=\frac{2\pi S_o}{(3\beta t)^{1/3}}\,Ai[(\alpha t-x)/(3\beta t)^{1/3}]. \qquad (2.2.19)$$

This is seen to correspond to a disturbance propagating with phase velocity α, a wavelength proportional to $(\beta t)^{1/3}$, and amplitude decreasing as $t^{-1/3}$. The limit of no dispersion ($\beta=0$) gives $\delta\rho(x,t)=S_o\delta(x-\alpha t)/2\pi$, that is a uniformly propagating compact pulse.

A question that naturally arises concerning dispersive waves is 'what happens to the energy?' To answer this question we consider the propagation of many modes, all with wave numbers centred about a value k_o, so that $S_o(k)$ is now peaked about this value. This is the antithesis of the previous example, where the source was peaked in x space, and hence flat in k space. In this context we will see that the more peaked our present pulse is in k space the less so it is in x space. In this case one may write $\omega(k)\simeq\omega(k_o)+\omega'(k_o)(k-k_o)$, with $\omega'(k)=d\omega/dk$ and so

$$\delta\rho(x,t)=\int_{-\infty}^{+\infty} S_o(k')e^{i(k'-k_o)x-i\omega'(k_o)(k'-k_o)t}dk'e^{ik_ox-i\omega(k_o)t} \qquad (2.2.20)$$

Taking $S_o(k)=\bar{S}\exp(-\beta(k-k_o)^2)$ gives

$$\delta\rho(x,t)=\bar{S}e^{ik_ox-i\omega(k_o)t}\left(\frac{\pi}{\beta}\right)^{\frac{1}{2}}e^{-(x-\omega'(k_o)t)^2/4\beta}. \tag{2.2.21}$$

This solution is of the form of a single mode where amplitude changes with space and time. In particular $|\delta\rho|^2$ is simply proportional to exp $(-(x-\omega' t)^2/2\beta)$ and since the energy is proportional to $|\delta\rho|^2$ we see that the energy propagates with velocity $\omega'(k_o)$. This is called the group velocity v_g, so by definition $v_g=d\omega/dk$. In a non-dispersive medium we have $v_g=v_{ph}$.

One might wonder why our wave envelope does not disperse. The reader will have no trouble in adding the next term, $\omega_o''(k-k_o)^2/2$ to the Taylor expansion of ω and finding an improved value for $\delta\rho$. For short times (2.2.21) is recovered, but is seen to disperse for longer times.

Thus, in conclusion, we see that in a general medium the propagation of any disturbance is governed by the dispersion function $D(\omega,k)$, $D=0$ relating the frequency ω to the wave number k. Associated with each mode is a phase velocity $v_{ph}=\omega/k$ which is the characteristic velocity for that mode, that is, an observer moving with that velocity sees the mode as a static disturbance. However, in a dispersive medium, the group velocity v_g is the more important velocity and this is the velocity of transfer of energy through the medium. We will see later on that inclusion of finite amplitude wave effects, ignored here, can complicate the concept of wave energy and its propagation. In particular, although we will be able to generalize the concept of group velocity, this will no longer be the velocity of energy propagation.

In the rest of this Chapter and in fact in most of the book, except Chapter 3, we shall study linear wave propagation in terms of normal modes. However, in Chapter 3 where we will discuss the extension of the concept of group velocity to unstable situations it will prove necessary to consider the complete response of the system as given by (2.2.16).

2.3 CMA diagrams

The dispersion relation for warm plasmas as given by (2.2.11) is simple and easily understood in terms of the k variation and the parameter variation (ω_p,α). However, if one considers a plasma as an electron gas in a magnetic field, not only is the parameter space enlarged to include ω_c, (the electron cyclotron frequency eB/m_ec), but the problem also becomes inherently three dimensional. The properties of the dispersion relation now become very complicated. A method of representation which immensely simplifies this study is to use a CMA diagram. These diagrams were first introduced by Clemmow, Mullaly and Allis, hence the CMA: see Clemmow and Mullaly (1955) and Stix (1962). More recently Infeld and Rowlands (1979a) have extended the concept to include nonlinear effects and this work is discussed fully in Chapter 8. Here, however, the basic idea behind the diagram representation will be introduced by considering a linear problem, namely that of a cold plasma in a constant external magnetic field.

The basic equations describing the plasma take the form

$$\frac{\partial \rho}{\partial t}+\nabla\cdot(\rho\mathbf{u})=0, \tag{2.3.1}$$

$$\frac{\partial u}{\partial t}+(\mathbf{u}\cdot\nabla)\mathbf{u}=\frac{e}{m}\left(\mathbf{E}+\frac{\mathbf{u}\wedge\mathbf{B}}{c}\right). \tag{2.3.2}$$

and the full set of Maxwell equations,

$$\nabla\wedge\mathbf{B}=\frac{1}{c}\frac{\partial\mathbf{E}}{\partial t}+4\pi\rho e\mathbf{u}/mc \tag{2.3.3}$$

$$\nabla\wedge\mathbf{E}=-\frac{1}{c}\frac{\partial\mathbf{B}}{\partial t}, \tag{2.3.4}$$

$$\nabla\cdot\mathbf{E}=\frac{4\pi e}{m}(\rho-\rho_o), \tag{2.3.5}$$

and $\nabla\cdot\mathbf{B}=0$. We consider an equilibrium state defined by $\rho=\rho_o$, $\mathbf{u}=0$, $\mathbf{E}=0$ and $\mathbf{B}=(B_o,0,0)$. Here again we treat the 'one component' model in which only electron dynamics are considered. As before, ρ and $\mathbf{u}$ are the electron density and velocity, whereas ρ_o, the ion density, is considered constant. Linearized normal mode analysis leads, in analogy to equation (2.2.10), to a set of coupled equations, which we write in matrix form as

$$\begin{pmatrix} \varepsilon_{11}-k_z^2c^2/\omega^2, & \varepsilon_{12}, & k_xk_zc^2/\omega^2 \\ -\varepsilon_{12}, & \varepsilon_{11}-(k_x^2+k_z^2)c^2/\omega^2, & 0 \\ -k_xk_zc^2/\omega^2, & 0, & \varepsilon_{33}-k_x^2c^2/\omega^2 \end{pmatrix}\begin{pmatrix}\delta E_x\\ \delta E_y\\ \delta E_z\end{pmatrix}=0, \tag{2.3.6}$$

where without loss of generality we have taken $k_y=0$. The dielectric coefficients ε_{ij}, known as the components of the dielectric tensor, are defined by

$$\varepsilon_{11}=S=(R+L)/2, \varepsilon_{12}=-\mathrm{i}D=-\mathrm{i}(R-L)/2, \tag{2.3.7}$$

and $\varepsilon_{33}=P$, where

$$R=1-\frac{\omega_p^2}{\omega(\omega+\omega_c)}, L=1-\frac{\omega_p^2}{\omega(\omega-\omega_c)}, \tag{2.3.8}$$

and $P=1-\omega_p^2/\omega^2$. Here ω_c is the electron cyclotron frequency ($\omega_{c_e}=eB_o/m_ec$) but we drop the subscript e, as ω_{ci} does not appear here. For a more complete discussion of the derivation of the above set of equations, see for example Bernstein and Trehan (1960) where ion dynamics are included. The equations have a non-trivial solution when the determinant is zero and this leads after a little manipulation to a dispersion relation, the analogue of equation (2.2.11),

$$A\eta^4-B\eta^2+C=0, \tag{2.3.9}$$

with

$$\begin{aligned} A&=S\sin^2\psi+P\cos^2\psi,\\ B&=RL\sin^2\psi+PS(1+\cos^2\psi),\\ C&=PRL \end{aligned} \tag{2.3.10}$$

with $k_x=k\cos\psi$, $k_z=k\sin\psi$ and $\eta=kc/\omega$. The nature of the problem can now be appreciated. One not only has to solve for η as a function of ψ, but also as a function of the two parameters ω_p and ω_c.

Waves can only propagate in the plasma if η is real, that is η^2 must be real and positive. The plus and minus signs for real η correspond to waves propagating in opposite directions (to the left and to the right). If $\eta^2<0$ or complex, then for real ω one would get complex k. Such waves are either growing or evanescent. The growing waves are not physical as there is no energy available to feed them. For evanescent waves if one stimulates the plasma by an external source at a frequency ω at some point, the disturbance decreases in amplitude exponentially away from the source. This behaviour arises because the change in sign of η^2 corresponds to a branch point in the integral over k appearing in the expression for $\delta\rho$ such as (2.2.16) (see for example Derfler and Simonen (1966)). Thus we are only interested in $\eta^2>0$. From (2.3.9) we have

$$\eta^2=[B\pm\sqrt{(B^2-4AC)}]/2A,$$

and it is readily shown that $(B^2-4AC)>0$. For a particular plasma, that is ω_p and ω_c given, η^2 will be some function of the angle ψ, with at most two values. This function will depend critically on the values of ω_p and ω_c. The idea behind a CMA diagram is to partition parameter space, in this case ω_{pe}, ω_c space, into regions where the η^2, ψ variation is topologically the same. For example, if R, L and P are all positive then (2.3.9) has two real solutions for all ψ and so a polar plot of η^2 as a function of ψ will consist of just two distinct closed curves centred about the origin. A critical line in this space is $C=0$, for along this line $\eta^2=B/A$, and hence it separates regions of two solutions from those of one, or one solution from no solution, depending on the signs of A and B. The conditions $R=0$, $L=0$ or $P=0$ are called cut-off conditions since, as one crosses such lines, one loses a solution. This is because a solution η^2 changes sign. This corresponds physically to a propagating wave ($\eta^2>0$) changing to an evanescent or non-propagating one ($\eta^2<0$). The other critical condition is $A=0$, which leads to just one solution $\eta^2=C/B$. Unlike the conditions $R=0$, $L=0$ or $P=0$, which only involve the plasma parameter, this condition is also a function of ψ. However, A can only be zero if S and P are of opposite signs so that the parameter conditions $S=0$ and $S=\infty$ are significant. Such a condition is called a resonance condition and corresponds physically to $\eta^2\to\infty$ (zero phase velocity).

Fig. 2.1. A CMA diagram for a cold plasma in a uniform magnetic field. The little polar plots schematically represent normalized phase velocity ω/ck as it depends on the propagation angle. The non-existence of two solutions at any angle means that evanescent waves exist in the region of parameter space. Polar plots are not to scale.

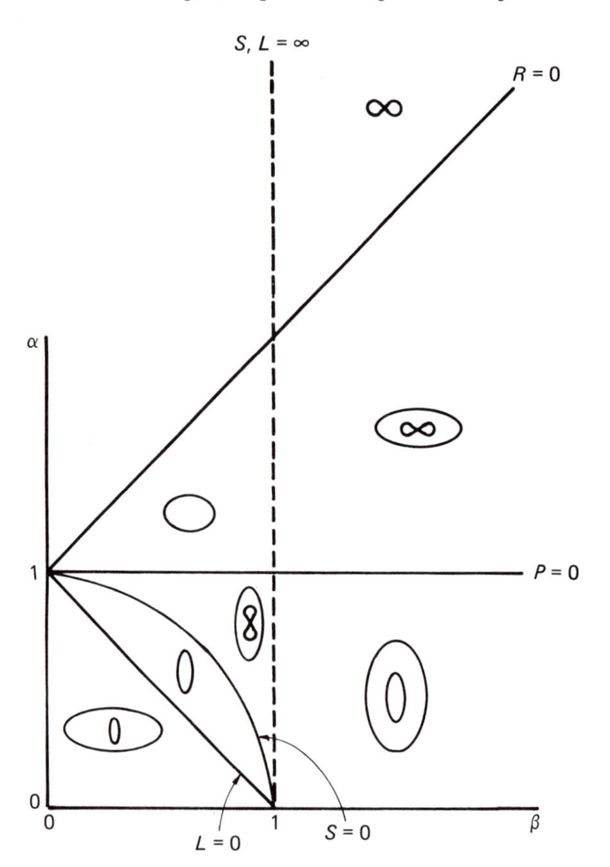

We are now in a position to construct a CMA diagram. We first consider a parameter space which in this case is an α, β space, where $\alpha = \omega_p^2/\omega^2$ and $\beta = \omega_c/\omega$. Then we divide this space into regions by using the conditions $P=0, R=0, L=0, S=0$ and $S=\infty$. This is illustrated in Figure 2.1. Now the next stage is to plot the polar diagrams of normalized phase velocity $1/\eta = \omega/kc$ against ψ in each of these regions. These polar plots will be topologically different in each region but in a particular region will retain the same general features over the whole region. Then one is able at a glance to appreciate the main qualitative features of wave propagation in each part of parameter space. This is also illustrated in Figure 2.1.

An alternative way of thinking about a CMA diagram is as representing a huge plasma pond in which the plasma density and external magnetic field (hence ω_p and ω_c) vary gradually. This variation is so slow that the plasma can locally be considered as homogeneous (ω_p and ω_c constant), such that the above theory applies. Changeovers to different topological regions then occur on the gradual scale. CMA diagrams have been used extensively to illustrate some of the main features of wave propagation in anisotropic media, particularly plasmas. For more detailed considerations see for example Stix (1962).

2.4 Instabilities

Though unstable waves exist in other media it was probably concern with energy production by nuclear fusion that led to the immense interest in instabilities, particularly of course, in plasmas where they are usually undesirable. Thus it seems inevitable that a study of instabilities should start with a study of the cold stream plasma instability, the prime example of many plasma instabilities.

In its simplest form we consider a plasma consisting of two cold electron plasma streams against a background of uniform positive charge created by stationary ions. The streams are characterized by a density and velocity which satisfy the equations (in one spatial dimension),

$$\frac{\partial \rho_i}{\partial t} + \frac{\partial}{\partial x}(\rho_i v_i) = 0, \tag{2.4.1}$$

and

$$\frac{\partial v_i}{\partial t} + v_i \frac{\partial v_i}{\partial x} = \frac{e}{m} E, \tag{2.4.2}$$

with the suffix i taking the values 1 and 2, thus representing the two counterstreaming electron beams. Poisson's equation

$$\frac{\partial E}{\partial x} = \frac{4\pi e}{m}(\rho_1 + \rho_2 - \rho_o) \tag{2.4.3}$$

completes the set of equations.

The equilibrium is chosen such that $\rho_1 = \rho_2 = \rho_o/2$, that is constant density and $v_1 = -v_2 = v_o$, where v_o is a constant velocity. One then again proceeds to linearize the above equations about this equilibrium and take all perturbed quantities to vary proportionally to $e^{ikx - i\omega t}$. Then, in complete analogy with previous wave propagation analyses, we obtain a dispersion relation which takes the form

$$D(\omega, k) = 1 - \omega_p^2 \left[\frac{1}{(\omega - kv_o)^2} + \frac{1}{(\omega + kv_o)^2} \right] = 0 \tag{2.4.4}$$

with $\omega_p^2 = 2\pi\rho_o e^2/m^2$. We may rewrite this as a quadratic equation in ω^2 which is readily solved to give

$$\omega^2 = \omega_p^2 + k^2 v_o \pm \sqrt{\left(\omega_p^4 + 4\omega_p^2 k^2 v_o^2\right)},$$

from which it is seen that one solution is negative for all k^2 such that

$$2\omega_p^2/v_o^2 > k^2 > 0. \tag{2.4.5}$$

If this condition is satisfied, one solution exists with $\omega^2 = -\gamma^2, \gamma$ real so that the perturbation changes with time proportionally to $e^{\gamma t}$. Such situations are said to be unstable, more correctly linearly unstable. The existence of an upper limit on k^2 for instability is characteristic of a wide range of problems.

At this stage, the study of instabilities and wave propagation are mathematically the same. A linearization of the equations is carried out about some assumed equilibrium and the condition that the linearized equations have a non-trivial solution leads to a dispersion relation. Because one restricts attention to bulk behaviour (infinite media) the wave number k must be real. Then the nature of the perturbation depends on ω. In Section 2.2 we discussed situations where ω was real, in which case the disturbance took the form of a constant amplitude wave. Above we have found solutions where $\omega = i\gamma$, γ real which for $\gamma > 0$ corresponds to an instability whilst $\gamma < 0$ corresponds to the disturbance being damped. However, all these examples were very simple. In general, the solution of a dispersion relation will be of the form $\omega = \omega_R + i\gamma$ with ω_R and γ both non-zero real quantities. Then for $\gamma > 0$ we have the case of a wave propagating with phase velocity ω_R/k and an amplitude which increases exponentially with time ($e^{\gamma t}$). For $\gamma<0$ the wave eventually damps away. When the growing and damped roots appear in pairs and the instability represented by the growing mode is physically justified (a corresponding energy source exists), the damped mode can simply be discarded as insignificant.

For $\gamma<0$ the linearization procedure, basic to all of the above considerations, is justified as long as the initial disturbance is sufficiently small. However, if $\gamma>0$, no matter how small this disturbance is initially, a time will come when, due to the exponential increase in amplitude, the disturbance will be comparable to the equilibrium values. When this happens the entire linearization procedure is bound to break down. Thus the study of instabilities using linear theory is justifiable for short times ($\gamma t<1$). The larger the value of γ, the shorter is this time.

The prime importance of the above γ analysis is in distinguishing between stable and unstable situations. For $t>1/\gamma$, the nonlinear terms so far neglected have to be taken into account. This is the subject of much of this book, but before studying such effects we consider in a little more detail some of the consequences of applying linear theory to unstable situations.

As already mentioned briefly, instabilities can only arise when there is some 'available' energy that can be transferred from the medium to the wave. Thus for media that are in thermodynamic equilibrium there is no available energy and so one does not expect instabilities to manifest themselves. Unfortunately, it has not been possible, in general at least, to quantify the concept of available energy, and hence the study of instabilities usually follows the analysis given above leading to a dispersion relation. The nearest approach to the idea of available energy has been the study of entities called Lyapunov functionals. This has led to some results, mainly in hindsight. For a review of this work see McNamara and Rowlands (1964), and for a specific example Benjamin (1972).

In the two stream instability described above, the available energy is simply the kinetic energy of the stream in equilibrium. This is not a thermodynamic equilibrium, but rather a stationary solution of the equations. It contains more kinetic energy than a static plasma would. For this model it will be seen from condition (2.4.5) that in a bulk plasma, where all k values are allowed, the system is always unstable. This is not necessarily the case for general non-thermodynamic equilibria. This important fact may be illustrated by warming the cold streams. This is done by introducing a pressure term in the momentum equation (2.4.2), that is, adding to the right hand side a term $-\dfrac{\alpha^2}{\rho_i}\dfrac{\partial \rho_i}{\partial x}$, where α^2 is an effective temperature. The resulting dispersion relation then takes the form

$$D=1-\omega_{\mathrm{p}}^2\left[\frac{1}{(\omega-kv_{\mathrm{o}})^2-k^2\alpha^2}+\frac{1}{(\omega+kv_{\mathrm{o}})^2-k^2\alpha^2}\right]=0,$$

$$=1-\omega_{\mathrm{p}}^2\psi(\omega). \tag{2.4.6}$$

Note that if one takes $v_{\mathrm{o}}=0$ this dispersion relation reduces to (2.2.11). This equation leads to a biquadratic for ω; Exercise 1. More elaborate models of the stream temperature have been introduced and some of them still lead to simple biquadratics for ω; Exercise 2.

However, a quick method that may be used to study the conditions under which a plasma is stable without solving (2.4.6) is a graphical one and is implemented as follows. The quantity $\psi(\omega)$, defined by (2.4.6), is *sketched* as a function of real ω. There is singular behaviour at $\omega=\pm kv_{\mathrm{o}}\pm k\alpha$, $\psi(\omega)\rightarrow$o for $\omega\rightarrow\pm\infty$ and $\psi(\omega)$ has a minimum at $\omega=0$ equal to $2/k^2(v_{\mathrm{o}}^2-\alpha^2)$. The dispersion relation is satisfied when $\psi(\omega)=1/\omega_{\mathrm{p}}^2$, which is equivalent to the horizontal line of height $1/\omega_{\mathrm{p}}^2$ intersecting the curve $\psi(\omega)$. Two distinct situations have to be considered. If $v_{\mathrm{o}}^2>\alpha^2$ then $\psi(\omega)$ is as shown in Figure

Fig. 2.2. The function $\psi(\omega)$ sketched for real ω: (*a*) For $v_{\mathrm{o}}^2>\alpha^2$ which can lead to instability; (*b*) For $v_{\mathrm{o}}^2<\alpha^2$, stable plasma. From Clemmow and Dougherty (1969).

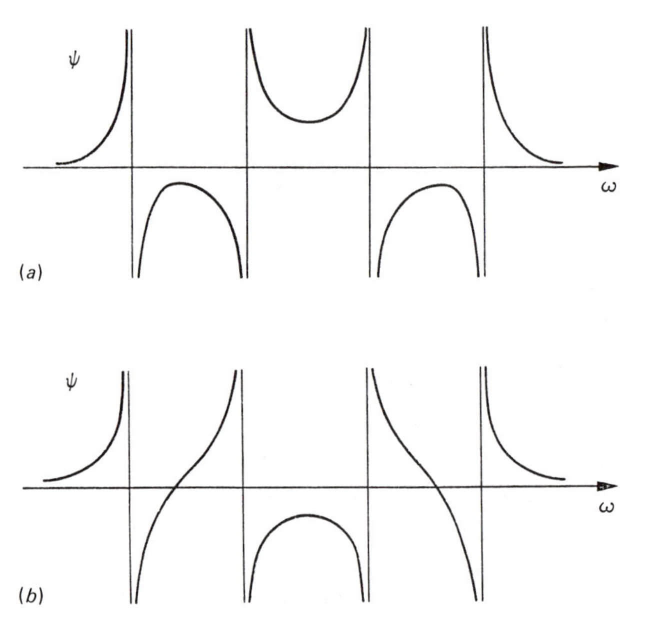

2.2(a) and if $2/k^2(v_o^2-\alpha^2)<1/\omega_p^2$, there are four intersections all corresponding to real values of ω, and, since $D(\omega,k)=0$ only has four solutions, all solutions must correspond to stable wave propagation. If $2/k^2(v_o^2-\alpha^2)>\omega_p^2$, there are only two intersections corresponding to two real values of ω. The other two must be complex conjugates, one of which will correspond to an unstable mode. Since all values of k are allowed we conclude that the plasma is unstable for $v_o^2>\alpha^2$. It is of course only necessary for one mode to be unstable for the plasma to be unstable as a whole. On the other hand, if $v_o^2<\alpha^2$ the variation of $\psi(\omega)$ with ω is as shown in Figure 2.2(b) and now we see that there are always four intersections and hence the dispersion relation has four real values of ω as solutions. The plasma is stable. Thus, unlike for the cold streams, we now have a stability criterion which depends on the properties of the plasma, namely $\alpha^2>v_o^2$ for stability. Physically, this is interpreted as saying that the effect of plasma pressure must be greater than the kinetic energy of the streams to promote stability (thermal energy greater than the kinetic energy of directed motion).

In summary, it can be seen that a study of whether a medium is unstable reduces to the mathematical problem of seeing if the dispersion relation $D(\omega,k)=0$ can have solutions with ω complex *and* with a positive imaginary part. For the two stream problem this could be studied graphically as shown above. Over the years, a number of different analytic techniques have been used to study the roots of $D(\omega,k)=0$. Perhaps the most sophisticated of these is the Nyquist complex plane diagram technique first used in electronic circuit theory, see Krall and Trivelpiece (1973). However, these techniques are now mainly of historical interest as the advent of high-speed computers has led to the introduction of fast algorithms for the numerical solution of such equations. For this reason the actual methods of solution of $D(\omega,k)=0$ will not be considered further here except to mention the analysis of Penrose (1960) who used the Nyquist diagram technique to great advantage in discussing hot plasmas, obtaining a useful stability criterion that will be given in Section 2.6.

Plasmas, in general, are not in thermodynamic equilibrium, and are thus prone to all manner of instabilities. Many have been catalogued by various authors (Rose and Clark (1961), Stix (1962), Montgomery and Tidman (1964), Clemmow and Dougherty (1969), Krall and Trivelpiece (1973), Cap (1976)) and reference should be made to these works for further details. In all cases, however, the basic technique is as expounded above, namely linearization, followed by a normal mode analysis leading to a dispersion relation whose solution reveals the conditions for stability.

2.5 The Vlasov equation

In the above we have described a plasma in terms of fluid-like equations relating macroscopic quantities such as density ρ and velocity v. A more detailed description of the plasma necessitates the introduction of a distribution function $f(\mathbf{x}, \mathbf{v}, t)$ where $f(\mathbf{x}, \mathbf{v}, t)\mathrm{d}\mathbf{x}\mathrm{d}\mathbf{v}$ is the probability of finding a particle in the 'volume' bounded by $\mathbf{x}$, $\mathbf{x}+\mathrm{d}\mathbf{x}$, and $\mathbf{v}$ and $\mathbf{v}+\mathrm{d}\mathbf{v}$. For a plasma where collective effects outweigh the effects of close collisions (remember, many particles in a Debye sphere), the function $f(\mathbf{x}, \mathbf{v}, t)$ satisfies the Vlasov equation. For the case where only electrostatic fields are important, and only electron dynamics are considered, the Vlasov equation is given by

$$\frac{\partial f}{\partial t}+\mathbf{v}\cdot\frac{\partial f}{\partial \mathbf{x}}+\frac{e}{m}\mathbf{E}\cdot\frac{\partial f}{\partial \mathbf{v}}=0 \tag{2.5.1}$$

and where $\mathbf{E}$ satisfies Poisson's equation which takes the form

$$\nabla\cdot\mathbf{E}=4\pi e\left[\int_{-\infty}^{+\infty} f(\mathbf{x}, \mathbf{v}, t)\mathrm{d}\mathbf{v}-\frac{\rho_0}{m}\right], \tag{2.5.2}$$

ρ_0 being the ion density considered constant in this model and f is for electrons. The above constitutes a closed set of equations. The justification of the Vlasov equation is not within the scope of this book but is discussed extensively elsewhere (see for example Montgomery and Tidman (1964), Ichimaru (1973)). However, regions of applicability of this equation were roughly outlined in Figure 1.3.

Physically, the Vlasov equation replaces the particle–particle Coulomb interaction by an averaged self-consistent electrostatic field which satisfies Poisson's equation, with the plasma represented by the charge density obtained by integrating $f(\mathbf{x}, \mathbf{v}, t)$ over all velocities. It is a good model equation when the plasma parameter g is small, see Section 1.3.

To study waves and instabilities in a Vlasov plasma we will now restrict our attention to one dimension. Consider an equilibrium independent of space and time, $f=f_0(v)$, with $E=0$. The only condition is that of charge neutrality,

$$\int_{-\infty}^{+\infty} f_0(v)\mathrm{d}v=\frac{\rho_0}{m}.$$

Now, unlike the case of thermodynamic equilibrium, $f_0(v)$ is not necessarily a Maxwellian distribution. (This is because a Maxwellian distribution is reached by collisional processes not included in the Vlasov model.) A simple linearization about this solution gives, on writing $f(x, v, t) = f_o(v) + \delta f(x, v, t)$,

$$\frac{\partial \delta f}{\partial t} + v\frac{\partial \delta f}{\partial x} = -\frac{e}{m}\delta E\frac{\mathrm{d}f_o}{\mathrm{d}v}$$

and

$$\frac{\partial \delta E}{\partial x} = 4\pi e \int_{-\infty}^{+\infty} \delta f(x, v, t)\mathrm{d}v.$$

This is best solved using a Fourier–Laplace transformation as applied in Section 2.2 to the simpler problems of a cold plasma. In this way one finds by analogy with (2.2.16) that

$$\delta E(x, t) = -\mathrm{i}\int_{\gamma} e^{pt}\mathrm{d}p \int_{-\infty}^{+\infty} \frac{S(p, k)e^{\mathrm{i}kx}\mathrm{d}x}{D(p, k)}, \tag{2.5.3}$$

where

$$S(p, k) = \int_{-\infty}^{+\infty} \frac{\delta f(x, v, t=0)\mathrm{d}v}{p + \mathrm{i}kv},$$

$$D(p, k) = k - \mathrm{i}\omega_\mathrm{p}^2 \int_{-\infty}^{\infty} \left(\frac{\frac{\mathrm{d}\hat{f}_o}{\mathrm{d}v}\mathrm{d}v}{p + \mathrm{i}kv}\right)$$

$$= k\left\{1 + \omega_\mathrm{p}^2 \int_{-\infty}^{+\infty} \frac{\hat{f}_o(v)\mathrm{d}v}{(p + \mathrm{i}kv)^2}\right\} \tag{2.5.4}$$

and we have written $f_o(v) = \rho_o \hat{f}_o(v)/m$. For long wavelength disturbance ($k \to 0$) it is tempting to expand the integrand in (2.5.4) in powers of k. If this is done the dispersion relation reduces to

$$1 + \frac{\omega_\mathrm{p}^2}{p^2}(1 - 3k^2\langle v^2\rangle/p^2 + \ldots) = 0$$

where $\langle v^2 \rangle = \int v^2 f_o(v)\mathrm{d}v$ and is a measure of the thermal spread or pressure. Then to lowest order one finds that

$$p^2 = -(\omega_\mathrm{p}^2 + 3k^2\langle v^2\rangle)$$

which is seen to be similar to (2.2.11), the result obtained for a warm plasma with $p = -\mathrm{i}\omega$. Thus in the long wavelength limit one can obtain from the Vlasov equation the results of the macroscopic theory. This, however, does not work the other way round and, as we shall see below, macroscopic treatment has excluded an important damping effect known as Landau damping. The two cold stream dispersion relation (2.4.4) is readily obtained from (2.5.4) by taking $\hat{f}_o=[\delta(v-v_o)+\delta(v+v_o)]/2$ and $p=-i\omega$.

For unstable plasmas the analysis is straightforward. One looks for solutions of $D(p,k)=0$ such that $\mathrm{Re}\, p>0$. In such a case the integral appearing in the definition of D exists and the whole procedure is consistent. A difficulty arises if one attempts to discuss solutions corresponding to purely oscillatory time dependence. One then expects a solution of the dispersion relation such that $p=-i\omega$ with ω real. However, if one makes this substitution into the integral in the definition of $D(p,k)$ then one sees in fact that the integrand becomes infinite for some $v(=\omega/k)$ and so is not really well defined. This difficulty was overcome by Landau who pointed out that in the expression for $\delta E(x,t)$, namely (2.5.3), $D(p,k)$ is defined for p's such that $\mathrm{Re}\, p>0$. (This is in virtue of the fact that the p integral is taken along the γ contour.) Thus, when one evaluates the p integral by the method of residues, it is necessary to analytically continue $D(p,k)$ into the complex p plane including the imaginary axis. The appropriate expression for $D(p,k)$ is the analytic continuation of the expression given by (2.5.4). A pleasing way of doing this, due to Landau, is to distance the contour of the v integration away from the real axis in such a manner as always to enclose the pole at $p=-ikv$. This new contour is called the Landau contour, and is such that the pole in the integrand at $p+ikv=0$ is always avoided. For unstable modes, $\mathrm{Re}\, p>0$, the introduction of the Landau contour makes no difference and the dispersion relation is as given by (2.5.4). For damped modes, $\mathrm{Re}\, p<0$, the distortion of the contour gives an extra contribution to the dispersion relation at the pole $p+ikv=0$. Thus for $\mathrm{Re}\, p<0$

$$D(p,k)=k-i\omega_p^2\left[\int\frac{\frac{d\hat{f}_o}{dv}dv}{p+ikv}+\frac{2\omega_p^2}{k}\frac{d\hat{f}_o}{dv}\right|_{v=ip/k}. \tag{2.5.5}$$

A most important consequence of this extra term in $D(p,k)$ is that plasma oscillations are damped and this is the Landau damping effect. This of course is distinct from the result obtained using macroscopic equations to describe the plasma as in Section 2.2. The detailed solution of (2.5.5) for this case, which depends on the form of f_o, will be discussed in Section 2.6.

For a more detailed treatment of the solution of the Vlasov equation, leading to the introduction of the Landau contour, see for example Clemmow and Dougherty (1969).

It is sufficient here to say that using the Laplace transform method to study the initial value problem leads inevitably to the necessity of an analytic continuation of $D(p,k)$ which can be accounted for by the introduction of the Landau contour. This procedure has always raised uneasiness, particularly among experimental physicists, and for years a consequence of this procedure, namely that plasma oscillations are damped (Landau damped) was not universally accepted. This was so until an experimental verification of Landau damping by Derfler and Simonen (1966) and Malmberg and Wharton (1966) finally convinced most plasma physicists. On the theoretical side, doubts were based on the fact that the Vlasov equation is time reversible, that is change of $t \to -t$ and $v \to -v$ does not change the equation. Thus, it was argued, one cannot have a damped mode without an unstable one. This paradox was finally removed when it was demonstrated that Landau damping is the result of the phase mixing of many modes and in fact the perturbed *distribution function* δf is not damped and retains all the information necessary for time reversal. It is this observation which lies behind the plasma echo experiments in which an apparently damped pulse reappears after a while. (Gould *et al.* (1967), Malmberg *et al.* (1968)). On the purely theoretical side, the phase mixing idea was first suggested by Bunemann and Dawson, see Dawson (1960). A more thorough treatment is given by Baldwin and Rowlands (1966), Rowlands (1969a) and Navet and Bertrand (1971). It should be stressed that the damping arises in a linear theory and has nothing to do with particles being trapped in the waves (which is an inherently nonlinear phenomenon and will be treated more fully in Section 6.3).

To illustrate Landau damping we consider a special case where the integral in (2.5.4) can be carried out analytically. We consider the 'resonance' distribution in velocity space

$$\hat{f}_o(v) = \frac{a/\pi}{a^2 + v^2}$$

which models a Maxwellian reasonably well for small v. Equation (2.5.4) then takes the form

$$D(p,k) = k\left\{1 + \frac{\omega_p^2 a}{\pi} \int_{-\infty}^{+\infty} \frac{dv}{(a^2+v^2)(p+ikv)^2}\right\}.$$

The integrand has poles at $v = \pm ia$ and $v = ip/k$. Now in (2.5.3), $D(p,k)$ is only needed for $\operatorname{Re} p > 0$, that is along the γ contour. If we deform the above contour in the lower half-plane we get a contribution only from the pole at $v = -ia$. Using the calculus of residues this then gives

$$D(p,k) = k\left\{1 + \frac{\omega_p^2}{(p+ka)^2}\right\}. \tag{2.5.6}$$

It is this expression which must be substituted into (2.5.3). The integral in (2.5.3) with respect to p can be performed using contour integration and completing the contour into the left half of the p complex plane. It is this step that needs the form of $D(p,k)$ in this part of the plane and necessitates the analytic continuation of $D(p,k)$. For the form of D as given by (2.5.6), analytic continuation is straightforward and D simply remains the same. Then the perturbed electric field will be proportional to e^{pt} where p is a solution of $D(p,k)=0$ which from (2.5.6) has solutions

$$p_{\pm}=-ka\pm i\omega_{p}.$$

These solutions correspond to *damped* plasma oscillations. For this case the analytic continuation is simple and quite acceptable. The replacement of (2.5.4) by (2.5.5) for $\mathrm{Re}\,p<0$ does not seem so simple, but of course is done using the same procedure.

In summary we can state that, with the one proviso that the dispersion relation has to be analytically continued, the normal mode analysis as applied to the Vlasov equation is the same as applied to the simpler plasma models discussed in Section 2.2.

It would seem that the above results could be obtained by an expansion in normal modes. For an unstable plasma this is allowed and leads to the same results as above. However, for a stable plasma, the set of normal modes is *not* complete and hence must be supplemented by terms of other types. Neglect of these extra terms in fact gives the solution as above with D defined for *all* p by (2.5.4). The extra terms over and above the normal mode terms are non-trivial and a complete analysis along these lines is much more complicated than using the Laplace transform to solve the problem (see Clemmow and Dougherty (1969), Section 8.5 on van-Kampen modes). Thus this alternative method would seem to be of little but historic interest now.

In the above we have illustrated the method of solution of the Vlasov equation by considering electrostatic effects in one dimensional plasmas. Of course real plasmas are infinitely more complicated. A major complication arises from anisotropic effects associated with the presence of equilibrium magnetic fields. Such effects arise since the behaviour of a charged particle, even in a uniform magnetic field, is very different in the direction of the field than in the plane perpendicular to the field. The inclusion of magnetic fields is easily incorporated into the Vlasov equation by extending the force term $e\mathbf{E}$ in (2.5.1) to the full Lorenz force $e(\mathbf{E}+\mathbf{v}\wedge\mathbf{B})$ where $\mathbf{B}$ is the magnetic field. The fields $\mathbf{E}$ and $\mathbf{B}$ satisfy Maxwell's equations with the charge density still given as in (2.5.2) and the current $\mathbf{J}$ by

$$\mathbf{J}(\mathbf{x},t)=e\int f(\mathbf{x},\mathbf{v},t)\mathbf{v}\,d\mathbf{v}$$

The simplest equilibrium situation is one where there is no spatial variation, $\mathbf{E}=0$ and $\mathbf{B}$ is a uniform field, B_0 say, in the x direction. The condition that the equilibrium state $f_o(\mathbf{v})$ satisfies the Vlasov equation is simply that $f_o(\mathbf{v})$ is any function of $v_\parallel$ and $v_\perp$, where the velocity has been written in cylindrical form $\mathbf{v}=(v_\parallel, v_\perp\cos\psi, v_\perp\sin\psi)$. The stipulation that no current flows (B_o is considered as an external field) places a weak condition on $f_o(v_\parallel, v_\perp)$.

The solution of the linearized Vlasov equation proceeds along the lines as outlined above and one finds that normal modes proportional to $e^{ik_\perp x + ik_\parallel z - i\omega t}$ exist provided $\omega, k_\perp$ satisfy a dispersion relation. The latter is of the form of a three by three determinant and is given in all its glory by Krall and Trivelpiece (1973). However; so as to illustrate some of the effects of anisotropy we consider the very special case of an electrostatic wave propagating perpendicular to B_o ($k_\parallel=0$ and magnetic perturbations neglected). The dispersion relation then reduces to

$$k_\perp^2 + 4\pi\omega_p^2\omega_c^2 \sum_{n=1}^{\infty} \frac{n^2}{(\omega^2 - n^2\omega_c^2)} \int J_n^2\left(\frac{k_\perp v_\perp}{\omega_c}\right) \frac{\partial f_o}{\partial v_\perp^2} \mathrm{d}\mathbf{v} = 0. \tag{2.5.7}$$

Here $J_n(x)$ is the Bessel function of the first kind. For a Maxwellian velocity distribution this may be written in the form

$$k_\perp^2 = \sum_{n=1}^{\infty} \frac{a_n}{\omega^2 - n^2\omega_c^2}, \tag{2.5.8}$$

where

$$a_n = \frac{2\omega_p^2\omega_c^2}{V_T^2} n^2 \mathrm{I}_n\left(\frac{k_\perp^2 v_T^2}{\omega_c^2}\right) \mathrm{e}^{-k^2 v_T^2/\omega_c^2}$$

from which it is seen that $a_n > 0$ for all n. In the above $v_T^2 = k_B T/m$ where T is the temperature in the Maxwellian and I_n a modified Bessel function. The existence of solutions of (2.5.8) is easily shown graphically. A plot of the right hand side as a function of ω^2, shown schematically in Fig. 2.3, reveals that one finds an infinite number of positive solutions of (2.5.8) for any value $k_\perp$ since the a_n's are all positive. Thus we have an infinite number of modes with real frequencies. These are the so-called Bernstein modes (Bernstein (1958)). They have been studied experimentally in some detail (Schmitt (1973)) and are a pleasing example of agreement between theory and experiment.

It may be noted that the above argument for all frequencies being real does not depend on the value of ω_c, thus seemingly leading to the result that electrostatic modes with real frequencies can propagate even in the limit of $\omega_c = 0$, that is no magnetic field. Where has Landau damping gone? The resolution of this paradox was given by Baldwin and Rowlands (1966), who showed that in the limit $\omega_c \to 0$, many real frequency modes are excited but they undergo phase mixing to produce damping. Thus Landau damping is recovered in this limit. Curiously, a dispersion relation of the form (2.5.8) applies for an electrostatic plasma described by what is known as a water-bag model (Section 6.5). The summation is over the number of water-bags and the cyclotron frequency is replaced by $k\Delta v$, where Δv is the difference in velocities between the water-bags (Rowlands (1969a)). Thus the solution of the dispersion relation always leads to a real frequency.

2.6 Weak instabilities

In most situations of unstable media, the essential features of the instability can be appreciated by considering the parameters designating the medium to be such that the instability is weak, that is the medium is almost stable. This restriction also allows one to make analytic progress. For example consider the two stream instability for warm plasmas. The dispersion relation is given by (2.4.6). It was shown in Section 2.4 that the plasma was stable for all wavelengths for $v_0^2 < \alpha^2$, that is for thermal effects larger than kinetic effects. The marginally stable mode thus corresponds to $v_0^2 = \alpha^2$ and hence we introduce a parameter ε such that $\alpha^2 = v_0^2(1+\varepsilon)$. Substitution of this into (2.4.6), followed by an expansion in ε, where it is assumed that $\omega^2 = 0(\varepsilon)$, gives to lowest order

$$\omega^2 = k^2\omega_p^2(\alpha^2 - v_0^2)/(\omega_p^2 + 2k^2 v_0^2),$$

Fig. 2.3. A plot of $k_\perp^2$ as a function of ω^2 illustrating the existence of an infinity of Bernstein modes for $k_\perp$ fixed. From Bernstein (1958).

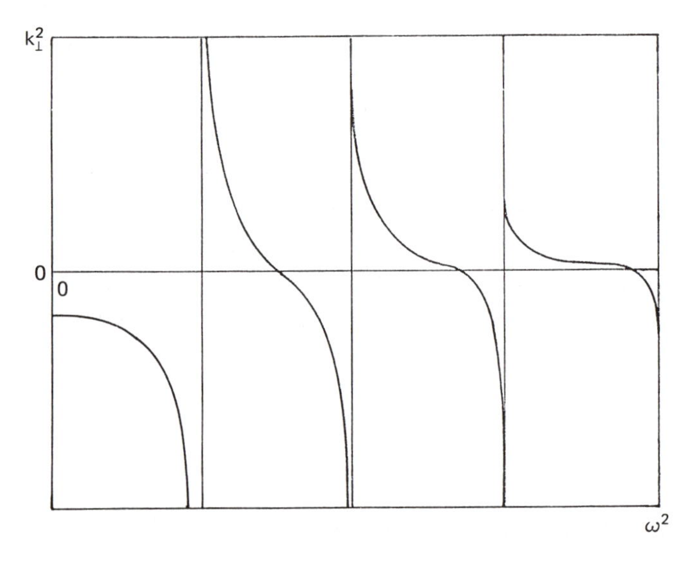

clearly showing the changeover from stability to instability as v_0^2 increases above α^2.

More formally, we can consider the dispersion relation of the form $D(\omega, k, \alpha)=0$ where α is some parameter designating the system. We now assume that for a particular value of α, α_c say, the system is marginally stable, that is ω is real and equal to some ω_c. Writing $\alpha=\alpha_c+\varepsilon$ and $\omega=\omega_c+\varepsilon\omega_1$, substituting into the dispersion relation and expanding in a Taylor series, give

$$\frac{\partial D}{\partial \alpha_c}+\frac{\partial D}{\partial \omega_c}\omega_1=0. \tag{2.6.1}$$

If D is a real function of ω, k and α, this merely gives a real shift in the frequency and is of little interest. Two distinct situations can, however exist which do lead to possible instabilities. The first is where $\partial D/\partial\omega_c=0$ so one makes a new ordering and writes $\omega=\omega_c+\sqrt{(\varepsilon\omega_1)}$ to give, to order ε

$$\partial D/\partial\alpha_c+\frac{\omega_1^2}{2}\partial^2 D/\partial\omega_c^2=0$$

or

$$\omega_1^2=-2\frac{\partial D/\partial\alpha_c}{\partial^2 D/\partial\omega_c^2}. \tag{2.6.2}$$

Now we see, depending on the sign of the right hand side, that one can have ω_1^2 of either sign, and if negative, it corresponds to an instability with $\omega_1 \sim \sqrt{|\alpha-\alpha_c|}$. An important example of the above approach to stability analysis is encountered when the dispersion relation is a real function of ω^2, as for example is (2.4.6). Then the medium goes unstable by ω^2 changing sign so that the marginal mode is $\omega=0$, and $\partial D/\partial\omega_c=0$ is automatically satisfied. Plasmas described by magnetohydrodynamic equations tend to lead to such dispersion relations.

The second instability mechanism arises when $D(\omega, k, \alpha)$ is itself a complex quantity. The classical example of this situation is the dispersion arising from the solution of the Vlasov equation. For simple electrostatic oscillations in one dimension, the appropriate dispersion relation has been obtained in Section 2.5 and given by (2.5.4) (with $p=-\mathrm{i}\omega$), which we now rewrite as

$$1+\frac{\omega_p^2}{k^2}\int_{-\infty}^{+\infty}\left(\frac{\dfrac{\mathrm{d}\hat{f}_0}{\mathrm{d}v}\mathrm{d}v}{(\omega/k-v)}\right)=0. \tag{2.6.3}$$

As stressed in Section 2.5 this form is only true if the real part of p and hence the imaginary part of ω is sufficiently positive. For a weak instability we consider this imaginary part to be small, write $\omega=\omega_R+i\omega_I$ and take $\omega_I \ll \omega_R$. Using the result that

$$\operatorname*{Lt}_{\eta\to 0}\int_{-\infty}^{+\infty}\frac{f(x)\mathrm{d}x}{x+i\eta}=P\int_{-\infty}^{+\infty}\frac{f(x)\mathrm{d}x}{x}-i\pi f(x)$$

and identifying η with ω_1 allows one to write (2.6.3) in the form

$$D(\omega,k)=D_R(\omega,k)+iD_I(\omega,k), \tag{2.6.4}$$

where

$$D_R=1+\frac{\omega_p^2}{k^2}P\int\frac{\frac{\mathrm{d}\hat{f}_o}{\mathrm{d}v}\mathrm{d}v}{(\omega/k-v)},$$

and

$$D_I=-(\omega_p^2\pi/k^2)\hat{f}_o'(\omega/k)$$

Note that both D_R and D_I are real functions of ω. Now treating D_I as of order ε gives, to lowest order

$$D_R(\omega_R,k)=0, \tag{2.6.5}$$

an equation for the real part of the frequency, and to next order

$$\omega_I=(\omega_p^2\pi/k^2)\hat{f}_o'(\omega_R/k)/\frac{\partial D_R}{\partial\omega_R}, \tag{2.6.6}$$

an equation for the imaginary part.

Most importantly, this latter result is also applicable to weakly damped waves in plasmas since the process of analytic continuation necessary in such cases is trivial in the present situation and (2.6.6) still applies.

It is now instructive to consider the long wavelength assumption ($k\to 0$). The integral defining D_R can be expanded in powers of k because, unlike the integral appearing in (2.5.6), it is a principal part integral. Then to lowest order we find

$$D_R\simeq 1-(\omega_p^2/\omega^2)(1+3k^2\langle v^2\rangle/\omega^2)$$

so that

$$\omega_R^2=\omega_p^2+3k^2\langle v^2\rangle$$

whilst from (2.6.6)

$$\omega_1=\tfrac{1}{2}(\omega_p^3\pi/k^2)\hat{f}_o'(\omega_R/k) \tag{2.6.7}$$

In particular for a Maxwellian distribution of velocities

$$f_o(v) = \rho_o(2\pi m K_B T)^{-\frac{1}{2}}\exp(-mv^2/2K_B T)$$

$$\omega_1 = -(\pi/8)^{\frac{1}{2}}\frac{\omega_p}{(k\lambda_D)^3}\exp\left(-\frac{1}{2k^2\lambda_D^2}\right)$$

$$\lambda_D^2 = mK_B T/4\pi\rho_o\rho e^2.$$

Here ω_I is the value of the Landau damping.

In summary we see that in the long-wave limit, whilst a treatment of plasma oscillations using fluid or macroscopic equations leads to the correct value of the real frequency, the phenomenon of Landau damping is an entirely microscopic effect. This is clear from (2.6.6) because there ω_I is given in terms of the microscopic distribution function $f_o(v)$.

The result (2.6.7) suggests that the condition for instability is that a value of v exists such that $\hat{f}_o'(v) > 0$. This is in fact a necessary condition. Physically this corresponds to the superimposition of a stream of particles on an otherwise monotonic decreasing function such as a Maxwellian and the instability is a manifestation of the two stream instability discussed in Section 2.4. For a necessary and sufficient condition for instability one must include the requirement that there can be a wave or waves with phase velocities ω/k in the region of velocities where $\hat{f}_o'(v) > 0$. Physically this corresponds to the condition that there are more particles moving slightly faster than the wave as compared to those that are slower than the wave. As these particles get carried along with the wave, particle energy is effectively lost and is fed into the waves.

For electrostatic waves the conditions for instability were first found by Penrose (1960) using a method based on the Nyquist diagram technique. He found the necessary and sufficient conditions for instability to be that (a) $f_o(v)$ has a minimum for some v, namely u; and (b) that

$$P\int_{-\infty}^{-\infty}\frac{(f_o(v)-f_o(u))}{(v-u)^2}\,dv > 0.$$

This expresses the familiar intuitive rule that instability occurs when the relative velocities of two plasma streams are in some sense larger than their thermal spreads.

Instabilities found using a Vlasov approach which do not follow from the macroscopic equations are called velocity space instabilities. Mathematically, they are seen to arise because of the presence of the resonant denominator in the expression for $D(p, k)$ as given by (2.5.4). Such a denominator implies that for the case of weak instabilities the dispersion relation is of the form of (2.6.4), that is, it contains both a real and an imaginary part.

In the case of electrostatic waves as discussed in detail above, the zero in the denominator is caused by a resonance between the phase velocity ω/k and the particle velocity v. For more complicated plasma situations one gets other types of resonant denominators which can then lead to further instabilities. A magnetized plasma (that is immersed in a uniform magnetic field) when treated using the Vlasov equation, gives rise to denominators of the form $\omega - n\omega_c - k_{\parallel}v_{\parallel}$. Here n is an integer, ω_c the cyclotron frequency and $k_{\parallel}$ and $v_{\parallel}$ are the components of wave number and velocity respectively along the magnetic field. The resonance is then a resonance between this particular component of velocity and the Doppler shifted phase velocity. It will be noted that this resonance condition does not involve the perpendicular components of either the velocity or the wave number. Thus for perpendicular propagation ($k_{\parallel}=0$) there are no resonant particles and hence no imaginary contribution to the dispersion relation.

It would be out of place here to attempt to catalogue the instabilities and waves that are possible in a magnetized plasma. Some are associated with the resonant condition discussed above so that the condition for instability involves $f'(v)$ being positive at some Doppler shifted phase velocity. These are intimately related to the two stream instability (cyclotron instabilities). However, another class occurs which does not depend on the resonant contribution but rather on the anisotropy of the equilibrium distribution function imposed by the presence of the magnetic field. In such a field the distribution function can be any function of $v_{\parallel}, v_{\perp}$, where $v_{\parallel}$ and $v_{\perp}$ are the parallel and perpendicular velocity components so $f_o = f_o(v_{\parallel}, v_{\perp})$. It can be shown that a single humped function of $v_{\parallel}^2 + v_{\perp}^2$ only (for example a Maxwellian), is stable in a uniform magnetic field. This suggests that the anisotropy of the distribution might be considered as a source of available energy which will drive an instability (Bernstein (1958)). A particularly important example is the loss-cone instability. Particles confined by a mirror type magnetic field, for example the Earth's field, must have their perpendicular velocity components sufficiently larger than their parallel components, as otherwise the particle would leak out of the mirror field. Consequently the distribution function is zero for $v_{\perp} < v_{\parallel}\tan\psi$ where ψ is the loss-cone angle. This gives rise to an instability. This instability, quite naturally called the loss-cone instability, was first studied by Rosenbluth and Post (1965). Another instability arises when the perpendicular pressure $p_{\perp}$ differs from the parallel pressure $p_{\parallel}$. Because this does not arise from a resonance condition it can be studied using macroscopic equations.

A further group of instabilities, the so-called universal instabilities, can be associated with yet another resonance. A charged particle in a magnetic field in the y direction which field is a function of z only, undergoes a drift in the y direction of magnitude $v_D = \frac{mcv_\perp}{2eB}\frac{dB}{dx}$ where $v_\perp$ is the perpendicular velocity. In the simple resonance conditions given above $\omega - \mathbf{kv}$ is now replaced by $\omega - k_\parallel v_\parallel - k_y v_D$ so that even for perpendicular propagation ($k_\parallel = 0$) the dispersion relation will have an imaginary contribution leading to the possibility of velocity space instabilities. It is found that even a Maxwellian velocity distribution function, which is stable in a uniform field, is now prone to instabilities. That is, a thermodynamic equilibrium velocity distribution can be unstable in a non-uniform magnetic field. The available energy in this case arises from the kinetic energy associated with the drift motion. Since non-uniform magnetic fields are expected in almost all realistic situations, be they laboratory experiments or space plasmas, such instabilities are called universal. However, the growth rates are usually small as compared to the time scale of the experiment and though indeed universal, they are not considered to be too damaging to plasma containment experiments.

In the above we have touched on some of the instabilities that can exist in a plasma. Some of these are discussed in the text books referenced above but a catalogue of them has been prepared (Cap (1976)) and this gives an excellent review of the vast amount of work that has been done in this area.

Exercises on Chapter 2

Exercise 1

Find how the equation for ω^2 following (2.4.4) is altered by inclusion of thermal effects in the streams as in (2.4.6).

Exercise 2

Find the two stream model that generalizes (2.5.6) as derived from two resonance distributions. Solve for ω (the solution is given in Chapter 3, equation 3.2.6).

Exercise 3

Extend the calculation of Exercise 2 to unequal temperatures of the beams, such that the denominators in the dispersion solution are $\omega - kv_1 + iku$ and $\omega - kv_2 + irku$. By expanding Imω in k or otherwise, show that the system is stable if

$$v_1 - v_2 \leq u(r+1)$$

and

$$\frac{\sqrt{3}+1}{\sqrt{3}-1} > r > \frac{\sqrt{3}-1}{\sqrt{3}+1}.$$

3

Convective and non-convective instabilities; group velocity in unstable media

3.1 Introduction

In Chapter 2 a linear theory of waves and instabilities was presented for the case of the propagation of waves in infinite and uniform media. Basically one considers the wave disturbance (in one dimension) to be proportional to $\exp(ikx - i\omega t)$ with ω, k satisfying the dispersion relation $D(\omega, k) = 0$. In many problems, for example infinite media, one must take k to be real, in which case one distinguishes between stable ($\mathrm{Im}(\omega(k) \leqslant 0$ and unstable disturbances for a particular k value $\mathrm{Im}\,\omega(k) > 0$).

Early experiments suggested that such an approach was not always sufficient. For example, experiments involving the interaction of charged particle beams with stationary plasmas consistently showed little sign of being unstable. On the other hand, all theoretical models based on the ideas outlined above unequivocably suggested the system to be unstable. In the case of plasma-beam experiments, the explanation which resolved this difficulty was given independently by Sturrock (1958) and by Fainberg, Kurilko and Shapiro (1961). The Soviet authors based their method on earlier work by Landau and Lifshitz (1959) Chapter 3. These latter authors were concerned with problems in fluid mechanics.

The resolution of this problem is based on the fact that it is not sufficient to treat the time development of a system by considering just a single k mode, but rather it is necessary to consider a spatial pulse or wave packet which is composed of a range of k values. Then unstable media can be classified into two distinct types. The instability could be *convective*, in which case a pulse would propagate away from any point sufficiently rapidly that the disturbance at that point would eventually decay with time. Alternatively it may be *non-convective* (or absolute), in which case the pulse may still propagate but the disturbance at any point in space would eventually grow with time. These two types of behaviour are illustrated in

Fig. 3.1. (Mathematically these characteristics arise out of a generalization of the concept of group velocity to unstable waves.)

From a practical point of view these two types of instabilities have to be considered separately. A convective instability in a finite medium may travel out of the medium before its amplitude has had sufficient time to grow significantly. In this case the medium may to all intent and purpose be considered as stable. This was the case of the early plasma-beam experiments. The reason why basically unstable solutions appeared to be

Fig. 3.1. Timelike packets in x, t space: (a) Timelike packet, also a spacelike packet. (b) Timelike packet, not a spacelike packet. From Sturrock (1960).

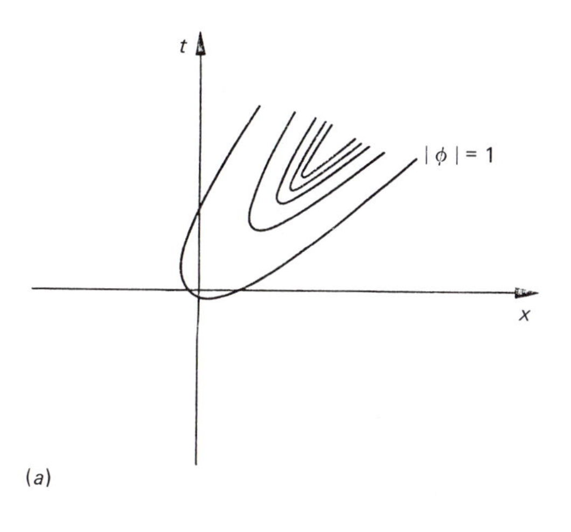

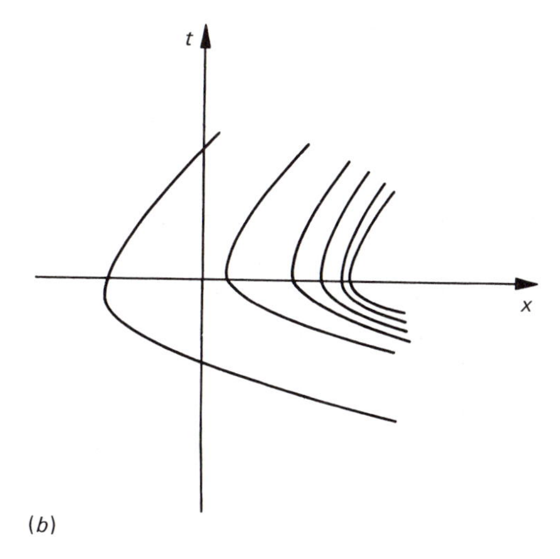

stable was that they were only convectively unstable and the unstable pulses were convected out of the measuring region before they had time to grow to a level that could be detected. (The basic plasma-beam instability was studied by designing the experimental device to ensure the instability to be non-convective by having beams entering the plasma from both ends.) For such situations one may associate with the unstable medium a critical length such that media shorter than this critical value may be considered stable. The instability does not have time to grow before it is convected out of the medium. Early magnetic mirror experiments were designed with such critical lengths in mind (Rosenbulth and Post (1965)).

A system with a non-convective instability can be associated with an oscillator whilst a convective instability is more closely associated with an amplifier.

A number of methods of distinguishing between convective and non-convective instabilities have been given over the last twenty years. The most rigorous and general treatment was given by Bers and Briggs for which see Briggs (1964)). This reference contains a complete list of earlier work; both theoretical and experimental. Since then the main emphasis has been in simplifying the technique, (see for example Baldwin and Rowlands (1970)) applying it to more realistic, but consequently more complicated situations (Callen (1968)), and elucidating the concept of the velocity of these unstable pulses (Hall and Heckrotte (1968)).

The emphasis of this Chapter will be on the basic understanding of pulse propagation in unstable media. This entails an extension of the concept of group velocity to such media.

It must be stressed that the analysis differentiating between convective and non-convective instabilities is based on *linear theory*, one which gives rise to unstable solutions ($\mathrm{Im}\omega(k)>0$) of the dispersion relation.

3.2 Kinematics of unstable wave packets

We assume waves to propagate in a one dimensional, infinite and uniform medium. The disturbance at a point x at time t may then be written in the form

$$\phi(x,t)=\int \mathrm{d}\omega \int_{-\infty}^{+\infty} \frac{S(k)\mathrm{e}^{\mathrm{i}(kx-\omega t)}\mathrm{d}k}{\mathrm{D}(\omega,k)}. \tag{3.2.1}$$

$S(k)$ specifies the shape of the initial pulse.

In the following we take $S(k)$ to be identically equal to unity, corresponding to an initial pulse proportional to $\delta(x)$, where $\delta(x)$ is the Dirac delta function. This is the antithesis of the usual approach to the study of instabilities where $S(k)$ is taken to be proportional to $\delta(k-k_o)$ so that the zeros of $D(\omega, k_o)$ determine the time dependence of ϕ. That is, we now study the Green function for the system rather than the time evolution of a normal mode.

The ω integration is along a contour which lies above any zeros of $D(\omega, k)$ for all real k. Since the above expression arises in solving an initial value problem, $D(\omega, k)$ is analytic above the contour and we shall assume it to be entire in both k and ω.

For $t>0$, the ω integration may be carried out by completing the contour in the lower half-plane and summing over residues to give

$$\phi(x,t)=\sum_n \int_{-\infty}^{+\infty} 2\pi \frac{e^{i(kx-\omega_n(k)t)}}{D'(\omega_n(k))} dk, \tag{3.2.2}$$

where

$$D(\omega_n(k), k)=0, \tag{3.2.3}$$

and $D'(\omega(k))=\partial D/\partial\omega$ for $\omega=\omega(k)$. The sum is over the zeros of $D(\omega, k)$ (assumed simple).

For those modes which are stable, that is $\mathrm{Im}\,\omega_n(k)\leqslant 0$ for all real k, the integrals appearing in (3.2.2) will remain bounded functions for all t. This is easily seen by using Schwartz's inequality. Thus we restrict our attention to unstable modes and consider at this stage, for the sake of simplicity, that only one such mode exists and, further, that it is only unstable for a range of k values ($k_a<k<k_b$). The significant contribution to (3.2.2) is then of the form

$$I(x,t)=2\pi \int_{k_a}^{k_b} \frac{e^{i(kx-\omega(k)t)}}{D'(\omega(k))} dk. \tag{3.2.4}$$

To evaluate the integral we analytically continue the integrand into the complex k plane in an attempt to find a contour along which $\mathrm{Im}(\omega/k)\equiv 0$. Two distinct possibilities arise. The first is where a path in the k plane connecting k_a and k_b exists and along which $\omega(k)$ is real. Then, by Schwartz's inequality, $I(x,t)$ decays asymptotically with time. In this case we have a linearly *unstable* medium but one where an initial pulse grows in amplitude for some time, but eventually decays asymptotically with time. Such an instability is called CONVECTIVE.

The second possibility is where no such path exists, that is, it is impossible to join k_a and k_b by a contour along which $\mathrm{Im}\,\omega(k)=0$. This means that along some part of any contour joining k_a and k_b, $\mathrm{Im}\,\omega(k)>0$. This in turn implies that $I(x,t)$ will increase exponentially with time for sufficiently long times. This type of instability is called NON-CONVECTIVE. Thus the classification of instabilities into convective or non-convective is simply related to the existence or not of a path in the complex k plane along which $\mathrm{Im}(\omega(k))=0$ and which connects two points on the real axis, k_a and k_b.

The possible vanishing of $D'(\omega(k))$ presents no problems, for this occurs only if two modes are equal at the point. But $D'(\omega)$ is of opposite sign for these modes and their contributions to the summation in (3.2.2) simply cancel. This is most easily seen as follows. Consider two modes $\omega_1(k)$ and $\omega_2(k)$. Then one may write $D(\omega,k)=A(\omega,k)(\omega-\omega_1(k))(\omega-\omega_2(k))$ where A is finite for $\omega=\omega_1(k)$ or $\omega=\omega_2(k)$. Then $D'(\omega_1(k),\,k)=A(\omega_1,k)(\omega_1(k)-\omega_2(k))$ whilst $D'(\omega_2(k),k)=A(\omega_2,k)(\omega_2(k)-\omega_1(k))$. Thus though D' vanishes for $\omega_1(k)=\omega_2(k)$ the net contribution to the summation in (3.2.2) is zero.

It is now convenient to introduce the transformation $z=\omega(k)$ and write (3.2.4) in the form

$$I=2\pi\int_C (1/D')\frac{dk}{dz}e^{i(kx-zt)}dz \tag{3.2.5}$$

The contour C is the mapping of $z=\omega(k)$ for k real and such that $k_a\leqslant k\leqslant k_b$. It starts on the real axis at $z_a=\omega(k_a)$, rises in the upper half of the z plane and ends on the real axis at $z_b=\omega(k_b)$. If the integrand is analytic in the area bounded by the contour C and the real axis, then the contour may be deformed to the real axis and again by Schwartz's inequality the integral I will at least remain bounded in time and thus we have a convective instability. If the integrand is not analytic, the deformation is not possible and we have a non-convective instability. Since, as pointed out above, points where $D'(z)=0$ are not relevant, the classification of instabilities into convective or non-convective reduces to an examination of the analyticity of dk/dz, that is the position of the zeros of $dz/dk(=d\omega/dk)$. An example corresponding to a non-convective instability is given as Fig. 3.2.

This basic result has been obtained in seemingly different ways by a number of different authors. (Fainberg, Kurilko and Shapiro (1961); Polovin (1961); Briggs (1964); Sudan (1965); Dysthe (1966); Hall and Heckrotte (1968); Baldwin and Rowlands (1970)). For example, it follows almost immediately by evaluating (3.2.4) using the method of stationary phase. The critical point in such an analysis is where $d\omega/dk=0$, and if $\mathrm{Im}(\omega_n(k))>0$ at this point, then $\phi(t)$ increases exponentially with time and the instability is non-convective.

The analysis given above can really be seen as *firstly* justifying the use of the method of stationary phase to the present problem, and *secondly* as reducing the area of search, namely to inside the C contour, for the critical point $d\omega/dk=0$. In what follows, we shall base our analysis on the method of stationary phase, implying that the extension of the C contour is straightforward.

For complicated dispersion relations the study of the analyticity of $\mathrm{d}k/\mathrm{d}z$ has to be carried out numerically (see Rognlien and Self (1972)). Again a number of different approaches have been suggested, all involving mapping quantities such as $\omega(k)$, $\mathrm{d}\omega/\mathrm{d}k$ for a range of complex values of k. These have been reviewed by Baldwin and Rowlands (1970) where an efficient procedure for the classification of instabilities is also suggested. However, to illustrate the general concepts we consider an example where most of the calculation can be carried out analytically. Treatment follows Infeld and Skorupski (1969).

For the case of two interpenetrating electron streams described by a resonance distribution function of the form

$$f(v)=\frac{Nu}{\pi}\left[\frac{1}{(v-v_1)^2+u^2}+\frac{1}{(v-v_2)^2+u^2}\right]$$

the dispersion relation is obtained by substituting the above form into (2.5.4) to give

$$\mathrm{D}(\omega,k)=1-\omega_\mathrm{p}^2\left[\frac{1}{(\omega-kv_1+\mathrm{i}kv)^2}+\frac{1}{(\omega-kv_2+\mathrm{i}kv)^2}\right]=0.$$

This may be solved to give

$$\omega/\omega_\mathrm{p}=\beta K+\mathrm{i}\surd[(1+4K^2)^{\frac{1}{2}}-1-K^2], \tag{3.2.6}$$

where $K=k(v_1-v_2)/2\omega_\mathrm{p}$, $\beta=(v_1+v_2-2\mathrm{i}u)/(v_1-v_2)$ and ω_p is the plasma frequency of one stream.

From the general analysis given above we know that the system is non-convectively unstable if the point ω_o in the ω complex plane, where

Fig. 3.2. Complex frequency plane showing $\omega(k)$, k real, for chosen branch. The point where $\mathrm{d}\omega/\mathrm{d}k=0$ is ω_0. From Infeld and Skorupski (1969).

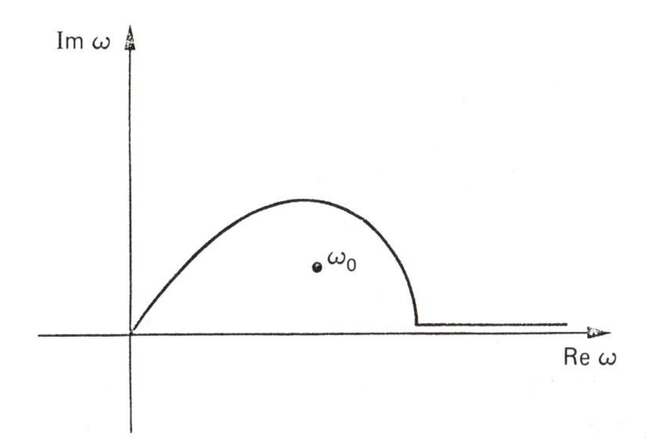

$d\omega/dk = 0$, is enclosed within the area bounded by the contour C and the real axis. The condition $d\omega/dk = 0$ reduces to an algebraic equation for the critical value of k, k_o say (in general complex) namely

$$8(1-\beta^2)^2(2k_o^3 - 3k_o^2) + 3(1-\beta^2)(3+5\beta^2)k_o - 2\beta^2(1+\beta^2) = 0.$$

This equation has been solved numerically and the value of k_o substituted into equation (3.2.6) to give the value of $\text{Im}\Omega_o = \text{Im}(\omega_o/\omega_p)$ at this critical point.

The results of this numerical investigation are shown in Fig. 3.3, where lines of equal growth (equal $\text{Im}(\omega_o/\omega_p)$) are plotted in a parameter space designated by $-v_2/v_1$ and u/v_1. In particular, the line $\text{Im}(\omega_o/\omega_p) = 0$ separates the convective and non-convective regions. Also shown is the line corresponding to $\text{Im}(\omega/\omega_p) = 0$ for *real* k, which separates the stable and unstable regions. For more details of this particular example see Infeld and Skorupski (1969).

Throughout the above discussion we have assumed that only one unstable mode (branch) exists in the range $k_a \leqslant k \leqslant k_b$. If two or more branches exist, each will make its own contributions to the sum in the expression for $\phi(x, t)$ given by (3.2.2). In particular, if one is non-convective, then the system is non-convectively unstable. However, if two are non-convective and $dz/dk = 0$ for the *same* values of k and z for each mode then

Fig. 3.3. Division of two plasma stream data into convective and absolute instabilities (non-convective) and the stable region. Lines of equal growth rate (equal $\text{Im}\Omega_o = \text{Im}\omega_o/\omega_p$) are indicated. From Infeld and Skorupski (1969).

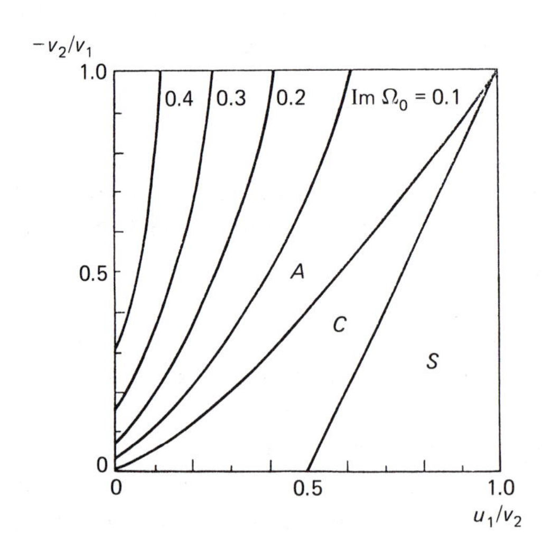

they make an equal but opposite contribution to the summation in equation (3.2.2). The system is then no longer non-convectively unstable.

Finally, to illustrate the time and space dependence of $\phi(x, t)$, we consider the case of an isolated singularity ($dz/dk = 0$) at the point (z_o, k_o). Near this point we may write

$$D(z, k) = (z - z_o) - a(k - k_o)^n$$

where n is some integer, since it has been assumed throughout that $D(\omega, k)$ is an entire function of ω and k. Here a may be taken to be constant. If this form is substituted into (3.2.5), then we have

$$\mathrm{I} \sim e^{i(k_o x - z_o t)} \int_c \frac{e^{i\left(\frac{z - z_o}{a}\right)^{1/n} x + i(z - z_o)t}}{(z - z_o)^{1 - 1/n}} \, dz$$

For large t, the contour may be chosen to pass around the point $z = z_o$ and end at $-i\infty$. Since for large t the contribution comes mainly from $z \simeq z_o$, the x dependence of the integrand may be neglected (for finite x) and then

$$\mathrm{I} \sim e^{i(k_o x - z_o t)}/t^{1/n}$$

Remembering that $z_o = \omega(k_o)$ and is generally complex, we write $z_o = \omega_R(k_o) + i\gamma(k_o)$ to give

$$\mathrm{I} \sim \frac{e^{\gamma(k_o)t}}{t^{1/n}} e^{i[k_o x - \omega_R(k_o)t]} \tag{3.2.7}$$

Thus at *fixed* x, ($x = 0$ say) the disturbance 'I' decreases exponentially with time for a convective instability ($\gamma(k_o) < 0$) but increases exponentially for a non-convective one ($\gamma(k_o) > 0$). When $\gamma = 0$ the disturbance remains finite and we call the instability marginally non-convective (see Exercise).

3.3 Moving coordinate systems

It will be appreciated that the distinction between convective and non-convective instabilities depends critically on the frame of reference. By considering this more explicitly one is able to obtain better insight into the propagation of these unstable modes.

A change to a moving coordinate system is readily carried out by replacing x by $x - vt$ in the above equations, in particular, equation (3.2.4). Here v is the relative velocity between the two systems. However, by defining $z = \omega(k) - kv$ this equation is transformed to the same form as (3.2.5) with x, z and C being replaced by x, z and C. The latter quantity is simply the mapping of z(k) for $k_a \leqslant k \leqslant k_b$. The analysis of the response $\mathrm{I}(t)$

is then entirely analogous to that for $I(t)$ and the critical point is simply $dz/dk=0$ or

$$\frac{d\omega}{dk}=v \tag{3.3.1}$$

This equation defines a critical value of k, k_o say, which may be used to calculate the corresponding value of ω at this point. We denote this value by $\omega_o(v)$ so as to make the v dependence explicit.

For large times, the major contribution to the integral comes from the region near k_o and so asymptotically we may write in analogy to (3.2.7), with x being replaced by $x+vt$, and taking $x=0$

$$I(t)\sim e^{i(k_o vt-\omega_o(v)t)}$$

Since $x=0$ we have $v=x/t$ and so

$$I(t)\sim e^{i(k_o x-\omega_o(x/t)t)}$$

The pulse is thus large where $\text{Im}\omega_o(x/t)>0$.

Now consider a convective instability, $\text{Im}\omega_o(v=0)<0$. Then for t, large but fixed, there exists a range of x values, including $x=0$, for which $\text{Im}\omega_o(x/t)<0$ and so I is exponentially small. For larger values of x there is a range where $\text{Im}\omega_o(x/t)>0$, in which case the pulse is large whilst for still larger values of x, there is a range where $\text{Im}\omega_o(x/t)<0$, and then the pulse is small again.

The maximum of the pulse will occur when $\text{Im}\omega_o$ is a maximum and this will occur for some critical value of v, v_o say such that

$$\frac{d\,\text{Im}(\omega)}{dk}(\omega_o)=0. \tag{3.3.2}$$

This condition taken together with (3.3.1), allows one to write

$$\frac{d\omega}{dk}(\omega_o)=v_o. \tag{3.3.3}$$

(Since the maximum value of $\text{Im}(\omega)$ lies on the contour C, k in the above equations must be real.) This velocity v_o is the velocity of propagation of the maximum of the pulse and although the shape of the pulse changes with time this velocity is a good characteristic for it.

The case of a non-convective instability is qualitatively the same. The maximum of the pulse may still move, but with a much reduced velocity. However, since $\text{Im}\omega_o(0)>0$, the point $x=0$ always lies in the region where the pulse grows with time. Note that in both cases the width of the pulse (region where $\text{Im}\omega(x/t)>0$) also grows with time.

It will be realized that the above discussion is in reality an extension of the concept of group velocity to unstable systems. (The velocity v_o defined by (3.3.3) is just the group velocity if $\mathrm{Im}\omega = 0$.) For a discussion of group velocity in a style somewhat similar to that given above see Lighthill (1978).

For a large number of problems, the system is only weakly unstable in the sense that if $\omega = \omega_R + i\omega_I$, then $\omega_I/\omega_R \ll 1$. Furthermore, if for real ω and k we write $D(\omega, k) = D_R(\omega, k) + iD_I(\omega, k)$ then the condition $D(\omega, k) = 0$ leads via a Taylor expansion in powers of ω_I/ω_R to

$$D_R(\omega_R(k), k) = 0$$

and

$$\omega_I = -\frac{D_I(\omega_R, k)}{\partial D_R/\partial \omega_R}. \qquad (3.3.4)$$

Similarly, by writing $k = k_R + ik_I$, one can write by analogy

$$D_R(\omega, k_R(\omega)) = 0$$

and

$$k_I(\omega) = -D_I(\omega, k_R)/\partial D_R/\partial k_R$$

If (3.3.4) is differentiated with respect to k we have

$$\frac{d\omega_R}{dk} = -\frac{\partial D_R}{\partial k} \bigg/ \frac{\partial D_R}{\partial \omega}$$

so that

$$\omega_I(k) = -\frac{d\omega_R}{dk} k_I(\omega)$$

This equation relates the growth rate in time to that in space in a simple manner.

As mentioned in the Introduction, the original need for the classification of nonlinear instabilities into convective or non-convective arose from the experimental observation that a particular system, supposedly unstable, showed little sign of instability (Sturrock (1958)). The explanation given by Sturrock was that the instability was convective and the unstable pulse propagated out of the system before it had time to grow to a significant amplitude. Of course, if the system had been longer the instability would have been seen. To incorporate these ideas into a study of a finite system of length L one can define an amplitude factor f such that

$$f = \frac{L}{v_o} M_k[\mathrm{Im}(\omega(k))]$$

where M_k stands for the maximum value for real k. Using the above relation between ω_I and k_I and the definition of v_o one can alternatively write

$$f = LM_{\omega}[\text{Im}(k(\omega))] \tag{3.3.5}$$

This may now be used as an effective criterion for instability for the case of a convective instability in a finite system. One chooses, somewhat arbitrarily perhaps, a value of f that may be tolerated experimentally, remembering that thermal noise will be amplified by a factor $\exp(f)$ by the convective instability traversing the system. This value together with (3.3.5) defines a critical length L_c, say, such that for a system with $L > L_c$ the thermal noise is amplified to such an extent as to cause concern.

Rosenbluth and Post (1965) have used the above concepts to obtain a critical size for a mirror-machine subject to the loss-cone instability. They show that this instability is convective and estimate the maximum value of $\text{Im}(k(\omega))$ to be used in (3.3.5). Their results show that this instability does impose realistic constraints on the design of mirror machines.

3.4 Higher dimensional systems

Throughout this Chapter we have considered the propagation of unstable waves in one-dimensional systems. The extension to more than one dimension is best done by using the method of stationary phase to evaluate the relevant integrals. By analogy with (3.2.4) we have

$$I(t) = \int_S \frac{e^{i(\mathbf{k}\cdot\mathbf{x} - \omega(\mathbf{k})t)}}{D'(\omega(\mathbf{k}), \mathbf{k})} d\mathbf{k},$$

where the limits of integration are such that $\text{Im}(\omega(\mathbf{k})) \geqslant 0$. This condition defines the surface S. In a moving coordinate system, where $\mathbf{x} = \hat{\mathbf{x}} + \mathbf{v}t$, the stationary phase point is defined, in analogy with (3.3.1), as

$$\mathbf{v} = \nabla\omega(\mathbf{k})$$

Now the criterion for a non-convective instability is that $\text{Im}\omega(\mathbf{k}) > 0$ for $\mathbf{k}$ such that $\nabla\omega(\mathbf{k}) = 0$. The characteristic velocity of the pulse v_o is now defined by (see (3.3.2) and (3.3.3))

$$\nabla[\text{Im}(\omega(\mathbf{k}))] = 0, \tag{3.4.1}$$

$$\nabla[R_e(\omega(\mathbf{k}))] = \mathbf{v}_o, \tag{3.4.2}$$

$\mathbf{k}$ being real.

Dysthe (1966) has considered the classification of instabilities in three dimensions and obtained the above conditions. His work gives the basic analysis implied in the above derivation in much more detail, though the importance of the contour C is not realized. When it comes to the task of finding the critical point defined by (3.4.1) and (3.4.2), the limitation imposed by the introduction of the contour C becomes of paramount importance. Briefly, it reduces the area of ω space where one needs to evaluate ω and its derivatives. Dysthe also considers some simple three-dimensional examples which can be treated analytically.

The concept of group velocity in stable multi-dimensional systems ($\mathrm{Im}\,\omega = 0$) has been discussed at length by Lighthill (1965). Not surprisingly, the group velocity is given by (3.4.2) with (3.4.1) redundant.

3.5 Summary

The concept of group velocity has been extended here to unstable systems. The method introduced also allows one to classify and hence distinguish between two types of instabilities, convective and non-convective.

The analysis is based on the representation of any disturbance $\phi(x, t)$ by a space–time transform of the form of (3.2.1). Such an equation itself is obtained by first converting partial differential equations in space and time to algebraic equations by introducing transforms. This procedure is discussed at length in Chapter 2. The results of the present Chapter are obtained from an expression such as (3.2.1) by considering the long time ($t \to \infty$) form of the relevant integral. However, the whole procedure of using transform techniques for this particular problem can be circumvented by applying asymptotic methods directly to the partial differential equations. This has been done for the Vlasov equation by Rowlands (1970) and not surprisingly the basic results, namely equations (3.3.2) and (3.3.3), are reproduced

Exercise on Chapter 3

Show that the velocity V_o of a moving frame of reference such that the linear instability around the state $w = 0$ of the Hohenberg–Swift–Pomeau–Manneville equation

$$\partial_t w = [\varepsilon - (\partial x^2 + 1)^2]w - w^3$$

is marginally non-convective, is

$$V_o = 4\varepsilon^{\frac{1}{2}} + 4\varepsilon^{3/2} + \ldots .$$

Hint Find $\omega_o(k)$ following from the equation, define

$$\omega = \omega_o - kV_o$$

and find the complex value of k and real V_o such that $\mathrm{d}\omega/\mathrm{d}k = 0$ for ω real. (The velocity V_o plays an important role in the theory of shock propagation, see Ben-Jacob *et al.* (1985).)

4

A first look at surface waves and instabilities

4.1 Introduction

One of the ironies of wave phenomena is that those waves which are most easily observed, such as waves on the surface of water, are more difficult to analyse theoretically than waves which propagate through a medium, such that their presence has to be inferred indirectly. Surface waves are more easily observed than bulk waves but *mathematically* it is more difficult to deal with them. The main reason for this is that for surface waves a boundary condition, such as continuity, must be satisfied at the surface of the wave which of course is not generally a simple plane. This was indicated briefly in the Introduction. Fortunately for *linearized* wave theory, the boundary condition has to be satisfied on the unperturbed surface which is usually planar. If the medium on either side of the boundary is homogeneous, the problem reduces to matching algebraic expressions at the interface. In this manner one obtains an algebraic dispersion relation for surface waves analogous to a bulk dispersion relation. However, in most practical situations the boundary is diffuse and this necessitates solving differential equations, with non-constant coefficients, for the behaviour perpendicular to the boundary. An algebraic dispersion relation is then replaced by an eigenvalue equation with a consequent increase in the difficulty of the problem.

The basic mathematical techniques are illustrated in Section 4.2 by considering the propagation of waves along the surface of a liquid. Because such waves are controlled by gravity, in that gravity tends to restore any perturbed surface to a flat one (whilst inertial effects tend to act in opposition), they are usually referred to as surface gravity waves. For such waves the medium, namely the liquid, is taken to be homogeneous, whilst the air above simply provides an atmospheric pressure.

In the real world the liquid is not usually homogeneous. Due to effects such as gradients in salinity or temperature, the density ρ, often has spatial variation. Depending on this density variation, which we assume to be in the vertical direction $\rho=\rho(z)$, the main disturbance of the wave may move away from the surface and be localized essentially in the region where $\rho^{-1}(z)\mathrm{d}\rho/\mathrm{d}z$ is a maximum. Then the wave is usually referred to as internal but is much more like a surface wave than a bulk one. For plasmas, which in many situations can be considered as fluids, the major density variation often occurs in the vicinity of what is usually taken as the surface. The propagation of internal waves or surface waves in inhomogeneous or stratified media can be discussed under the same mathematical umbrella and this is done in Section 4.2.

In Nature the density variation is usually such that the light fluid lies on top of the heavier fluid. In such a case internal waves can propagate confined to a region near the interface. However, if an inversion occurs where the dense fluid is above the light fluid then the mathematical analysis which leads to wave propagation in the usual case now leads to an instability. This is the Rayleigh–Taylor instability. That such a situation is unstable is not surprising but the form it takes is not so obvious.

For plasma systems, where real gravitational effects may be neglected, it can be shown to a good approximation that the effect of magnetic field curvature is equivalent to an effective gravitational force.

This may best be seen by comparing the drift motion of the guiding centre of a charged particle in a gravitational field **g**, namely

$$\mathbf{v}_\mathrm{g}=\frac{m}{e}(\mathbf{g}\wedge\mathbf{B})/B^2,$$

where **B** is the magnetic field and e, m the charge and mass of the particle, with the drift due to curvature effects and magnetic gradients

$$\mathbf{v}_\mathrm{c}=\frac{m}{e}(v_\parallel^2+\tfrac{1}{2}v_\perp^2)\frac{\mathbf{R}\wedge\mathbf{B}}{(RB)^2}.$$

Here **R** is the radius of curvature of the field line whilst $v_\parallel$ and $v_\perp$ are the parallel and perpendicular components of velocity with respect to the direction of **B**. Unlike other possible drifts, such as that produced by an electric field **E**, namely

$$\mathbf{v}_\mathrm{e}=\mathbf{E}\wedge\mathbf{B}/B^2$$

the mass dependence of $\mathbf{v}_\mathrm{g}$ and $\mathbf{v}_\mathrm{c}$ are identical. For thermal plasmas one usually takes a velocity average and then defines an effective gravitational constant $\mathbf{g}_\mathrm{m}$ by

$$\mathbf{g}_\mathrm{m}=(kT/\rho)\mathbf{R}/R^2, \qquad (4.1.1)$$

where T is the temperature and ρ the density. For a more detailed discussion of particle drifts see Chen (1984) whilst a good physical discussion of this analogy is in Rosenbluth and Longmire (1957).

Thus the phenomenon of inversion is now significant for the appropriate sign of the magnetic curvature. Unfortunately the simplest of plasma confinement arrangements, that of a region of high density plasma contained by an envelope of high intensity magnetic field, is equivalent to the inversion situation. Thus simple plasma confinement systems are almost all prone to Rayleigh–Taylor instabilities. This instability is considered in Section 4.3.

Another case where the same mathematical analysis is applicable is that of a non-homogeneous fluid subjected to an acceleration such as is the situation during an explosion. When this happens, the gravitational acceleration $\mathbf{g}$ is simply replaced by the actual acceleration. It was in fact in this context that Taylor (1950) made his contribution to the analysis of the instability which now bears his name.

A second, equally important, class of instabilities in inhomogeneous media is the Kelvin–Helmholtz instability. In this case the equilibrium is that of a stratified fluid where different layers are in relative transverse motion. Here the free energy needed to drive the instability comes from the kinetic energy associated with the relative velocity between the layers. (This is in contrast to the Rayleigh–Taylor instabilities where the free energy arises because of the density inversion layer in the gravitational field.) Such instabilities are the subject of Section 4.4.

Of course, in our finite world all media must have a surface, and though for many phenomena the presence of a surface need not be considered (bulk waves), for others it is of paramount importance (surface waves). Besides the simple fluids and plasmas we have so far considered there are many more media which exhibit surface wave phenomena. For example consider the waves often seen propagating through a corn field driven by the wind in what is essentially a Kelvin–Helmholtz mechanism. Perhaps the destructive effects of these waves, which lead to the corn being laid flat to the ground in periodic intervals across the field in a direction perpendicular to that of the wind, could be alleviated by a stabilizing mechanism suggested by analogy with plasma physics. One could perhaps plant a periodic row of trees at right angles to the prevailing wind thus introducing a fixed perpendicular component of spatial variation, an effect known to lead to stabilizing effects in laboratory plasmas and discussed for example in Section 4.4. In large fields one might need more than one row.

However, here we choose to illustrate these wider fields of application by considering the waves on the interface between a solid and its melt. It is well known that in certain cases this interface will go unstable and spontaneously deform from a flat surface to a contorted one where nonlinear effects take over to control the final outcome. This phenomenon is known as the Mullins–Sekerka instability and will be discussed in Section 4.5. However, this is but a single example of surface waves and associated instabilities that abound in our exotic world. We discuss it in the hope that it may give the impetus for others to use the methods outlined in this Chapter and apply them to a far greater range of phenomena.

4.2 Simple surface waves

To help contrast the study of surface waves to that of bulk waves, which make up most of this book, we consider the example of waves on the surface of a simple fluid such as water. The fluid is described by a density $\rho(\mathbf{x},t)$, velocity $\mathbf{v}(x,t)$ and pressure $p(\mathbf{x},t)$ and is subject to a gravitational force. Thus inside the fluid we may write

$$\frac{\partial\rho}{\partial t}+\boldsymbol{\nabla}(\rho\mathbf{v})=0, \tag{4.2.1}$$

$$\frac{\partial v}{\partial t}+(\mathbf{v}\cdot\boldsymbol{\nabla})\mathbf{v}=-\mathbf{g}-\frac{1}{\rho}\boldsymbol{\nabla}p, \tag{4.2.2}$$

the usual fluid equations used to study bulk waves. Surface effects enter only through boundary conditions.

The pressure at the surface must be equal to the atmospheric pressure, which is taken to be a linear function of the height of the surface. Thus if we denote by $\xi(\mathbf{x}_\perp,t)$ the vertical displacement of the surface about a horizontal plane which is usually taken as the unperturbed surface of the fluid ($\mathbf{x}_\perp$ is the pair of coordinates in the horizontal plane) then neglecting the density of air compared to the density of the fluid (strictly speaking, ρ is the difference in densities)

$$p(\mathbf{x}_\perp,\xi,t)=p_\mathrm{a}+\rho g\xi, \tag{4.2.3}$$

Here p_a is the atmospheric pressure at the chosen unperturbed surface and is a constant. This condition is applied at the actual surface of the fluid which, when a wave propagates, takes a complicated geometric form. However, if we restrict attention to a small amplitude disturbance and are content with a linearized treatment, then we may apply the above condition not at the actual boundary but at the unperturbed one so that (4.2.3) is replaced in a linear treatment by

$$\delta p(\mathbf{x}_\perp,0,t)=+g\rho_\mathrm{o}\delta\xi(\mathbf{x}_\perp t). \tag{4.2.4}$$

Here δ denotes perturbed quantities and ρ_o is the unperturbed density. The complications of a nonlinear theory are indicated in Section 4.6 and treated more fully in Chapters 5, 6 and 8.

A further complication arises in practice. In the above we have implicitly assumed that the atmospheric pressure is continuous across the fluid surface so that a change of height of the surface must be accompanied by a change of fluid pressure. This must be modified in general because of the restoring effects of surface tension. Such effects produce an extra force which can only be balanced by a further change in pressure so that (4.2.4) is replaced by

$$\delta p = +g\rho_o \delta\xi - T\nabla_\perp^2 \xi. \tag{4.2.5}$$

Here T is the surface tension and $\nabla_\perp$ the two dimensional Laplacian. The spatial variation of the perturbed quantities in the horizontal plane is generally proportional to $\exp(i\mathbf{k}_\perp \cdot \mathbf{x}_\perp)$, in which case the effect of surface tension is to introduce an effective gravitational potential $\bar{g} = g + k_\perp^2 T/\rho_o$. Thus for the sake of simplicity we will not explicitly introduce the effect of surface tension but will simply remember that such effects can be included merely by replacing g by $\bar{g}$.

A second boundary condition derives from the fact that the velocity of the surface must just be the velocity of the fluid at the surface. It is sufficient to consider the vertical component only, so that

$$\frac{\partial \xi}{\partial t} + (\mathbf{v} \cdot \nabla)\xi = \mathbf{v} \cdot \mathbf{i}_z, \tag{4.2.6}$$

where $\mathbf{i}_z$ is the unit vector in the vertical direction. The velocity is evaluated at the surface of the fluid. In the linear approximation this reduces to

$$\frac{\partial \xi}{\partial t}(\mathbf{x}_\perp, t) = \mathbf{i}_z \cdot \delta\mathbf{v}(\mathbf{x}_\perp, 0, t), \tag{4.2.7}$$

with $\delta\mathbf{v}$ evaluated at $z=0$. Conditions (4.2.4) and (4.2.7) may be used to eliminate ξ to give a final boundary condition in the form

$$\frac{\partial}{\partial t}\delta p = +g\rho_o \mathbf{i}_z \cdot \delta\mathbf{v}(\mathbf{x}_1, 0, t). \tag{4.2.8}$$

Finally, we must say something about conditions at the bottom of the fluid. At the bottom we introduce a further boundary condition, namely that the normal component of velocity is zero. In the linear regime and for the case where the unperturbed fluid is of uniform depth h, this takes the form

$$\mathbf{i}_z \cdot \delta\mathbf{v}(\mathbf{x}_\perp, -h, t) = 0, \tag{4.2.9}$$

since we have chosen the origin of the z coordinate to be at the unperturbed surface.

We are now in a position to study surface waves in the linearized approximation. In the unperturbed state $\mathbf{v}=0$, the density ρ_o may be taken to be an arbitrary function of $\mathbf{x}$, and the vertical pressure gradient balances the gravitational force such that

$$g=-\frac{1}{\rho_o}\frac{\mathrm{d}p_o}{\mathrm{d}z}. \tag{4.2.10}$$

We will consider ρ_o to be a function of z but not of the horizontal coordinates $\mathbf{x}_\perp$. The equations are linearized in the usual manner and because of the assumed homogeneity in the horizontal direction we may write

$$\mathbf{v}(\mathbf{x},t)=\delta\mathbf{v}(z)\mathrm{e}^{-\mathrm{i}\omega t+\mathrm{i}k_\perp \mathbf{x}_\perp}. \tag{4.2.11}$$

However, unlike for the case of bulk waves, we cannot assume a normal mode solution in the z direction. It is in fact this difference that leads to the greater mathematical complexity of surface as compared with internal waves. An approximation usually made is to assume the fluid to be incompressible. That is, when moving with the fluid, no changes in the density are observed (this is distinct from the assumption $\rho=$constant throughout, a case which will be considered separately). Thus

$$\frac{\mathrm{d}}{\mathrm{d}t}\rho\equiv\left(\frac{\partial}{\partial t}+\mathbf{v}\cdot\nabla\right)\rho=0, \tag{4.2.12}$$

or, in its linearized form

$$-\mathrm{i}\omega\delta\rho+\delta\mathbf{v}\cdot\boldsymbol{\nabla}\rho_o=0. \tag{4.2.13}$$

For most fluids this is a good approximation but it sometimes needs to be amended for plasmas. Equations (4.2.1) and (4.2.12) further imply that $\nabla\cdot\mathbf{v}=0$, that is

$$\mathrm{i}\mathbf{k}_\perp\cdot\delta\mathbf{v}+\frac{\mathrm{d}}{\mathrm{d}z}\delta v_z=0. \tag{4.2.14}$$

The linearized momentum equation (4.2.2) reduces to

$$-\mathrm{i}\omega\delta\mathbf{v}=-\mathbf{g}\frac{\delta\rho}{\rho_o}-\frac{1}{\rho_o}\boldsymbol{\nabla}\delta p. \tag{4.2.15}$$

By using standard elimination procedures the above equations may be reduced to the single ordinary differential equation

$$\frac{1}{\rho_o}\frac{d}{dz}\left(\rho_o\frac{d\delta v_z}{dz}\right)-k_\perp^2\left(1+\frac{g}{\omega^2}\frac{1}{\rho_o}\frac{d\rho_o}{dz}\right)\delta v_z=0. \tag{4.2.16}$$

This equation has to be solved subject to the boundary conditions, (4.2.8) and (4.2.9). In the present case these can be written in terms of δv_z only, namely

$$\frac{d\delta v_z}{dz}=+\frac{gk_\perp^2}{\omega^2}\delta v_z \tag{4.2.17}$$

at $z=0$ and $\delta v_z=0$ at $z=-h$.

A simpler boundary condition that is often used and is in fact more applicable for the case of a plasma–vacuum boundary is that $\delta v_z=0$ on this boundary as well as at $z=-h$.

The simplest but non-trivial case is where the unperturbed fluid is homogeneous, that is ρ_o is constant. Then the solution of (4.2.16) is simply (we treat $k_\perp>0$ in what follows)

$$\delta v_z=A\mathrm{e}^{-k_\perp z}+B\mathrm{e}^{k_\perp z}.$$

This solution, together with the boundary conditions (4.2.17), gives the dispersion relation

$$\omega^2=gk_\perp\tanh(k_\perp h). \tag{4.2.18}$$

This is the classic ω, $k_\perp$ relation for surface waves on a homogeneous fluid of arbitrary depth. For deep layers ($k_\perp h\gg 1$), $\omega^2=gk_\perp$. It will be noted that, unlike bulk fluid waves ($\omega=kc$), surface waves are dispersive. This property is lost in the shallow water, long wave limit, $\omega^2=ghk^2$. Furthermore, the disturbance δv_z is confined to a surface layer of width of order $1/k_\perp$.

Even this simple treatment can be used to explain much of the fascinating behaviour of sea waves that impinge on our coasts. For a more detailed account of these aspects, reference should be made to Lighthill (1978).

A particular case where the density is non-uniform but can still be treated analytically is where the density variation is of the form

$$\rho_o=\bar{\rho}_o{}^{-z/\lambda}.$$

Equation (4.2.16) now reduces to an equation with constant coefficients whose general solution is of the form

$$\delta v_z=Ae^{l+z}+Be^{l+z}$$

where

$$\frac{1\pm[1+4k_\perp^2\lambda^2(1-g/\omega^2\lambda)]^{\frac{1}{2}}}{2\lambda} \tag{4.2.19}$$

Application of the boundary conditions (4.2.17) gives the following dispersion relation

$$\omega^2 = +gk_\perp^2 \frac{e^{(l_+ - l_-)h} - 1}{l_+ e^{(l_+ - l_-)h} - l_-} \tag{4.2.20}$$

The homogeneous medium result is obtained in the limit $\lambda \to \infty (l_+ \to k_\perp, l_- \to -k_\perp)$.

A somewhat surprising result is obtained in the deep fluid limit ($h \to \infty$). In this case (4.2.20) reduces to $\omega^2 = gk_\perp^2 l$, a seemingly different dispersion relation to the case of a homogeneous medium ($\omega^2 = gk_\perp$). However, l is by definition a function of ω and simple algebraic manipulation reveals that, even for the inhomogeneous medium, $\omega^2 = gk_\perp$. This is independent of $\lambda_\parallel$, the inverse density gradient. This result suggests that for deep fluid systems, a continuous density variation is not too important. In fact, if we consider the deep fluid limit from the start and impose the boundary condition $\delta v_z \to 0$ as $z \to -\infty$, we have

$$\delta v_2 = A e^{gk_\perp^2 z/\omega 2}.$$

This is an exact solution of (4.2.16) and satisfies the boundary condition (4.2.17) if $\omega^2 = gk_\perp$. This result is independent of the particular density variation. This simplicity does not survive if we use the plasma boundary condition $\delta v_z = 0$ at $z = 0$.

Another approximation frequently made in studying inhomogeneous media is to consider the density to be discontinuous at some value of z, $\bar{z}$ say such that

$$\begin{aligned} \rho_o &= \rho_1 z > \bar{z}, \\ &= \rho_2 z < \bar{z}. \end{aligned}$$

Note that since the true surface is at $z = 0$, we have $-\bar{z} < 0$. Then for $z = \bar{z}$, (4.2.16) reduces to a very simple equation which has exponential solutions.

For $z > \bar{z}$ we write

$$\delta v_z = A e^{k_\perp z} + B e^{-k_\perp z} \tag{4.2.21}$$

and use the boundary condition (4.2.17) to reduce this to

$$\delta v_z = A\{\cosh(k_\perp z) + (gk_\perp/\omega^2)\sinh(k_\perp z)\}. \tag{4.2.22}$$

Similarly, for $z < \bar{z}$ the solution satisfying the boundary condition at $z = -h$ is

$$\delta v_z = \bar{A} \sinh(k_\perp(h + z)). \tag{4.2.23}$$

To obtain a unique solution we use the fact that (4.2.16) is true for all z between 0 and $-h$. Thus we first multiply by ρ_o and integrate it between the limits $\bar{z}-\varepsilon$ and $\bar{z}+\varepsilon$ to obtain

$$\rho_o \frac{\mathrm{d}\delta v_z}{\mathrm{d}z}\Big|_{\bar{z}-\varepsilon}^{\bar{z}+\varepsilon} = k_\perp^2 \int_{\bar{z}-\varepsilon}^{\bar{z}+\varepsilon} \rho_o \left(1+\frac{g}{\omega^2 \rho_o}\frac{\mathrm{d}\rho_o}{\mathrm{d}z}\right)\delta v_z \mathrm{d}z$$

$$= \frac{k_\perp^2 g}{\omega^2} \rho_o \delta v_z \Big|_{\bar{z}-\varepsilon}^{\bar{z}+\varepsilon}$$

$$+ k_\perp^2 \int_{\bar{z}-\varepsilon}^{\bar{z}+\varepsilon} \rho_o \left(\delta v_z - \frac{g}{\omega^2}\frac{\mathrm{d}\delta v_z}{\mathrm{d}z}\right)\mathrm{d}z.$$

Now, though ρ_o and $\mathrm{d}\delta v_z/\mathrm{d}z$ are discontinuous at $z=\bar{z}$, the integrand remains bounded. In the limit as $\varepsilon \to 0$ the integral goes to zero. In this case we have

$$\rho_1 \frac{\mathrm{d}\delta v_z}{\mathrm{d}z}\Big|_+ - \rho_2 \frac{\mathrm{d}\delta v_z}{\mathrm{d}z}\Big|_- = +\frac{k_\perp^2 g}{\omega^2}(\rho_1-\rho_2)\delta v_z(\bar{z}) \tag{4.2.24}$$

where $|_+, |_-$ denote the values of the quantity on either side of $z=\bar{z}$. We have used the condition that δv_z is continuous at $z=\bar{z}$. This condition is obtained from (4.2.7) since ξ, the displacement of the surface, must be continuous (otherwise the fluid would come apart at the surface $z=\bar{z}$).

This procedure of replacing an inhomogeneous medium by one with discontinuous behaviour is quite widespread. The major advantage is that the differential equation with spatially varying coefficients is replaced by one with constant coefficients in the different regions. The differential equations are easily solved, but this is at the expense of introducing extra boundary conditions of the form (4.2.24). These are obtained from the original differential equation by the process elucidated above. This whole procedure reduces the problem to the solution of algebraic equations rather than finding the eigenvalues of differential equations.

Using these conditions in conjunction with (4.2.22) and (4.2.23) gives the dispersion relation

$$(\omega^2/k_\perp g)^2(\rho_2-\rho_1 T_h T_z)+\rho_2(T_z-T_h)(\omega^2/k_\perp g)+T_z T_h(\rho_1-\rho_2)=0, \tag{4.2.25}$$

where $T_h=\tanh(k_\perp(h+z))$ and $T_z=\tanh(k_\perp z)$.

An important limit is for the case where the fluid–air interface ($z=0$) is far removed from the discontinuity at $\bar{z}$. Since $\bar{z}<0$ this is the limit $\bar{z}\to-\infty$ so that $T_z\to-1$ but such that $h+\bar{z}=\hat{h}$ remains finite, where $\hat{h}$ is now the distance from the bottom of the fluid to the interface $z=\bar{z}$. The dispersion relation reduces to

$$(\omega^2/k_\perp g)^2(\rho_2+\rho_1 T_h)-\rho_2(1+T_h)(\omega^2/k_\perp g)-T_h(\rho_1-\rho_2)=0. \tag{4.2.26}$$

This may be factorized to give

$$\omega^2=k_\perp g(\rho_2-\rho_1)T_h/(\rho_2+\rho_1 T_h), \tag{4.2.27}$$

and $\omega^2/k_\perp g=+1$. The second root corresponds to a surface wave on the surface $z=0$ in a medium of infinite depth. The first solution corresponds to a wave localized horizontally by the discontinuity at $\bar{z}$ which is at height $\hat{h}$ above the bottom of the medium with density ρ_2. The medium with density $\rho_1(<\rho_2)$ extends infinitely upwards. Thus this wave may be considered as an internal wave localized about $z=\bar{z}$ or else as a surface wave propagating on the surface between the two media. Finally, if we consider the case of $\hat{h}\to\infty$, that is, the surface $z=\bar{z}$ is now well away from the bottom and the top of the fluid, (4.2.27) reduces to ($T_h\to 1$)

$$\omega^2=k_\perp g\frac{(\rho_2-\rho_1)}{(\rho_2+\rho_1)}. \tag{4.2.28}$$

This is of course the result for two semi-infinite homogeneous media of densities ρ_1 and ρ_2, having a common interface (at $z=\bar{z}$, but now $\bar{z}$ is purely arbitrary and could be taken to be zero). This result, that ω^2 depends critically on the density variation is to be contrasted with the result obtained above for the exponentially varying density. In that case ω^2 was found to be independent of the density variation, at least for the case of an infinitely deep fluid ($h\to\infty$). Note that for small ρ_1/ρ_2, (4.2.28) also becomes density independent.

In the above we have illustrated most of the basic ideas behind the treatment of surface waves, in particular the difference between this approach and that used for bulk waves. Many interesting phenomena have been ignored. Two examples are: the effect of various forms of attenuation, and the changes brought about by the bottom surface, here taken to be horizontal at $z=-h$, being dependent on position, $\mathbf{x}_\perp$. These are treated in detail in more specialized texts such as Lighthill (1978).

4.3 The Rayleigh–Taylor instability

The simple result (4.2.28), which describes the propagation of a wave ($\omega^2>0$) on the interface between a light fluid of density ρ_1 on top of a heavy one of density $\rho_2(\rho_2>\rho_1)$, opens up a Pandora's box of instabilities. Suppose the heavier fluid lies on top of the lighter one, as when a beaker of water is inverted, then $\rho_1>\rho_2$ and $\omega^2<0$, that is instead of wave propagation we have an instability. In plasma physics, as we have seen in Section 4.1, the quantity g should be considered as an effective gravitational field, depending on magnetic field curvature. For many magnetic configurations $g<0$. Now even for $\rho_2>\rho_1, \omega^2<0$ we again have an instability. Such instabilities plague the controlled thermonuclear reactor programme and have received considerable attention. It is not the purpose of this book to discuss this aspect in great detail, but in the following few paragraphs we outline some of the salient points. For a thorough discussion of the Rayleigh–Taylor instability, reference should be made to Chandrasekhar (1961). The review by Wesson (1981) gives more modern results particularly relevant to the nuclear fusion programme.

For plasma physics applications it is more appropriate to use the plasma boundary condition $\delta v_z=0$ at $z=0$ in place of (4.2.17). However, the result (4.2.28) still applies since the boundary $z=0$ has effectively been pushed an infinite distance away from the discontinuity. In fact (4.2.25) is replaced by

$$\omega^2=\frac{k_\perp g(\rho_2-\rho_1)T_h}{\rho_2-\rho_1 T_h/T_z}, \tag{4.3.1}$$

so that in the limit $\bar{z}\to-\infty, T_z\to-1$, the above reduces to (4.2.27) which itself reduces to (4.2.28) for $h\to\infty$.

A method that has been suggested to stabilize the instability is to introduce a magnetic field B at right angles to the z direction. The net result is to replace (4.2.16) by

$$\frac{1}{\rho}\frac{\mathrm{d}}{\mathrm{d}z}\left(\rho\frac{\mathrm{d}\delta v_z}{\mathrm{d}z}\right)-k_\perp^2\left(1+\frac{g}{\omega^2}\frac{1}{\rho}\frac{\mathrm{d}\rho_o}{\mathrm{d}z}\right)\delta v_z=0, \tag{4.3.2}$$

where $\rho=\rho_o-(\mathbf{k}_\perp\cdot\mathbf{B}(z))^2/\omega^2$. For the case of a discontinuity in the density, but not in $B(z)$, (4.2.24) still holds. Furthermore, if $B(z)$ is taken to be independent of z, the analysis proceeds by analogy with that given in the last Section. In particular (4.2.28) is replaced by

$$\omega^2=\{k_\perp g(\rho_2-\rho_1)+2(\mathbf{k}_\perp\cdot\mathbf{B})^2\}/(\rho_1+\rho_2), \tag{4.3.3}$$

showing that the introduction of the field can be a stabilizing effect. Unfortunately, disturbances propagating at right angles to $\mathbf{B}$, that is $\mathbf{k}_\perp\cdot\mathbf{B}=0$, are unaffected by this stabilization.

In an attempt to overcome this disappointment one is naturally led to the idea of introducing shear into the magnetic field. The idea is to allow $\mathbf{B}$ to be a function of z so that $\mathbf{k}\cdot\mathbf{B}(z)\neq 0$ except possibly at one value of z, z_c say. It is then necessary to reconsider the full equation (4.3.3), when it will be noted that it is singular at the point $\rho\omega^2=0$, that is

$$\omega^2=(\mathbf{k}_\perp\cdot\mathbf{B}(z))^2/\rho_o.$$

Since the right hand side is positive definite, such a singularity can only occur for $\omega^2>0$, a stable configuration. Unfortunately this includes the marginally stable case $\omega^2=0$, for which the equation is singular for $z=z_c$. The plane $z=z_c$ is called the resonant plane and for more general configurations the surface defined by $\mathbf{k}\cdot\mathbf{B}=0$ is called a resonant surface. Such surfaces play an important role in plasma dynamics, as is to be expected since on such surfaces any possible stabilization due to shear is minimal. It is found that the perturbations such as δv_z become localized around such surfaces as if trying to eliminate the stabilizing effects of shear.

4.4 The Kelvin–Helmholtz instability

This instability arises when stratified layers of fluid are in relative motion. For example, think of a wind blowing over a water surface. Here the stratified layers are the water and the air with the wind causing the relative motion. In this case the instability manifests itself in waves on the water surface. A very similar situation occurs in the upper atmosphere where the wind velocity and density both vary with height above the Earth's surface. In this case the secondary effect of cloud formation in regions of higher density results in the instability being made visible by a cloud pattern, the herring-bone cloud formation mentioned before (see Fig. 1.1).

The basic equations are still (4.2.1) and (4.2.2), but the initial state or equilibrium is such that the fluid is streaming so that there is a non-zero velocity v (in a direction perpendicular to the z direction) which may be an arbitrary function of z. The density may also be a function of z. Assuming the perturbed motion to be incompressible, the linearized equations of motion reduce to

$$\frac{1}{\rho_o}\frac{d}{dz}\left\{\rho_o(\omega-\mathbf{k}_\perp\cdot\mathbf{u})\frac{d\delta v_z}{dz}\right\}+\frac{1}{\rho_o}\frac{d}{dz}\left(\rho_o\frac{d}{dz}(\mathbf{k}_\perp\cdot\mathbf{u})\delta v_z\right)$$
$$-k_\perp^2\left[(\omega-\mathbf{k}_\perp\cdot\mathbf{u})+\frac{\dfrac{g}{\rho_o}\dfrac{d\rho_o}{dz}}{(\omega-\mathbf{k}_\perp\cdot\mathbf{u})}\right]\delta v_z=0. \tag{4.4.1}$$

For $u=0$ or $\mathbf{k}_\perp\cdot\mathbf{u}=0$ this reduces to (4.2.16). The boundary condition (4.2.17) is changed to

$$\frac{d}{dz}\{(\omega-\mathbf{k}\cdot\mathbf{u})\delta v_z\}=\frac{gk_\perp^2\delta v_z}{(\omega-\mathbf{k}_\perp\cdot\mathbf{u})}$$

at $z=0$, whilst the condition $\delta v_z=0$ at $z=-h$ remains unchanged.

The simplest example to illustrate some of the properties of the Kelvin–Helmholtz instability is to consider the case where ρ and $\mathbf{u}$ are discontinuous at some value of z, $\bar{z}$ say. The condition that the fluid does not separate at this surface is equivalent to demanding that the displacement $\delta\xi$ is continuous. This requires, upon application of the linearized version of (4.2.6), that $\delta v_z/(\omega-\mathbf{k}_\perp\cdot\mathbf{u})$ be continuous. The second condition is obtained in an analogous manner to that used to derive (4.2.24), namely by integrating the basic equation (4.4.1) over a small region on each side of $z=\bar{z}$. Thus the condition which replaces (4.2.24) becomes

$$\rho_1(\omega-\mathbf{k}_\perp\cdot\mathbf{u}_1)^2\frac{\mathrm{d}\delta v_z}{\mathrm{d}z}\bigg|_+ -\rho_2(\omega-k_\perp\cdot\mathbf{u}_2)^2\frac{\mathrm{d}\delta v_z}{\mathrm{d}z}\bigg|_- =k_\perp^2 g(\rho_1-\rho_2)\delta u(\bar{z}). \quad (4.4.2)$$

where $\delta u(z)=\delta v_z/(\omega-k_\perp\cdot\mathbf{u})$ and is continuous at $z=\bar{z}$. For $z\neq\bar{z}$ where both ρ_o and u are constants (4.4.1) reduces to $d^2\delta v_z/\mathrm{d}z^2-k_\perp^2\delta v_2=0$, so that δv_z is readily found. The solution must satisfy the boundary conditions at $z=0$ and $z=-h$. Here we consider the simple case where these boundaries are far away from $z=\bar{z}$ and it is then sufficient to simply demand that δv_z tend to zero as z changes from $\bar{z}$. Thus we write

$$\begin{aligned}\delta v_z&=A\mathrm{e}^{-k_\perp(z-\bar{z})} \qquad && z>\bar{z},\\ &=B\mathrm{e}^{k_\perp(z-\bar{z})} && z<\bar{z}.\end{aligned}$$

The continuity of δu (but not of δv_z) gives the relation

$$A/(\omega-\mathbf{k}_\perp\cdot\mathbf{u}_1)=B/(\omega-\mathbf{k}_\perp\cdot\mathbf{u}_2).$$

Finally, using (4.4.2) gives the dispersion relation

$$\rho_1(\omega-\mathbf{k}_\perp\cdot\mathbf{u}_1)^2+\rho_2(\omega-\mathbf{k}_\perp\cdot\mathbf{u}_2)^2=k_\perp g(\rho_2-\rho_1) \quad (4.4.3)$$

This reduces to (4.2.28) for $\mathbf{k}_\perp\cdot\mathbf{u}=0$.

Solving for ω, we find

$$\omega=\frac{(\rho_1\mathbf{k}_\perp\cdot\mathbf{u}_1+\rho_2\mathbf{k}_\perp\cdot\mathbf{u}_2)\pm\sqrt{[k_\perp g(\rho_2^2-\rho_1^2)-\rho_1\rho_2(\mathbf{k}_\perp\cdot\mathbf{u}_1-\mathbf{k}_\perp\cdot\mathbf{u}_2)^2]}}{(\rho_1+\rho_2)}, \quad (4.4.4)$$

so that the condition for instability is

$$\rho_1\rho_2(\mathbf{k}_\perp\cdot\mathbf{u}_1-\mathbf{k}_\perp\cdot\mathbf{u}_2)^2>k_\perp g(\rho_2^2-\rho_1^2). \quad (4.4.5)$$

If we denote by ψ the angle between $\mathbf{k}_\perp$ and the velocities $\mathbf{u}$ then the above may be written in the form

$$k_\perp>\frac{g(\rho_2^2-\rho_1^2)}{\rho_1\rho_2(u_1-u_2)^2\cos^2\psi},$$

which can always be satisfied for all values of the relative density. However, in practice, short wavelength disturbances (large $k_\perp$) are particularly vulnerable to the stabilizing influence of surface tension and viscosity.

It was shown in Section 4.2 that the effect of surface tension could be incorporated into the theory simply by replacing the gravitational acceleration g by an effective value $\bar{g}=g+k_\perp^2 T/\rho$. Here ρ is the difference in densities in the two adjacent media. Thus if in (4.4.5) g is replaced by $\bar{g}$ with $\rho=\rho_2-\rho_1$ then the condition for instability now becomes (we put $\psi=0$ as this is the easiest direction for the instability to grow)

$$\rho_1\rho_2(u_2-u_1)^2>(\rho_2^2-\rho_1^2)\left\{\frac{g}{k_\perp}+\frac{k_\perp T}{(\rho_2-\rho_1)}\right\}$$

The right hand side takes its minimum value for $k_\perp=g(\rho_2-\rho_1)/T$ giving the condition for instability as

$$(u_2-u_1)^2>2\frac{(\rho_1+\rho_2)}{\rho_1\rho_2}\sqrt{[Tg(\rho_2-\rho_1)]}, \tag{4.4.6}$$

a result first obtained by Kelvin. Thus, as expected, the introduction of surface tension limits the value of $k_\perp$ which then imposes a minimum value on the velocity difference before the surface will go unstable.

For a wind blowing over the sea, the critical velocity difference is 650 cm/s. In practice, as already pointed out by Kelvin, much smaller velocities will ruffle the sea surface and this can only be due to effects omitted in the above theory. However, this instability has been demonstrated under laboratory conditions and the experimental results agree well with (4.4.6) (see Chandrasekhar (1961), Chapter II).

A more interesting result arises when, due to some imposed condition, the perturbation must have a specific wavenumber in a direction perpendicular to the velocity **u**. That is to say, if the velocity **u** is in the x direction, k_y is a fixed quantity whereas k_x is arbitrary. The instability condition may then be written as

$$(u_1-u_2)^2>\frac{(\rho_1+\rho_2)T}{\rho_1\rho_2}F(k_x,k_y,S) \tag{4.4.7}$$

Where $S=g(\rho_2-\rho_1)/T$ and

$$F=\frac{(k_x^2+k_y^2)^{\frac{1}{2}}}{k_x^2}(S+k_x^2+k_y^2) \tag{4.4.8}$$

We now minimize F with respect to k_x and find that the minimum value is given by

$$2k_x^2=S+k_y^2+\sqrt{\left[(S+k_y^2)(S+9k_y^2)\right]} \tag{4.4.9}$$

For $k_y \equiv 0$, we find $k_x^2 = S, F = 2\sqrt{S}$ and then (4.4.7) reduces to (4.4.6). However, this is the minimum value of F and thus any finite value of k_y will increase the right hand side of (4.4.7) over the value given by (4.4.6), thus increasing the velocity difference necessary for instability. Thus the imposition of a finite value of k_y is a stabilizing effect. In Section 4.1 it was suggested that fields of corn are prone to the Kelvin–Helmholtz instability. The above result suggests that the destructive power of the wind blowing over a field of corn could be limited by planting trees periodically in a direction perpendicular to the prevailing wind, if known. This would impose a value of k_y equal to $2\pi/l$ where l is the distance between the trees.

The Kelvin–Helmholtz instability has been discussed in the above under the assumption that the density ρ and velocity u are discontinuous at some surface $z = \bar{z}$. A discussion of the instability for the case where the variation in u and ρ is continuous leads to some interesting mathematical problems and, as these are more general than just in the present context, they are worth discussing further. Equation (4.4.1) can be rewritten in the form

$$\mathrm{D}(z)\mathrm{L}\delta v_z = 0 \tag{4.4.10}$$

where $\mathrm{D}(z) = \omega - \mathbf{k}_\perp \cdot \mathbf{u}(z)$ and the linear operator L takes the form

$$\mathrm{L}\delta v_z \equiv \frac{1}{\rho_o}\frac{\mathrm{d}}{\mathrm{d}z}\left(\rho_o \frac{\mathrm{d}\delta v_z}{\mathrm{d}z}\right) - \left[k_\perp^2 + \frac{\left(g + \frac{1}{2}\frac{\mathrm{d}}{\mathrm{d}z}\mathrm{D}^2\right)}{\rho_o \mathrm{D}^2}\frac{\mathrm{d}\rho_o}{\mathrm{d}z} + \frac{\mathrm{d}^2\mathrm{D}}{\mathrm{d}z^2}\frac{1}{\mathrm{D}}\right]\delta v_z = 0 \tag{4.4.11}$$

It seems natural from the form of (4.4.10) to think that by solving $\mathrm{L}\delta v_z = 0$ one has arrived at the complete solution. However, as first pointed out by Case (1959), the complete solution must also include solutions of the inhomogeneous equation

$$\mathrm{L}\delta v_z = \delta(\mathrm{D}(z)),$$

where δ is the Dirac delta function. This is most readily appreciated by noting that the equation $xf(x) = 0$, in addition to the obvious solution $f(x) = 0$, also has the solution $f(x) = \delta(x)$, since $x\delta(x) = 0$.

This procedure has been discussed in detail for the case $u(z) = u_o z/d$ where u_o and d are constants. It is somewhat surprisingly found that the system is stable. However, the inclusion of a more general variation for $u(z)$ and hence the inclusion of the last term in the coefficient of δv_z in (4.4.11) showed the existence of the basic instability.

We conclude with two observations which in fact have a wide range of relevance. Firstly, the form of the basic eigenvalue equation (4.4.10) is quite common, in which case the inclusion of the solutions of the corresponding inhomogeneous equations must also be taken into account. Secondly, the simplifying procedure of replacing spatially varying coefficients by ones having discrete steps is extremely useful but should be treated with some degree of caution.

4.5 Solid–liquid interface instabilities

The very practical need to grow single crystals of various materials with an extremely high degree of homogeneity has led to much research into a particular method, known as growth from the melt. Experimentally it is found that simple metals cooled sufficiently rapidly will not form simple crystals with smooth surfaces, but rather these surfaces will have a honeycombed structure (somewhat reminiscent of the Giants' Causeway in Ireland). This of course means a loss in homogeneity and hence is to be avoided. This honeycombed structure is the final configuration of a linear instability that is controlled by nonlinear effects. Like many other examples in Nature, whilst progress has been made in an understanding of an instability in a linear regime, the final state, obviously controlled by nonlinear effects, is much less well understood. This is of course regrettable from a practical point of view but in this case also from an aesthetic one. The nonlinear effects, as revealed experimentally, lead to beautiful dendritic structures, with perhaps the common snowflake as an exceptionally fine example.

However, in the context of the rest of this Section it is the linear stability analysis that is relevant. This is based on the work of Mullins and Serkerka. For an introduction to this area of research and for some fine illustrations of dendritic growth see the book by Woodruff (1973).

The following is based on a paper by Sekerka (1967) in which a careful examination is made of the linear stability of a planar interface between an alloy and its melt. The solid–liquid interface is assumed on average to be moving with a constant velocity V. Then, in a frame of reference relative to this interface, the diffusion of the solute in the liquid phase is governed by the 'flowing' diffusion equation

$$\frac{\partial C}{\partial t} = v\frac{\partial C}{\partial x} + D_c \nabla^2 C, \tag{4.5.1}$$

where x is the direction perpendicular to the moving boundary. The velocity v is in reality the velocity of the interface averaged over the perpendicular plane. The temperature T_l satisfies a similar equation but with D_c replaced by the thermal diffusivity D_t. In the solid, concentration diffusion is ignored but the temperature T_s again satisfies an equation of the above form but now with D_c replaced by the appropriate thermal diffusivity D_s.

The complexity of the present problem arises because of the nonlinearity, not of the governing equations, but of the boundary conditions. As in all cases studies in this Chapter, the boundary is in general non-planar and we express its position in the form $x=\phi(y,t)$ so that the boundary velocity $v(y,t)=V+\partial\phi/\partial t$. (For simplicity we assume no variation in the z direction.) Now on this surface we must have continuity of temperature, that is $T_1=T_s$. The other usual temperature condition at a boundary, of the continuity of thermal currents, must in the present context be supplemented to include the latent heat L produced at the interface by the solidification process. Thus at the boundary we have

$$vL=K_s\frac{\partial T_s}{\partial x}-K_1\frac{\partial T_1}{\partial x}, \tag{4.5.2}$$

where the thermal conductivities K are given by $K=D/\rho S$ and ρ is the appropriate density and S the specific heat. It is also assumed that the solute concentration at the boundary is in thermodynamic equilibrium. This gives

$$v=\frac{D_c}{C_o(k-1)}\frac{\partial C}{\partial x}. \tag{4.5.3}$$

Here C_o is the equilibrium solute concentration in the solid and C_o/k the concentration in the liquid at the phase boundary. Finally, the temperature of the interphase must also be the thermodynamic temperature of melting T_m modified due to the presence of the solute so that

$$T_1=T_m+mC. \tag{4.5.4}$$

We have neglected a capilliary contribution in (4.5.4).

The same linearization procedure as discussed in Section 4.2 is now invoked. In particular, the boundary conditions are applied at the unperturbed boundary, which in the moving reference frame is simply $x=0$. A major simplification that occurs in the present problem compared to those considered in Section 4.2 arises because the defining equations are linear (though not the boundary conditions). Thus, although the steady state temperatures and concentrations (T_s, T_c and C) are spatially dependent (functions of x), the linearized equations only involve partial derivatives with constant coefficients. Thus the perturbing spatial variation in the y direction may be accounted for by assuming variation proportional to $\exp(iky)$ and the time dependence to $\exp(-i\omega t)$.

The treatment given by Sekerka (1967) involves use of a Laplace–Fourier transform to deal with the time and y dependence though, as is apparent from the analysis described in Chapter 2, it is sufficient to use a normal mode analysis.

The various perturbed quantities must also decay exponentially with x away from the boundary. After considerable algebra an algebraic dispersion relation is obtained which can be written in the usual form $\mathrm{D}(\omega, k)=0$. This is given by Sekerka (1967) equation 53 with his $\omega \equiv k$ and his $s=-\mathrm{i}\omega$. A considerably simplification can be made by noting that both D_l and D_t are typically a thousand times larger than D_c. This means that the temperature relaxes very quickly to a *local* thermodynamic value, which itself varies through the time variation of C because of the boundary condition (4.5.4). With this simplification Sekerka was able to derive a necessary and sufficient condition for stability, that is that the planar boundary remain planar no matter what the wavelength of the perturbation.

4.6 A first look at gravity wave instabilities

Up till now we have been using simple linear theory to describe rather complex surface configurations. This same linear theory can also be used to suggest when to *expect* instabilities when we move on to nonlinear theory. To illustrate this, we will introduce a general technique by looking at gravity waves on a water surface again. The technique will offer us our first glimpse of things to come (large amplitude waves), and can be used for a wide range of problems.

We have seen how gravity waves on a water surface can be described by a set of differential equations and boundary conditions. Small but finite amplitude waves will be considered in some detail in Chapter 5, where an understanding of how these waves develop will help us formulate general weakly nonlinear theory. By a combination of simple algebra and a presentation of more advanced results without a derivation, we will now try to gain some insight into the instabilities that plague finite amplitude water waves. Depth will be taken to be arbitrary. The details can be found in McLean (1982a, b) and, for the deep water limit, in Saffman and Yuen (1985). Extensions from simple waves, as considered here, to modulated wavetrains on a water surface, again of arbitrary depth, can be found in Infeld *et al.* (1987).

4.6.1 The small amplitude onset of wave instability

As found in Section 4.2, the dispersion relation for gravity waves over a finite depth h is

$$\omega=\sqrt{[k\tanh(hk)]}. \tag{4.6.1}$$

Here we have taken the gravitational acceleration to be equal to 1.0, dropped the subscript in k, and chosen the plus sign when extracting the square root in (4.2.18). Thus the wave propagates to the right. We will use knowledge of this simple dispersion relation to obtain information about the onset of instability brought on by finite wave amplitude (at first glance this might seem to be outside the scope of linear theory).

Suppose we abandon our rest frame, in which (4.6.1) holds, and just sit on the wave it describes. Furthermore, we will measure lengths in units of reciprocal wavenumber $k^{-1}=\lambda/2\pi$ and take the x axis along $\mathbf{k}$. Thus we are in fact moving with velocity $\sqrt{(\tanh h)}$ along x. Any other linear wave (any $\mathbf{K}$) will now obey the Doppler shifted dispersion relation

$$\Omega_{\pm} \sqrt{[K\tanh(hK)]}-cK_x \tag{4.6.2}$$

$$c=\sqrt{(\tanh h)}. \tag{4.6.3}$$

So far there is little difference between our basic or carrier wave ω, $\mathbf{k}$ and any secondary waves that might be propagating with Ω, $\mathbf{K}$, all being described by linear theory. We will, however, try to think of the former wave as the vanishing amplitude limit of a nonlinear wave structure. If the amplitude A were increased, linear theory would no longer apply and the dispersion relation (4.6.1) would cease to hold. However, the resulting nonlinear wave would still be describable by a wavenumber k, albeit dependent on A. We would still insist on normalizing lengths to $k(A)^{-1}$. The amplitude A will be treated as a control parameter in what follows.

Simple algebra will now tell us something (though by no means everything) about the possible onset of instability when the amplitude of the carrier wave, A, is small. Methods for extending this approach will be developed progressively in Chapters 5 and 8. Now that the amplitude of the carrier wave is finite, all secondary waves will feel its presence, in particular its periodicity.

It is a mathematical fact of life, following from Floquet–Bloch theory, that, with the conventions introduced above (and even when the amplitude of the carrier wave is not small), all secondary waves travelling in the medium are of the form (η is the elevation of the water surface above the mean)

$$\eta=e^{i(K_x x+K_y y-\Omega t)}P(x), \tag{4.6.4}$$

where $P(x)$ is a periodic function with period 2π. However, when the amplitude of the carrier wave *is* small, the value of Ω is one of

$$_{\pm}=\pm\sqrt{[K\tanh(hK)]}-c(K_x+m) \tag{4.6.5}$$

$$K^2=(K_x+m)^2+K_y^2. \tag{4.6.6}$$

The difference with respect to (4.6.2) is in the altered form of K. But why the extra integer m? Well, if $P(x)$ is periodic, period 2π, so will $e^{imx}P(x)$ be. Therefore, the function $P(x)$ can be freely exchanged for $e^{imx}P(x)$ in (4.6.4). Thus, from the point of view of the generalization of Fourier analysis as expressed by (4.6.4), there is in fact a degeneracy. All m modes are valid. This degeneracy is introduced by the *very existence* of the carrier wave mode. The value of m need not be the same for Ω_+ and Ω_- and so from now on it will be designated by m_+ or m_-.

Let us now recapitulate the situation. We are happily sitting on a wave, which we consider to be basic, and its amplitude can, in theory, be increased from a very small value. We also observe other waves that are *always* small amplitude.

When can the whole system go unstable? There are at least two possible answers to this question in physical terms (hopefully giving the same conditions). One is to treat our basic wave as a material property of the medium, and ask when Ω can become complex in this altered medium. (This is in the spirit of the fact that the basic wave is static in our coordinate system.) Complex Ω can be expected when a degeneracy takes place and Ω has a double root, for example

$$\Omega_+(\mathbf{K}, m_+) = \Omega_-(\mathbf{K}, m_-). \tag{4.6.7}$$

It is natural to expect two roots to go complex by first merging into a double, real root. This 'going complex' can be expected to occur when a control parameter increases. In our problem the carrier wave amplitude A is just this control parameter. Thus, in our picture, increasing A will further alter the medium. This (increase of A) will remove the degeneracy (4.6.7) and could build $\mathrm{Im}\Omega$ up from zero. (In fact, when it does, the *maximum* growth rate proves to be proportional to $A^{32m_+ - m_-|}$.)

Those readers who are familiar with the concept of wave–wave coupling will appreciate a second way of looking at this onset of instability. We now think of the carrier wave and the two secondary waves as a triad in which energy can leak from two components to a third if the Manley–Rowe conditions (e.g. Davidson (1972), Chapter 6):

$$\begin{aligned} &\mathbf{K}_+ = \mathbf{K}_- + M\mathbf{k} \\ &\Omega_+ = \Omega_- + M\omega \\ &\mathbf{K}_\pm = \mathbf{K} + m_\pm \mathbf{i}_x,\ M = m_+ - m_-,\ \mathbf{k} = (1, 0) \end{aligned} \tag{4.6.8}$$

are satisfied for some integer M. The three waves are thus treated on a more equal footing in this reasoning.

As $\omega = 0$ in our coordinate system, these two approaches lead to the same condition, as expected. We will concentrate on $m_+ = N, m_- = -N$; and

$m_+ = N, m_- = -N-1$, covering all interesting cases if the degeneracy in K_x is considered. We thus introduce the following classification for resonances so obtained:

$$\text{Class I}\quad \sqrt{[K_+ \tanh(hK_+)]} + \sqrt{[K_- \tanh(hK_-)]} = 2Nc = Mc \tag{4.6.9}$$

$$\mathbf{K}_\pm = (K_x \pm N, K_y)$$

$$\text{Class II}\quad \sqrt{K_+ \tanh(hK_+)]} + \sqrt{[K_- \tanh(hK_-)]} = (2N+1)c = Mc$$
$$\mathbf{K}_+ = (K_x + N, K_y)$$
$$\mathbf{K}_- = (K_x - N - 1, K_y).$$

It should be stressed that we have only found necessary conditions for the onset of instability. Corresponding resonance curves as follow from (4.6.9) and (4.5.10) in the K_x, K_y plane for three values of h are given in Fig. 4.1. We expect that, as the amplitude of the basic wave is increased from zero, unstable regions will appear around at least some segments of these resonance curves. Numerical calculations (obtained in the papers referenced above and in the captions), fully confirm this expectation (Fig. 4.2).

Some analytical methods for finding the unstable regions and corresponding growth rates of instabilities for small K will be outlined in Chapters 5 and 8. Work is also proceeding on the case $K_x = \frac{1}{2}$, and K_y in the corresponding unstable Class II regions. Though we will not review this work we will try to indicate why this case is of special interest.

Fig. 4.1. Resonance curves from the linear dispersion relation. N is the order of the resonance: (*a*) $kh = 0.5$; (*b*) $kh = 2$; (*c*) $kh = \infty$ (deep water). (*a*) and (*b*) from McLean (1982b), (*c*) from Saffman and Yuen (1985).

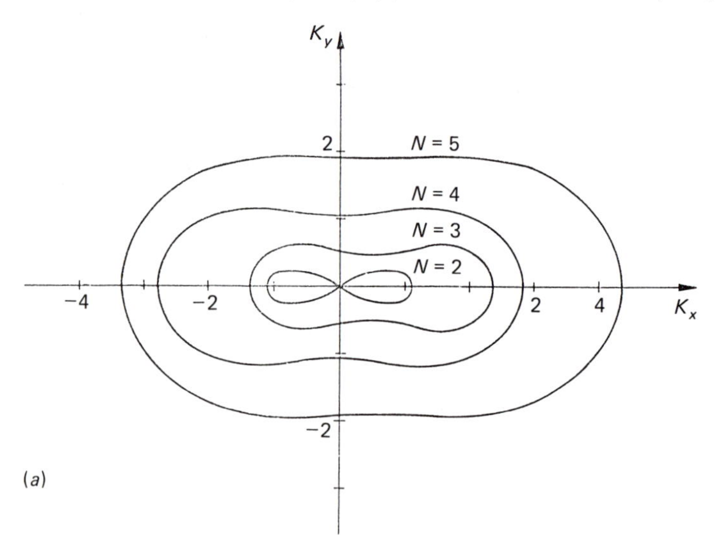

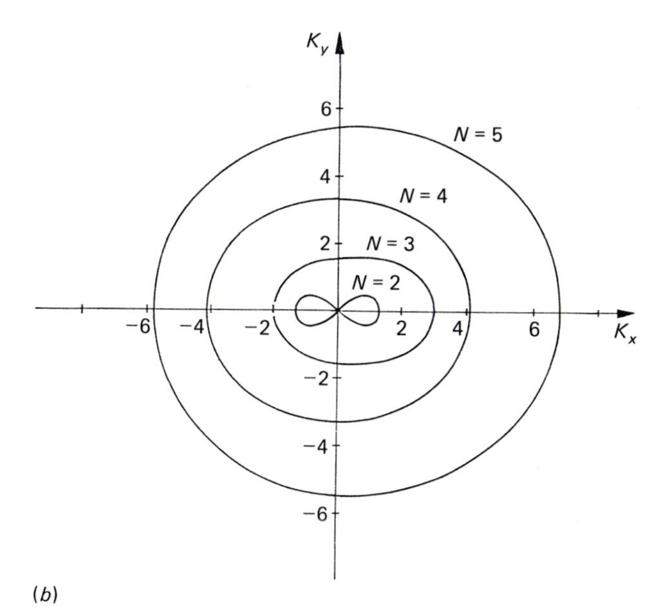

(*b*)

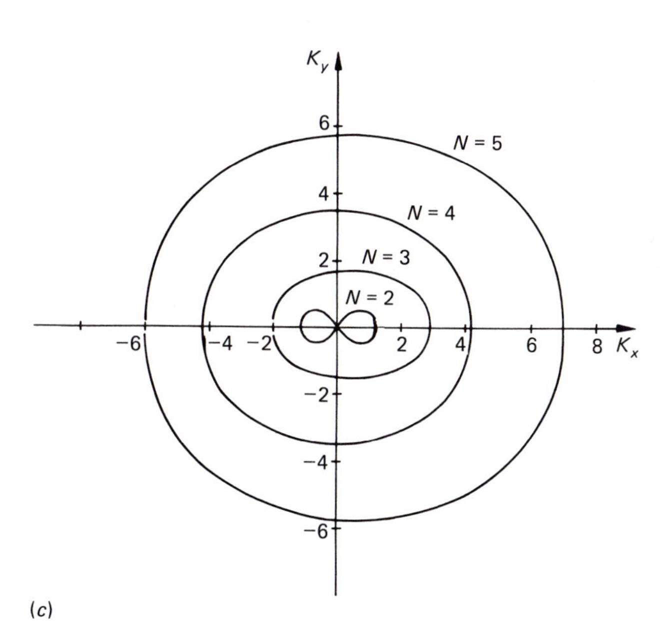

(*c*)

Fig. 4.2. Bands of instability for $M=2$, $M=3$, and (*a*) $kh=0.5$ and $kA=0.1$ and 0.16; (*b*) $kh=2$ and $kA=0.2$ and 0.35; (*c*) $kh=\infty$ (deep water) and $kA=0.3$ and 0.41. Here A is the wave amplitude. Broken lines correspond to zero amplitude. (*a*) and (*b*) from McLean (1982b); (*c*) from Saffman and Yuen (1985).

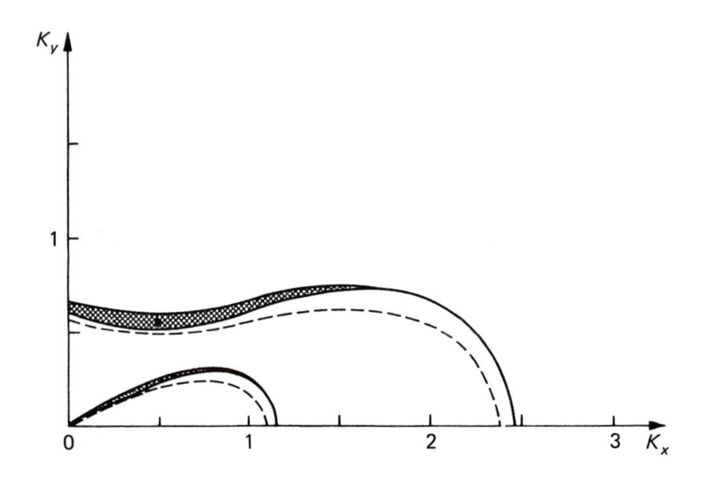

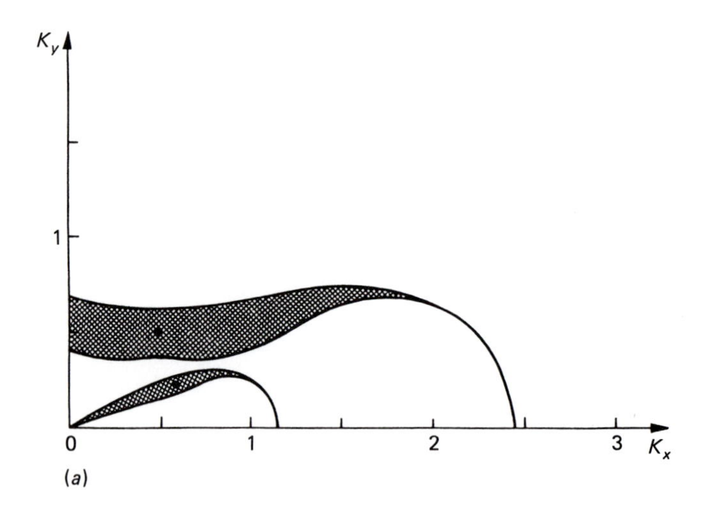

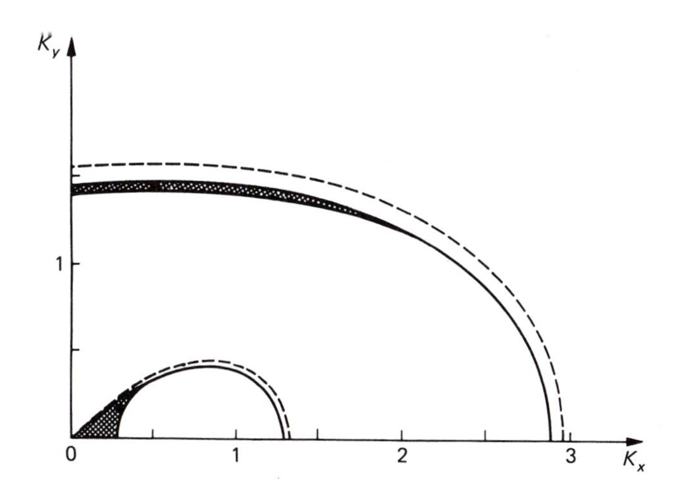

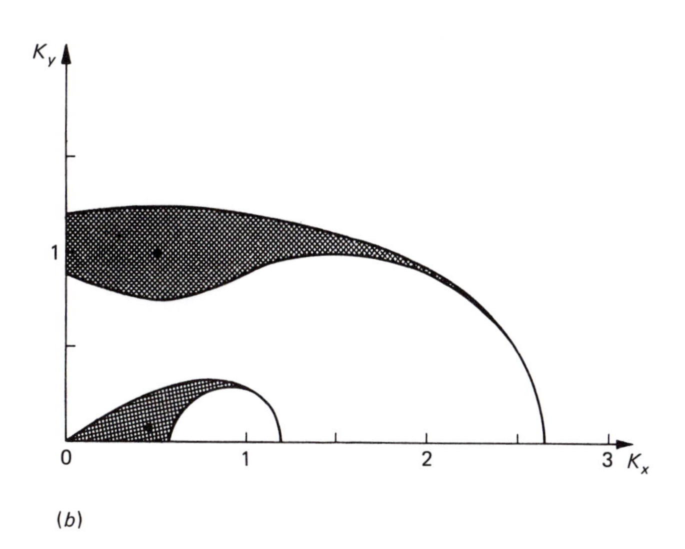

(b)

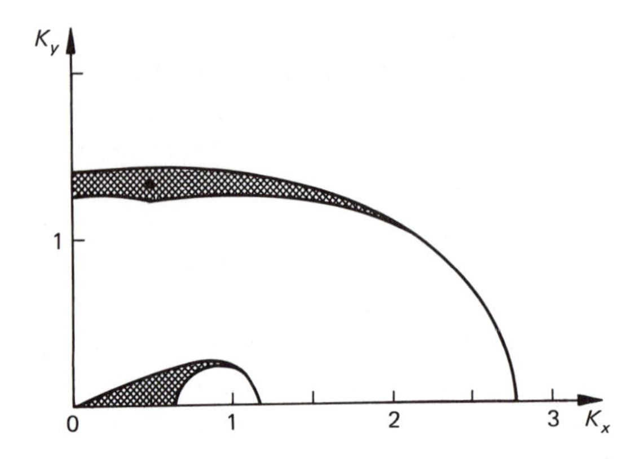

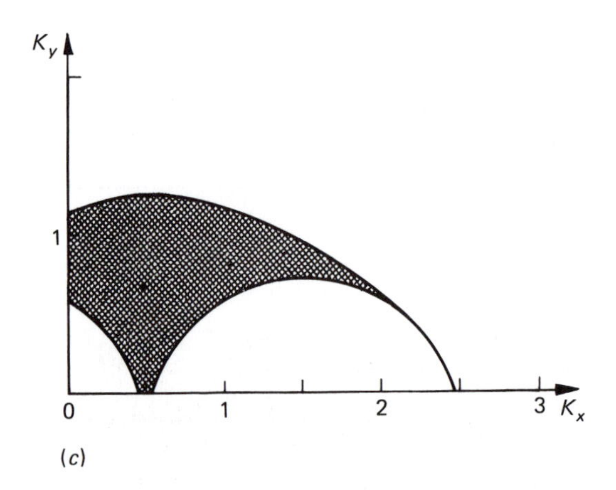

(c)

4.6.2 Further numerical results

As maximum growth rates, corresponding to **K** values denoted by dots on Fig. 4.2, are proportional to $A^{|M|}$ when A is small, the lowest, $M=2$, Class I instability will dominate for small amplitude carrier waves (the instability of Fig. 1.4(*b*) is an example of this). For larger amplitudes the $M=3$, Class II instability becomes equally important and even dominant for steep enough waves. Note from Fig. 4.2 that the maximum growth rate of the $M=3$ instability corresponds to $K_x=\frac{1}{2}$, the axis of symmetry of all Class II resonance curves, and to a non-zero K_y. This instability will thus try to double the wavelength of the carrier wave (the more so as it transpires that $\mathrm{Re}\Omega=0$ for it and so it is co-moving with the carrier wave). It will also introduce a third dimension to the wave structure. This phenomenon sometimes takes computer experts by surprise, as $K_x=\frac{1}{2}$ perturbations are difficult to bar from a two-dimensional simulation in which an even number of wavelengths is considered. The end product can look something like Fig. 4.3. When Class I and Class II instabilities couple, the end result can be a beautiful pattern like that illustrated by Fig. 4.4. References to experiments in which similar structures have been produced and observed on a water surface can be found in the Saffman and Yuen article (1985), the most extensive being Su *et al.* (1982). See also part I of Debnath (1983), and van Dyke (1982) for photographs.

Fig. 4.3. A bifurcated three-dimensional wave in which wavelength doubling has occurred. From Saffman and Yuen (1985).

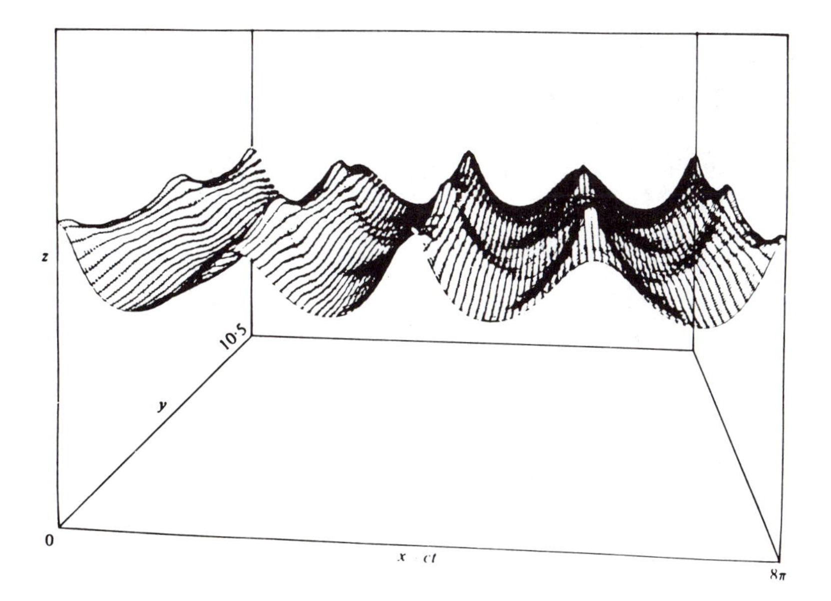

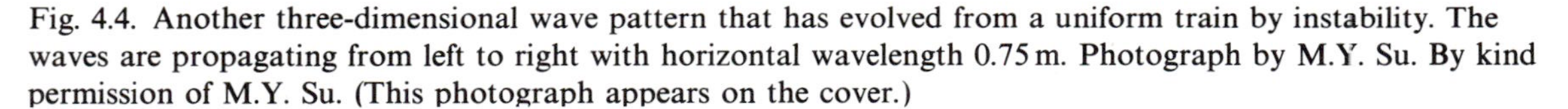

Fig. 4.4. Another three-dimensional wave pattern that has evolved from a uniform train by instability. The waves are propagating from left to right with horizontal wavelength 0.75 m. Photograph by M.Y. Su. By kind permission of M.Y. Su. (This photograph appears on the cover.)

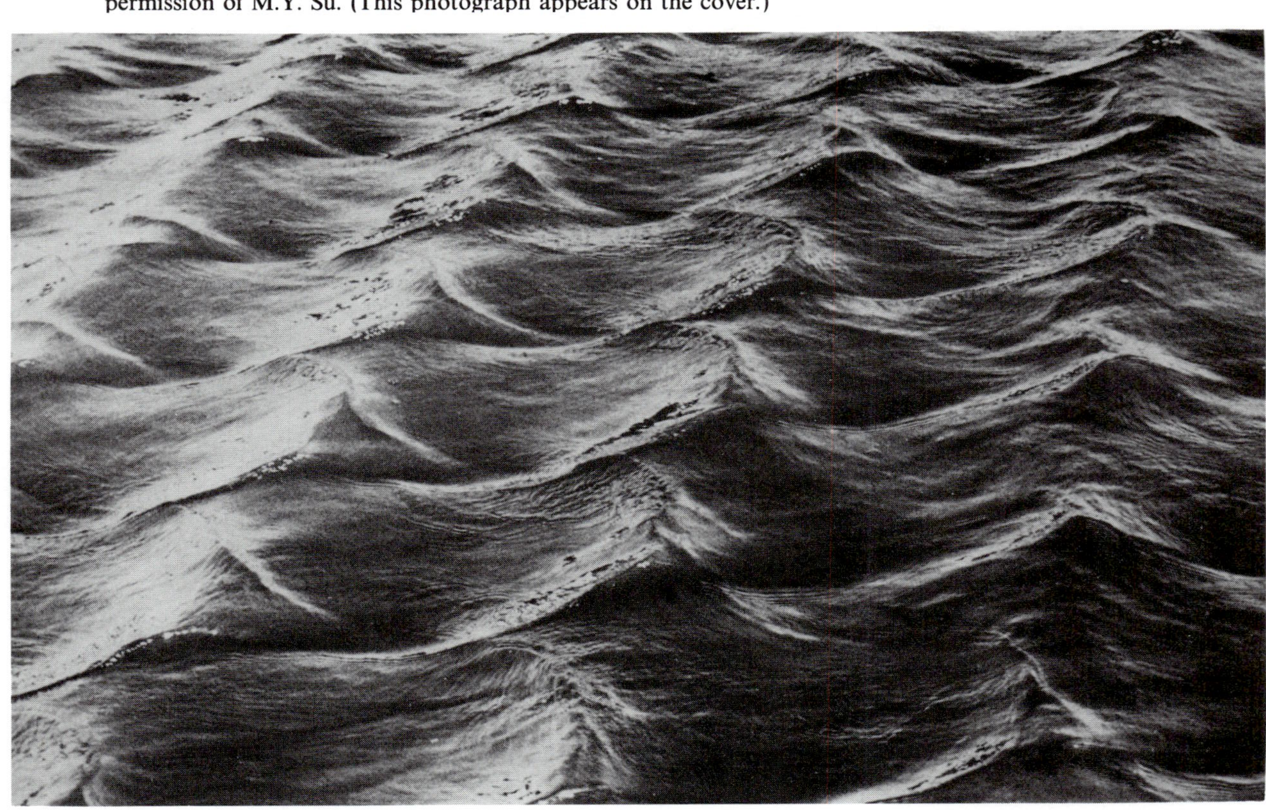

We will return to the problem of nonlinear water wave stability in Chapters 5 and 8, where we will concentrate on two h limits (h small and $h \to \infty$). We will also briefly review about a dozen or so model equations describing water surfaces in Section 5.4.

4.7 Summary

When reconsidering the first five sections of this Chapter, it is gratifying to see the universality of the methods discussed in Section 4.2. Of course, it is the linearization that makes the calculations tractable, although this by itself still leaves problems. For the fluids one still has to solve differential equations which, though linear, can have spatially varying coefficients.

Though this difficulty does not arise in other problems (such as the solid–liquid interface problem) the algebraic complexity of the dispersion relation makes up for this simplification.

It would be out of context here to dwell on nonlinear aspects. It should, however, be said that solid–liquid interface instabilities and subsequent dendritic growth could be a fruitful area to persue for a better understanding of nonlinear effects. This is in virtue of the relative ease of carrying out controlled experiments, particularly with alloy systems, where the growth rates are relatively low. Recent progress in this area is reviewed by Langer (1980).

Finally, we have seen in Section 4.6 how simple linear analysis can help us look for the possibility of nonlinear wave instabilities. Although this is not conclusive it can be a useful clue. More serious analyses will come in Chapters 5 and 8.

Exercises on Chapter 4

Exercise 1

Draw the $M=2$ and $M=3$ resonance curves for the Kadomtsev–Petviashvili equation (1.2.13) and compare them with those obtained from (4.5.9) and (4.5.10) in the $h \to 0$ limit. Comment on the similarity.

Hint: use the identity
$$K = |K|, K>0$$
$$= -|K|, K<0.$$

Exercise 2

Draw the lowest order resonance curves for the Kadomtsev–Petviashvili equation with positive dispersion, (1.2.13) with a minus sign in front of the last term. (In contradistinction to the water surface wave problem, this equation admits class II, $M=1$ instabilities.)

5

Model equations for small amplitude waves and solitons; weakly nonlinear theory

5.1 Introduction

5.1.1 Some physical equations ask for surgery

Classical physicists usually agree on their equations. In this respect they are very fortunate and can feel rewarded for not working in more fashionable fields such as the frontiers of high energy physics. However, these established equations often compound many different physical effects and can be difficult to solve.

Once we have derived as realistic a set of equations as possible for a given situation, we can try to reduce the number of terms or otherwise simplify by some logical process. Only very good scientists can get away with formulating equations that model chosen phenomena from the start, say by ignoring some physical effects or chopping off terms they consider to be insignificant. Lesser mortals are well advised to develop a systematic scheme for simplifying model equations. To do this one should look for at least one small, dimensionless parameter and use it as a surgical tool.

The above remarks concern a theoretical treatment. Computer scientists, on the other hand, will increasingly welcome elaborate mathematical models, embracing more and more rather than less and less physics.

There can be two broad justifications for introducing a small parameter scheme to simplify a system of physical equations (other than being fed up with not being able to solve it). One is that a dimensionless parameter is always small (an example of such a parameter is the ratio of the centre of mass velocity of motion of a massive heavenly body to c, the velocity of light). The second is somewhat more refined: we concentrate on a *class of phenomena* for which certain dimensionless parameters are small. An example is furnished by the amplitude of a water surface wave divided by its wavelength, often seen to be small for water waves.

The first type of small parameter expansion leads to reasonable models in which various numerical values, such as periods of motion, radii of trajectories etc., are off by small fractions of the values obtained. In the second type we obtain a theory for a restricted class of phenomena which may break down even if it was initially consistent (e.g., the water wave amplitude of our example grows too large after a while).

Surgery performed on a system of physical equations is not the only way to make life simpler. A second method is to keep the equations in all their complexity, but look for *solutions* that are restricted from the start. One example is furnished by small amplitude, almost cosine waves, another by arbitrary travelling wave functions of one variable $\theta = \mathbf{k} \cdot \mathbf{x} - \omega t$ for which partial differential equations become ordinary differential equations (Chapter 6).

Finally, we can first simplify the equations and then restrict the class of solutions.

In this Chapter, we will illustrate these procedures for dealing with some difficult wave equations and then give a partial assessment of the results. The theme of a critical assessment of the simplification procedures that can be applied to model equations will be continued in Chapter 8.

5.1.2 Examples

One example of an important set of equations in hydrodynamics that is universally accepted and quite horrid is that which governs water surface waves (gravity waves). Some of the physical assumptions involved were described in Section 1.4 where the equations were derived. However, we will write them out again, reinstating surface tension which will be of interest here. We assume the velocity field to be irrotational, and so a velocity potential ϕ can be introduced

$$\mathbf{v} = \nabla\phi, \tag{5.1.1}$$

satisfying (1.4.1)

$$\nabla \cdot \mathbf{v} = \nabla^2\phi = 0. \tag{5.1.2}$$

We choose the z coordinate to be the height measured from the bottom, assumed to be flat. The normal fluid velocity should vanish for $z = 0$,

$$\frac{\partial\phi}{\partial z} = 0 \text{ on } z = 0. \tag{5.1.3}$$

The equation of the surface is $z = h + \eta(x, y, t)$ (here h is the average height) and the velocity of the fluid is thus the velocity of the surface. This yields

$$\eta_t + \phi_x \eta_x + \phi_y \eta_y = \phi_z \text{ on } z = h + \eta. \tag{5.1.4}$$

Finally, pressure balance on the fluid surface gives, (T/ρ is the surface tension):

$$\phi_t + \tfrac{1}{2}[\phi_x^2 + \phi_y^2 + \phi_z^2] + g\eta - (T/\rho)(\eta_{xx} + \eta_{yy})(1 + \eta_x^2 + \eta_y^2)^{-3/2} = 0$$
$$\text{on } z = h + \eta. \tag{5.1.5}$$

Viscous effects have been neglected. The reader unfamiliar with these equations may find them frightening, especially as (5.1.4) and (5.1.5) are to be satisfied on the *unknown* surface $z = h + \eta$. In any case, we count on his moral support in trying to simplify (5.1.2)–(5.1.5).

Another example, perhaps not quite so convoluted but still formidable, is furnished by the set of equations governing a volume plasma wave mode similar to sound waves in air (ion acoustic wave in a two component electron collision dominated plasma – see Section 1.3). We will consider a hydrogen plasma composed of electrons and ions of equal density $n_o, n_e = n_i = n_o$, and neglect electron inertia and ion temperature $(m_e/m_i \to 0, T_i/T_e \to 0)$. In the presence of an external magnetic field

$$\mathbf{B}_o = B_o \mathbf{x}, \tag{5.1.6}$$

the ion dynamics are governed by

$$\partial_t n + \nabla \cdot (n\mathbf{v}) = 0 \tag{5.1.7}$$

$$\partial_t \mathbf{v} + (\mathbf{v} \cdot \nabla)\mathbf{v} + \nabla \phi + \Omega_c \mathbf{x} \wedge \mathbf{v} = 0 \tag{5.1.8}$$

$$\boldsymbol{\nabla}^2 \phi = \mathrm{e}^{\phi} - n. \tag{5.1.9}$$

(Davidson (1972), Chapter 2). Here ϕ is the electric field potential defined by $\mathbf{E} = -\boldsymbol{\nabla}\phi$, and normalized to $(K_B T_e)/e$, n and $\mathbf{v}$ the ion density and velocity in units of n_o and $(K_B T_e/m_i)^{\frac{1}{2}}$ henceforth referred to as the ion sound velocity. All lengths have been normalized to $(K_B T_e/4\pi n_o e^2)^{\frac{1}{2}}$, the Debye length (Section 1.3), and Ω_c is a normalized measure of the strength of the magnetic field:

$$\Omega_c = \mathrm{B}_o/\sqrt{(4\pi \mathrm{n}_i m_i)}c = V_A/c.$$

Thus $B_o/n_i^{\frac{1}{2}}$, rather than B_o, dictates the physics of these waves. The velocity V_A is known as the Alfvén velocity. (It is comparable to c only in very dilute, strongly magnetized plasmas.)

Before embarking on various schemes for making life simpler, there is one problem that can always be solved painlessly: that of linear waves (vanishing amplitude waves) if they exist. We will quote the results as they will prove useful in what follows. The reader will have no difficulty in generalizing the calculation of Section 4.2 to show that a small amplitude ripple on the water surface assumed of the form

$$\eta = a\cos(k_o x - \omega_o t) \tag{5.1.10}$$

satisfies equations (5.1.2)–(5.1.5) if

$$\omega_o^2 = (gk_o + Tk_o^3)\tanh(k_o h) \tag{5.1.11}$$

(Exercise 1). (The reason for the zero subscript will be given in Section 5.3.) For very small k_o there is little dispersion ($\omega_o/k_o \simeq$ const). A similar calculation based on (5.1.7)–(5.1.9) with

$$\eta = a\cos(\mathbf{k}_o \cdot \mathbf{x} - \omega_o t) \tag{5.1.12}$$

yields the dispersion relation

$$k_o^2 + 1 - \frac{k_{oy}^2 + k_{oz}^2}{\omega_o^2 - \Omega_c^2} - \frac{k_{ox}^2}{\omega_o^2} = 0. \tag{5.1.13}$$

The $\Omega_c = 0$ case, for which

$$\omega_o^2 = k_o^2/(1 + k_o^2) \tag{5.1.14}$$

will be of interest. For small k_o, ion-acoustic waves in a magnetic field-free plasma are also seen to propagate almost without dispersion ($\omega_o/k_o \simeq$ const).

In Section 5.2, we will derive several model equations from (5.1.2)–(5.1.5) and (5.1.7)–(5.1.9). Surprising as it might seem at this stage, the same model equations will be obtained in both contexts, the diversity of the two physical situations notwithstanding.

In Section 5.3, we will see what can be accomplished by assuming 'a' in (5.1.10) small but finite and treating wave *envelope* dynamics (deriving equations for 'a' dynamics). Next, a family tree of various models obtained in Sections 5.2 and 5.3, is followed by a brief section on a general amplitude approach that painlessly extends the small 'a' formalism (Sections 5.4 and 5.5). In Section 5.6 we will investigate the long-time behaviour of instabilities in systems with periodic boundary conditions. Some general conclusions on the relations between model equations and the real world wind up the present Chapter (Section 5.7).

5.2 A few model equations as derived by introducing a small parameter

We will now give an example of how a complicated set of equations can be simplified by introducing a parameter that *may* be small in some physical situations.

5.2.1 Shallow water, weak amplitude gravity waves

It will prove convenient to rewrite (5.1.2)–(5.1.5) in dimensionless form:

$$\nabla^2\phi=0 \tag{5.2.1}$$

$$\phi_z=0 \text{ on } z=0 \tag{5.2.2}$$

$$\eta_t+\phi_x\eta_x+\phi_y\eta_y=\phi_z \text{ on } z=1+\eta \tag{5.2.3}$$

$$\phi_t+\tfrac{1}{2}[\phi_x^2+\phi_y^2+\phi_z^2]+\eta-S(\eta_{xx}+\eta_{yy})(1+\eta_x^2+\eta_y^2)^{-3/2}=0$$
$$\text{on } z=1+\eta. \tag{5.2.4}$$

Here length is measured in units of the depth h and velocities in units of $(gh)^{\frac{1}{2}}$. A fluid velocity measured in units of $(gh)^{\frac{1}{2}}$ is known as a Froude number. The dimensionless constant S derives from surface tension:

$$S=T/(g\rho h^2).$$

The value of S for water is 1 when $h=0.27$ cm.

We will now treat a restricted class of wave phenomena such that the wave 'feels' the bottom. Thus, the wavelength will be assumed to be *longer* than the depth. The wave amplitude will be taken to be *much* smaller than the depth. As the depth is 1 in our scaling, these relations will be expressed by

$$\text{wavelength: depth: amplitude as } \varepsilon^{-\frac{1}{2}}:1:\varepsilon \tag{5.2.5}$$

($\varepsilon^{\frac{1}{2}}$ being the basis of the scaling is merely a convention).

The linear wave dispersion relation, (5.1.11) in the shallow water, small hk_o limit can furnish some hints as to how to proceed. However, they will only be hints, as our ambition is to derive an equation capable of describing more than just cosine shape waves. Bearing this in mind, we can check from (5.1.11) that for small k_o, linear waves have both phase and group velocities $\simeq(gh)^{\frac{1}{2}}$, or 1 in our system. Fortunately nonlinear, non-cosine shape waves of small but finite amplitude also move with approximately this velocity (the general type of reasoning follows from Exercise 4, though the physical context is different there). If the nonlinear wave propagates, say, along the x direction, it would thus seem natural to take $x-t$ or $x+t$ as the basis for a new coordinate. We will in fact concentrate on the first choice, always remembering that by doing so, we have specified a basic wave structure moving from left to right along x. This and (5.2.5) suggest taking

$$\xi=\varepsilon^{\frac{1}{2}}(x-t) \tag{5.2.6}$$

as a new coordinate. The t scaling is also easy to derive, once again referring back to linear wave modes. Taking (5.1.11) a step further we see that the frequency in our moving coordination system, ω_o^*, is given by

$$\omega_o^* = \omega_o - k_o \sim k_o^3, \tag{5.2.7}$$

suggesting that in our coordinate system, in which frequencies become ω_o^* due to (5.2.6), we take

$$\tau = \varepsilon^{3/2} t. \quad (5.2.8$$

The second statement implicit in (5.2.5) suggests

$$\eta = \varepsilon\eta^{(1)} + \varepsilon^2\eta^{(2)} + \ldots \tag{5.2.9}$$

The remaining scalings are less obvious:

$$\sigma = \varepsilon y \tag{5.2.10}$$

$$\phi = \varepsilon^{\frac{1}{2}}\phi^{(1)} + \varepsilon^{3/2}\phi^{(2)} + \ldots \tag{5.2.11}$$

but ensure inclusion of the essential physics in the second order equation, as we will presently see.

The procedure outlined above is known as variable stretching (e.g. see Taniuti and Wei (1968); Su and Gardner (1969), Jeffrey and Kakutani (1972)). It assumes the possibility of introducing new coordinates and variables such that the slowness of coordinate dependence and smallness of some of the physical variables can be taken out in a uniform way, (thus, in the new variables, the rather anaemic wave profile of Fig. 5.1 can become a

Fig. 5.1. Schematic illustration to the derivation of the Kadomtsev–Petviashvili equation for water wave equations. Waves are assumed to be small amplitude (A/h small) and the water shallow (h/λ small), but not necessarily near-cosine in shape.

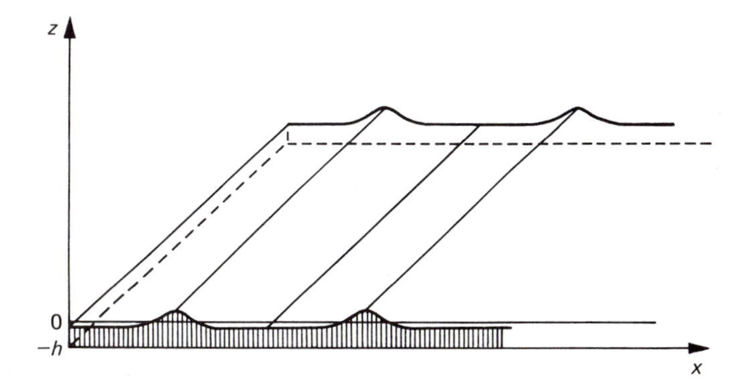

full blooded nonlinear wave, as high as it is long, as it were). Knowledge of ε is then the key to recovering a specific physical situation.

We now look for solutions to (5.2.1) in the form

$$\phi = \sum_{n=0}^{\infty} z^n f_n(x, y, t) \tag{5.2.12}$$

yielding

$$\sum_{n=0}^{\infty} [n(n-1)z^{n-2} f_n + \nabla^2 f_n z^n] = 0. \tag{5.2.13}$$

Thus

$$f_{n+2} = \frac{-\nabla^2 f_n}{(n+1)(n+2)}. \tag{5.2.14}$$

Equation (5.2.2) implies $f_1 = 0$ and so all odd f_n are zero. Now

$$\phi = \sum_{m=0}^{\infty} (-1)^m \frac{z^{2m}}{(2m)!} \nabla^{2m} f \tag{5.2.15}$$

and the zero subscript is dropped in f_o. In our stretched coordinates

$$\nabla^2 = \varepsilon \partial_\xi^2 + \varepsilon^2 \partial_\sigma^2 \tag{5.2.16}$$

$$\phi = \varepsilon^{\frac{1}{2}} \left(f + \sum_{m=1}^{\infty} (-1)^m \frac{z^{2m}}{(2m)!} [\varepsilon \partial_\xi^2 + \varepsilon^2 \partial_\sigma^2]^m f \right), \tag{5.2.17}$$

where f is an as yet undetermined function of ξ, σ, τ. Both (5.2.1) and (5.2.2) are now satisfied. We must also satisfy the boundary conditions on $z = 1 + \eta$. To do this we will use the operator identity

$$\partial_t = -\varepsilon^{\frac{1}{2}} \partial_\xi + \varepsilon^{3/2} \partial_\tau, \tag{5.2.18}$$

following from (5.2.6) and (5.2.9). The lowest order expression for ϕ is

$$\phi \simeq \varepsilon^{\frac{1}{2}} f - \varepsilon^{3/2} \frac{z^2}{2} f_{\xi\xi}, \tag{5.2.19}$$

implying

$$\phi_z \simeq -\varepsilon^{3/2} z f_{\xi\xi}. \tag{5.2.20}$$

From now on we need only worry about satisfying the two boundary conditions on $z = 1 + \eta$: (5.2.3) and (5.2.4). The first of these yields, to lowest order ($\varepsilon^{3/2}$):

$$f_{\xi\xi} - \eta_\xi^{(1)} = 0 \tag{5.2.21}$$

$$\eta^{(1)} = f_\xi = \phi_\xi^{(1)} \tag{5.2.22}$$

and (5.2.4) adds nothing new. The next order ($\varepsilon^{5/2}$) contribution from (5.2.3) is, when $z=1+\eta$ is used,

$$-\eta^{(2)}_{\xi}+\phi^{(1)}_{\xi\tau}+\phi^{(2)}_{\xi\xi}+2\phi^{(1)}_{\xi}\phi^{(1)}_{\xi\xi}-\tfrac{1}{6}\phi^{(1)}_{\xi\xi\xi\xi}+\phi^{(1)}_{\sigma\sigma}=0. \tag{5.2.23}$$

Here $\eta^{(1)}$ has been expressed as $\phi^{(1)}_{\xi}$. The $\varepsilon^{5/2}$ terms in condition (5.2.4) yields, upon differentiation by ξ

$$\eta^{(2)}_{\xi}+\phi^{(1)}_{\xi\tau}-\phi^{(2)}_{\xi\xi}+\phi^{(1)}_{\xi}\phi^{(1)}_{\xi\xi}+(\tfrac{1}{2}-S)\phi^{(1)}_{\xi\xi\xi\xi}=0. \tag{5.2.24}$$

If we now add the two above equations, (5.2.23) and (5.2.24), we obtain

$$\phi^{(1)}_{\xi\tau}+\tfrac{3}{2}\phi^{(1)}_{\xi}\phi^{(1)}_{\xi\xi}+\tfrac{1}{6}(1-3S)\phi^{(1)}_{\xi\xi\xi\xi}+\tfrac{1}{2}\phi^{(1)}_{\sigma\sigma}=0. \tag{5.2.25}$$

This is known as the Kadomtsev–Petviashvili equation (1970).

Other forms are sometimes more convenient. If we wish to express (5.2.25) in terms of the velocity components

$$v=\phi^{(1)}_{\xi},\ u=\phi^{(1)}_{\sigma},$$

we obtain

$$v_{\tau}+\tfrac{3}{2}vv_{\xi}+\tfrac{1}{6}(1-3S)v_{\xi\xi\xi}+\tfrac{1}{2}u_{\sigma}=0 \tag{5.2.26}$$

$$v_{\sigma}=u_{\xi}. \tag{5.2.27}$$

Note how the number of independent variables has been reduced from four to three (depth no longer appears as an independent variable), and of dependent variables from two to one. The above derivation follows Infeld and Rowlands (1979b). Kinematic viscosity would add a $v_{\xi\xi}$ term (Bartucelli *et al.* (1985)).

If we are considering a very thin film of water, $h<0.47$ cm, the coefficient of the third term becomes negative. When this happens, it is more customary to write (5.2.26) in the form

$$v_{\tau}+\tfrac{3}{2}vv_{\xi}+\tfrac{1}{6}(3S-1)v_{\xi\xi\xi}-\tfrac{1}{2}u_{\sigma}=0,\ S>1/3 \tag{5.2.28}$$

obtained by taking $\xi\to-\xi, v\to-v$. This equation, together with (5.2.27), which is unaltered, is known as the Kadomstev–Petviashvili equation with positive dispersion. However, its validity in this physical context could be questioned due to the enhanced importance of the discarded frictional effects at the bottom when the depth is so minute (Miles (1976)). Further doubts centre on the introduction of small scale dimples on the wave profile due to the surface tension. This may invalidate the long wave hypothesis, see Hunter and Vanden-Broeck (1983a).

The Kadomstev–Petviashvili equation (KP) will play an important role in this book. The ξ,τ version (without perpendicular variable dependence) has been known since 1895 as the Korteweg–de Vries equation (including the effect of surface tension). Notice how modest the inclusion of perpendicular dynamics seems to be. We will see, however, that this one additional term in (5.2.25) (as compared with KdV), opens the door to a wealth of physical effects (Sections 7.3, 7.4 and 8.4).

It is important to remember that head-on collisions of various wave structures cannot be described by the KP equation (5.2.26)–(5.2.27).

5.2.2 Weak amplitude ion acoustic waves in an unmagnetized plasma

We now proceed to another example of small parameter surgery. This time we will not operate on the equations themselves. It will prove simpler to attack the Lagrangian, thus expanding just one entity in the small parameter.

Equations (5.1.7)–(5.1.9) with $\Omega_c = 0$ and $\mathbf{v} = \nabla\psi$ (no vorticity) can be derived from a Lagrangian density

$$L = \tfrac{1}{2}n(\nabla\psi)^2 + n\psi_t + n\phi - e^{\phi} - \tfrac{1}{2}(\nabla\phi)^2 + 1. \tag{5.2.29}$$

We will now consider two space coordinate dependence: x, y, leaving a physical discussion of this restriction to Chapter 8.

We now proceed as in Subsection 5.2.1. Again, the dispersion relation (5.1.14) is of the form $\omega_o - k_o \sim k_o^3$ for small k_o and suggests

$$\xi = \varepsilon^{1/2}(x - t) \tag{5.2.30}$$

$$\tau = \varepsilon^{3/2}t. \tag{5.2.31}$$

We also take

$$\sigma = \varepsilon y \tag{5.2.32}$$

$$n = 1 + \varepsilon n^{(1)} + \ldots \tag{5.2.33}$$

$$\psi = \varepsilon^{1/2}\psi^{(1)} + \ldots \tag{5.2.34}$$

$$\phi = \varepsilon\phi^{(1)} + \ldots \tag{5.2.35}$$

A fuller exposition of linear arguments justifying (5.2.30)–(5.2.35) is given in Appendix 1. However, the linear dispersion relation may strike the reader as a somewhat flimsy basis for an expansion leading to a model that still prides itself on being nonlinear, as were the initial equations (5.1.7)–(5.1.9). For arguments supporting all expansions other than (5.2.32), and based on the properties of *solitons*, we once more suggest Exercise 4. The two references together (Appendix 1 and Exercise 4) should give some feeling for the physical basis for what we are trying to accomplish.

We now expand (5.2.29). The first non-trivial term is $L^{(2)}$ (as $L^{(1)} = -\psi_\xi^{(1)}$):

$$L^{(2)} = -n^{(1)}\psi_\xi^{(1)} + n^{(1)}\phi^{(1)} - \tfrac{1}{2}\phi^{(1)^2} + \tfrac{1}{2}\psi_\xi^{(1)^2} + \psi_\tau^{(1)} - \psi_\xi^{(2)}, \qquad (5.2.36)$$

yielding the following identities as Euler–Lagrange equations

$$\delta_n{}^{(1)}: \quad \psi_\xi^{(1)} = \phi^{(1)} \qquad (5.2.37)$$

$$\delta\phi^{(1)}: \quad n^{(1)} = \phi^{(1)} \qquad (5.2.38)$$

$$\delta\psi^{(1)}: \quad n^{(1)} = \psi_\xi^{(1)} \qquad (5.2.39)$$

The next order Lagrangian, when expressed in terms of $\psi^{(1)}$, is

$$L^{(3)} = \psi_\tau^{(1)}\psi_\xi^{(1)} + \tfrac{1}{3}\psi_\xi^{(1)^3} - \tfrac{1}{2}\psi_{\xi\xi}^{(1)^2} + \tfrac{1}{2}\psi_\sigma^{(1)^2}$$
$$+\text{perfect differentials}, \qquad (5.2.40)$$

yielding, as the Euler–Lagrange equation when superscripts are omitted,

$$\delta\psi: \quad \psi_{\xi\tau} + \psi_\xi\psi_{\xi\xi} + \tfrac{1}{2}\psi_{\xi\xi\xi\xi} + \tfrac{1}{2}\psi_{\sigma\sigma} = 0. \qquad (5.2.41)$$

The Kadomstev–Petviashvili equation again!

Inclusion of electron–ion collisions, usually the most important dissipative mechanism, would add new terms to (5.2.41) without damaging its general form of an equation for ψ with quadratic nonlinearities only. Nor is the order of the highest derivative raised. Depending on how we order parameters that did not need to be specified before, such as $T_i m_i^{\frac{1}{2}}/T_e m_e^{\frac{1}{2}}$, we obtain additionally either a Galilean-invariant term

$$-\nu\partial_\xi(\psi_{\xi\tau} + \psi_\xi\psi_{\xi\xi}), \qquad (5.2.42)$$

where ν is a measure of electron–ion collisional dissipation (Kakutani and Kawahara (1970)), or

$$-\nu\partial_\xi^3\psi \qquad (5.2.43)$$

(Tagare and Shukla (1977)). The papers quoted derive these additional terms for equations in one space dimension ξ, but no new terms appear in the two-dimensional version with our ordering in ξ, σ, τ. Not surprisingly, a Lagrangian will no longer exist for either model.

Returning to (5.2.41), it is generally simpler to expand just one entity, namely the Lagrangian, though historically this equation was first laboriously derived from (5.1.7)–(5.1.9) (Kako and Rowlands (1976)), and then, somewhat later, from the Lagrangian as above (Infeld (1981b)).

However, now that we do have this useful tool, we might as well further exploit it to derive the simplest conservation equations using Noether's theorem, before we resume the main theme of this Section.

To complicate matters, the Lagrangian (5.2.40) contains a second derivative term, and so it is impossible to apply Noether's treatment in its original form. Therefore, we introduce a new variable χ

$$\chi = \psi_{\xi\xi} \tag{5.2.44}$$

and use the equivalent Lagrangian

$$L = \psi_{\tau\xi}\psi_{\xi} + \tfrac{1}{3}\psi_{\xi}^{3} + \psi_{\xi} + \psi_{\xi}\chi_{\xi} + \tfrac{1}{2}\chi^{2} + \tfrac{1}{2}\psi_{\sigma}^{2} \tag{5.2.45}$$

yielding (5.2.44) as a Euler–Lagrange equation. The resulting equation for energy conservation will take the slightly more elaborate form than usual

$$\begin{aligned} &\partial_{\tau}(\psi_{\tau}L_{\psi\tau} + \chi_{\tau}L_{\chi\tau} - L) + \partial_{\xi}(\psi_{\tau}L_{\psi_{\xi}} + \chi_{\tau}L_{\chi\xi}) \\ &+ \partial_{\sigma}(\psi_{\tau}L_{\psi\sigma} + \chi_{\tau}L_{\chi\sigma}) = 0, \end{aligned} \tag{5.2.46}$$

for which we obtain, for our particular L (5.2.45):

$$\begin{aligned} &\partial_{\tau}(\tfrac{1}{3}\psi_{\xi}^{3} - \tfrac{1}{2}\psi_{\xi\xi}^{2} + \tfrac{1}{2}\psi_{\sigma}^{2}) + \partial_{\xi}(\psi_{\xi\tau}\psi_{\xi\xi} - \psi_{\tau}^{2} - \psi_{\tau}\psi_{\xi}^{2} - \psi_{\tau}\psi_{\xi\xi\xi}) \\ &+ \partial_{\sigma}(-\psi_{\tau}\psi_{\sigma}) = 0. \end{aligned} \tag{5.2.47}$$

The two equations for momentum conservation are

$$\partial_{\tau}(\psi_{\xi}^{2}) + \partial_{\xi}(\tfrac{2}{3}\psi_{\xi}^{3} + \psi_{\xi}\,\psi_{\xi\xi\xi} - \tfrac{1}{2}\psi_{\xi\xi}^{2} - \tfrac{1}{2}\psi_{\sigma}^{2}) + \partial_{\sigma}(\psi_{\xi}\psi_{\sigma}) = 0, \tag{5.2.48}$$

and

$$\begin{aligned} &\partial_{\tau}(\psi_{\xi}\psi_{\sigma}) + \partial_{\xi}(\psi_{\tau}\psi_{\sigma} + \psi_{\sigma}\psi_{\xi}^{2} + \psi_{\sigma}\psi_{\xi\xi\xi} - \psi_{\sigma\xi}\psi_{\xi\xi}) \\ &+ \partial_{\sigma}(\tfrac{1}{2}\psi_{\xi\xi}^{2} + \tfrac{1}{2}\psi_{\sigma}^{2} - \tfrac{1}{3}\psi_{\xi}^{3} - \psi_{\tau}\psi_{\xi}) = 0. \end{aligned} \tag{5.2.49}$$

Finally, (5.2.41) itself can be written in conservation form

$$\partial_{\tau}\psi_{\xi} + \partial(\tfrac{1}{2}\psi_{\xi}^{2} + \tfrac{1}{2}\psi_{\xi\xi\xi\xi}) + \partial_{\sigma}(\tfrac{1}{2}\psi_{\sigma}) = 0. \tag{5.2.50}$$

There is also an infinity of conservation laws in which powers of τ, ξ, and σ appear explicitly; and also a set of non-local laws (Infeld (1981a), Infeld and Frycz (1983); Zakharov and Shulman (1980)).

5.2.3 Weak amplitude ion acoustic waves in a magnetized plasma

In this derivation, which is another example of a small parameter expansion, we will consider ion acoustic waves in very strong external magnetic fields, such as waves described by (5.1.7)–(5.1.9), with Ω_c of order one.

There is now no Lagrangian, at least none analogous to (5.2.29), and we must expand all the equations. Now the velocity **v** is not curl-free and no velocity potential can be introduced. However, there is nothing to stop us from taking variables to be z independent (perpendicular dependence reduced to one coordinate). In contradistinction to the case of Subsection

5.2.2, it will prove convenient to expand x and y uniformly. The small $\mathbf{k}_o$ expansion of the dispersion relation (5.1.13) is once again of the form $\omega_o - k_o \sim k_o^3$ and so t stretching will be as before (once again, see Appendix 1 for a more complete argument). We have

$$\xi = \varepsilon^{\frac{1}{2}}(x-t) \tag{5.2.51}$$

$$\sigma = \varepsilon^{\frac{1}{2}} y \tag{5.2.52}$$

$$\tau = \varepsilon^{3/2} t \tag{5.2.53}$$

$$n = 1 + \varepsilon n^{(1)} + \ldots \tag{5.2.54}$$

$$v_x = \varepsilon v_x^{(1)} + \varepsilon^2 v_x^{(2)} + \ldots \tag{5.2.55}$$

$$v_y = \varepsilon^2 v_y^{(1)} + \varepsilon^3 v_y^{(2)} + \ldots \tag{5.2.56}$$

$$v_z = \varepsilon^{3/2} v_z^{(1)} + \varepsilon^{5/2} v_z^{(2)} + \ldots \tag{5.2.57}$$

$$\phi = \varepsilon \phi^{(1)} + \varepsilon^2 \phi^{(2)} + \ldots \tag{5.2.58}$$

This expansion scheme only differs from that of (5.2.30)–(5.2.35) in the treatment of perpendicular dynamics. The expansion of v_y and v_z will follow from that of v_x and the fact that there is no z dependence. In lowest order in each equation we have

$$n^{(1)} = \phi^{(1)}, \tag{5.2.59}$$

$$\partial_\xi n^{(1)} = \partial_\xi v_x^{(1)}, \tag{5.2.60}$$

$$v_z^{(1)} = \Omega_c^{-1} \partial_\sigma \phi^{(1)}, \tag{5.2.61}$$

$$v_y^{(1)} = \Omega_c^{-1} \partial_\xi v_z^{(1)} \tag{5.2.62}$$

In the next order we obtain

$$n^{(2)} = \phi^{(2)} - (\partial_\xi^2 + \partial_\sigma^2)\phi^{(1)} + \tfrac{1}{2}\phi^{(1)^2}, \tag{5.2.63}$$

$$-\partial_\xi n^{(2)} + \partial_\tau n^{(1)} + \partial_\xi (n^{(1)} v_x^{(1)}) + \partial_\xi v_x^{(2)} + \partial_\sigma v_y^{(1)} = 0, \tag{5.2.64}$$

$$-\partial_\xi v_x + \partial_\tau v_x^{(1)} + v_x^{(1)} \partial_\xi v_x^{(1)} + \partial_\xi \phi^{(2)} = 0. \tag{5.2.65}$$

If we now add the last two equations using (5.2.63), and express all in terms of $n^{(1)}$ we obtain

$$\partial_\tau n^{(1)} + n^{(1)} \partial_\xi n^{(1)} + \tfrac{1}{2}\partial_\xi^3 n^{(1)} + \tfrac{1}{2}(1 + \Omega_c^{-2})\partial_{\xi\sigma\sigma} n^{(1)} = 0. \tag{5.2.66}$$

This is the Zakharov–Kuznetsov equation (1974). It has retained much of the essential physics, including the $\mathbf{E} \wedge \mathbf{B}$ drift along z and the polarization drift along y (Appendix 1). Like many successful classical model equations, it has turned up in other physical contexts, such as vortex soliton theory (Nozaki (1981)).

Oddly enough, (5.2.66) is derivable from a Langrangian that does not seem to have an antecedent giving the initial equations (5.1.7)–(5.1.9):

$$L=\psi_\tau\psi_\xi+\tfrac{1}{3}\psi_\xi^3-\tfrac{1}{2}\psi_{\xi\xi}^2-\tfrac{1}{2}(1+\Omega_c^{-2})\psi_{\xi\sigma}^2 \tag{5.2.67}$$
$$n^{(1)}=\psi_\xi.$$

Again we must remove the second derivative terms before conservation laws can be obtained as in Subsection 5.2.2. Take the equivalent Langrangian

$$L=\psi_\tau\psi_\xi+\tfrac{1}{3}\psi_\xi^3+\psi_\xi\chi_\xi+\tfrac{1}{2}\chi^2+\tfrac{1}{2}(1+\Omega_c^{-2})(2\psi_\sigma\rho_\xi+\rho^2) \tag{5.2.68}$$

where

$$\delta_\chi:\ \chi=\psi_{\xi\xi}$$
$$\delta\rho:\ \rho=\psi_{\xi\sigma},$$

lead to the conserved densities (dropping the superscript):

$$n^2 \text{ and } \tfrac{1}{3}n^3-\tfrac{1}{2}n_\xi^2-\tfrac{1}{2}(1+\Omega_c^{-2})n_\sigma^2$$

(momentum and energy; see Exercise 5). Of course n is also a conserved density, as follows directly from equation (5.2.66).

These forms of conserved densities are somewhat less surprising if we remember that to lowest order $v_x=n$ *and so* n^2 can be written nv_x and n^3 as nv_x^2 etc.

5.3 Weakly nonlinear waves

We have seen that equations can be simplified, leading to more tractable models. As indicated in Subsection 5.1, another approach would consist of leaving the equations alone, but working on a special class of solution, such as a small amplitude, nearly cosine wave solution. When this is our programme, we can expand in the amplitude 'a'. This procedure leads to an interesting mathematical structure, linear theory comprising the ground floor. New physical effects are usually introduced at the a^2 level. The procedure of this Section can be used whenever a nonlinear wave equation (or set of equations) has been solved in the linear limit; $a\to 0$. A reference for some of the mathematical background is the book by Krylov and Bogolyubov (1947), but our presentation will be self-contained.

5.3.1 Spreading, splitting and instabilities

As a simple example, take the Korteweg–de Vries equation in the form

$$u_t+uu_x+u_{xxx}=0 \tag{5.3.1}$$

We will now look for approximate though nonlinear solutions to this equation. They will look like linear waves on the small scale (say one or two wavelengths), but their amplitudes, wavenumbers and frequencies will be modulated on a larger scale (many wavelengths). We will try to find an equation that will describe this large-scale modulation. In view of what was said above, we take

$$u = \frac{a}{2} e^{i\theta_o} + \text{c.c.} \tag{5.3.2}$$

where 'a' is small and,

$$\partial_x \theta_o = k_o, \ \partial_t \theta_o = -\omega_o, \ \omega_o = -k_o^3. \tag{5.3.3}$$

In the course of this calculation, which will be an 'a' expansion of almost everything in sight, ω_o and k_o will nevertheless be 'a' independent. They will be the frequency and wavenumber a zero-amplitude wave would have. Later on it will, in some calculations, prove expedient to use ω and k, the physical frequency and wavenumber found by taking a stop-watch and a ruler to the wave. These quantities will in general depend on the amplitude. However, for the moment we will retain ω_o and k_o. They are no longer physical quantities (but are useful tools) when $a > 0$, but we do have $\omega(a \to 0) = \omega_o$; $k(a \to 0) = k_o$.

We will now take a closer look at a, k_o, ω_o, which will cease to be constants in next order.

As 'a' is our expansion parameter, u given by (5.3.2) is a first order quantity. All fast variations are assumed to be described by θ_o, and thus dependence of a, ω_o and k_o on x and t will be slow, differentiation raising the order. We will try to improve on the theory of (5.3.2) and (5.3.3) so as to describe a, ω_o, k_o dynamics on a slow scale. We also expect higher harmonics in u. Thus, if subscripts denote the order in an 'a' expansion, and a^* the complex conjugate of 'a';

$$\frac{\partial a}{\partial t} = A_2(a, a^*) + A_3 + \ldots \qquad \frac{\partial k_o}{\partial t} = M_1 + M_2 + \ldots \tag{5.3.4}$$

$$\frac{\partial a}{\partial x} = B_2(a, a^*) + B_3 + \ldots \qquad \frac{\partial k_o}{\partial x} = N_1 + N_2 + \ldots$$

$$u = u_1 + u_2(\theta_o, x, t) + \ldots \qquad \frac{\partial \omega_o}{\partial x} = \frac{\partial \omega_o}{\partial k_o}(N_1 + N_2 + \ldots)$$

As k_o and ω_o are 'a' independent, this expansion entails the additional assumption that $\partial k_o/\partial x$ etc. are of order 'a'. Here A_2, B_2 etc. are as yet unknown functions and will determine the dynamics of ω_o, k_o, a. Higher order $u(u_2, u_3$ etc.) depend on x, t on the fast scale through θ_o only. All other dependence is slow, and differentiation by x or t raises the order. Thus, for example, $\partial u_2/\partial\theta_o$ is second order, but $\partial a^2/\partial x$ is third order. It takes a moment's reflection to see what order the contribution of a given term will be. For example, the operator $\partial/\partial x$ will not raise the order when acting on θ_o, but will when acting on 'a'. The second order terms in the KdV equation are

$$(-\omega_o\partial_{\theta_o}+k_o^3\partial_{\theta_o}^3)u_2+\tfrac{1}{2}A_2\mathrm{e}^{\mathrm{i}\theta_o}-\tfrac{3}{2}k_o^2B_2\mathrm{e}^{\mathrm{i}\theta_o}-\tfrac{3}{2}k_oaN_1\mathrm{e}^{\mathrm{i}\theta_o}$$

$$+\tfrac{1}{2}\partial_x\left(\frac{a^2}{4}\mathrm{e}^{2\mathrm{i}\theta_o}\right)+\text{c.c.}=0. \tag{5.3.5}$$

We will now think of (5.3.5) as an equation to determine u_2. If secular terms such as $\theta_o\mathrm{e}^{\mathrm{i}\theta_o}$, which are physical nonsense, are to be avoided in u_2, all $\mathrm{e}^{\mathrm{i}\theta_o}$ terms in (5.3.5) must cancel. This yields via (5.3.4)

$$\frac{\partial|a|^2}{\partial t}+\frac{\partial}{\partial x}(v_g|a|^2)=0,\ v_g=\frac{\partial\omega_o}{\partial k_o}=-3k_o^2. \tag{5.3.6}$$

This is the law of conservation of wave energy. It will be seen in Section 5.5 and more generally in Chapter 8 to be only an approximation and will not survive a more exact calculation. Equation (5.3.6), together with

$$\frac{\partial k_o}{\partial t}+v_g\frac{\partial k_o}{\partial x}=0 \tag{5.3.7}$$

which follows from (5.3.3), give a crude theory of small 'a' wavetrain behaviour. There is one degenerate characteristic velocity v_g (for $|a|^2$ and k_o propagation), two Riemann variables, $|a|^2$ and k_o, one of which is in fact a Riemann invariant (k_o). If we were to follow a segment of a wavetrain with wavevector k_o we would find that, after a long time, it was to be found at a point such that x and t would satisfy

$$x/t=v_g(k_o) \tag{5.3.8}$$

(this is a result from linear wave theory which, if forgotten, can be found in Whitham (1974), p. 374). Along the characteristic that follows constant k_o (5.3.7), we have from (5.3.6) and (5.3.8)

$$\left(\frac{\partial}{\partial t}+v_g\frac{\partial}{\partial x}\right)|a|^2=\frac{\mathrm{d}|a|^2}{\mathrm{d}t}=-\frac{\partial v_g}{\partial x}|a|^2=-|a|^2/t, \tag{5.3.9}$$

and this is solved to give

$$|a|^2\sim t^{-1}.$$

The wave amplitude will diminish as $t^{-\frac{1}{2}}$ for large t. We will now see that the total $|a|^2$ integral between characteristics is conserved. From (5.3.6) (x_1 and x_2 are points on two characteristics), we have

$$\frac{\partial}{\partial t}\int_{x_1}^{x_2}|a|^2\mathrm{d}x=\int_{x_1}^{x_2}\frac{\partial}{\partial t}|a|^2\mathrm{d}x+v_{g_2}|a|_2^2-v_{g_1}|a|_1^2$$

$$=-\int_{x_1}^{x_2}\frac{\partial}{\partial x}[v_g|a|^2]\mathrm{d}x+v_{g_2}|a|_2^2-v_{g_1}|a|_1^2=0. \qquad (5.3.10)$$

The two above statements ($|a|\sim t^{-\frac{1}{2}}$ and total $|a|^2$ conservation between characteristics) imply that the wavetrain between x_1 and x_2 will have no alternative but to spread out. Another way of looking at this spreading is to observe that group lines corresponding to slightly different k_o, say k_{o_1} and k_{o_2}, move out from x_1 and x_2 with slightly different velocities x_1/t and x_2/t, but the energy between them remains constant. These findings are illustrated in Fig. 5.2(*a*). As k_o is conserved on characteristics and the wavetrain spreads, new crests will perforce appear. We have thus been able to describe the asymptotic (large t) behaviour. We will not explore this feature, but instead will consider a different ordering of parameters, leading to different wave dynamics.

Indeed, a quite different nonlinear behaviour follows if k_o *variation* is as slow as 'a' variation, even though $k_o \gg a$. In this case in (5.3.4)

$$M_1=N_1=0,$$

and up to the order considered equation (5.3.5) will be altered but still gives (5.3.6), though now with v_g constant. Now (5.3.6)–(5.3.7) describe a wave that does not spread, both k_o and $|a|^2$ being invariants. Interesting dynamics are thus postponed to the next order.

It is sometimes claimed that linear theory predicts spreading, whereas nonlinear theory does not. This calculation, however, demonstrates that both cases can follow from an 'a' expansion and the choice depends on the ordering of $\partial a/\partial x$ and $\partial k_o/\partial x$, which does not follow from the equations and must be postulated or taken from observation. We must somehow know how the small parameters compare before proceeding with the theory. The relevance of simple models in wave dynamics to real physical situations depends on the interplay of small parameters rather than on our approach.

Going back to the case $M_1=N_1=0$ and anticipating the conclusions of the following pages, the next order calculation will lead to a splitting of characteristics for stable wavetrains. In physical terms, this implies that a modulation on a uniform wavetrain will propagate on both characteristics and will eventually split in two (Fig. 5.2). In Chapter 8, an argument will be

Fig. 5.2. A nonlinear wavetrain will spread and its amplitude diminish if $\partial a/\partial x \sim a\partial k_o/\partial x$, a small, k_o remains constant on a characteristic (*a*). If on the other hand $\partial a/\partial x$ and $\partial k_o/\partial x$ are of the same order or else $\partial a/\partial x \gg \partial k_o/\partial x$, the wavetrain will not immediately spread, but any modulation will split (in the case of stable waves). The two emerging modulations will eventually spread after a while, but this is not shown (*b*). Here $t_2 > t_1$.

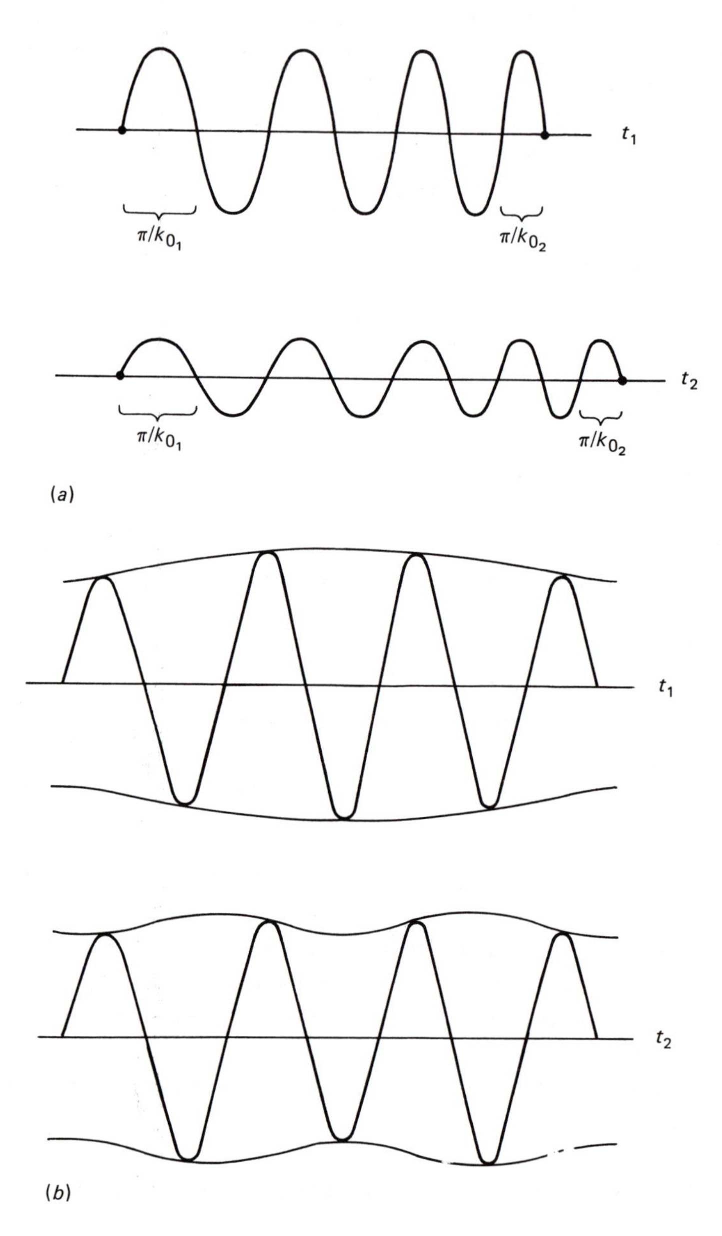

given to the effect that each of the two resulting modulations will spread (always assuming the initial wave to be stable).

Thus the fate of a stable, almost uniform wavetrain will be either *spreading* if $\partial a/\partial x \sim a\partial k_o/\partial x$; or *modified wavetrain splits in two→spreading of each modified component* if $\partial a/\partial a \sim \partial k_o/\partial x$. There is also a possibility of intermediate situations that we will not go into.

Later on in this Section we will see what the implications of the above theoretical results are for entire wavepackets (as opposed to almost uniform wavetrains considered here).

Proceeding with $M_1 = N_1 = 0$, we obtain $A_2/B_2 = -v_g$ and

$$u_2 = \frac{a^2}{24k_o^2} e^{2i\theta_o} + \text{c.c.} + C(a, a^*). \tag{5.3.11}$$

The third order equation is

$$(-\omega_o \partial_{\theta o} + k_o^3 \partial\theta_o^3)u_3 + \tfrac{1}{2}(A_3 - 3k_o^2 B_3 + 3ik_o B_{2x} - 3k_o^2 a N_2)e^{i\theta_o}$$
$$+ \text{c.c.} + C_a A_2 + C_a^* A_2^* + \partial_x(u_1 u_2 + \tfrac{1}{2}u_1^2) = 0 \tag{5.3.12}$$

where

$$C_a = \frac{\partial C}{\partial a}.$$

This time we have two non-secularity conditions; all constant and all $e^{i\theta_o}$ terms must cancel. The first condition yields

$$C = -\frac{aa^*}{12k_o^2} + \text{const}, \tag{5.3.13}$$

the second, expressed in terms of the amplitude a; is:

$$i\left(\frac{\partial a}{\partial t} + v_g \frac{\partial a}{\partial x}\right) + \tfrac{1}{2}\frac{\partial^2 \omega_o}{\partial k_o^2}\frac{\partial^2 a}{\partial x^2} - \omega_2 a^2 a^* + \gamma a = 0 \tag{5.3.14}$$

$$\frac{\partial^2 \omega_o}{\partial k_o^2} = -6k_o, \omega_2 = -\frac{1}{24k_o}, \tag{5.3.15}$$

and γ is a so-far arbitrary constant (in this order). We have arrived at the nonlinear Schrödinger equation with a cubic nonlinearity. This in fact results when the above procedure is applied to most nonlinear wave equations in one variable.

The simplest non-zero solution to (5.3.14) is a constant a_o such that

$$a_o^2 = \gamma/\omega_2, \tag{5.3.16}$$

and for the moment we will concentrate on linear perturbations on (5.3.16) (other non-zero solutions, such as soliton or shock shaped envelopes, will be treated in Chapter 6, and their stability in Chapter 8). Thus:

$$a = a_o + \delta a_1 \mathrm{e}^{\mathrm{i}(K_x - \Omega t)} + \delta a_2 \mathrm{e}^{-\mathrm{i}(Kx - \quad {}^{*}t)}, \tag{5.3.17}$$

equation (5.3.14) will give, upon linearization in δa:

$$\begin{aligned} &(\Omega - Kv_g - \tfrac{1}{2}\omega_{o_{kk}} K^2 - \omega_2 a_o^2)\delta a_1 - \omega_2 a_o^2 \delta a_2^* = 0 \\ &-\omega_2 a_o^2 \delta a_1^* + (-\Omega^* + Kv_g - \tfrac{1}{2}\omega_{o_{kk}} K^2 - \omega_2 a_o^2)\delta a_2 = 0. \end{aligned} \tag{5.3.18}$$

The characteristic determinant for equations in $\delta a_1 + \delta a_2^*$ and $\delta a_1 - \delta a_2^*$, obtained by adding the complex conjugate of the second component to the first, vanishes when

$$\Omega/K = v_g \pm \sqrt{[(\omega_2 \omega_{o_{kk}} a_o^2 + \tfrac{1}{4}\omega_{o_{kk}}^2 K^2)]}. \tag{5.3.19}$$

This is a generalization of Lighthill's theorem, in which the K^2 term is absent, Whitham (1974). If $\omega_2 \omega_{o_{kk}} > 0$, as is the case for KdV (5.3.15), Ω/K (and hence a characteristic) splits and so also will modulations on a uniform wavetrain (Fig. 5.2(*b*)). Splitting will be illustrated by experimental results later in this Section, though the context will not be exactly the same as in this calculation (Fig. 5.5). A general amplitude theory will be presented in Section 5.5 and Chapter 8.

For $\omega_2 \omega_{o_{kk}} < 0$ we obtain instability in a K band

$$0 < K < 2a_o \sqrt{|\omega_2/\omega_{o_{kk}}|} = K_c, \tag{5.3.20}$$

in which the growth rate is

$$\Gamma = \sqrt{|\omega_2 \omega_{o_{kk}} a_o^2 K^2 - \tfrac{1}{4}\omega_{o_{kk}}^2 K^4|}, \tag{5.3.21}$$

and the maximum growth rate for given a_o is achieved at $K = K_c/\sqrt{2}$.

We have up to now concentrated on the amplitude 'a' of the nonlinear wave (5.3.2) and have formulated a theory of its slow variations. However, as we saw in Section 4.6, the basic wavenumber k_o will also have its say in all this: it will introduce a degeneracy in the perturbation wavenumber K. The perturbed mode with growth rate Γ given by (5.3.21) will appear periodically, as seen in Fig. 5.3. A more mathematical argument for this degeneracy will be given in Section 8.2, but it should not come as a great surprise if we think of the basic wave as a background entity, introducing a periodicity into the space in which the perturbation (5.3.17) travels.

We will see later in this Section how the stability chart of Fig. 5.3, obtained from small 'a' theory, stands up to numerical confirmation when this restriction is lifted for a specific example (capillary water waves). Agreement will be better than we have any right to expect!

5.3.2 The story of deep water waves

Study of the dynamics of deep water gravity waves has led scientists to a better understanding of the role played by the NLS in weak amplitude wave theory.

For $h \to \infty$ and neglect of surface tension ($T=0$), equations (5.2.1)–(5.2.4) simplify to (reinstating the g)

$$\nabla^2 \phi = 0 \tag{5.3.22}$$

$$\left.\begin{aligned} &\eta_t + \phi_x \eta_x + \phi_y \eta_y = \phi_z \qquad (5.3.23) \\ &\phi_t + \tfrac{1}{2}(\nabla \phi)^2 + g\eta = 0 \qquad (5.3.24) \end{aligned}\right\} \quad z = \eta$$

The undisturbed water surface is now taken to be $z=0$. The linear dispersion relation (5.1.11) becomes

$$\omega_o^2 = g k_o \tag{5.3.25}$$

We now drop y dependence (general y dependent formulas will be given at the end of this Section). Proceeding as for the KdV equation, we define

$$\frac{\partial a}{\partial t} = A_2(a, a^*) + A_3 + \ldots \tag{5.3.26}$$

$$\frac{\partial a}{\partial x} = B_2(a, a^*) + B_3 + \ldots \tag{5.3.27}$$

Fig. 5.3. Stability regions in a, K plane for unstable waves as obtained from a^2 theory. S = stable region, U = unstable. There is a degeneracy in K causing behaviour to be periodic in K. The loci of maximum instabilities are indicated by broken lines. Growth rates are zero for $K = nk_o$.

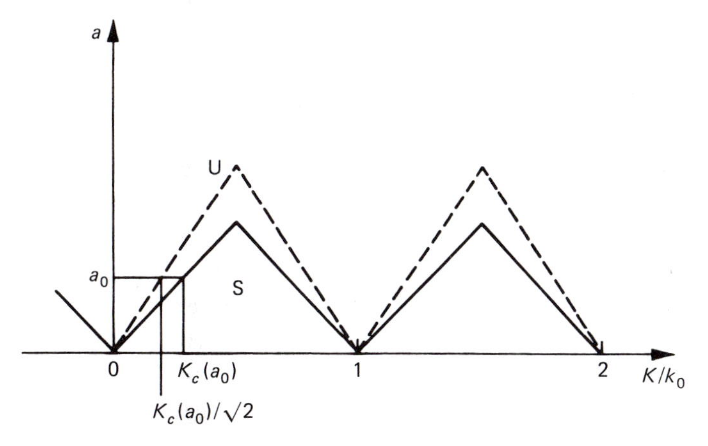

Taking $\partial k_o/\partial x$ to be of order 'a' would once more lead to (5.3.6) and wave spreading. Rather than repeat the ensuing discussion we just take

$$\frac{\partial k_o}{\partial t}=M_2+\ldots \tag{5.3.28}$$

$$\frac{\partial k_o}{\partial x}=N_2+\ldots \tag{5.3.29}$$

from the start.

Linear theory, Section 5.1 yields

$$\eta_1=\frac{a}{2}\mathrm{e}^{\mathrm{i}\theta_o}+\text{c.c.} \tag{5.3.30}$$

$$\phi_1=\frac{\omega_o a}{2\mathrm{i}k_o}\mathrm{e}^{\mathrm{i}\theta_o}\mathrm{e}^{k_o z}+\text{c.c.} \tag{5.3.31}$$

As before, all fast x, t dependence is through θ_o and

$$\partial_x\theta_o=k_o, \partial_t\theta_o=-\omega_o. \tag{5.3.32}$$

The second order equations following from (5.3.22)–(5.3.24) are

$$(k_o^2\partial_{\theta_o}^2+\partial z^2)\phi_2=-2k_o\partial_{\theta_o x}\phi_1 \tag{5.3.33}$$

and, at $z=0$:

$$-\omega_o\partial_{\theta_o}\eta_2-\partial_z\phi_2=-\partial_t\eta_1+\eta_1\partial_z^2\phi_1-k_0^2(\partial_{\theta_o}\phi_1)(\partial_{\theta_o}\eta_1)$$
$$-\omega_o\partial_{\theta_o}\phi_2+g\eta_2=-\partial_t\phi_1+\omega_o\eta_1\partial_{\theta_o}z\phi_1-\tfrac{1}{2}k_o(\partial_{\theta_o}\phi_1)^2-\tfrac{1}{2}(\partial_z\phi_1)^2.$$

Here θ_o and z dependence were assumed fast, x and t dependence slow and differentiation of either of the latter pair raised the order. The $z=0$ conditions were obtained by expanding ϕ around $\eta=0$:

$$\phi(\eta)=\phi(\mathrm{o})+\eta_z^{\phi}(\mathrm{o})+\tfrac{1}{2}\eta^2\phi_{zz}(\mathrm{o})+\ldots \tag{5.3.34}$$

The non-secularity condition is

$$A_2+v_g B_2=0$$

leading to

$$\frac{\partial|a|^2}{\partial t}+\frac{\partial}{\partial x}(v_g|a|^2)=0, \tag{5.3.35}$$

v_g being constant in this equation, and

$$\eta_2=\tfrac{1}{4}k_o a^2\mathrm{e}^{2\mathrm{i}\theta_o}+\text{c.c.}+\text{const.} \tag{5.3.36}$$

$$\phi_2=\tfrac{1}{2}\left[\frac{a_t}{k_o}\mathrm{e}^{\mathrm{i}\theta_o}+\frac{\omega_o a_x}{k_o^2}(1-k_o z)\mathrm{e}^{\mathrm{i}\theta_o}\right]\mathrm{e}^{k_o z}+\text{c.c.} \tag{5.3.37}$$

When these expressions are used in the third order equations analogous to (5.3.33), the consistency condition again becomes the NLS equation:

$$\mathrm{i}\left(\frac{\partial a}{\partial t}+v_{\mathrm{g}}\frac{\partial a}{\partial x}\right)+\tfrac{1}{2}\frac{\partial^2\omega_{\mathrm{o}}}{\partial k_{\mathrm{o}}^2}\frac{\partial^2 a}{\partial x^2}-\omega_2 a^2 a^*+\gamma a=0 \tag{5.3.38}$$

$$\frac{\partial^2\omega_{\mathrm{o}}}{\partial k_{\mathrm{o}}^2}=-\tfrac{1}{4}\sqrt{(g/k_{\mathrm{o}}^3)} \text{ and } \omega_2=\tfrac{1}{2}\sqrt{(gk_{\mathrm{o}}^5)}. \tag{5.3.39}$$

(Exercise 6).

In view of what was said above about stability, and the fact that $\omega_{\mathrm{o}_{kk}}\omega_2$ is negative, small amplitude, nonlinear deep water gravity waves are unstable. This was mentioned in Chapter 4. Although this instability is one of the basic phenomena of hydrodynamics, it was only properly described in 1967 (Benjamin (1967), especially the plate reproduced here in the Introduction in Fig. 1.4(*b*), Benjamin and Feir (1967), Lighthill (1967)). We will give a historical survey of this problem in the introduction to Chapter 8, where large amplitude waves will also be considered.

5.3.3 Mystery of the missing term

In the above calculation, ω_{o} and k_{o} were the frequency and wavenumber our wave would have in the vanishing amplitude limit and so they satisfied the linear dispersion relation throughout. This was seen to force the 'a' dynamics into the NLS equation (5.3.14). Although two cases were considered in detail, there should by now be little doubt that calculations will be similar, and a NLS equation obtained, for a wide class of wave equation.

Another approach would be to use the *physical* frequency ω and wavenumber k. In general, amplitude dependence is expected and so we anticipate

$$\omega=\omega_{\mathrm{o}}(k)+\omega_2(k)|A|^2, \tag{5.3.40}$$

where A is the wave amplitude, henceforth taken to be real. A new phase θ will correspond to our altered ω and k:

$$k=\theta_x, \omega=-\theta_t. \tag{5.3.41}$$

To avoid confusion with the ω_{o} of previous subsections, we will write $\omega_{\mathrm{o}}(k)$ wherever this confusion might arise. Our definition of phase will in general imply an altered form of the amplitude, now denoted by A. Since the nonlinear wave is the same in both conventions, we have

$$a\mathrm{e}^{\mathrm{i}\theta_{\mathrm{o}}}=A\mathrm{e}^{\mathrm{i}\theta}, \tag{5.3.42}$$

and we will now expand in A.

The basic equations are (5.3.40) and

$$k_t + \omega_x = 0, \tag{5.3.43}$$

which follows from (5.3.41); and wave energy conservation in A^2 theory:

$$\frac{\partial A^2}{\partial t} + \frac{\partial}{\partial x}\left(\frac{\partial \omega}{\partial k} A^2\right) = 0. \tag{5.3.44}$$

Once more we will consider $\partial k/\partial x \sim \partial A/\partial x$. If we now write out our equations in expanded form:

$$\frac{\partial k}{\partial t} + [\omega_{o_k} + \omega_{2_k} A^2]\frac{\partial k}{\partial x} + \omega_2 \frac{\partial A^2}{\partial x} = 0 \tag{5.3.45}$$

$$\frac{\partial A^2}{\partial t} + \omega_{o_k}\frac{\partial A^2}{\partial x} + \omega_{o_{kk}} A^2 \frac{\partial k}{\partial x} + \frac{\partial}{\partial x}\left(A^2 \frac{\partial}{\partial k}\omega_2 A^2\right) = 0 \tag{5.3.46}$$

and apply our ordering in A, we see that (5.3.45) up to third order is

$$\frac{\partial k}{\partial t} + \omega_{o_k}\frac{\partial k}{\partial x} + \omega_2 \frac{\partial A^2}{\partial x} = 0. \tag{5.3.47}$$

The fourth order term can be discarded, as it is of higher order than the rest and does not introduce any new physics. Equation (5.3.46) is third, fourth and higher order. The next to last term, which is the only fourth order term but introduces a coupling, will be retained. The last term is sixth order and will be discarded. Adding (5.3.47) multiplied by $\frac{1}{2}\sqrt{\omega_{o_{kk}}/\omega_2}$ to what remains of (5.3.46) times $(2A)^{-1}$, we obtain

$$\frac{\partial R_+}{\partial t} + (\omega_{o_k} + \sqrt{(\omega_2 \omega_{o_{kk}} A^2)})\frac{\partial R_+}{\partial x} = 0 \tag{5.3.48}$$

$$R_+ = \tfrac{1}{2}\int \sqrt{[(\omega_{o_{kk}})/\omega_2]}\,\mathrm{d}k + A. \tag{5.3.49}$$

This is a characteristic equation with characteristic velocity

$$\frac{\mathrm{d}x_+}{\mathrm{d}t} = \omega_{o_k} + \sqrt{(\omega_2 \omega_{o_{kk}} A^2)}. \tag{5.3.50}$$

Subtraction leads to a characteristic equation with Riemann invariant

$$R_- = \tfrac{1}{2}\int \sqrt{[(\omega_{o_{kk}})/\omega_2]}\,\mathrm{d}k - A, \tag{5.3.51}$$

and the two characteristic velocities can be written jointly as

$$\frac{\mathrm{d}x}{\mathrm{d}t}\pm = \omega_{o_k} \pm \sqrt{(\omega_2 \omega_{o_{kk}} A^2)}. \tag{5.3.52}$$

This would be similar to the result obtained via the NLS equation (5.3.19) if not for the additional term

$$\tfrac{1}{4}\omega_{o_{kk}}K^2 \tag{5.3.53}$$

in the square root in (5.3.19) but missing here in (5.3.52) (the difference between a^2 and A^2 would be higher order). Is it possible to introduce this term here? Perhaps we will be able to find the answer by trying to reproduce the NLS equation from (5.3.40), (5.3.43), (5.3.44). We would expect the NLS to govern the behaviour of the complex amplitude 'a' defined by

$$a = A e^{i(\theta - \theta_o)} \tag{5.3.54}$$

$$\partial_x(\theta - \theta_o) = k - k_o,\ \partial_t(\theta - \theta_o) = -\omega + \omega_o. \tag{5.3.55}$$

However, when we derive an equation for the x, t dynamics of 'a', we fail to obtain the NLS equation (Exercise 7). This is the second result to fall short of expectations, (5.3.52) being the first. No doubt the two failures are somehow connected. Presumably one of (5.3.40), (5.3.46) and (5.3.47), which are all approximate, will prove to be the culprit.

Such was the state of affairs in the mid-seventies, when Yuen and Lake (1975) suggested taking another look at the dispersion relation (5.3.40) in the hope of finding a missing term that might set things straight. (They considered deep water gravity waves and we will also revert to this simple case in Exercise 7.) The idea was excellent, but their calculation suffers from some important mistakes which cancel out to give (almost) the right answer. This is often true of pioneering work. Fornberg and Whitham (1978) produced the formula in correct form without deriving it. It is

$$\omega = \omega_o(k) + \omega_2(k)A^2 - \tfrac{1}{2}\omega_{o_{kk}}A_{xx}A^{-1}, \tag{5.3.56}$$

and this is one of the most important equations of the theory. The procedure used to derive (5.3.56) is known as Whithams's averaged Lagrangian calculation and will be described briefly in Subsection 5.5 and in some detail in Chapter 8. By comparing it with (5.3.40) we find our missing term!

If we now look for an equation describing 'a' dynamics, using (5.3.56) instead of (5.3.40), and (5.3.54)–(5.3.55), we obtain the much desired NLS equation (Exercise 7). To complete the argument by recovering the K^2 term we must rederive (5.3.19) directly from (5.3.56) and (5.3.43)–(5.3.44) see Exercise 8.

Note that the Yuen–Lake correction to the dispersion relation of (5.3.40) is zero when A is constant. Thus the small amplitude dispersion relation (5.3.40), found in most books, is not strictly incorrect, but it is limited in relevance to constant amplitude waves. This is worth remembering, even when the rest of this Section is forgotten.

5.3.4 Dynamics of a wavepacket

Up to now, we have been treating idealized situations in which small modulations on uniform wavetrains were investigated, as in Fig. 5.2(*b*). We will now use our main result, equation (5.3.56), to obtain at least a rough qualitative picture of the dynamics of a whole wavepacket (localized pulse). This will be possible if we know both the linear dispersion relation and the sign of ω_2. Suppose we do, and that

$$\omega_2 < 0,\ \omega_{\text{o}kk} < 0,\ \omega_{\text{o}kkk} < 0, \qquad (5.3.57)$$

as is the case for KdV waves (5.3.3), and they are stable and split modulations. In the k_x of order or smaller than A_x theory, equation (5.3.56) holds. If we model the wavepacket at $t = t_\text{o}$ by uniform k and

$$\begin{aligned} A &= A_\text{o}\cos\kappa(x - x_\text{B}), \quad |x - x_\text{B}| \leqslant \pi/2\kappa \\ &= 0 \qquad\qquad\qquad\quad , \ \text{other } x, \end{aligned} \qquad (5.3.58)$$

we have a rough representation of the wavepacket of Fig. 5.4(*a*). This is consistent with $k_x \leqslant A_x$ theory. It is of course just a model, but will be seen to cover the essential dynamics.

The third term in (5.3.56) will be

$$-\tfrac{1}{2}|\omega_{\text{o}kk}|\kappa^2 \qquad (5.3.59)$$

and will be independent of x (neglecting end effects at A and C). The second term, however, will be proportional to $-A^2$ and so the phase velocity ω/k

Fig. 5.4. Schematic illustration to the theory of wavepacket evolution in $A_x \sim k_x$ theory. V_p = phase velocity, V_g = group velocity.

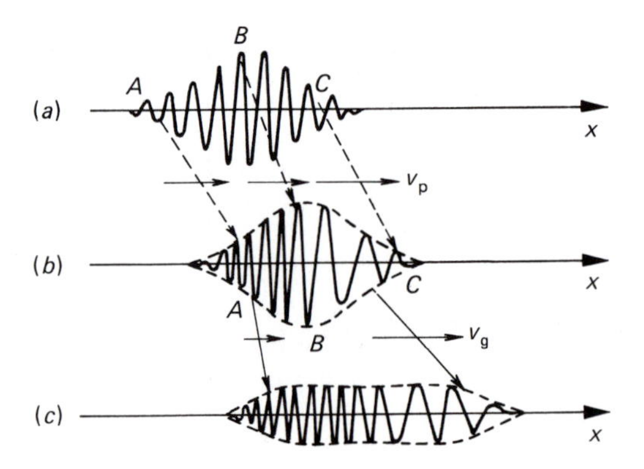

will be smaller at B than at A and C. After a while the wavepacket will resemble Fig. 5.4(b) and will lose its symmetry. There are two new developments:

1. The amplitude is now asymmetric and can roughly be modelled by

$$\begin{aligned} A &= A_0 \cos\kappa(x - x_B) \qquad -\pi/2\kappa \leqslant x - x_B \leqslant 0 \\ &= A_0 \cos\kappa(x - x_B) \qquad 0 \leqslant x - x_B \leqslant \pi/2\kappa \\ &= 0 \qquad \text{other } x, \end{aligned} \tag{5.3.60}$$

   and $\kappa_1 > \kappa_2$.

2. The wavenumber k is no longer uniform. The increment Δk is positive between A and B, where the wavepacket has been squeezed, and negative between B and C, where it has been

Fig. 5.5. Seven stages in the evolution of an ion acoustic wavepacket. The first four resemble those of Fig. 5.4. This and the previous Figure are taken from Ikezi *et al.* (1978).

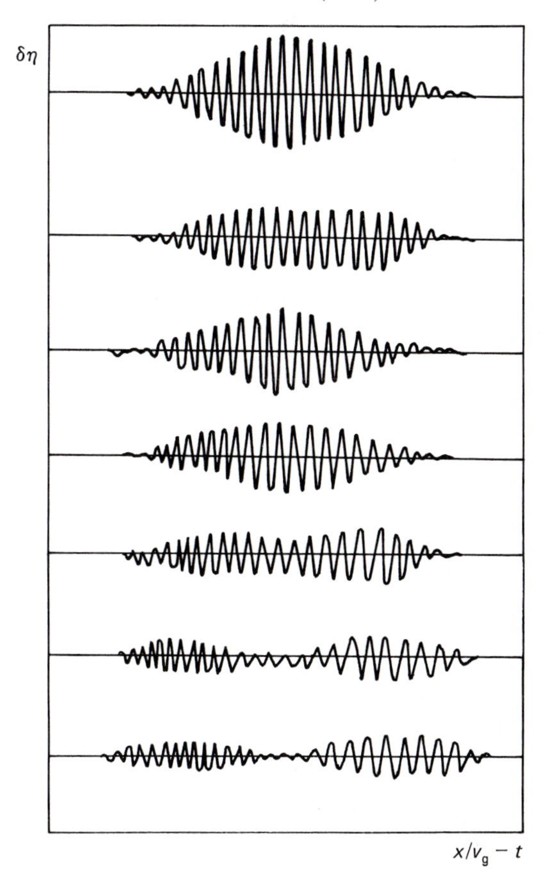

stretched. It is now time to calculate the group velocity from (5.3.56):

$$v_g = \mathrm{d}\omega/\mathrm{d}k = \omega_{o_k} + \omega_{o_{kk}}\Delta k + \omega_{2k}A^2 + \tfrac{1}{2}\omega_{o_{kkk}}\kappa_{1,2}^2 \tag{5.3.61}$$

Now

$$v_g(x > x_B) > v_g(x < x_B) \tag{5.3.62}$$

as both $\omega_{o_{kk}}$ and $\omega_{o_{kkk}}$ are negative, and Δk and κ^2 are larger in the A–B region than in the B–C region. The envelope in the B–C region will therefore move faster. As a result, the envelope expands and flattens (Fig. 5.4(*c*)).

Finally, as time proceeds, two distinct wavepackets will emerge, each travelling at its own group velocity. This stage, however, is outside the scope of the theory as represented by (5.3.56). The behaviour described above, including the splitting, was observed by Ikezi *et al.* for ion acoustic waves (1978); Fig. 5.5). Notice that k is indeed larger for the signal emerging on the left.

5.3.5 Some generalizations

We will now give some useful generalizations of formulae derived or quoted above.

Two space dimensions

We will start with an equation governing the amplitude of a planar wavetrain perturbed at an angle ψ to the direction of propagation x. It generalizes (5.3.14). The Schrödinger equation for 'a' dynamics becomes:

$$i\left(\frac{\partial a}{\partial t} + \omega_{o_{\mathbf{k}}}\cdot\frac{\partial a}{\partial \mathbf{x}}\right) + \tfrac{1}{2}\omega_{o_{\mathbf{kk}}}: a_{\mathbf{xx}} - \omega_2(\psi)(a^2a^* - a_o^2 a) = 0 \tag{5.3.63}$$

(We must remember to first calculate $\omega_{o_{kk}}$ and *then* take $\mathbf{k}$ along x. In the language of the perturbation wavevector $\mathbf{K}$, $\cos\psi = \mathbf{k}\cdot\mathbf{K}/kK$). Three-dimensional perturbations thus herald new possibilities of instabilities even when the basic wave was one-dimensionally stable, and conversely of stabilization for non-zero ψ when a one-dimensional analysis gave instability as the result.

We will now return to the idealized situation described before (Subsection 5.3.1); that of a uniform wavetrain perturbed slightly. In three dimensions, the perturbation becomes

$$\delta a_1 \mathrm{e}^{\mathrm{i}(\mathbf{K}\cdot\mathbf{x} - \Omega t)} + \delta a_2 \mathrm{e}^{-\mathrm{i}(\mathbf{K}\cdot\mathbf{x} - \Omega^* t)} \tag{5.3.64}$$

and (5.3.19) becomes

$$\Omega/K = \omega_{o_{\mathbf{k}}}\cdot\mathbf{n} \pm \sqrt{[a_o^2\omega_2\omega_{o_{\mathbf{kk}}}:\mathbf{nn} + \tfrac{1}{4}(\omega_{o_{\mathbf{kk}}}:\mathbf{nn})^2K^2]} \tag{5.3.65}$$

where

$$\mathbf{K} = K\mathbf{n}.$$

To the order considered, ω_o, k_o, a_o and ω, k, A_o are interchangeable within the square root. Equation (5.3.19) is recovered for $\mathbf{n}$ along $\mathbf{x}$. Finally, (5.3.56) generalizes to

$$\omega = \omega_o(k) + \omega_2(k, \psi)A^2 - \tfrac{1}{2}\omega_{o\mathbf{kk}} : A_{\mathbf{xx}}A^{-1}. \tag{5.3.66}$$

We now propose to illustrate these formulae by extending the examples considered earlier in this Section to more space dimensions.

For the deep water case, ω_2 is ψ independent and $\omega_o^2 = gk_o$ yields, for $\mathbf{k}_o = k_o\mathbf{i}_x$, which should be substituted after performing all differentiations:

$$\omega_{o\mathbf{kk}} : \mathbf{n}\,\mathbf{n} = -\tfrac{1}{4}\sqrt{[(g/k_o^3)]}(1 - 2\tan^2\psi)\cos^2\psi \tag{5.3.67}$$
$$\cos\psi = \mathbf{k}\cdot\mathbf{K}/kK.$$

Thus the deep water wave instability is confined to $\psi < \tan^{-1}\sqrt{2}$ for all wavelengths $k_o \gg h^{-1}$ (Fig. 5.6, deep water end). The ambitious reader is encouraged to see Exercise 6 in this context.

Similarly, if KdV (5.3.1) is replaced by the Kadomstev–Petviashvili equation in the form

$$u_t + uu_x + u_{xxx} + v_y = 0 \tag{5.3.68}$$
$$v_x = u_y$$

more convenient for calculation of ω_2 (Exercise 9), we obtain

$$\omega_2 = -\frac{1}{24k_o^2}\frac{1 - \tan^2\psi/3k_o^2}{1 + \tan^2\psi/3k_o^2}. \tag{5.3.69}$$

This might suggest the possibility of an instability, as a change of sign occurs in ω_2 when $\psi = \tan^{-1}(\sqrt{(3)}k_o)$. However, the second factor in Lighthill's theorem comes to the rescue:

$$\omega_{o\mathbf{kk}} : \mathbf{nn} = -6k_o[1 - \tan^2\psi/3k_o^2]\cos^2\psi, \tag{5.3.70}$$

and no instability is introduced! All that happens at $\psi = \tan^{-1}\sqrt{(3)}k_o$ is that two Ω/K values merge (in the $K \to \infty$ limit). This is a residue of narrow angle instabilities following from the full parent equations, both in the water wave problem (5.2.1)–(5.2.4), see the dotted line in Fig. 5.6 (shallow water end); and in the magnetic field free ion acoustic wave problem (5.1.7)–(5.1.9), $\Omega_c = 0$ (Fig. 8.15).

Water over finite depth with surface tension included

The equations for water waves with surface tension (5.1.2)–(5.1.5) in x, t space lead to a Schrödinger equation (Kawahara (1975a, b)). In the first reference effects of a variable depth are also studied; see also Santini (1981) and Turpin *et al.* (1983). The coefficients are (ρ has been absorbed in T):

$$\omega_o^2 = (gk_o + Tk_o^3)\sigma, \sigma = \tanh(k_o h) \tag{5.3.71}$$

$$v_g = \frac{1}{2\omega_o}[(g+3k_o^2 T)\sigma + (g+Tk_o^2)k_o h(1-\sigma^2)] \tag{5.3.72}$$

$$\omega_2 = \frac{k_o^2}{16\sigma\omega_o}\left[\frac{k_o}{g\sigma^2 + Tk_o^2(\sigma^2-3)}\right]\{g^2(9\sigma^4 - 10\sigma^2 + 9)$$

$$+ gTk_o(15\sigma^4 - 44\sigma^2 + 30)$$

$$+ T^2k_o^4(\sigma^2-3)(6\sigma^2-7)\} + \frac{2(g+Tk_o^2)}{v_g{}^2 - gh}$$

$$\{6(g+Tk_o^2)\sigma - 2(g+3Tk_o^2)\sigma^3 + 3(g+Tk_o^2)k_o h(1-\sigma^2)^2\}] \tag{5.3.73}$$

Fig. 5.6. Unstable angles ψ for perturbations on a uniform, small amplitude wavetrain on the surface of a body of water of depth h. Shallow water waves to the left, deep water waves to the right (here $k_o h$ is the measure of the depth). Three models are considered: the full set (5.2.1)–(5.2.4) denoted by a continuous line; Boussinesq (5.4.2), broken lines; and Kadomtsev–Petviashvili, dotted line (the unstable region has shrunk to a curve).

The limitation on the instability angle $\psi_c \langle \tan^{-1}(2^{-\frac{1}{2}})$ obtained from the deep water NLS equation is seen on the right.

This comparison illustrates how information can be lost if we first derive a model equation for general shape waves, and then a NLS equation for near-cosine waves. From Infeld (1980a).

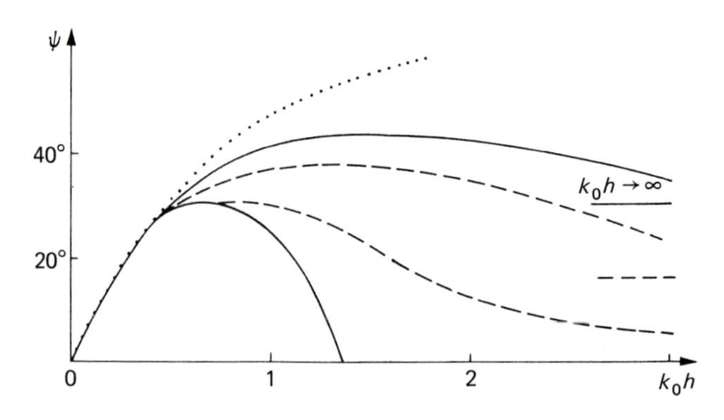

We will first consider $T=0$. Our previous unstable deep water values are recovered for $T=0, h\to\infty$. For finite h, ω_2 changes sign when $k_o h=1.36$ and this is the water depth limit of the instability (Benjamin (1967)). The three-dimensional (but still $T=0$) generalization can be found in Hayes (1973) and Davey and Stewartson (1974), and the former is the basis of the solid curve of Fig. 5.6. This curve is the locus of all vanishing $\omega_2\omega_{o\mathbf{kk}}:\mathbf{nn}$. For a more general analysis of the three-dimensional $T=0$ case, see Infeld *et al.* (1987).

When surface tension *dominates* gravitational effects (capillary waves of a few mm wavelength), (5.5.71)–(5.3.73) yield stability in the very shallow

Fig. 5.7. Stability chart (*a*) and loci of maximal growth rates (*b*) for finite amplitude capillary waves as found from a 51 term expansion applied to an exact solution (Crapper 1957). Small A^2 theory predicts the broken lines (Fig. 5.3), and the expansion the continuous lines. The critical value of A corresponds to streamlines that are vertical at a point. From Ma (1984).

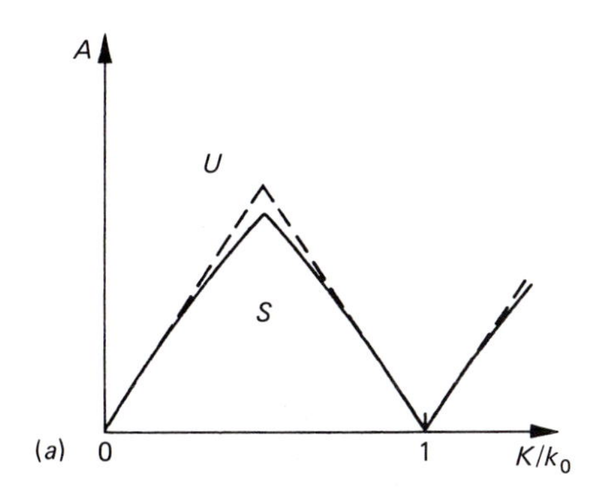

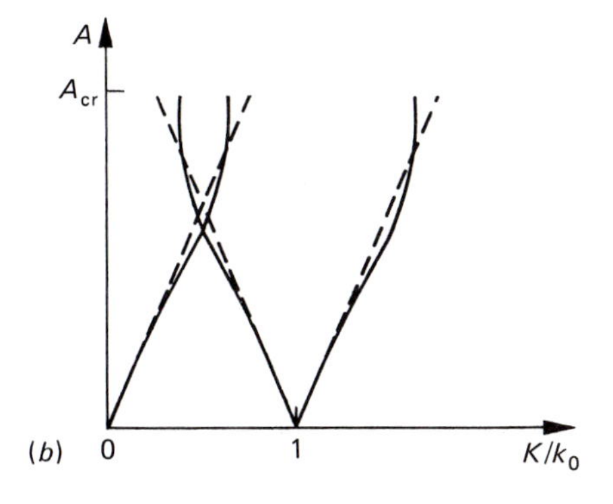

water limit, as the KdV equation is obtained. In the deep water limit we have

$$\omega_o = [Tk_o^3]^{\frac{1}{2}}, v_g = \omega_{ok} = \tfrac{3}{2}[Tk_o]^{\frac{1}{2}}, \omega_{okk} = \tfrac{3}{4}[T/k_o]^{\frac{1}{2}} \tag{5.3.74}$$
$$\omega_2 = -\tfrac{1}{16}[Tk_o^7]^{\frac{1}{2}}$$

and $\omega_{okk}\omega_2 < 0$. Instability again! This instability has been investigated by more exact methods (expansion of the stream function in 51 terms), and this furnishes a check on the small amplitude theory, which proves to be surprisingly good (Fig. 5.7; Ma (1984)).

Finally, Fig. 5.8 gives the division of parameter space into stable and unstable regions for arbitrary T/g. Equations (5.3.71) correspond to the upper right-hand corner. There are five changeovers from stability to instability, two due to the denominators in ω_2. For large T/g, an instability appears in the shallow water limit. It was lost in the KdV approximation, (though once again, two roots merge in this model and thus remind us of the instability in the fuller model).

The similarity of two of the continuous curves of Fig. 5.8 to the continuous line curves of Fig. 5.6 is fortuitous.

Fig. 5.8. Regions of modulational stability (shaded) for capillary-gravity waves. The KdV model always gives stability, the unstable region having shrunk to the line $T/gh^2 = \frac{1}{3}$ (broken line). However, KdV may be unphysical for points above this line (Section 5.2). From Kawahara (1975a) with $T/gh^2 = \frac{1}{3}$ line added. See also Ablowitz and Segur (1979).

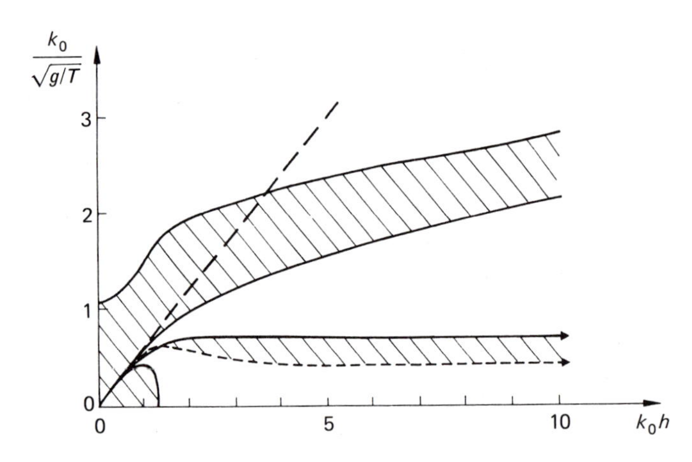

5.4 A general look at two families of model equations

In Section 5.2 we were able to simplify some rather formidable equations by assuming various things about variable dependence, weakness of amplitude etc., but allowing the shape of solutions to remain quite general. In Section 5.3 we started with a solution that was small amplitude and near-cosine in shape and then proceeded to find equations governing the amplitude as it *gradually* varied in space and time. These two approaches are essentially different in the sense that one can first simplify the equations as in Section 5.2 and then look at the dynamics of the amplitude of cosine shaped solutions as in Section 5.3, but of course not vice versa.

Figure 5.9 shows how various equations, derived from the basic fluid set

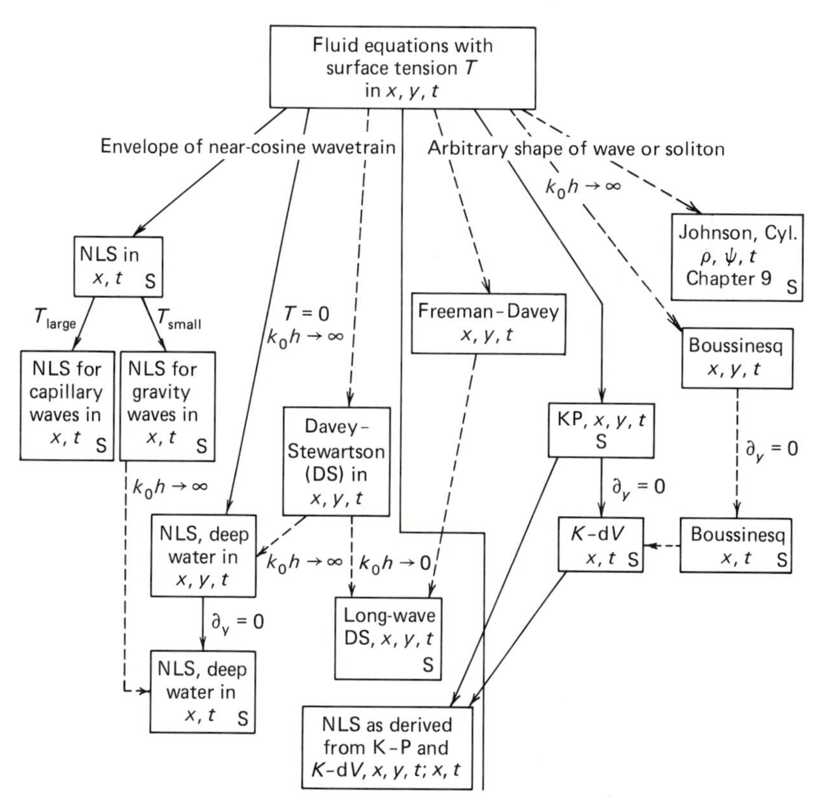

Fig. 5.9. Interdependence of some fluid dynamic equations. The dividing line down the middle separates equations for arbitrary shape waves on the right, from equations for the envelopes of locally near-cosine shaped waves on the left. Small parameters that enable us to move down the figure are not generally indicated, but some can be found in the text. S = completely solvable equation.

(5.2.1)–(5.2.4), are interrelated. General shape expansions are to the right of the dividing line, those in which the lowest order wave is $a\exp(i[k_o x-\omega_o t])$ to the left. The dividing line can only be crossed from right to left. Roughly speaking, derivations that appear in this book are indicated by continuous arrows, others by broken line arrows. Figure 5.9 may seem a bit heavy at first glance, but the reader might like to trace the derivations of Section 5.2 and Section 5.3 down the figure.

The shallow water Boussinesq equation (1871), when generalized to include y dependence and made Galilean-invariant, appears in two versions:

$$h_t+(uh)_x+(vh)_y=0 \tag{5.4.1}$$

$$u_t+uu_x+gh_x\begin{cases}+\frac{1}{3}gh_o^2h_{xxx}=0 & (5.4.2)\\ -\frac{1}{3}h_o(u_{xt}+uu_{xx})_x=0 & (5.4.3)\end{cases}$$

$$v_x=u_y, \tag{5.4.4}$$

where u, v are the x and y components of the velocity, averaged over the depth (Infeld (1980a)). Here h is the depth at a point and h_o the average depth. These equations were derived from the full water wave equations (5.2.1)–(5.2.4) as Kadomtsev–Petviashvili was, but are more general in that wave propagation in both directions is possible. The two forms are equivalent to the order considered, but a stability analysis for an almost uniform wavetrain as in Section 5.3 yields different results for each form. The first version (5.4.2) gives a stability chart as shown in Fig. 5.6 (broken line). It approximates the full equation result very well for $h_o k_o<0.5$, but fails to predict the known one-dimensional destabilization at $h_o k_o=1.36$ (Section 5.3).

The second version (5.4.3), on the other hand, predicts one-dimensional destabilization for $h_o k_o=1.5$, not very different from the true value of 1.36. However, this model breaks down completely for $h_o k_o=1.73$, as can be seen just by deriving the linear dispersion relation. (Exercise 11 and Infeld (1980a)).

The diversity of the two models is not surprising, as they only really claim to be valid for small $h_o k_o$, just like Kadomstev–Petviashvili (see Section 5.2 and Fig. 5.1). The surprising thing is that they still yield reasonable stability results for $h_o k_o$ of order 1.

The other equations dragged in from the cold in Fig. 5.9 (Davey–Stewartson (1974) and Freeman–Davey (1975)) are both pairs of equations that are intermediate between the original set and those derived in Section 5.2. They will not be written out in full. It is extremely interesting, however, that the long wave Davey–Stewartson pair of equations for wave envelope dynamics contains second order space differentiation only through the operators: $\partial_x^2+\partial_y^2$ in one equation and $\partial_x^2-\partial_y^2$ in the other, Anker and Freeman (1978), Zakharov and Shulman (1980). This renders the set completely solvable. This is more than can be said of even the simplest x, y, t version of the NLS equation (one equation!). This solvability was first conjectured by Ablowitz and Segur (1979) and then demonstrated by Zakharov and Shulman (1980).

Figure 5.10 presents a similar family tree for the ion acoustic plasma wave problem (5.1.7)–(5.1.9) and its derivative equations as treated in this book. So far two model equation derivations have been given (Section 5.2); the Kadomstev–Petviashvili equation for $\Omega_c=0$, and the Zakharov–Kuznetsov equation for large Ω_c. The others will appear in Chapter 9. In a way, the story of Fig. 5.10 is simpler than that of its predecessor, as volume equations breed volume equations and no artifacts are required to eliminate unknown surface shapes. All equations of Fig. 5.10 are valid for the entire wave (not the wave envelope). However, many of these equations can be reduced to the NLS equation with various coefficients if solutions are assumed proportional to $a\exp[i(k_0x-\omega_0t)]$ to lowest order (see, for instance, Exercise 10).

Fig. 5.10. Interdependence of some plasma ion acoustic equations. Model equations to be derived in future chapters are indicated. S = completely solvable equation.

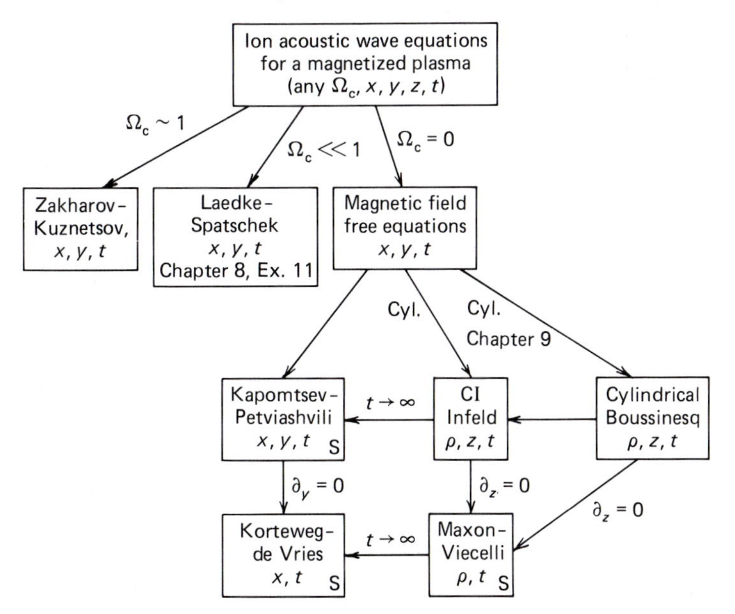

In Chapter 8 we will develop the theme of a critical assessment of these models, and extensions to large amplitude waves will be considered. Most models represented in Fig. 5.10 will in some respects prove to be better than could be expected (as Boussinesq was just seen to be).

Figures 5.9 and 5.10 should be used as references for this Chapter and the remainder of the book. At the moment they may seem somewhat formidable, but the reader who decides to investigate the subject more fully will find no trouble in adding to them from time to time.

5.5 A natural extension to finite amplitude waves due to Hayes

Some of the results of Section 5.3 can naturally be extended to arbitrary amplitude of the nonlinear wave (in addition to the amplitude no longer being small, the shape ceases to be a cosine). The form of the equations will be similar to those of (small) A^2 theory. It will be more convenient to use A, ω, k, as we are now far from the linear limit and ω_o, k_o lose their usefulness (this is also true of a, which was their creature). The functions A, ω, k will still be slowly varying in space and time, all fast variations appearing in the phase θ, (still assumed to increase by 2π over one wavelength and to yield $k=\theta, \omega=-\theta_t$). This is the idea behind geometrical optics. We will try to generalize Lighthill's theorem (5.3.52), which will be seen to describe fully nonlinear waves more adequately than weakly nonlinear waves (in (5.3.19) the amplitude is no longer small, but K^2 still is and can be dropped). The following treatment is largely based on Hayes (1973).

Suppose we know the form of a Lagrangian for a system of equations governing wave propagation. When considering locally plane waves with slowly varying properties (amplitude, wavelength etc.), we will find it useful to treat this Lagrangian averaged over one wavelength. Just what form the condition of small variations of ω, k, A will take when this averaging is performed will be explained in Chapter 8 in some detail. Here we merely outline a reasoning that leads to a natural extension of the results of Section 5.3 to large amplitude, slowly varying plane waves.

We now take a Langrangian as given. Such a Lagrangian was found for the plasma wave problem in Section 5.2 (5.2.29). For the deep water wave equations it will be given by the density

$$L=\int_{-\infty}^{\eta}[\phi_t+\tfrac{1}{2}(\phi_x^2+\phi_y^2+\phi_z^2)+gz]\mathrm{d}z, \tag{5.5.1}$$

ϕ variation yielding (5.3.22) and (5.3.23), and η variation the condition (5.3.24), see Luke (1967).

Thus we assume the Lagrangian to be known and the variational principle will be the usual (we suppress y dependence in the following)

$$\delta\int\int L\mathrm{d}x\mathrm{d}t=0. \tag{5.5.2}$$

Now suppose we average our Lagrangian over one wavelength to obtain

$$\mathscr{L}=\frac{1}{2\pi}\int L\mathrm{d}\theta, \tag{5.5.3}$$

and $\mathscr{L}$ depends on ω, k, A, but not their derivatives. This follows from the assumed slowness of variations in ω, k, A and can be put in a more rigorous form (Sections 8.1 and 8.2). Whitham (1974) has shown that gradual changes in ω, k, A, perceptable only on the scale of many wavelengths, will be governed by the variational principle

$$\delta\int\int\mathscr{L}(\omega,k,A)\mathrm{d}x\mathrm{d}t=0. \tag{5.5.4}$$

Integration is taken over a *large* number of wavelengths and a time much *longer* than ω^{-1}. Variation (5.5.4) and the identities $\omega=-\theta_t, k=\theta_x$ yield the two Euler–Lagrange equations and one equality of mixed partial derivatives

$$\mathscr{L}_A=0 \tag{5.5.5}$$

$$\frac{\partial\mathscr{L}_\omega}{\partial t}-\frac{\partial\mathscr{L}_k}{\partial x}=0 \tag{5.5.6}$$

$$\frac{\partial k}{\partial t}+\frac{\partial\omega}{\partial x}=0. \tag{5.5.7}$$

(so we are not entirely rid of the *gradients* of θ in spite of the integration (5.5.3)!).

Choosing the new variable

$$I=\mathscr{L}_\omega \tag{5.5.8}$$

to eliminate A, our dependent variables become ω, k, I, and I has taken over from A as a measure of the amplitude.

Equation (5.5.5) takes the form

$$\mathscr{L}_I=0. \tag{5.5.9}$$

At this point we rather unglamorously introduce the Hamiltonian function as a constant of integration of (5.5.8) when integrated over ω:

$$\mathscr{H}(k,I)=\omega I-\mathscr{L} \tag{5.5.10}$$

In view of (5.5.9) and the identity that follows it, we have

$$\omega = \mathscr{H}_I \tag{5.5.11}$$

$$\mathscr{L}_k = -\mathscr{H}_k. \tag{5.5.12}$$

We now have all we need to eliminate ω and promote k and I to be a complete set of dependent variables. Equations (5.5.6) and (5.5.7) become

$$\frac{\partial k}{\partial t} + \frac{\partial}{\partial x}\mathscr{H}_I = 0 \tag{5.5.13}$$

$$\frac{\partial I}{\partial t} + \frac{\partial}{\partial x}\mathscr{H}_k = 0. \tag{5.5.14}$$

The characteristic equations are

$$\frac{\mathrm{d}R_\pm}{\partial t} + (\mathscr{H}_{Ik} \pm \sqrt{(\mathscr{H}_{kk}\mathscr{H}_{II})}\frac{\partial R_\pm}{\partial x} = 0$$

$$R_\pm = \int \sqrt{(\mathscr{H}_{kk}/\mathscr{H}_{II}I)}\mathrm{d}k \pm \int I^{-\frac{1}{2}}\mathrm{d}I \tag{5.5.15}$$

Thus $R_\pm$ are constant, along characteristics moving with velocities

$$\frac{\mathrm{d}X_\pm}{\mathrm{d}t} = \mathscr{H}_{Ik} \pm \sqrt{(\mathscr{H}_{kk}\mathscr{H}_{II})} \tag{5.5.16}$$

respectively.

It is easily checked that $\mathrm{d}\mathscr{H}_k/\mathrm{d}I$ on a characteristic is equal to the companion characteristic velocity. Thus it is x_+ on C_- and x_- on C_+ (Exercise 12: Hayes seems to have missed this). If (5.5.14) is interpreted as energy conservation, the first characteristic velocity is thus the velocity at which the *increment* of energy flows when measured on the second characteristic. Thus, if we were travelling with a characteristic, say C_-, we could learn the velocity of the other characteristic C_+ just by measuring $\mathrm{d}\mathscr{H}_k/\mathrm{d}I$. In contradistinction to lowest order theory (Section 5.3) the characteristic velocities are not velocities at which the total energy flows (this velocity is $\mathscr{H}_k/I$). We will return to this problem in Chapter 8.

Equations (5.5.15) and (5.5.16) are reminiscent of (5.3.48)–(5.3.52). We would of course expect to obtain these formulae in the small A limit. To this end, assume A small and

$$\mathscr{L} = A^2 D(\omega, k) + \mathrm{O}(A^4)$$
$$D = \omega - \omega_o(k).$$

To lowest order (A^2):

$$\mathscr{L}_\omega = I = A^2$$
$$\mathscr{L}_k = -\mathscr{H}_k = -A^2\omega_{ok}.$$

This is the stage at which we introduce k and I as our sole dependent variables, obtaining

$$\mathscr{H}_{kk} = I\omega_{okk} \tag{5.5.17}$$

$$\mathscr{H}_{II} = \omega_I = \omega_2 \tag{5.5.18}$$

$$\mathscr{H}_{Ik} = \omega_{ok}, \tag{5.5.19}$$

and (5.5.15) yields (5.3.49) and (5.3.51) in this limit. (The Riemann invariants $R\pm$ are recovered by first substituting (5.5.17)–(5.5.19) in (5.5.15), then integrating over k and I, and finally using $I = A^2$.)

The three-dimensional extension of (5.5.15) and (5.5.16) is

$$R_{\pm} = \int \mathscr{H}_{\mathbf{kk}} : \mathbf{n}\mathrm{d}\mathbf{k} / \sqrt{(\mathscr{H}_{II}\mathscr{H}_{\mathbf{kk}} : \mathbf{nn}I)} \pm \int I^{-\frac{1}{2}}\mathrm{d}I = 0 \tag{5.5.20}$$

constant on characteristics moving with velocities

$$\mathbf{X}_{\pm} = \frac{\mathrm{d}\Omega +}{\mathrm{d}\mathbf{k}} = \mathscr{H}_{I\mathbf{k}} \pm \mathscr{H}_{II}\mathscr{H}_{\mathbf{kk}} \cdot \mathbf{n} / \sqrt{(\mathscr{H}_{\mathbf{kk}} : \mathbf{nn})} \tag{5.5.21}$$

where $\mathbf{n}$ is the normal along the perturbation vector $\mathbf{K}$ and must be specified. Up to now we were able to obtain equations free of $\mathbf{K}$. Even now, neither the modulus K nor Ω, the perturbation frequency, appear in (5.5.21). This is due to the assumption that K and Ω are small. In particular, the term proportional to K^2 present in (5.3.65), does not appear in (5.5.21). This deserves an explanation. In equation (5.3.65), the whole correction term within the square root was small, A^2 (or a^2) and K^2 being comparable small quantities. Now the amplitude is large (zero order) and all components of (5.5.21) or (5.5.16) are of the same order. The small K^2 correction has been outclassed and dropped. On the level of the Lagrangian, space derivatives in the slow scale, such as A_x, k_x etc. are absent in $\mathscr{L}$ (and hence in $\mathscr{H}$).

For unstable, large amplitude waves, equation (5.5.15), covers the regions $\mathbf{K} \simeq m\mathbf{k}$ in Fig. 5.3, but for arbitrary amplitudes. The slope $\partial\Gamma/\partial K$ is given in these regions by

$$\frac{\partial\Gamma}{\partial K} = \sqrt{(|\mathscr{H}_{II}\mathscr{H}_{\mathbf{kk}} : \mathbf{nn}|)}$$

where $\mathbf{n}$ is the unit vector along $\mathbf{K} - m\mathbf{k}$.

In many problems (though not for deep water waves), mean flow effects can complicate this theory. We will see how to cope with these effects in Chapter 8.

A comprehensive picture of nonlinear wave dynamics for a given class of waves is best gained by performing both calculations: for A^2 and K^2 small and competitive, as in Section 5.3 and Fig. 5.2; and also for large A, small K

as outlined in this Section and treated in more detail in Chapter 8 (where some approaches that are more practical than Hayes', if less elegant, are used). The information obtained is usually complementary.

General A, K analysis is difficult and numerics usually enter the calculations before long. The field is fairly young (Figs. 5.1, 8.12(*b*), Martin *et al.* (1980), McLean (1982a and b), Laedke and Spatschek (1982b), Kuznetsov *et al.*, (1984), Ma (1984), Saffman and Yuen (1985) and papers quoted there, Infeld and Frycz (1987)). Some examples appeared in Section 4.6, others will be encountered briefly in Chapter 8. The special case of perpendicular perturbations ($\theta=\pi/2$) is a favourite, as calculations tend to simplify.

5.6 Temporal development of instabilities and wave–wave coupling

We saw in Section 5.3 that a uniform wavetrain solution of the nonlinear Schrödinger equation would be unstable whenever the signs of the highest derivative term and the nonlinear term were the same. This was in fact the case for one surface dimensional deep water gravity waves. The theory presented in Section 5.3 yielded a cutoff K_c in perturbation wavenumber, above which instabilities would not appear. Thus K must satisfy

$$0<K<K_c=2\sqrt{|\omega_2/\omega_{okk}|}\,a_o(0) \tag{5.6.1}$$

for instability (the slight change in notation will prove convenient in what follows). All this is no doubt very interesting, but a natural question to ask is: what happens next? A uniform wavetrain will presumably be destroyed, energy being pumped into the perturbation as long as the analysis leading to (5.6.1) holds, but obviously this cannot go on for ever and the exponential growth of the perturbation must surely give way to a different behaviour?

This question was answered numerically by Yuen and Ferguson (1978a). (For more of the same and numerical details, see Herbst *et al.* (1985).) They took the nonlinear Schrödinger equation (5.3.14) in a coordinate frame in which the $\partial a/\partial x$ term vanishes and chose $\omega_{okk}=-\frac{1}{4}, \omega_2=\frac{1}{2}$, the values obtained for deep water waves when $\omega_o=k_o=1$. The actual values are less important than their relative signs. Thus the equation they considered was just

$$i\frac{\partial a}{\partial t}-\frac{1}{8}\frac{\partial^2 a}{\partial x^2}-\tfrac{1}{2}a^2a^*+\tfrac{1}{2}a_o^2(0)a=0. \tag{5.6.2}$$

The initial condition was taken to be

$$a(x,0)=a_o(0)+\frac{a_o(0)}{10}\cos Kx \tag{5.6.3}$$

and periodicity in x was assumed:

$$a(x,t)=a(x+2\pi/K,t). \tag{5.6.4}$$

All this is essentially consistent with (5.3.17) and the theory of Section 5.3 should apply initially. Sure enough, growth of the perturbation (second term in 5.6.3) was found at the onset, and also generation of higher (mK) modes, not included in the theory of Section 5.3. Furthermore, when

$$\tfrac{1}{2}K_c<K<K_c=2^{3/2}a_o(0), \tag{5.6.5}$$

the initially unstable mode ($m=1$) levelled off and then relinquished its energy to the $m=0$ mode, the higher harmonics slavishly following suit (Figs. 5.11(*a*) and 5.12(*a*)). After a while, the system returned to its initial state (5.6.3). The effect was cyclic and the system was really a nonlinear clock (Fermi–Ulam–Pasta recurrence, or FUP). This interesting behaviour had also been observed experimentally for water waves (Yuen *et al.* (1977)). When $K<K_o/2$, higher m *unstable* modes are generated by nonlinear coupling and the phenomenon loses its simplicity, each unstable mode trying to force its own behaviour on the system. Thus much more time will elapse before the initial state (5.6.3) is reproduced. Even when this does occur, it will only be an approximate repetition of the initial data (Fig. 5.11(*b*)). Periodicity has been downgraded to ergodicity (see Chapter 10). A general mathematical, epsilon delta type theory of both the periodic and ergodic cases has been suggested by Thyagaraja, (1979) and (1983). The

Fig. 5.11. Sum of deviations of Fourier amplitudes from initial values

$$\varepsilon=N^{-1}\sum_{j=1}^{N}|a_j(t)-a_j(0)|$$

where N is the number of grid points used in the computation. Two cases are shown: (*a*) $\tfrac{1}{2}K_c<K<K_c$, leading to exact periodicity (FUP recurrence). (*b*) $0<K<\tfrac{1}{2}K_c$, leading to an almost perfect repetition of the initial conditions after a long time interval. From Yuen and Ferguson (1978a).

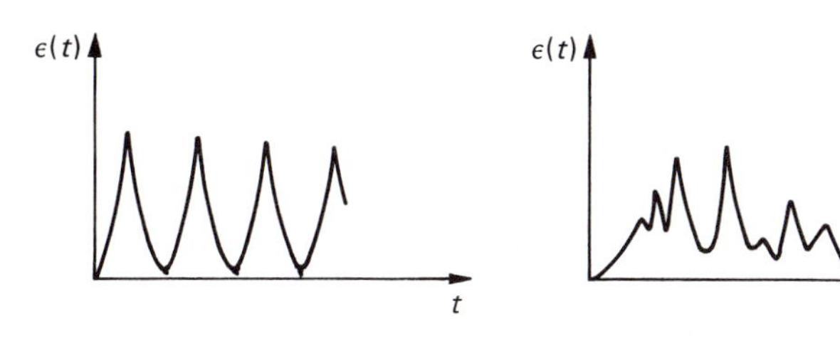

essence of his theory is that the system in question, subject to space-periodic boundary conditions, behaves as if its number of modes m was limited. Another way of looking at this is that the higher m modes are enslaved to the lower modes, Rowlands (1980). When a limited number of modes interact, a periodic or ergodic recurrent motion results.

We now concentrate on the (5.6.5) case and give a quantitative theory of the numerical results of Yuen and Ferguson, Figs. 5.11(a) and 5.12(a).

Fig. 5.12. Energy in lowest modes as a function of time: (a) Found numerically by Yuen and Ferguson for $a_0(0)=1$, $a_1(0)=0.05$, $K=2$; (b) Found from the calculation of Infeld (1981c); and (c) The period T for reproduction of initial data as follows from numerical calculation (dots) as compared with the theory of this Section (solid). Notice how much better agreement is in the right half of the Figure. Broken lines correspond to the theory of (5.6.16). From Infeld (1982).

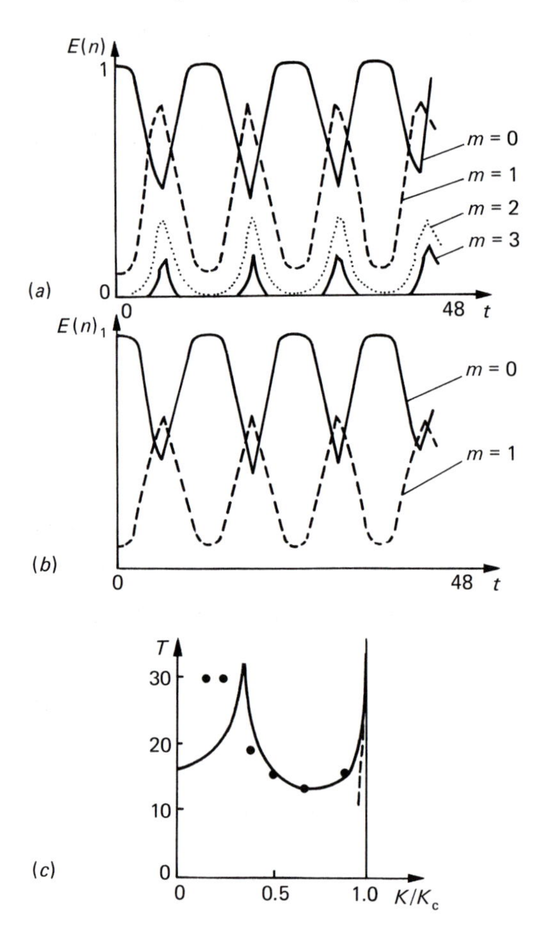

Treatment follows Infeld (1981c). We will look for solutions to (5.6.2) in the form

$$a = a_o(t) + t) + a_{-1}(t)e^{-iKx} + a_1(t)e^{iKx} \tag{5.6.6}$$
$$a_{-1}(0) = a_1(0)$$

Here we neglect the generation of all $m \geqslant 2$ modes. This procedure is in the *spirit* of Thyagaraja's Theory, though more modes would have to be taken for that theory to apply directly. This three mode theory (5.6.6) is used in fluid dynamics (Benney (1962), Bretherton (1964), Fornberg and Whitham (1978)). If we were to linearize in a_1, a_{-1}, we would of course just recover the instability of Section 5.3. However, we now proceed from (5.6.6) with no further simplification. Equations (5.6.2) and (5.6.6) yield, for the time development of the $m = 0$ mode,

$$i\dot{a}_o + \tfrac{1}{2}a_o^2(o)a_o = \tfrac{1}{2}a_o^2 a_o^* + a_{-1}a_1 a_o^* + a_{-1}a_{-1}^* a_o + a_o a_1 a_1^*, \tag{5.6.7}$$

and two similar equations for the development of a_{-1} and a_1. All three can be obtained from the Lagrangian

$$L = \frac{\sqrt{-1}}{2} \sum_{l=-1}^{1} (a_l^* \dot{a}_l - \dot{a}_l^* a_l + \sum_l (\tfrac{1}{2}a_o^2(0) + \tfrac{1}{8}K^2 l^2)|a_l|^2 \tag{5.6.8}$$

$$-\tfrac{1}{4}\sum_l |a_l|^4 - \sum_{i<j} |a_i|^2 |a_j|^2 - \tfrac{1}{2}F,$$

$$F = a_o^{*2} a_{-1} a_1 + a_o^2 a_{-1}^* a_1^*,$$

in particular

$$F(0) = 2a_o^2(0)a_1^2(0).$$

Time invariance and Noether's theorem give

$$\sum_{l=-1}^{1} (\dot{a}_l L\dot{a}_l + \dot{a}_l^* L_{\dot{a}l^*} - L) = \text{const}, \tag{5.6.9}$$

whereas invariance under the phase transformation $a_e \rightarrow a_e e_l$ yields the equation

$$\sum (\varepsilon_l a_l L_{\dot{a}l} - \varepsilon_l a_l^* L_{\dot{a}l}) = \text{const}. \tag{5.6.10}$$
$$2\varepsilon_o - \varepsilon_1 - \varepsilon_{-1} = 0,$$

So as to obtain conservation laws from (5.6.10) we will consider two separate cases:

$$\varepsilon_o = \varepsilon_1 = \varepsilon_{-1};\ \varepsilon_o = 0, \varepsilon_1 = -\varepsilon_{-1}. \tag{5.6.11}$$

Altogether, we now have three equations that are free of time differentiation. They are, for our initial conditions,

$$|a_{-1}|^2 = |a_1|^2 = a_1(0)^2 + \tfrac{1}{2}X;$$
$$X = a_o^2(0) - |a_o|^2 \qquad (5.6.12)$$
$$F = \tfrac{3}{4}X^2 + (\tfrac{1}{4}K^2 - a_o(0)^2 + a_1(0)^2)X + 2a_o(0)^2 a_1(0)^2.$$

If we now multiply (5.6.7) by a_o, subtract its complex conjugate, square the result, and express all quantities in terms of X and the known constants, we obtain

$$X^2 = 4|a_o|^4|a_1|^4 - F^2 = \tfrac{7}{16}(X - X_1)(X - X_2)(X - X_3)(X - X_4) \qquad (5.6.13)$$
$$X_1 = (K^2 - 4a_1(0)^2 - \sqrt{\Delta})/2, \Delta = (K^2 - 4a_1(0)^2)^2 + 64a_o(0)^2 a_1(0)^2$$
$$= X_2 = 0$$
$$X_3 = \tfrac{4}{7}(2a_o^2 - 3a_1(0) - \tfrac{1}{4}K^2)$$
$$X_4 = (K^2 - 4a_1(0)^2 + \sqrt{\Delta})/2$$
$$X_2 < X < X_3 \text{ when } X_3 > 0.$$

Equation (5.6.12) can be solved to give

$$X = X_1 X_3 sn^2(m, \alpha t)/[X_3 sn^2(m, \alpha t) + X_1 - X_3] \qquad (5.6.14)$$
$$m = X_3(X_4 - X_1)/X_4(X_3 - X_1)$$
$$\alpha^2 = 7X_4(X_3 - X_1)/64.$$

The solution is periodic with period $T = 2K(m)/\alpha$, where $K(m)$ is the complete elliptic integral, and the maximum deviation of $|a_o|^2$ is X_3. For $K^2 > K_c + 2a_1(0)^2$ the recurrence will not occur. This generalizes the (squared) cut-off of linear theory K_c^2. If we now take the numerical values for which Yuen and Ferguson gave the most detail (including Fig. 5.11(*a*)):

$$a_o(0) = 1, a_1(0) = 0.05, K_c = 2^{3/2} \qquad (5.6.15)$$

and use (5.6.13) to generate the time dependence of the $m = 0$ and $m = 1$ modes in our theory, we obtain Fig. 5.12(*b*). The accuracy of T, as well as of the time development of the $m = 0$ mode, are staggering. This is due to enslavement (following of the $m = 1$ modes by those of higher m when only $m = 1$ corresponds to linear instability). The maximal energy content in the $m = 1$ modes, on the other hand, is down by about 30% as compared with the numerical results of Yuen and Ferguson. This is due to neglect of the interplay with higher modes. Theoretical and numerical values of the period T for various K, as well as a comparison with earlier theories based on a small $K_c - K$ calculation, yielding

$$T \simeq (K_c - K)^{-\frac{1}{2}} \qquad (5.6.16)$$

(Rowlands (1980), Janssen (1981)), are given in Fig. 5.12(*c*).

The above theory is surprisingly successful even though no small parameter directly justifies the model. The approach has limitations. When the initial conditions differ considerably from the form of (5.6.3), and the perturbation of $a_0(0)$ is a more general periodic function of x, more modes should be included in the calculation. Even five modes (as opposed to our three) are difficult to treat, though some very special cases have been attempted (McEwan *et al.* (1972), Falk (1982), Romeiras (1982)). Further developments and their relation to modern theories of turbulence will be taken up in Chapter 10.

Wave–wave coupling in one space dimension plays an important role in most branches of classical physics and even appears in studies of population fluctuations etc. Most of the important references prior to date of publication can be found in the book by Weiland and Wilhelmsson (1977), one of the standard sources on the subject.

The problem of generalizing the above to higher spatial dimensions has also been considered (Yuen and Ferguson (1978b), Martin and Yuen (1980)). For the elliptic form of the nonlinear Schrödinger equation ($\partial_x^2 + \partial_y^2$), instability is still confined to **K** satisfying

$$K_x^2 + K_y^2 < K_c^2. \tag{5.6.17}$$

Further considerations are similar to those for the one-dimensional one, exact periodicity being confined to **K** such that (5.3.65):

$$(mK_x)^2 + (nK_y)^2 < K_c^2 \tag{5.6.18}$$

is satisfied by exactly one non-zero (m, n) pair. For the hyperbolic form of NLS appearing in deep water wave theory with $\partial_x^2 - 2\partial_y^2$, modes are unstable in the open region (5.3.65):

$$0 < (mK_x)^2 - 2(nK_y)^2 < K_c^2, \tag{5.6.19}$$

and there can be infinitely many (m, n) pairs yielding instability. We can no longer even hope for periodicity, as energy can now leak to higher and higher modes. This simple theoretical picture is consistent with the numerical results quoted above.

Thus we see that for times larger than one e folding, an unstable perturbation to a uniform wavetrain will either cyclically exchange its energy with the system (FUP recurrence), or a non-periodic recurrence is obtained (the system is represented by ergodic motion in phase space). Including more than one space dimension can change the picture completely, adding to the above the further possibility of a slow thermalization. One space dimensional analysis is seen to be too restrictive for some problems, such as deep water waves. This can be true even when the initial nonlinear structure was planar.

5.7 Concluding remarks

We have seen how a small parameter expansion can be useful in extracting models from frightening sets of equations. When possible, expanding Lagrangians is usually simpler than struggling with all the equations. The models obtained are particularly useful when they admit solution by the inverse scattering method (which will be presented in Chapter 7).

Stability considerations can be a check on the merits of a model. If we find that stability charts following from the full set and the model are similar, more power to the model. Those considered here gave reasonably good stability charts for waves. Some further comparisons will be presented in Chapter 8, where they will be seen to further improve for solitons. The authors have no ready explanation for why so many model equations should be more physical for solitons than for waves, but somehow this does seem to be the case. Perhaps one reason is that some of the trouble involved in passing from a full set of equations to a model is tied up with the periodicity of a nonlinear wave and the symmetries involved (see Section 8.4).

In all the examples treated here and in Chapter 8, the stretching parameter ε can be taken to be at least $\frac{1}{2}$ before predictions are very far off. They often still make some sense for $\varepsilon = 1$!

There can sometimes be a physical objection to using a particular model for a given situation. This objection may not be covered by our analysis. An example is furnished by the shallow water models treated here when $k_0 h$ is very small. When the depth is appreciable, neglect of bottom friction may not be very important to gravity waves on the surface. This approximation will, however, become increasingly more objectionable as we take $k_0 h$ to be smaller. Thus, one should take a second look at the physical assumptions introduced when a model equation was derived. One may have unwittingly violated one or more of them in the process of deriving the model.

Finally, we saw in Section 5.6 that a small parameter expansion is not the only possible vehicle for finding a good model. Retention of the lowest order modes proved extremely successful when investigating the Fermi–Ulam–Pasta recurrence in a one-dimensional system with periodic boundary conditions. This was true in spite of the absence of an ε in our procedure. Why the method works so well is not fully understood and deserves attention.

Exercises on Chapter 5

Exercise 1
Assume $\phi = af(z)\sin(k_o x - \omega_o t)$ and (5.1.10) to solve (5.1.2)–(5.1.5) in the $a \to 0$ limit. Show that

$$f(z) = (\omega_o/k_o)\frac{\cosh(k_o z)}{\sinh(k_o h)}$$

and $\omega_o(k_o)$ is given by (5.1.11).

Exercise 2
Assuming $n, \mathbf{v}, \phi$ all proportional to $ae^{i(\mathbf{k}_o \mathbf{x} - \omega_o t)}$, derive (5.1.13) in the vanishing 'a' limit (without loss of generality, assume $k_{o_z} = 0$).

Exercise 3
Derive equation (5.2.31) from (5.1.8) for $\partial \mathbf{v}/\partial z = 0$ and $\Omega_c = 0$.

Exercise 4
Find a solution of (5.1.7)–(5.1.9) depending on $x - Mt$ only, and such that $\mathbf{v}_\perp = 0$, $\delta M = M - 1$ is small, and ϕ, v_x and $n - 1$ all tend to zero for $|x - Mt| \to \infty$. Show that this solution is

$$\phi = 3\delta M \operatorname{sech}^2[(\tfrac{1}{2}\delta M)^{\frac{1}{2}}(x - Mt)] + 0(\delta M)^2.$$

Explain how the form of this soliton solution suggests all the expansions in (5.2.32)–(5.2.37) except that of y (this Exercise is based on the treatment in Davidson (1972), Subsection 2.2.2).

Exercise 5
Find the complete differential forms of the conservation laws (momentum and energy) of the Zakharov–Kuznetsov equation, either straight from the equation or using the Lagrangian (5.2.68).

Exercise 6
Find the third order equation analogous to (5.3.33). By demanding that all $e^{i\theta_o}$ terms cancel on the right hand side of the two boundary conditions, derive the NLS equation (5.3.38) with coefficients given by (5.3.39).

Repeat the calculation in x, y, t and show that the only resulting change is $\partial_x^2 \to \partial_x^2 - 2\partial_y^2$ in NLS.

Exercise 7
Using (5.3.56) in (5.3.43) and (5.3.44) and defining

$$\Delta\theta = \theta - \theta_o, \partial_x \Delta\theta = \tilde{k}, \partial_t \Delta\theta = -\tilde{\omega}$$

obtain for deep water waves, for which $\omega_o(k) = k^{\frac{1}{2}}, \omega_2 = \frac{1}{2}k^{5/2}$,

$$A_t + \tfrac{1}{2}(\omega_o/k_o)A_x - (\omega_o/8k_o^2)(\Delta\theta_{xx}A + 2\Delta\theta_x A_x) = 0,$$

and

$$\Delta\theta_t + \tfrac{1}{2}(\omega_o/k_o)\Delta\theta_x - (\omega_o/8k_o^2)(\Delta\theta_x^2 - A_{xx}A^{-1}) + \tfrac{1}{2}\omega_o k_o^2 A^2 = \text{const.}$$

Show that these two equations combine to give the NLS equation in the form,

$$i[a_t+\tfrac{1}{2}(\omega_o/k_o)a_x]-(\omega_o/8k_o^2)a_{xx}-\tfrac{1}{2}\omega_o k_o^2 a^2 a^* + \gamma a = 0,$$

for $a=Ae^{i\Delta\theta}$. Would this work with (5.3.40) instead of (5.3.56)? Explain.

Exercise 8
How will (5.3.50) be altered when the proper dispersion relation (5.3.56) is used in place of (5.3.40)? Take

$$A=A_o+\delta A e^{i(Kx-\Omega t)}+\text{c.c.}$$
$$k=k_o+\delta k e^{i(Kx-\Omega t)}+\text{c.c.}$$

and derive

$$\Omega/K=\frac{\partial\omega_o}{\partial k_o}\pm\sqrt{(\omega_2\omega_{okk}A_o^2+\tfrac{1}{4}\omega_{okk}^2K^2)}.$$

Exercise 9
Show that the linear dispersion equation following from (5.3.68) is

$$\omega_o=-k_{ox}^3+k_{oy}^2k_{ox}^{-1}.$$

Find the 2×2 tensor $\omega_{o\mathbf{kk}}$ for $k_{oy}=0$ and show that $\omega_{\mathbf{k}}\cdot\mathbf{n}=-3k_o^2\cos\psi$ and

$$\omega_{o\mathbf{kk}}:\mathbf{nn}=-6k_o\cos^2\psi[1-\tan^2\psi/3k_o^2].$$

By analogy with the KdV calculation of Section 5.3, show that

$$\omega_2=\omega_2(\psi=0)(1-\tan^2\psi/3k_o^2)/(1+\tan^2\psi/3k_o^2).$$

Draw a polar plot of $\Omega/K(\psi)$ for $K\to0$ (use the above results and (5.3.65)).

Exercise 10
Find ω_2 and $\omega_{\mathbf{kk}}$ for the reduction of the Zakharov–Kuznetsov equation

$$v_t+vv_x+\partial_x(v_{xx}+v_{yy})=0$$

to NLS in the small a theory. Find $\Omega/K(\psi)$ and draw a polar plot for $K\to0$. (This is a failure of small a^2 theory, as it misses an instability for $\psi\simeq\pi/2$, see Chapter.)

Exercise 11
Derive the linear dispersion relations following from both variants of (5.4.1)–(5.4.4). What occurs when $h_ok_o=\sqrt{3}$ if the (5.4.3) version is considered?

Exercise 12
Derive (5.5.21) from the three-dimensional generalizations of (5.5.13)–(5.5.14):

$$\frac{\partial\mathbf{k}}{\partial t}+\nabla\mathscr{H}_I=0$$

$$\frac{\partial I}{\partial t}+\nabla\cdot\mathscr{H}_{\mathbf{k}}=0$$

and assuming perturbed quantities $\delta\mathbf{k}$, $\delta I\sim e^{i(\mathbf{K}\cdot\mathbf{x}-\Omega t)}$. Show that $d\mathscr{H}_{\mathbf{k}}/dI$ is equal to the 'companion' characteristic velocity on a characteristic and that

$$\frac{\partial \mathscr{H}_{\mathbf{k}}}{\partial I} \cdot \dot{\mathbf{x}} = \mathscr{H}_{I\mathbf{k}}^2 - \mathscr{H}_{II} \mathscr{H}_{\mathbf{kk}} : \mathbf{nn}$$

and is the same on the two characteristics. Show that a second expression having this property is

$$(\mathscr{H}_{I\mathbf{k}} - \dot{\mathbf{x}})^2.$$

What are the values of these two characteristic-invariant expressions in the $A \to 0$ limit?

6

Exact methods for fully nonlinear waves

6.1 Introduction

In this Chapter we will deal in some detail with the mathematical methods that can be used to treat fully nonlinear wave problems.

The best known single development of the last twenty years is the discovery of a method for solving the initial value problem for a limited class of nonlinear partial differential equations. This is the inverse scattering method (ISM). This method, however, proves difficult to extend to general initial conditions and is more useful for solitons than for waves. It will be presented in Chapter 7. There are, however, several other methods for dealing with nonlinear waves (often solitons also). They deserve notice in their own right. Some of them have been developed fairly recently. Few are limited to equations solvable by ISM. This Chapter will concentrate on these methods as applied to nonlinear waves. We hope to give an idea of how rich the family of known nonlinear waves now is.

Thus the next two chapters (6 and 7) will concentrate on methods as illustrated by simple plasma physics and fluid dynamics problems. We will find the shapes of nonlinear waves (Chapter 6) and solitons (Chapter 7) without *in principle* assuming small amplitude of the wave. Where we do restrict considerations to small amplitude (Section 6.5) it will be done in the hope of extracting more information out of an exact method (Lagrangian description of fluid flow) than would otherwise be forthcoming. Yes, by restricting a formalism (small in place of general amplitude) we will learn more. What will be observed is the folding over of contours in x, v phase space. In more physical terms, this is the development of two streams where there was only one. As a rule, we will leave the problem of stability of the newly found nonlinear waves (and solitons of Chapter 7) with respect to small perturbations, to Chapter 8.

Upon reading this Chapter, the reader may well ask whether contemporary nonlinear wave analysis is a story of investigating new phenomena, as a branch of physics should be, or else is it perhaps becoming a series of exercises in developing new mathematical methods? At present the answer would probably be that it is becoming a hybrid discipline based on both lines of approach. It is one of the aims of this book to try to bring the two together.

In Section 6.2, we present the method of phase plane analysis for determining the shape of a stationary wave structure (function of $x-v_0t$ only) without actually solving the equations. Nonlinear waves both with and without dissipative effects as well as two in-depth soliton systems, are treated as examples. Section 6.3 is on BGK (Bernstein–Kruskal–Greene) waves, a class of stationary waves the form of which follows from the constancy of energy in the statistical equations that describe a plasma in phase space (Vlasov and Poisson). Recent extensions, based on other constants of motion, are reviewed. Section 6.4 describes how exact wave profiles can be found by introducing a Lagrangian description that follows a fluid or plasma element. Contrary to views held initially, stationary BGK waves can sometimes be recovered from this method and we will see how to do this. Finally, Section 6.5 introduces a small amplitude expansion into the *exact* method of Section 6.4 so as to obtain analytically the folding of contours in x, v space, hitherto observed only in numerical calculations.

6.2 Phase plane analysis

Stationary waves, be they periodic, soliton or shock like, constitute an important and easily researched class of classical wave modes. However, direct solution of the ordinary differential equations that govern their behaviour is not always the optimal procedure. Phase plane analysis is often both simpler and more informative. This is best illustrated by treating some familiar examples.

Phase plane analysis is also a key to understanding chaotic behaviour in dynamical systems, see Helleman (1984) and his references.

6.2.1 One stationary wave in a dissipationless medium

We will first look at ion acoustic plasma waves as described by (5.1.7)–(5.1.9) with $\Omega_c=0$. We will concentrate on structures that depend on $\xi=x-v_0t$ only. Denoting ξ differentiation by a dash, we find

$$-v_0n'+(nv)'=0 \tag{6.2.1}$$

$$-v_0v'+vv'+\phi'=0 \tag{6.2.2}$$

$$\phi''=e^{\phi}-n. \tag{6.2.3}$$

If we now introduce $w=v-v_o$ so as to observe the world from the coordinate system of the wave, and then integrate, we find (the third equation is multiplied by ϕ' before integration):

$$nw=M \tag{6.2.4}$$

$$\tfrac{1}{2}w^2+\phi=E=\tfrac{1}{2}M^2 \tag{6.2.5}$$

$$\tfrac{1}{2}\phi'^2=e^{\phi}+M(M^2-2\phi)^{\frac{1}{2}}-C. \tag{6.2.6}$$

the C and M are positive constants of integration. Their number has been reduced from three to two by a similarity transformation (Exercise 1), and they were chosen such that the electrostatic potential of the soliton wave (single pulse) vanishes asymptotically with $|\xi|\to\infty$:

$$\mathrm{Lt}\phi=\mathrm{Lt}(n-1)=0 \tag{6.2.7}$$

When this is the case, we also have

$$\mathrm{Lt}w=M$$

from (6.2.4).

For the soliton, or single pulse wave, ϕ_ξ will also tend to zero in the assymptotic regions. Thus, values of C and M corresponding to a soliton should satisfy (6.2.6) with both ϕ and $\phi\xi$ zero, yielding

$$C_s=1+M^2.$$

If we were to draw ϕ_ξ^2 as a function of ϕ, (6.2.6), the curve would resemble a cubic with a negative coefficient multiplying the cube term. Such a curve descends from the upper left hand corner of the graph to a minimum and proceeds through a maximum to decrease once more. However, our curve (6.2.6) differs from a cubic in that it is aborted at $\phi_\xi^2=\frac{1}{2}M^2$, where the square root in (6.2.6) ceases to be real. For solitons, $C=C_s$, the minimum is at the origin, where $\partial\phi_\xi^2/\partial\phi=0$ and $\partial^2\phi_\xi^2/\partial\phi^2>0$. There is a positive ϕ_ξ^2 segment to the right of this minimum. The curve is then aborted to the right of this segment at a value of $\phi(M^2/2)$ for which ϕ_ξ^2 is negative.

We will defer a detailed analysis, yielding values of C and M corresponding to physical situations, to Chapter 8 and Fig. 8.13. It can be checked that, once we have fixed the value of M, the right hand side of (6.2.6) will take non-negative values for a finite ϕ interval provided that C satisfies

$$C_{\mathrm{LW}}(M)\geqslant C\geqslant C_s=1+M^2.$$

Calculation of $C_{\mathrm{LW}}(M)$ is more difficult than of C_s and is deferred to Chapter 8. This value corresponds to the $\phi_\xi^2(\phi)$ curve just touching the positive ϕ axis from below (maximum of ϕ_ξ^2 at $\phi>0$, $\phi_\xi^2=0$). Thus the two limiting values of C correspond to extrema of the $\phi_\xi^2(\phi)$ curve touching the ϕ axis (a minimum for the soliton; a maximum for the linear wave limit). Intermediate C for which the minimum is below and the maximum above the axis correspond to nonlinear waves. Values of C smaller than C_s or greater than C_{LW} no longer correspond to physical entities (no finite ϕ interval corresponds to non-negative ϕ_ξ^2).

We will now take a closer look at the condition that the square root in (6.2.6) be real, or $\phi < \frac{1}{2}M^2$. This can be seen to imply that there are no solitons for M larger than the root of

$$1 + M^2 - e^{M^2/2},$$

which is $M = 1.5852$. All our present considerations are limited to M smaller than this value.

The details of the reasoning outlined above are not important at this stage, but the method is. Thus, when looking for travelling wave solutions to a set of nonlinear differential equations, it is often possible to arrive at one equation of the form

$$(\phi')^2 = f(\phi, C_i),$$

where the C_i are constants. Simple restrictions, such as $(\phi')^2 \geqq 0$, ϕ bounded and real will then yield the region in C_i parameter space corresponding to physically interesting solutions. If we introduce the convention that $\phi \to 0$ at infinity, it follows that solitons can only exist if real C_i can be found such that

$$f(0, C_i) = 0$$

and the $\phi'^2(\phi)$ curve will touch the origin such as to be tangential to the ϕ axis from above ($\partial\phi'^2/\partial\phi = 0$, $\partial^2\phi'^2/\partial\phi^2 > 0$ at the origin).

In Chapter 8 we will require detailed knowledge of the existence region in C, M parameter space for our problem (6.2.1)–(6.2.3). This region is shown in Fig. 8.13. Here we just treat this problem as an illustration of the phase plane method.

We now proceed to draw ϕ_ξ (rather than ϕ_ξ^2) against ϕ (henceforth referred to as phase curves) for values of C and M that have been checked to fall within the existence region described above. These curves are shown in Fig. 6.1, where M is fixed. It will be more natural to vary C from the linear wave value $C_{LW}(M)$ to the soliton value $C_s = 1 + M^2$ in our comments on Fig. 6.1.

The phase curve for a linear wave has shrunk to a point and is the centre of the whole family of phase curves. The closed contours going around this centre and perpendicular to the ϕ axis correspond to periodic, nonlinear waves. Each 'nonlinear wave' phase curve is travelled clockwise as ξ is increased, once round corresponding to one wavelength. The largest curve corresponds to a soliton and is travelled just once (clockwise) for the entire range $-\infty \leqslant \xi \leqslant \infty$. The phase point emerges from the origin, reaching the ϕ axis again from above at the soliton ϕ maximum, and then re-entering the origin from below as ξ tends to $+\infty$. It leaves and re-enters the origin at acute angles such that $\phi_\xi = d\phi$ for large magnitude negative ξ and $\phi_\xi = -d\phi$ as ξ tends to $+\infty (d>0)$. (This is in contradistinction to the 'nonlinear wave' phase curves, which are all perpendicular to the ϕ axis.) This soliton behaviour can be better understood if we refer back to known formulas for solitons such as (1.2.9), all of which give exponential behaviour in the far field:

$$\phi \sim e^{-d|\xi|}, \quad |\xi| \to \infty.$$

This behaviour indeed implies $\phi\xi \simeq \pm d\phi$ when $|\xi|$ is large and ϕ small.

In section (6.2.3) KdV will be investigated and phase curves will be seen to form a family very similar to that of Fig. 6.1, once again vindicating this simple model in the plasma wave context.

With a little practice, one can read off stationary wave properties from phase diagrams such as Fig. 6.1, learning quite a lot as a result of very little work. Doubts about the existence or otherwise of soliton waves in a given situation can quite often be resolved by a similar phase plane analysis, (Infeld and Rowlands (1979c)).

Fig. 6.1. Phase plane curves $\phi_\xi(\phi)$ for M fixed ($\leqslant 1.5852$) and C varying from the linear wave limit (centre) to the soliton value $1+M^2$ (largest phase curve).

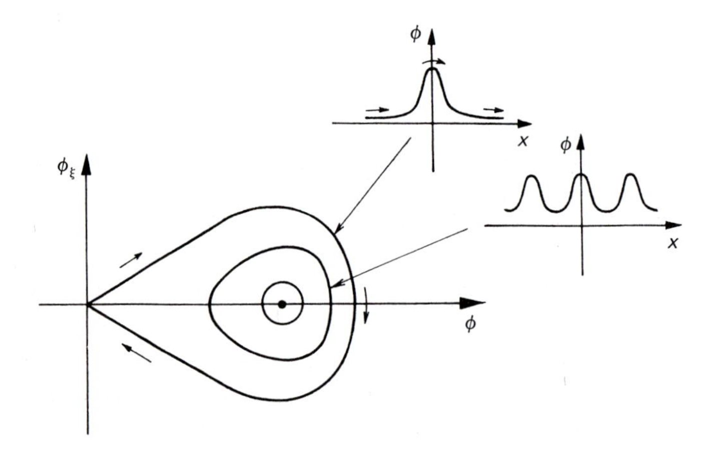

The nonlinear Schrödinger equation is particularly interesting to phase analyse. Without loss of generality

$$c = \pm 1 \tag{6.2.8}$$

and

$$i\partial_t \phi + \bar{b}\phi + c\partial_x^2 \phi + |\phi|^2 \phi = 0. \tag{6.2.9}$$

This equation plays an important role in the theory of nonlinear waves, as signalled in Chapter 1 and seen in Chapters 5 and 8. A phase plane analysis is relevant when considering stationary waves, functions of $\xi = x - v_D t$. In fact, due to the rather limited role played by the phase in (6.2.7), more general waves of the form

$$\phi_o(x - v_D t)e^{i[v_D(x - v_D t)/2c + \omega_o t]} \tag{6.2.10}$$

where ϕ_o is real, also lead to an ordinary differential equation for the function ϕ_o:

$$\begin{aligned} &c\phi_{o\xi\xi} + b\phi_o + \phi_o^3 = 0 \\ &b = \bar{b} - \omega_o - v_D^2/4c. \end{aligned} \tag{6.2.11}$$

Multiplying (6.2.11) by $\phi_{o\xi}$ and integrating, we obtain

$$c(\phi_{o\xi})^2 = B - b\phi_o^2 - \tfrac{1}{2}\phi_o^4, \tag{6.2.12}$$

where B is a constant of integration. Families of phase curves for various signs of b and c are given in Fig. 6.2.

As only even powers of ϕ_o and $\phi_{o\xi}$ appear in (6.2.12), these phase curves are symmetrical about both axes. Fig. 6.2(*a*), for $c = 1$ and $b < 0$, is seen to have centres at $\phi_o = \pm |b|^{\frac{1}{2}}$. For $B = -\frac{1}{2}|b|^2$, the whole phase curve shrinks to a point at one of these centres. The small ellipses around, say, $\phi_o = |b|^{\frac{1}{2}}$, depict small amplitude, almost cosine waves, (6.2.1) being satisfied for

$$\phi_{o\xi}^2 + |b|(\phi_o - |b|^{\frac{1}{2}})^2 \simeq \text{const.} \tag{6.2.13}$$

solved by

$$\phi_o = |b|^{\frac{1}{2}} + a\cos[|b|^{\frac{1}{2}}(\xi - \xi_o)], \quad \phi_{o\xi} = -a|b|^{\frac{1}{2}}\sin[|b|^{\frac{1}{2}}(\xi - \xi_o)],$$

an ellipse in parametric form. As we move out from the centre, the contours get larger and less like ellipses, the corresponding waves becoming more and more nonlinear. They are still periodic and once round corresponds to one wavelength in physical space. The separatrix to the right of the origin (right hand half of the figure of eight) passes through the saddle point at $(0,0)$, where $(\phi_o, \phi_{o\xi}) \to (0,0)$ as $\xi \to \pm\infty$. Once more, we have a soliton described in phase space.

So far we have encountered a family of solutions quite like the earlier, ion acoustic wave family. Now, however, some new entities enter the scene.

The left hand half of the separatrix corresponds to a rarefaction soliton, ϕ_o being negative. The two soliton solutions are given by

$$\phi_o = \pm|2b|^{\frac{1}{2}}\operatorname{sech}[|b|^{\frac{1}{2}}(\xi-\xi_o)]$$

(these entities become more interesting after (6.2.10) is applied). The closed curves embracing the separatrix also correspond to nonlinear waves, though for them ϕ_o takes on both positive and negative values. Fig. 6.2(b) is

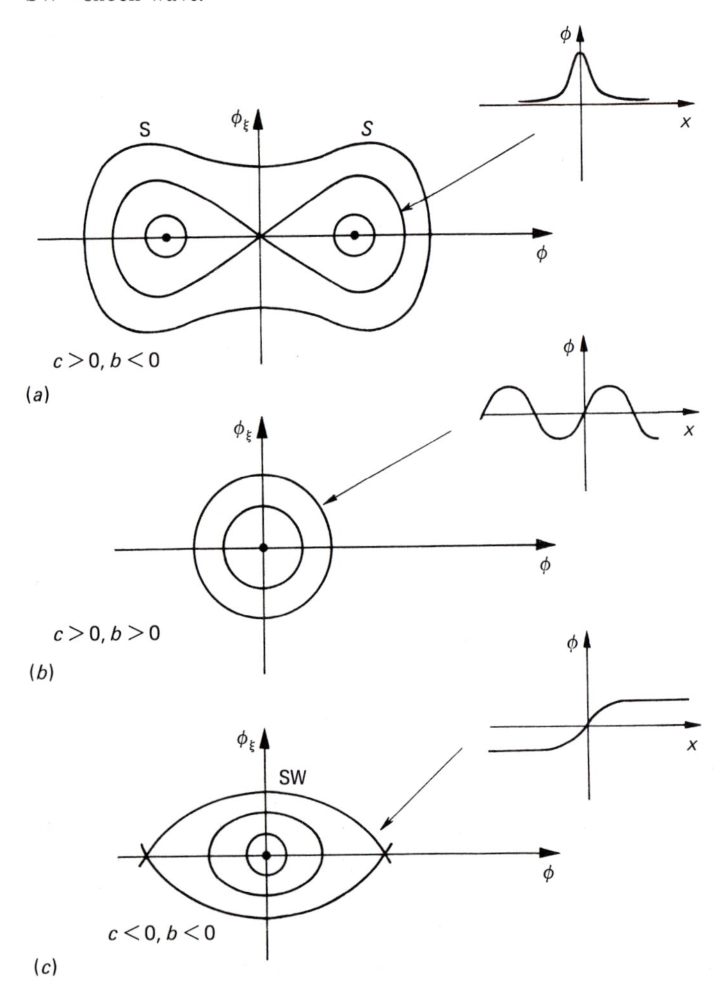

Fig. 6.2. Phase diagrams for solutions to the nonlinear Schrödinger equation (6.2.12). S is the separatrix corresponding to a soliton, SW = shock wave.

comprised entirely of nonlinear wave contours, whereas Fig. 6.2(c) depicts both nonlinear waves and a new kind of phase curve SW, connecting two saddle points, for which (top branch)

$$(\phi_o, \phi_{o\xi}) \rightarrow (-|b|^{\frac{1}{2}}, 0) \text{ for } \xi \rightarrow -\infty$$
$$\rightarrow (|b|^{\frac{1}{2}}, 0) \text{ for } \xi \rightarrow +\infty.$$

thus the entity it represents will be a shock wave of strength $2|b|^{\frac{1}{2}}$ in physical space (ϕ, ξ). Its mirror image is obtained from the bottom branch. In Fig. 6.2(c), saddle points and centres have changed places as compared to Fig. 6.2(a).

Thus the cubic nonlinear Schrödinger equation yields a rich variety of nonlinear wave structures, namely solitons, rarefaction solitons, several kinds of periodic nonlinear waves, and a pair of shocks. Indeed, as a result of this overabundance, scientists are not sure that all these solutions correspond to physical waves. However, some have been shown to be physical, see Yuen and Lake (1975), Saffman and Yuen (1985), and Infeld *et al.* (1987).

We will now look at the modified KdV equation, appearing in electric circuit theory (Scott (1970)), double layer theory (Torvén (1981), (1986)) and for a model of solitons in a multicomponent plasma (Verheest (1987)):

$$u_t + u^2 u_x + u_{xxx} = 0.$$

Solutions corresponding to stationary waves $u(x - u_o t)$ satisfy

$$u'^2 = u_o u^2 - \tfrac{1}{6} u^4 + \alpha u + \beta.$$

We will now concentrate on $u_o > 0$.

When phase curves are drawn for $\alpha_c > \alpha > 0$, $\alpha_c = \frac{4}{3} v_o^{3/2}$, we find that $u'^2(u)$ has two maxima; a larger, positive maximum, at a positive value of u and a smaller maximum at a negative value of u. Thus, as β is increased from β_1, where the larger maximum is zero, we find (for β_2 the smaller maximum is zero, and for β_3 the minimum is zero).

I. $\beta_1 < \beta \leqslant \beta_2$: one family of nonlinear waves

II. $\beta_2 < \beta < \beta_3$: two families of nonlinear waves

III. $\beta = \beta_3$: a separatrix (two solitons).

IV. $\beta_3 < \beta$: nonlinear wave phase curves that embrace all previous closed curves. Phase curves similar to the all-embracing curve of Fig. 6.2(a), though asymmetric with the larger section on the right.

We will now say just a few words about the stability of these nonlinear wave structures with respect to small amplitude, long wavelength perturbations; phase plane analysis helps classify the results.

It was generally known that weakly nonlinear, small amplitude waves satisfying the modified KdV equation, corresponding to the innermost phase curves of class I, are unstable. This can be seen to follow from an analysis similar to that of Subsection 5.3.1 resulting in a negative ω_{okk} but a positive ω_2 (Fornberg and Whitham (1978)). On the other hand, solitons must be stable, as the equation is solvable by inverse scattering (Chapter 7).

One might wonder just when the amplitude limitation of the instability (as measured by β) appears as β is increased from β_1 to β_3. The answer is rather pleasing, as it is just when the second wave family appears at $\beta = \beta_2$. Thus we have instability of the first family of waves for $\beta_1 \leqslant \beta < \beta_2$, and stability for $\beta_2 \leqslant \beta \leqslant \beta_3$. The second family is stable and the all-embracing waves $\beta > \beta_3$ are unstable (our remarks about stability are based on an analysis similar to that of Subsection 8.2.3, see Murawski and Infeld (1988)).

We will now consider some cases for which the basic ideas of a phase plane analysis, outlined above, will still serve, though extensions will prove necessary.

6.2.2 A two-fluid layer soliton pair

These methods can in principle also be extended to more numerous dependent variables. However, the analysis tends to lose its pictorial simplicity and, furthermore, only gives *limitations* on possible solutions, rather than the complete physical information of the one component case. The method is best illustrated by a simple example, as all these diagram analyses tend to be more easily done than said.

Su (1984) investigated a two-fluid layer system for the possible existence of soliton pairs (one in each fluid for stratified fluid see Liv (1984)). The fluids flow over a flat bed and the average thicknesses of the two components are equal (h). The bottom fluid (1) is much heavier than the top one (2). If the two fluid equations following (5.2.1)–(5.2.4) and $T=0$ are expanded according to the scheme (notice the by now familiar unequal stretching):

$$
\begin{aligned}
h_1 &= h(1+\varepsilon^2 j_1^{(1)} + \ldots) & \rho_2/\rho_1 &= \varepsilon^2 \\
h_2 &= h(1+\varepsilon j_2^{(1)} + \ldots) & \xi &= \varepsilon^{\frac{1}{2}}(x - v_o t)/h \\
v_1 &= v_o(\varepsilon^2 v_1^{(1)} + \varepsilon^3 v_1^{(2)} + \ldots) & \tau &= \varepsilon^{3/2} v_o t/h \\
v_2 &= v_o(\varepsilon v_2^{(1)} + \varepsilon^2 v_2^{(2)} + \ldots), & &
\end{aligned}
\tag{6.2.14}
$$

one obtains for the normalized excess heights in first order, j_1 and j_2, see Kakutani and Yamasaki (1978), Miles (1979):

$$
\begin{aligned}
&2\partial_\tau j_1 + \partial_\xi(j_2 + \partial_\xi^2 j_1) + 0 \\
&2\partial_\tau j_2 + \partial_\xi(j_1 + \tfrac{3}{2} j_2^2 + \tfrac{1}{3}\partial_\xi^2 j_2) = 0.
\end{aligned}
\tag{6.2.15}
$$

The ratio of the two fluid densities was taken to be ε^2. In the real world this quantity is rarely much smaller than 10^{-1}, at least at room temperature (e.g. water above mercury). There is thus little leeway for $\varepsilon^{\frac{1}{2}}$ to be small enough to justify our expansion (6.2.14) and so the physical utility of this model is an open question. Nevertheless, equations (6.2.15) are good material for a phase plane analysis.

We now consider modes propagating with a constant speed common to j_1 and j_2, taking for convenience

$$\eta=\sqrt{3}\left(\xi-\frac{d}{2}\tau\right) \tag{6.2.16}$$

as our sole independent variable. We obtain

$$j_{1\eta\eta}-dj_2+j_2=0 \tag{6.2.17}$$
$$j_{2\eta\eta}-dj_2+j_1+\tfrac{3}{2}j_2^2=0. \tag{6.2.18}$$

This system conserves energy

$$E=j_{1_\eta}^2+j_{2_\eta}^2 V(j_1,j_2), \tag{6.2.19}$$

Fig. 6.3. $V=0$ curve (AB) with solutions trajectories emanating from it for $E=0$, $d=-1.5$. No soliton pair. (The trajectories emanating from the left continue into the right hand region.)

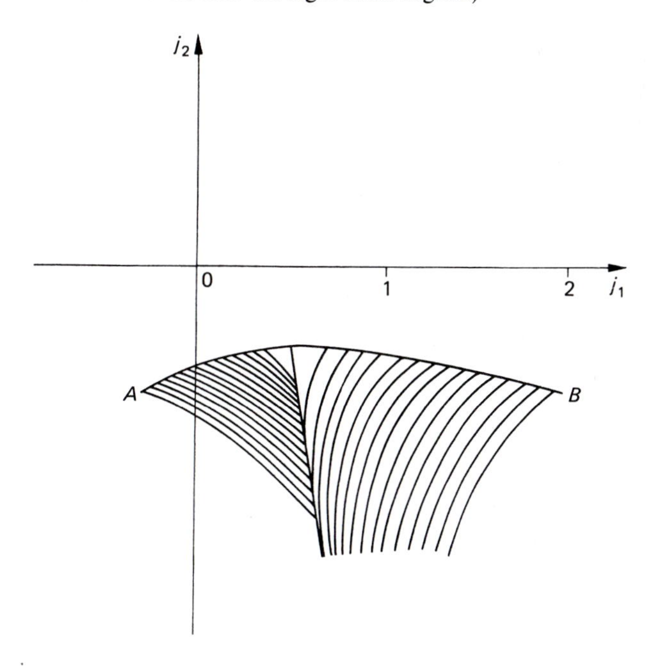

where

$$V = j_2^3 + 2j_1j_2 - d(j_1^2 + j_2^2). \tag{6.2.20}$$

Further analysis, in which proof of existence of soliton, periodic and quasi-periodic waves will be given, concentrates on $E=0$.

For a double soliton solution to exist, the phase curve in the j_1, j_2 plane should connect some point on $V=0$ with the origin, which is an isolated point of $V=0$ if $|d|>1$. This is due to the fact that $j_{1\eta}=j_{2\eta}=0$ both at the crests and at infinity, where the fluid is at rest and $j_1=j_2=0$. As η increases

Fig. 6.4. $V=0$ curve (AB) for $E=0$, $d=1.5$. S=soliton pair solution.

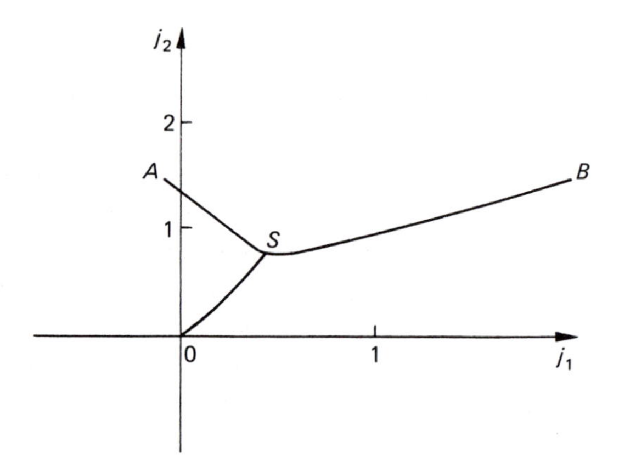

Fig. 6.5. $V=0$ curve (AB) for $E=0$, $d=-0.5$. S=soliton pair solution (depression in first fluid, positive soliton in second). PQ=nonlinear wave solution (elevation of second fluid riding a depression of first, both periodic).Regions of finite thickness, but otherwise resembling PQ, appear and represent quasi-periodic waves, see Su (1984), which reference is the basis of these three figures.

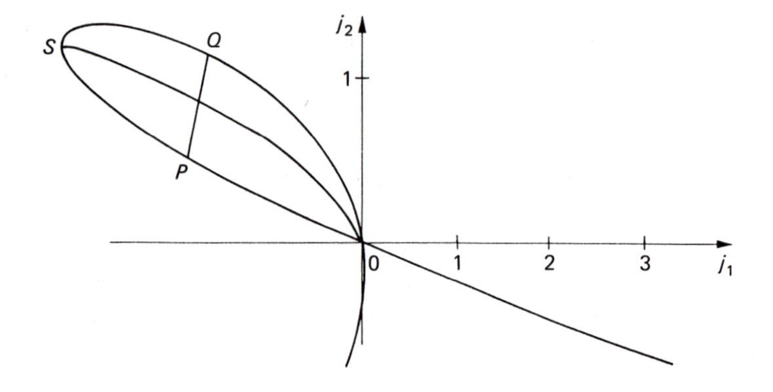

from $-\infty$ to $+\infty$, a phase point will emerge from the origin, reach $V=0$ and travel back again (back and forth along the same path if the two solitons are symmetrical).

Fig. 6.3 shows $V=0$, below which curve all $E=0$ solutions must lie. Also shown are some of the solution curves emanating from the $V=0$ curve not following from a simple phase plane analysis, for $d=-1.5$. Clearly no curve can connect with the origin. Su found more generally that there were no solutions corresponding to soliton pairs for $d \leqslant -1$ or $0 \leqslant d \leqslant 1$.

The next two phase diagrams, Figs. 6.4 and 6.5, show examples of phase curves corresponding to soliton pair solutions. In the first example both solitons are elevations riding together in the two fluids, in the second a soliton (j_2) rides a depression (j_1). Periodic waves (PQ), as well as quasi-periodic waves, not shown here, also appear for the case illustrated by Fig. 6.5.

Going back to equation (6.2.14) we see that the solitons (or depressions) are of very small amplitude in the heavy fluid at the bottom, generating much larger (but still small) solitons in the lighter fluid, which 'overreacts'. The same is of course the case for periodic nonlinear waves following from our expansion scheme. A phase plane analysis without the solution curves (which Su found numerically) would still tell us quite a lot. For example, the $d=-1.5$ case would be seen *not* to support solitons even if the details below the AB curve were not known. Likewise, $d=1.5$ and $d=-0.5$ would be seen to be *likely* to admit soliton solutions. Thus the soliton addict can be saved a lot of time by starting his investigations with a phase plane analysis for two dependent variables when reduction to one *independent* variable is possible.

6.2.3 Weak ion acoustic shock waves in a collisional plasma

Phase plane analysis can also be useful even when dissipative terms complicate matters. We will illustrate this by a dissipative shock wave problem.

When electron–ion collisions are important, the KdV equation for waves in a warm electron, cold ion plasma as obtained in Section 5.2 can be extended to

$$\frac{\partial u}{\partial \tau}+u\frac{\partial u}{\partial \xi}+\tfrac{1}{2}\frac{\partial^3 u}{\partial \xi^3}-\nu\frac{\partial}{\partial \xi}\left(\frac{\partial u}{\partial \tau}+u\frac{\partial u}{\partial \xi}\right)=0, \qquad (6.2.21)$$

where ν is a dimensionless measure of the dissipation due to electron–ion collisions, see Kakutani and Kawahara (1970) for derivation and most of the following analysis. This form of the dissipative equation is still Galilean-invariant.

A phase plane analysis is now complicated by the fact that $u_\xi(u)$ is not obtainable as an algebraic function of u, even for a travelling wave solution $u(\xi - u_o t)$. Instead, one obtains upon integration,

$$\frac{d^2u}{d\xi^2} = 2\nu(u-u_o)\frac{du}{d\xi} - (u-u_1)(u-u_2), \tag{6.2.22}$$

where u_1 and u_2 are constants depending on the constant of integration of (6.2.21), A:

$$u_{1,2} = u_o \pm (u_o^2 + A)^{\frac{1}{2}}. \tag{6.2.23}$$

Introducing

$$u = \alpha(v-1) + u_o \tag{6.2.24}$$

$$\frac{dv}{dx} = w$$

$$\xi = x/\sqrt{\alpha}$$

$$\alpha = (u_o^2 + 2A)^{\frac{1}{2}}, \delta = 2\nu\alpha^{\frac{1}{2}},$$

(6.2.22) simplifies to

$$\frac{dv}{dx} = w \tag{6.2.25}$$

$$\frac{dw}{dx} = \delta(v-1)w - v(v-2). \tag{6.2.26}$$

An analysis similar to that of Subsection 6.2.1 reveals that the point $(v, w) = (0, 0)$ is a saddle point and $(v, w) = (2, 0)$ is a centre when $\delta = 0$. For $\delta > 0$ algebra is no longer sufficient and (6.2.25)–(6.2.26) had to be solved numerically. For $0 < \delta < 2$, the centre becomes an unstable focus (spiral point, Fig. 6.6(*a*)), and for $\delta > 2$ a nodal point (the phase curve connects this point with the origin without spiralling). Note the lack of continuity of the solutions in phase space when $\delta \to 0$.

Two basic kinds of shock wave in physical space (v, x) follow. When dissipation is small the profile is oscillatory $(0 < \delta < 2)$, but ceases to be so for larger δ. Kakutani and Kawahara stress the unusual feature that increase of dissipation δ tends to increase the shock steepness (in addition to smoothing out the ripples). The additional term in (6.2.21) as compared with KdV is a profile steepening term (*usually* dissipation tends to smooth large gradients out).

Although (6.2.25) and (6.2.26) had to be solved numerically to obtain Fig.

6.6, phase plane arguments can help us understand and interpret it. We can write (6.2.25) and (6.2.26) in the form

$$\frac{dE}{dx} = \frac{d}{dx}\left(\tfrac{1}{2}v_x^2 + \tfrac{1}{3}v^3 - v^2\right) = \delta(v-1)v_x^2 \tag{6.2.27}$$

Fig. 6.6. (*a*) Phase diagram of (v, w) for KdV (dashed lines are E=const) and extended KdV, δ=0.5 for stationary shock. Note how similar the KdV contours are to those of Fig. 6.1. (*b*) Wave profile of the oscillatory shock wave for δ=0.5. For $\delta \geqslant 2$ the oscillations disappear. From Kakutani and Kawahara (1970).

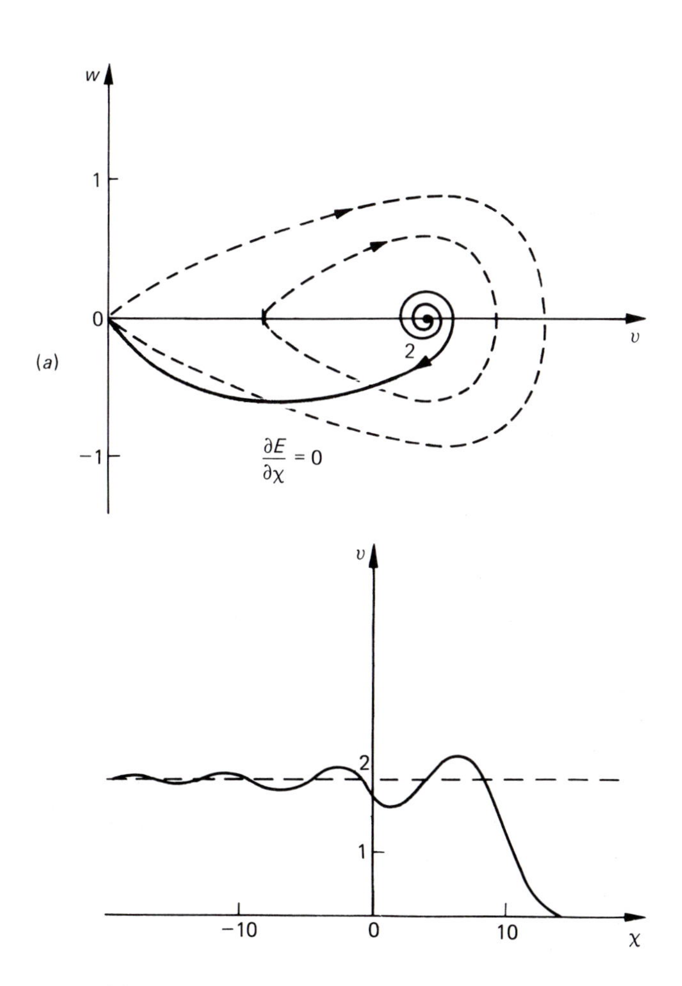

and E should increase with x for $v>1$ and decrease for $v<1$. Thus we move out from $(2,0)$ to larger and larger E contours until we reach $v=1$ where E begins to decrease (Fig. 6.6(a)). This explains the behaviour quantitatively and tells us that the sense of the phase curve is from $(2,0)$ to $(0,0)$ as x increases from $-\infty$ to $+\infty$. Notice how dissipation has deprived the phase curves of their $w\to -w$ symmetry.

Shock waves of ion acoustic type have indeed been found in plasma experiments in which $T_i/T_e \ll 1$, see Anderson *et al.* (1967) (in which experiment this ratio was $\frac{1}{8}$).

6.3 Bernstein–Greene–Kruskal waves

The differential equations appearing in this book admit certain symmetries involving both the independent and the dependent variables (Lie symmetries). In general, self-similar solutions can be associated with one or more of these symmetries. Here we will present an example of such a class of solutions taken from the statistical theory of a two component plasma. The symmetry is time translation, and the solutions describe nonlinear, stationary waves. Some recent developments, based on other symmetries, will be mentioned briefly.

6.3.1 Statistical description of a plasma and BGK waves

A good model for a fully ionized two component plasma is the Vlasov equation:

$$\frac{\partial}{\partial t}f(x,v,t)+v\frac{\partial}{\partial x}f(x,v,t)+\frac{e}{m}E(x,t)\frac{\partial}{\partial v}f(x,v,t)=0 \qquad (6.3.1)$$

$$\frac{\partial E}{\partial x}=\sum 4\pi en=\sum 4\pi e\int \mathrm{d}v f(x,v,t). \qquad (6.3.2)$$

Here $f(x,v,t)$ is the distribution function for one species of a hydrogen plasma (electron or ion). Summation is over the two components. Equation (6.3.1) carries Vlasov's name and has retained the Liouville form of a conservation of the number of particles in a phase space volume element, the E term being an acceleration in this interpretation. Equation (6.3.2) is just Poisson's equation. The region in n, T parameter space where (6.3.1) and (6.3.2) can describe a plasma reasonably well is indicated roughly in Fig. 1.3 (for plasmas $n_e \simeq n_i$).

It is easily seen by inspection that *any* function of the energy

$$W=\tfrac{1}{2}mv^2+e\phi(x), \qquad (6.3.3)$$

where

$$E(x) = -\partial\phi/\partial x, \tag{6.3.4}$$

and ϕ is time independent, satisfies (6.3.1).

Bernstein, Greene and Kruskal (1957) developed a theory of waves in plasmas in which distributions were functions of W only. Once we assume this we can forget about Vlasov's equation and concentrate on solving (6.3.2). It will prove convenient to introduce two W's in place of v as variables of integration in (6.3.2). For each species

$$dv = \pm dW_j/2m_j[W_j - e_j\phi]^{\frac{1}{2}} \tag{6.3.5}$$
$$j = \text{e,i (electron, ion)}.$$

We also propose to introduce the symbol $\vec{f}_j(W_j)$ for charges moving from left to right (and $\overleftarrow{f_j}$ for right to left). We find, from (6.3.2) and (6.3.5)

$$\frac{\partial^2\phi}{\partial x^2} = -\sum_j 4\pi e_j \int dW_j \frac{[\vec{f_j}(W_j) + \overleftarrow{f_j}(W_j)]}{[2m_j(W_j - e_j\phi)]^{\frac{1}{2}}}. \tag{6.3.6}$$

The ϕ waves considered here fall into two broad categories: those for which the potential ϕ does not trap any particles, and those for which a certain number of electrons, ions, or both are prisoners of ϕ. We will first consider the no-trapped particle solutions which might be expected to be simpler.

6.3.2 No trapped particles

Consider a stationary wave travelling against a cold hydrogen plasma with velocity $-v_o$ (from right to left). Without loss of generality we can assume $\phi(0) = 0$. Then, in the coordinate system of the wave,

$$W_{e,i} = \tfrac{1}{2}m_{e,i}v^2 + e\phi(x), \tag{6.3.7}$$
$$W_{o_{e,i}} = \tfrac{1}{2}m_{e,i}v_o^2.$$

Since no particles are to be trapped, we have for all x,

$$m_i v_o^2/2e \geqslant \phi(x) \geqslant -m_e v_o^2/2e. \tag{6.3.8}$$

The cold particle distribution functions are

$$\vec{f}_{e,i} = n_o v_o \delta(W_{e,i} - W_{o_{e,i}}) \tag{6.3.9}$$
$$\overleftarrow{f}_{e,i} = 0.$$

Equations (6.3.6) and (6.3.9) yield

$$\frac{\partial^2\phi}{dx^2} = -4\pi e n_o[(1 - 2e\phi/m_i v_o^2)^{-\frac{1}{2}} - (1 + 2e\phi/m_e v_o^2)^{-\frac{1}{2}}]. \tag{6.3.10}$$

The solution to (6.3.10) is straightforward if laborious

$$x(\phi)=\frac{y_o^2}{eb^2}[2(m_e-m_i)(\tfrac{1}{3}N(\phi)^{\frac{3}{2}}-\tfrac{1}{3}N_o^{\frac{3}{2}}-aN(\phi)^{\frac{1}{2}}+aN_o^{\frac{1}{2}}) \tag{6.3.11}$$

$$-\tfrac{1}{3}\mathrm{sgn}(\phi)(m_e m_i)^{\frac{1}{2}}([2b]^{-\frac{1}{2}}[(2b^2-8ab)F(r,\alpha)+16abE(r,\alpha)$$

$$-4N^{\frac{1}{2}}(b^2-[N-a]^2)^{\frac{1}{2}})],$$

where

$$a+b=\left(\frac{\mathrm{d}\phi}{\mathrm{d}x}\right)_o^2=N_o, b=8\pi n_o v_o^2(m_e+m_i) \tag{6.3.12}$$

$$N(\phi)=a+8\pi n_o m_i v_o^2\left(1-\frac{2e\phi}{m_i v_o^2}\right)^{\frac{1}{2}}+8\pi n_o m_e v_o^2\left(1+\frac{2e\phi}{m_e v_o^2}\right)^{\frac{1}{2}} \tag{6.3.13}$$

$$F(r,\alpha)=\int_o^\alpha \mathrm{d}w/(1-r^2\sin^2 w)^{\frac{1}{2}}, E(r,\alpha)=\int_o^\alpha \mathrm{d}w(1-r^2\sin^2 w)^{\frac{1}{2}} \tag{6.3.14}$$

$$r=\left(\frac{a+b}{2b}\right)^{\frac{1}{2}}, \alpha=\arcsin\left(\frac{N_o-N}{N_o}\right)^{\frac{1}{2}}.$$

Strictly speaking, this formula gives $x(\phi)$ for the monotonic segment of ϕ that passes through $x=0$. It corresponds to half a wavelength, from minimum to maximum of ϕ. (Without loss of generality we assume $(\mathrm{d}\phi/\mathrm{d}x)_o>0$.) This would correspond to the upper half of a closed contour in a phase diagram such as Fig. 6.1. The next segment of $\phi(x)$ (from maximum to minimum), which is its mirror image, combines with our segment to give one wavelength. The complete ϕ function is periodic.

6.3.3 Various limits

The small amplitude limit is easier to derive from (6.3.10) than from (6.3.11):

$$\frac{\mathrm{d}^2\phi}{\mathrm{d}x^2}\simeq -4\pi n_o e^2 v_o^2(m_e^{-1}+m_i^{-1})\phi \tag{6.3.15}$$

solved by

$$\phi=\phi_{\max}\sin(2\pi x/\lambda) \tag{6.3.16}$$
$$\lambda/2\pi=v_o/\sqrt{[4\pi n_o e^2(m_e^{-1}+m_i^{-1})]}\simeq v_o/\sqrt{(4\pi n_o m_e^{-1}e^2)}.$$

The wavelength λ looks deceptively like the Debye length, introduced for electrons in Section 1.3. However, the physical meaning of λ is different, as v_o is the phase velocity of the nonlinear wave, not the thermal velocity of the plasma particles.

It would be disappointing to introduce all this machinery just to obtain a small amplitude sine wave, so we will now try another limit. The 'one component' model is obtained for small m_e/m_i. Now (6.3.11) becomes

$$\omega_{pe}x = v_o[\arcsin(y/A) + A - (A^2 - y^2)] \tag{6.3.17}$$

$$A = (d\phi/dx)_o/v_o(4\pi n_o m_e)^{\frac{1}{2}} \geq 0 \tag{6.3.18}$$

$$\omega_{pe}^2 = 4\pi n_o e^2/m_e \tag{6.3.19}$$

$$y = (1 + 2e\phi/m_e v_o^2)^{\frac{1}{2}} - 1, \tag{6.3.20}$$

and the no-trapping condition (6.3.20) gives $A \leqslant 1$. Fig. 6.7 shows wave profiles for three A values including $A = 1$, the maximum amplitude. This nonlinear wave is exceptional in that its wavelength $2\pi v_o/\omega_p$ (but not the distance between neighbouring nodes) is independent of the amplitude A, (6.3.17) being valid for

$$-\frac{\pi v_o}{2\omega_{pe}} + \frac{v_o A}{\omega_{pe}} \leqslant x \leqslant \frac{\pi v_o}{2\omega_{pe}} + \frac{v_o A}{\omega_{pe}}, \tag{6.3.21}$$

in which interval

$$A(A-2)|\phi_{min}| \leqslant \phi \leqslant A(A+2)|\phi_{min}| = \frac{m_e v_o^2}{2e}$$

The importance of waves for which $\partial\lambda/\partial A = 0$ was noted in Chapter 5 and will be taken up again in Section 8.2.

Up to now we have considered waves against a uniform plasma background. BGK waves can, however, also coexist with beam-plasma or multiple beam configurations. Two equal and opposite electron beams

Fig. 6.7. Wave profiles as given by (6.3.17) for $A = 0.3$ ——; 0.6 – – –, 1——, ϕ in units of $|\phi_{min}|$.

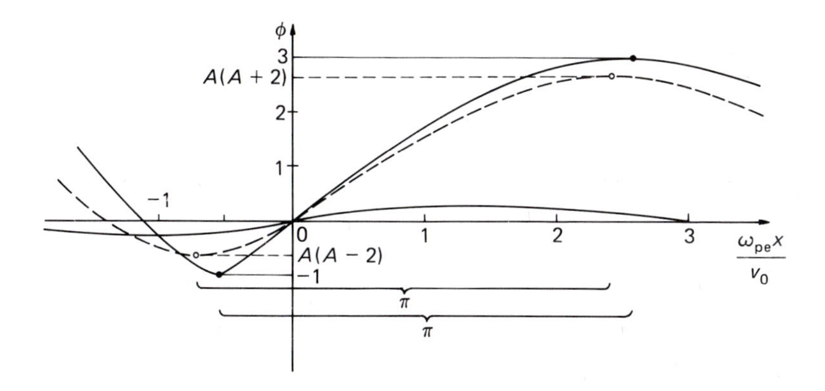

against a cold ion background is an example of this type of set-up. Different f functions will be involved, but ϕ need not be affected!

In view of the simple sum in the integrand of (6.3.6), $\phi(x)$ generated by these two equal, opposite electron beams with energies $\pm W_{oe}$ and half the previous density each, will be identical to (6.3.17) (Rowlands (1969b)). Other related exact solutions are known, e.g. the relativistic generalization of (6.3.17) (Infeld (1972)).

6.3.4 Trapped particle equilibria

Little more of interest can be said about the shapes of no-trapped particle, cold plasma BGK equilibria. It is of course possible to introduce 'half way house' f_i (step functions, truncated Maxwellians etc. in f_i), having compact W support and coexisting with $\phi(x)$ other than the above, that still trap no particles. However, life really becomes interesting when we do admit trapped particles. As these prisoners of ϕ will travel back and forth in a potential well, their distributions in a stationary situation must satisfy

$$\vec{f}_j(W_j^{TR}) = \overleftarrow{f}_j(W_j^{TR}),\ (e_j\phi)_{max} \geqslant W_j \geqslant (e_j\phi)_{min} \qquad (6.3.22)$$

(particles that are only reflected once in an aperiodic potential will also satisfy (6.3.22)). Bernstein *et al.* discovered that an arbitrary wave $\phi(x)$, as well as arbitrary ion and untrapped electron distributions, can in principle be supported by a suitably chosen trapped electron distribution (the words 'electron' and 'ion' can be interchanged in this statement). If we separate the trapped electron distribution in (6.3.6), taking all other quantities as given, we have

$$\frac{d^2\phi}{dx^2} = -4\pi e\left[\int_{e\phi} \frac{[\vec{f}_i(W_i) + \overleftarrow{f}_i(W_i)]dW_i}{[2m_i(W_i - e\phi)]^{\frac{1}{2}}} - \int_{-e\phi_{min}}^{\infty} dW_e \right.$$

$$\left. \frac{[\vec{f}_e(W_e) + \overleftarrow{f}_e(W_e)]dW_i}{[2m_e(W_e + e\phi)]^{\frac{1}{2}}} - g(e\phi)\right] \qquad (6.3.23)$$

$$g(e\phi) = \int_{-e\phi}^{-e\phi_{min}} \frac{f_e^{TR}(W_e)dW_e}{[2m_e(W_e + e\phi]^{\frac{1}{2}}},\ f_e^{TR} = 2\vec{f}_e^{TR}. \qquad (6.3.24)$$

This equation gives $g(e\phi)$ but not $f_e^{TR}(W_e)$. However, (6.3.24) can be solved to give

$$f_e^{TR}(W_e) = \frac{(2m_e)^{\frac{1}{2}}}{\pi} \int_{e\phi_{min}}^{-We} \frac{dg(W)}{dW} \frac{dW}{[-W_e - W]^{\frac{1}{2}}},\ W_e < -e\phi_{min}, \qquad (6.3.25)$$

subject to $g(e\phi_{\min})=0$. To see this, substitute (6.3.25) into (6.3.24) and change the order of integration with the appropriate limits in the integrals. Derivation of f_e is simplest when ϕ_{xx} is a reasonably simple function of ϕ. We will now investigate two such cases (ϕ_{xx} linear and quadratic binomial in ϕ, leading to a sine wave and hyperbolic secant squared pulse respectively). Both waves will be physically important. (It will now prove convenient to define constants at $\phi=\phi_{\min}$.)

If we wish $\phi(x)$ to be

$$\phi(x)=\phi_o \sin kx, \tag{6.3.26}$$

with ϕ_o *as large as we like*, and the plasma to be cold except for the trapped particle population, we again take $\overleftarrow{f}_{e,i}$ zero for untrapped particles and

$$\vec{f}_e(W_e)=n_e^u[2m_e(W_- - e\phi_o)]^{\frac{1}{2}}\delta(W_e - W_-),\ \ W_- > e\phi_o \tag{6.3.27}$$

$$\vec{f}_i(W_i)=n_i^u[2m_i(W_+ + e\phi_o)]^{\frac{1}{2}}\delta(W_i - W_+),\ \ W_+ > e\phi_o, \tag{6.3.28}$$

where $n_e^u v, n_i^u$, W_- and W_+ are positive constants. The densities are

$$n_e^u(x)=\hat{n}_e^u\left[\frac{W_- - e\phi_o}{W_+ - e\phi_o}\right]^{\frac{1}{2}} \tag{6.3.29}$$

$$n_i^u(x)=\hat{n}_i^u\left[\frac{W_+ + e\phi_o}{W_+ - e\phi}\right]^{\frac{1}{2}}. \tag{6.3.30}$$

From (6.3.23),

$$g(e\phi)=-\frac{k^2}{4\pi e}\phi+n_i^u(x)-n_e^u(x). \tag{6.3.31}$$

The distribution of trapped electrons is found from (6.3.25) and (6.3.31) to be

$$f^{TR}(W_e)=\frac{(2m_e)^{\frac{1}{2}}}{\pi}(e\phi_o - W_e)^{\frac{1}{2}}\left(-\frac{k^2}{2\pi e^2}+\frac{\hat{n}_i^u}{W_+ + W_e}+\frac{\hat{n}_i^u}{W_- - W_e}\right),$$

$$e\phi_o > W_e > -e\phi_o = 0, \text{ other } W_e. \tag{6.3.32}$$

The requirement that $g(e\phi_{\min})=g(-e\phi_o)=0$ yields

$$\frac{k^2}{4\pi e}\phi_o=\hat{n}_e^u-\hat{n}_i^u. \tag{6.3.33}$$

The remaining two conditions, that $f^{TR}(W_e)\geqslant 0$ and vanishing of the total current, furnish two further restrictions on the constants. For details see Exercise 4 and Davidson (1972). In the linear limit $\phi_o \to 0$, the trapped particle population vanishes, though k is not uniquely determined as it was in (6.3.16). This is due to the fact that $g(e\phi)$, the correction to (6.3.10), is linear in ϕ_o, and at no stage is (6.3.15) true. One sometimes loses information by taking the small amplitude limit of a nonlinear wave, as opposed to having ϕ small and linear from the start. Thus nonlinearity can be a vehicle for introducing arbitrariness in waves. We will return to this theme in Chapter 8.

For an example of a pulse-like ϕ, take the by now familiar form:

$$\phi = 3\eta^2 \operatorname{sech}^2(\eta x), \quad \phi_{\min} = 0, \quad \phi_{\max} = 3\eta^2, \tag{6.3.34}$$

Fig. 6.8. Distribution functions and wave profiles for three cases described in the text: (*a*) no trapped particles; (*b*) trigonometric ϕ and trapped electron population; (*c*) soliton shaped ϕ, also supported by a trapped electron population.

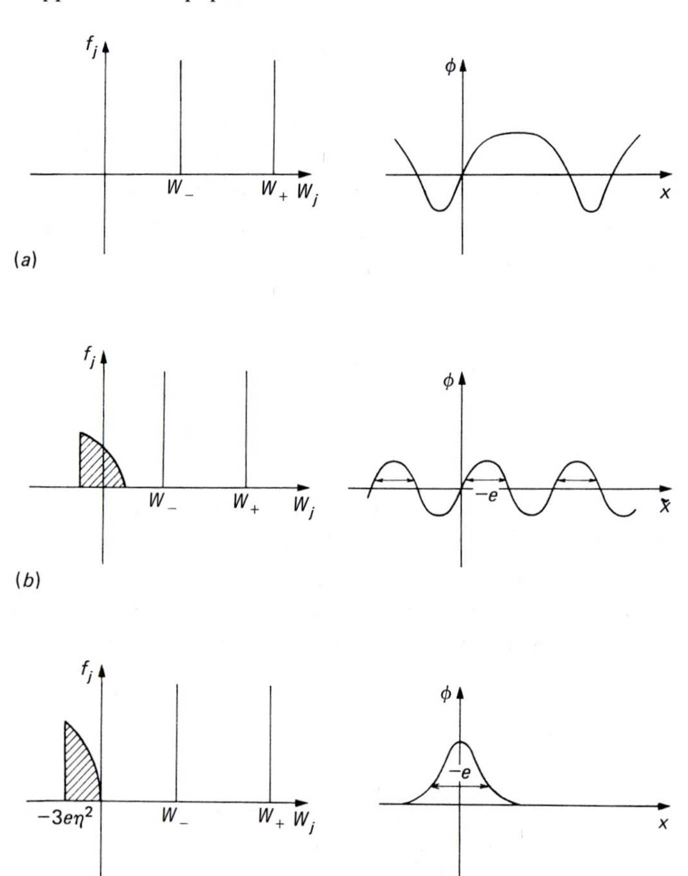

untrapped populations as in (6.3.27), (6.3.28) but with $\phi_{\min}=0$, and from (6.3.23)

$$g(e\phi)=\frac{1}{4\pi e^2}[4\eta^2 e\phi-2(e\phi)^2/e]+n_i(x)-n_e^u(x)], \tag{6.3.35}$$

$$\hat{n}_e^u(x)=\hat{n}_e^u\left[\frac{W_-}{W_-+e\phi}\right]^{\frac{1}{2}}$$

$$\hat{n}_i^u(x)=\hat{n}_i^u\left[\frac{W_+}{W_+-e\phi}\right]^{\frac{1}{2}},$$

leading to

$$f^{\mathrm{TR}}(W_e)=\frac{(2m_e)^{\frac{1}{2}}\sqrt{-W_e}}{\pi}\left(2\left[\eta^2-\frac{4W_e}{3e}\right]\frac{1}{\pi e^2}+\frac{n_i^u}{W_++We}+\frac{n_e^u}{W-We}\right), \tag{6.3.36}$$

The requirement $g(e\phi_{\min})=g(0)=0$ yields $n_e=n_i$. For the remaining two conditions see Exercise 4.

Fig. 6.8 plots the distribution functions f_i v. W_j for our three cases. The left hand components of Figs. 6.8(*b*) and (*c*) are deceptively similar. It should be remembered, however, that the trapped particle regions appear periodically in *configuration space* in case (*b*), and just once in case (*c*). Thus we must also know $\phi(x)$ to complete the picture.

6.3.5 Stability; subsequent developments

Historically, interest in BGK waves was first focused on finding new solutions, e.g. Montgomery and Joyce (1969), Smith (1970a, b), Schamel (1972), (the original BGK paper gave no examples). Once people realized that this is fairly easy, interest shifted to problems of stability and interpreting numerical and laboratory experiments.

Stability of the cold, 'one component' plasma wave (6.3.17) was demonstrated by Jackson (1963). Infeld (1972) found the dispersion relation for a linear perturbation around (6.3.17) by introducing

$$X=\int^{x}\frac{\mathrm{d}x}{1+y(x)}$$

as a new space coordinate (y given by (6.3.20). This dispersion relation is, in the new space coordinate X

$$(\Omega-Kv_o)^2=\omega_{\mathrm{pe}}^2, \tag{6.3.37}$$

the same as in x space without the wave, if the Galilean shift is retained. A small K analysis developed by Rowlands (1969b) for the two electron beam problem mentioned at the end of Subsection 6.3.3, ϕ still given by (6.3.17), yielded, for the same X,

$$\Omega^2 = \frac{3-2(1-A^2)^{3/2}-[9-8(1-A^2)^{3/2}]^{\frac{1}{2}}}{2[1-(1-A^2)^{3/2}]} v_0^2 K^2 \leqslant 0. \qquad (6.3.38)$$

The presence of the two electron beams generates the instability. In the zero amplitude limit, $A \to 0$, we obtain $\Omega^2 = -v_0^2 K^2$, the well-known two stream instability treated in Section 2.4. The BGK wave has a stabilizing effect and the growth rate (6.3.38) tends to zero as $A \to 1$. This calculation, however, was limited to small K, Ω and should not be taken to imply that BGK waves are generally stabilizing (e.g. this same wave *destabilizes* for perturbations such that $\frac{1}{2}\omega_{pe} v_0^{-1} < K < \omega_{pe} v_0^{-1}$, Infeld and Rowlands (1975)).

The above results have been confirmed numerically by Bertrand *et al.* (1971) and extended to more than one dimension by Infeld and Rowlands (1973) and (1975).

Configurations in which trapped particles play an important part were investigated for stability by Butler and Gribben (1968), Goldman (1970), Kruer and Dawson (1970), Liu (1972), Bloomberg (1974), Infeld and Rowlands (1977), Schwarzmeier *et al.* (1979), Rasmussen (1982), Gribben (1983) and many others. Some further references are given in Franklin's review (1977). One could summarize the results of these investigations by recognizing that, as a rule, instabilities tend to pose a serious problem when setting up a BGK wave. There are two general types of instability. One is the type that perseveres in the $A \to 0$ limit, such as (6.3.38), and is not entirely the 'fault' of the BGK wave. However, these instabilities are in general either amplified or diminished by the presence of the wave and this effect is important. The second type of instability owes its very existence to the wave. An example is the sideband or trapped particle instability, a two-stream effect caused by particles trapped by the wave resonating with the main body of the plasma in a collective manner; Goldman (1970).

Originally, BGK waves were considered by many physicists to be mathematical curiosities, but some numerical and laboratory experiments have subsequently shown similar time-independent states to evolve from various initial conditions (e.g. Armstrong and Montgomery (1967), Wharton *et al.* (1968), Morse and Nielson (1969a, b), Rasmussen (1982)). However, the predominance of theoretical papers over numerical and experimental results is still somewhat embarassing, even when compared with other fields of nonlinear classical physics.

Bernstein–Greene–Kruskal waves as described here have been quite painlessly extended to include magnetic fields by Laird and Knox (1965), Bell (1965), Lutomirski and Sudan (1966), Abraham-Shrauner (1968), Infeld (1974), and Abraham-Shrauner and Feldman (1977), to name a few examples.

It has recently proved possible to further extend the BGK formalism by various mathematical methods, see Abraham–Shrauner (1984a, b), (1985), Lewis and Symon (1984).

BGK wave solutions were formed by considering invariance of the Vlasov equation with respect to time translation. The most systematic way to extend this is to find the whole Lie symmetry group for (6.3.1)–(6.3.2) and then generate related exact solutions. The Lie group is

$$\begin{aligned} &X\to x+\varepsilon\xi^{x} && \xi^{x}=(2a_4-a_5)x+a_3t+a_2 \\ &t\to t+\varepsilon\xi^{t} && \xi^{t}=(a_4-a_5)t+a_1 \\ &v\to v+\varepsilon\xi^{v} && \xi^{v}=a_4v+a_3 \\ &f_j\to f_j+\varepsilon\eta^{j} && \eta^{j}=(2a_5-3a_4)f_j \\ &E\to E+\varepsilon\eta^{E} && \eta^{E}=a_5E, \end{aligned} \tag{6.3.39}$$

where the a_k are constants. The first two constants, a_1 and a_2, represent shifts in t and x, a_3 a Galilean transformation: and a_4, a_5 a general rescaling (Problem 5).

To obtain new solutions to (6.3.1)–(6.3.2), one takes a subgroup of (6.3.39) in which some combinations of the a_k are zero e.g. $2a_5-3a_4$, and integrates along the characteristics spanning a surface defined by the generators ξ, η of the Lie group. The original BGK waves follow when only a_1 is non-zero (Problem 5).

So far, the number of interesting and new solutions obtained by this procedure is slightly disappointing. More is forthcoming in the 'one component' model (static ion background), which admits a general function of time, $g(t)$, in the Lie group (generalized, Galilean transformation with time dependent velocity):

$$\begin{aligned} &\xi^{x}=(2a_4-a_5)x+a_3t+a_2+g(t) \\ &\xi^{v}=a_4v+a_3+\dot{g} \\ &\eta^{E}=a_5E-\frac{m_e}{e}\ddot{g}, \end{aligned} \tag{6.3.40}$$

the remaining three generators $\xi^{t}, \eta^{e}, \eta^{i}$ being unaltered (Roberts (1985)). Future investigations are expected to reveal new solutions for both the one and two component cases.

6.4 Lagrangian methods

Though the BGK waves discussed in the last Section (6.3) are exact solutions of the relevant set of nonlinear equations, they are restricted in that they are stationary solutions (static in a uniformly moving coordinate system). That is, though the original equation involved x and t dependence, the BGK solutions are functions of $x-v_0t$ only, where v_0 is the constant velocity of the moving frame. Such solutions, if found to be stable, can be candidates to represent the final state reached by a plasma evolving from some unstable but simpler state. Thus such solutions could well arise in practice and in fact do, as many experiments confirm (Section 6.3). Unfortunately, the method can say nothing about the 'dynamics', that is the time evolution of the system. In particular it can give no information about the initial state from which the BGK mode evolved. (There may exist certain invariants of the dynamical motion which may be useful but such quantities have only been found to date for much simpler systems. These are briefly discussed in Chapter 7.)

Thus any method which allows one to study the time evolution would be extremely useful. One such method is based on the introduction of Lagrangian variables and will be discussed below as applied to the evolution of plasma waves. Unfortunately, though extremely powerful in giving an explicit solution to the time evolution problem, it can only be made to perform so well for the simplest of cases. However, this method is worth considering, as it does give an explicit solution and illustrates the role of BGK modes in this more general problem. For early work see Konyukov (1959) and Petviashvili (1967).

We now follow Davidson and Schram as described in Davidson (1972) and Albritton and Rowlands (1975).

The basic equations for a one component, cold electron fluid, in which the ions are treated as a neutralizing background, are

$$\frac{\partial \rho_e}{\partial t}+\frac{\partial}{\partial x}(\rho_e v_e)=0 \tag{6.4.1}$$

$$\frac{\partial v_e}{\partial t}+v_e\frac{\partial v_e}{\partial x}=-\frac{e}{m}E \tag{6.4.2}$$

$$\frac{\partial E}{\partial t}+v_e\frac{\partial E}{\partial x}=\frac{4\pi\rho_0}{m}ev_e. \tag{6.4.3}$$

The third equation was taken in a form that will prove convenient, as we will go over to a system in which the convective derivative $\partial_t+v_e\partial_x$ simplifies. This form was obtained from Poisson's and Maxwell's electrostatic equations (the constant background ion density is ρ_o):

$$\frac{\partial E}{\partial x} = -4\pi e(\rho_e - \rho_o)/m$$

$$\frac{\partial E}{\partial t} = 4\pi e v_e \rho_e/m.$$

The method we now introduce is based on a transformation from the usual, Eulerian variables x and t to Lagrangian variables that follow the fluid, $\bar{x}, \tau$. They are defined such that $\tau = t$ but, dropping the e subscript

$$\bar{x} = x - \int_o^\tau v(\bar{x}, \tau')\mathrm{d}\tau' \tag{6.4.4}$$

so that $\bar{x}$ is a function of both x and t, but $\bar{x}$ and τ are treated as independent variables. Importantly, the convective derivative $\partial/\partial t + v\partial/\partial x$ becomes $\partial/\partial\tau$ as promised, and (6.4.1) and (6.4.2) yield

$$\frac{\partial v}{\partial \tau} = -eE/m, \quad \partial E/\partial\tau = 4\pi\rho_o ev/m$$

giving the basic equation

$$\frac{\partial^2}{\partial t^2} E + \omega_p^2 E = 0, \quad \omega_p^2 = 4\pi\rho_o e^2/m^2, \tag{6.4.5}$$

and an identical equation for v. These are solved to give

$$E(\bar{x}, \tau) = A(\bar{x})\cos(\omega_p\tau) + B(\bar{x})\sin(\omega_p\tau) \tag{6.4.6}$$

$$v(\bar{x}, \tau) = \frac{e}{\omega_p m_e}[-A\sin(\omega_p\tau) + B\cos(\omega_p\tau)].$$

The functions A and B are obtained from the initial conditions since for $t=0$, we have simply $\bar{x}=x$. Thus $A=E(\bar{x},0), B=\omega_p m_e v(\bar{x},0)$. It is now possible to write out the transformation from x to $\bar{x}$ explicitly:

$$x = \bar{x} + \frac{v(\bar{x},0)}{\omega_p}\sin(\omega_p\tau) - \frac{eE(\bar{x},0)}{m\omega_p^2}[1-\cos(\omega_p\tau)]. \tag{6.4.7}$$

The form of the density is a little more complicated. The transformed version of (6.4.1) may be written

$$\frac{\partial}{\partial\tau}[\rho(\bar{x},\tau)\psi(\bar{x},\tau)] = 0$$

where

$$\psi = 1 + \int_0^\tau \frac{\partial v(\bar{x}, \tau')}{\partial \bar{x}} \mathrm{d}\tau' \tag{6.4.8}$$

and

$$\frac{\partial}{\partial x} = \psi^{-1} \frac{\partial}{\partial \bar{x}},$$

so that

$$\rho(\bar{x}, \tau) = \rho(\bar{x})/\psi(\bar{x}, \tau),$$

where $\rho(\bar{x}) = \rho(\bar{x}, t=0)$ and $\psi(\bar{x}, 0) = 1$ from (6.4.8).

In principle we now have an exact solution for $E(x, t)$ for all $t > 0$, though $\bar{x}$ is given as an implicit function of x and t. With the proviso that the physical quantities must be single valued functions of x and t, considerations so far were very general. As a simple example, consider $v(\bar{x}, 0) = 0$ and

$$E(\bar{x}, 0) = a\sin(k\bar{x}).$$

Thus, in this case, in view of $t = \tau$:

$$E(\bar{x}, t) = a\sin(k\bar{x})\cos(\omega_\mathrm{p} t) \tag{6.4.9}$$

$$kx = k\bar{x} - \alpha(t)\sin(k\bar{x}), \ \alpha(t) = \frac{eak}{m\omega_\mathrm{p}^2}[1 - \cos(\omega_\mathrm{p} t)]. \tag{6.4.10}$$

It is important to realize at this stage that all physical quantities such as the electric field, velocity etc. are *single* valued functions of x. This imposes the conditon that $\bar{x}$ be a single valued function of x for all time. It is easily seen from (6.4.10) that this implies

$$|\alpha(t)| < 1$$

leading to

$$2ea/m\omega_\mathrm{p}^2 < 1.$$

This is a condition on the amplitude and raises the question of what happens if one tries to propagate a larger amplitude wave in a cold plasma (experimentally or computationally). This would in fact lead to a multi-stream flow not describable by the Lagrangian formalism, which is tailored to just one stream.

To solve equation (6.4.10) we follow Jackson (1960) and first note that $\sin(k\bar{x})$ must be a periodic function of x with period $2\pi/k$. Thus we may write

$$\sin(k\bar{x}) = \sum_{n=1}^{\infty} b_n \sin(nkx)$$

(b_o can be seen to be zero by integrating both sides of the equation over x and using (6.4.10)). We have the usual inverse relation

$$b_n = \frac{k}{\pi} \int_0^{2\pi/k} \sin(k\bar{x})\sin(nkx)\mathrm{d}x.$$

Changing variables to $y = k\bar{x}$ gives, together with (6.4.10)

$$b_n = \frac{1}{\pi} \int_0^{2\pi} \mathrm{d}y \sin y(1-\alpha\cos y)\sin(n[y-\alpha\sin y]).$$

Integration by parts gives

$$b_n = \frac{1}{n\pi} \int_0^{2\pi} \cos y \cdot \cos(ny - n\alpha\sin y)\mathrm{d}y$$

$$= \frac{1}{2n\pi} \int_0^{2\pi} \mathrm{d}y[\cos\{(n+1)y - n\alpha\sin y\} + \cos\{(n-1)y - n\alpha\sin y\}].$$

Such integrals can be expressed in terms of Bessel functions. In particular

$$\frac{1}{2\pi} \int_0^{2\pi} \cos(ny - \beta\sin y))\mathrm{d}y = J_n(\beta)$$

and from the identity $J_{n+1}(\beta) + J_{n-1}(\beta) = \dfrac{2nJ_n(\beta)}{\beta}$ it follows that

$$b_n = \frac{2J_n(n\alpha(t))}{n\alpha(t)}. \tag{6.4.11}$$

Thus we finally obtain an explicit solution

$$E(x,t) = 2a\cos(\omega_p t) \sum_{n=1}^{\infty} \frac{J_n(n\alpha(t))}{n\alpha(t)} \sin(nkx)$$

and using Poisson's equation

$$\rho = \rho_o - (amk/2\pi e)\cos\omega_p t) \sum_{n=1}^{\infty} \frac{1}{\alpha(t)} J_n(n\alpha)\cos(nkx) \tag{6.4.12}$$

It will be noted that $E(x,t)$ and ρ are periodic functions of time with period $2\pi/\omega_p$, a value independent of the amplitude of the nonlinear plasma wave! This is of course just the period of small amplitude plasma waves discussed in Chapter 2 and obtained using a linearization procedure. Cold plasma waves are one of the few examples where the time period is in fact independent of amplitude. The significance of such waves is commented on in Section 8.2.

The complexity of the above solution arises not from the differential equations but from the solution of the coordinate transformation from Lagrangian to Eulerian coordinates. A case where this transformation is much simpler is where we take

$$\rho(x,0)=\rho_o\{1+af(x)\}$$

where $f(x)$ is periodic, period $2\pi/k$ and defined in the 0, $2\pi/k$ interval by

$$f(x)=\begin{cases}+1, & \dfrac{\pi}{2K}>x\geqslant 0\\[2ex] -1, & \dfrac{3\pi}{2K}>x\geqslant \pi/2K\\[2ex] +1, & \dfrac{2\pi}{K}\geqslant x\geqslant \dfrac{3\pi}{2K}\end{cases}$$

and $a<\frac{1}{2}$. The transformation from Lagrangian to Eulerian coordinates is now straightforward and the form of $\rho(x,t)$ as a function of x for three t values is given in Fig. 6.9.
(An illustration of (6.4.12) would resemble Fig. 6.9 somewhat, but with the corners rounded off, see Davidson (1972), p. 39.)

The solution (6.4.12) found by Davidson and Schram is not of the form of a BGK wave, namely a function of $x-v_ot$ only. However such solutions to

Fig. 6.9. Three profiles for step function densities in a cold plasma (initial profile; $\frac{1}{4}$ period; $\frac{1}{2}$ period).

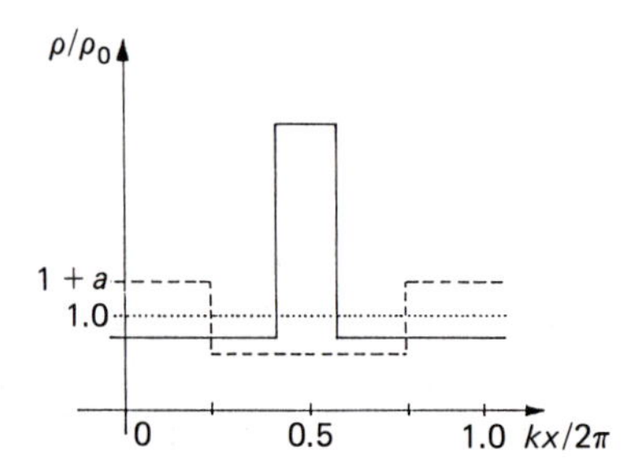

the basic equations can also be found by considering a different class of initial conditions. Consider in (6.4.6), $A(\bar{x})=b\sin(ky)$ and $B(\bar{x})=-b\cos(ky)$, where y is yet undetermined. Then it is readily seen that

$$E(\bar{x},t)=b\sin(ky-\omega_p t), \tag{6.4.13}$$

and from (6.4.7)

$$x=\left(\bar{x}-\frac{b}{4\pi\rho_o e}\sin ky\right)+\frac{b}{4\pi\rho_o e}\sin(ky-\omega_p t).$$

We now choose

$$\bar{x}-\frac{b}{4\pi\rho_o e}\sin ky=y$$

so that

$$kx-\omega_p t=(ky-\omega_p t)+\frac{bk}{4\pi\rho_o e}\sin(ky-\omega_p t).$$

This is now of the form of (6.4.10) with $x=kx-\omega_p t$, $\bar{x}=ky-\omega_p t$ and $\alpha=-b/4\pi\rho_o e$, so that using the result (6.4.11) we have

$$E(x,t)=\frac{4\pi\rho_o e}{k}\sum_{\mu=1}^{\infty}(-)^{n+1}\left(\frac{2J_n(nkb/4\pi\rho_o e)}{n}\right)\sin(n(kx-\omega_p t)) \tag{6.4.14}$$

This differs from $a\sin(kx)$ at $t=0$ as it should do, since we do not in general expect two different solutions to agree at a given moment in time. $E(x,t)$ is now a function of $x-v_o t$ $(v_o=\omega_p/k)$ only and so is of the form of a BGK mode. However, it will be noted that to obtain this particular wave solution the wave has to be set up in the first instance (at $t=0$). It does not evolve from some general initial condition. The relationship between the Lagrangian method and the BGK modes deserves a little more attention. Assuming in equation (6.4.1), (6.4.2) and (6.4.3) that all quantities are functions of $\eta=x-v_o t$ only, then they reduce to $\rho v=\rho_o v_o, v^2=c+2e\phi/m$ where $E=-\mathrm{d}\phi/\mathrm{d}\eta, v=v-v_o$, c is an integration constant and

$$\frac{\mathrm{d}^2\phi}{\mathrm{d}\eta^2}=4\pi e\rho_o\left[\frac{v_o}{\left(c+\dfrac{2e\phi}{m}\right)^{\frac{1}{2}}}-1\right].$$

This equation may be integrated to obtain essentially (6.3.17)–(6.3.20). This solution was illustrated by Fig. 6.7 for various values of A. This is now of the standard form of a Lagrangian type solution and so we see that the BGK method gives, as it should, the same result as the Lagrangian method.

However, it must be stressed that the Lagrangian method gives one a complete solution of the full time dependent problem whereas the BGK method gives a particular solution corresponding to particular initial conditions. These are the *only* initial conditions that will propagate uniformly in space without change of shape. (Any other initial conditions will subsequently distort.)

In summary it is seen that the Lagrangian method gives explicit solutions for the problem of time evolution of nonlinear cold plasma waves. Furthermore, for a special choice of initial conditions the plasma disturbances propagate at a constant phase velocity and have the form of a BGK mode.

Unfortunately, the above method is not easily extendable, though some progress has been made. For example consider the case of a hot plasma in which case the momentum conservation equation (6.4.2) has an extra pressure term $-(1/m\rho)\partial p/\partial x$ on the right hand side. We will now present a new family of waves in this context. Taking an equation of state of the form

$$\left(\frac{\partial}{\partial t}+v\frac{\partial}{\partial x}\right)(p/\rho^{\gamma})\equiv\frac{\partial}{\partial \tau}(p/\rho^{\gamma})=0$$

where γ is some constant, gives instead of (6.4.5)

$$\frac{\partial^2 v}{\partial \tau^2}+\omega_{\rm p}^2 v=-\frac{1}{m\rho}\frac{\partial^2}{\partial \tau \partial x}(p/\psi^{\gamma})$$

where $p(x)=p(x,t=0)$. From the definition of ψ, we see that $\partial\psi/\partial\tau\equiv\partial v/\partial x$ which, used in conjunction with the above equation, gives

$$\frac{\partial^2 \psi}{\partial \tau^2}+\omega_{\rm p}^2\psi=(-1/m)\frac{\partial}{\partial x}\frac{1}{\rho}\frac{\partial}{\partial x}(p/\psi^{\gamma})+\omega_{\rm p}^2(x) \tag{6.4.15}$$

where $\omega_{\rm p}^2(x)=4\pi e^2\rho(x)/m^2$. The last term on the right hand side is an integration constant. This equation itself may be simplified for the special class of initial conditions where P/ρ^{γ} is constant (independent of x), a situation that would be implied for example if the equation of state was $p(x,t)/\rho^{\gamma}(x,t)=$constant. If this is true at $t=0$ it will be true for all time (see the equation following (6.4.14)). Then with $\psi(\bar{x},\tau)=\psi(\bar{x},\tau)/\omega_{\rm p}^2(x)$, (6.4.15) reduces to

$$\frac{\partial^2 \psi}{\partial \tau^2}+\omega_{\rm p}^2\psi=1+\lambda\frac{\partial}{\partial z}\left(\frac{1}{\psi^{\gamma+1}}\frac{\partial \psi}{\partial z}\right), \tag{6.4.16}$$

where $\lambda=\gamma/4\pi e^2$ and the new independent variable z is defined by

$$z=\int_0^{\bar{x}}\rho(\bar{x}',t=0)\mathrm{d}x'.$$

Not surprisingly, equation (6.4.16), though somewhat simplified, is still nonlinear and little is known about its solution in general. However, there is a fundamental difference between the general case and the cold plasma one, ($\lambda=0$). In the latter case the solution, as obtained above, is of the form

$$\psi = 1/\omega_p^2 2 + A(\bar{x})\cos(\omega_p\tau) + B(\bar{x})\sin(\omega_p\tau) \tag{6.4.17}$$

and unless $A=B=0$, ψ can never relax to a final time-independent state. Even this state ($A=B=0$) corresponds to the trivial case of $\rho=\rho_o, v=E=0$. For the warm plasma ($\lambda\neq 0$) the situation is different and this difference can be seen by considering linear analysis. The introduction of the spatial derivative essentially introduces dispersion. This is most readily seen by writing $\psi=1/\omega_p+\delta\psi(z,t)$ and linearizing to give

$$\frac{\partial^2\delta\psi}{\partial\tau^2} + \omega_p^2\delta\psi = \lambda\omega_p^2(\gamma+1)\frac{\partial^2\delta\psi}{\partial z^2}. \tag{6.4.18}$$

A normal mode analysis of this equation, following the general procedures discussed in Chapter 2, gives rise to a solution proportional to $e^{ikz-i\omega(k)t}$ where $\omega(k)$ satisfies the dispersion relation

$$\omega^2 = \omega_p^2 + \lambda(\gamma+1)\omega_p^2 k^2. \tag{6.4.19}$$

Thus we now expect that initial disturbances of a pulse-like nature will disperse, giving rise possibly to a stationary solution that pervades all space. This stationary solution, if it exists, will be a function of $\xi=z-v_o\tau$ and satisfy the equation

$$v_o^2 - \frac{d^2\psi}{d\xi^2} + \omega_p^2\psi = 1 + \lambda\frac{d}{d\xi}\left(\frac{1}{\psi^{\gamma+1}}\frac{d\psi}{d\xi}\right).$$

By its very nature this is just a BGK mode and such modes have been discussed by Infeld and Rowlands (1979c) in the original Eulerian coordinates (Exercise 7). The waves do in fact exist and are stable to long wavelength perturbations.

However, one expects more interesting solutions to exist where the dispersive effects are balanced by the nonlinear effects. (In fact this balancing of different types of effects is the main theme of Chapter 7.) A particular class of such solutions has recently been obtained by the authors (Infeld and Rowlands (1987a, b)) in the limit of small λ. Using multiple time perturbation theory a solution to (6.4.16) is found of the form

$$\psi(z,\tau) = 1 + A(z)\cos(\omega\tau+\chi),\ A \leqslant A_1$$

where $\omega=\omega_p(1+\xi)$ and χ, A_1 and ξ are constants, the first two determining the third. The function $A(z)$ satisfies a second order nonlinear ordinary differential equation which is shown to have periodic solutions. In contrast to the cold plasma case the above solution is given in terms of two arbitrary *constants* χ and A_1 whereas for the cold case ($\lambda \equiv 0$) the solution involves arbitrary *functions* ($v(x,0)$ and $E(x,0)$). The variation of the density ρ for a warm plasma ($\lambda \neq 0$), as a function of the Euler coordinate x for three t values, is illustrated in Fig. 6.10. Note certain similarities (and also differences) with Fig. 6.9.

In this Section we have applied the method of Lagrangian coordinates and seen how it simplifies the basic set of equations. For a cold plasma it has been possible to obtain an explicit solution for large amplitude waves. For a warm plasma the method is useful in that the original set of three coupled nonlinear equations for ρ, v and E is reduced to a relatively simple scalar nonlinear equation (6.4.16).

6.5 Lagrangian interpolation

In this section we will depart slightly from the theme of the Chapter title, and the wave amplitude will be assumed small in some considerations. The Section might therefore be considered to belong in Chapter 5. However, it follows logically from the exact method of the previous Section (6.4) and therefore will be included here. (This is in the spirit of our introduction of Hayes' method, which is exact but follows from ideas gained from an amplitude expansion in Chapter 5.) We have seen in Section 6.4 how powerful Lagrangian techniques are when applied to the cold plasma problem and how a complete analytic solution of the time dependent

Fig. 6.10. Oscillations of a warm plasma. Again three stages in the time development of the density profile are shown. From Infeld and Rowlands (1987a).

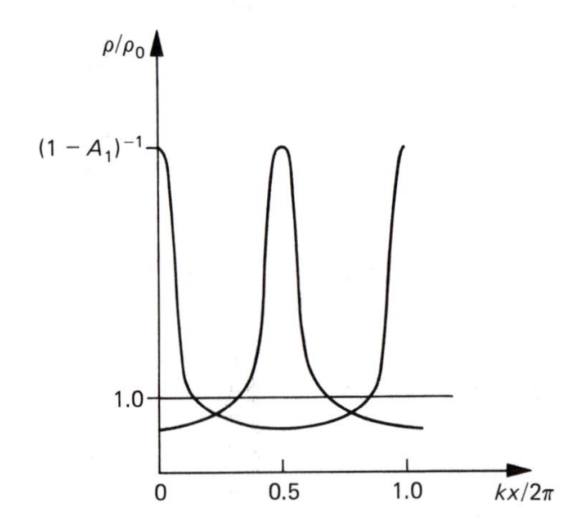

problem can be obtained. However, the method relies heavily on the introduction of a new and somewhat complex frame of reference, defined by (6.4.4), and this only makes sense if the state of the system can be adequately described by just one velocity at a given point. Of course, this was the case for the cold plasma. Even for the warm plasma, the introduction of a Lagrangian transformation reduced the complexity of the equations immensely, but unfortunately still left one with a scalar nonlinear partial differential equation to solve.

Thus it might seem that the method could not be usefully applied to discuss the simplest of instabilities, namely the two-stream instability, as each stream has its own velocity at every point of the system. However, from a more mathematical point of view, the simplification in the solution, for example the form of $E(x,t)$ as given by (6.4.9) is due to the *implicit* equation (6.4.10) relating the space variable x and the introduced variable $\bar{x}$. It is the purpose of this Section to introduce an approximate scheme where the perturbed quantities are expressed relatively simply in terms of a variable such as $\bar{x}$ which is then made to satisfy an approximate equation of the form of (6.4.10). The procedure is best illustrated by considering a perturbative solution to the cold plasma problem. Using conventional perturbation techniques and proceeding to third order gives the solution to (6.4.1), (6.4.2) and (6.4.3) in the form

$$E(x,t) = -2\pi A\sin\theta - \frac{\pi k A^2}{2e\rho_o}\sin 2\theta - \frac{3\pi k^2 A^3}{16\rho_o^2 e^2}\sin 3\theta + 0(A^4) \tag{6.5.1}$$

where A is the amplitude, assumed small, and determined by initial conditions and $\theta = kx - \omega_p t$. This should be compared with the exact solution obtained by the Lagrangian method namely (6.4.13) or (6.4.14). In particular we get exact agreement with the terms in (6.4.14) to $0(A^4)$ if we identify

$$A = \frac{-b}{2\pi}\left[1 - \tfrac{1}{8}\left(\frac{kb}{4\pi\rho_o e}\right)^2\right].$$

Alternatively, to $0(A^4)$, we may write $E(x,t)$, following (6.4.13), as

$$E(x,t) = b\sin\theta,$$

where b is given as a function of A above, or what is equivalent to this order

$$b = -2\pi A\left[1 + \frac{1}{32}\left(\frac{kA}{\rho_o e}\right)^2\right], \tag{6.5.2}$$

and θ satisfies a Lagrangian type equation

$$\theta = \theta + (bk/2\pi\rho_o e)\sin\theta.$$

If the perturbative theory was carried out to higher orders in A one could still express the solution in Lagrangian form but the relationship between A and b would change. In fact, by comparison of (6.4.14) and (6.5.1) it is seen that the exact relationship is simply

$$A = -\frac{4\rho_o e}{k} J_1\left(\frac{kb}{4\pi\rho_o e}\right). \tag{6.5.3}$$

Thus, in conclusion, we see that using conventional perturbation theory and assuming a Lagrangian form we can get the exact form of the solution as a function of x and t but the relationship between the Lagrangian amplitude b and the physical amplitude A is only obtained approximately. In this approach the Lagrangian variable has been introduced as part of an extrapolation scheme rather than having any direct physical context. It is now possible to apply this technique to other physical situations and in particular to the two-stream situation.

We choose to describe the plasma by the water-bag model given in Berk and Roberts (1967), in which case the velocity distribution function $f(x, v, t)$ is of the form

$$\begin{aligned} f(x, v, t) = {} & A_1[H(v-\phi_1) - H(v-\phi_2)] \\ & + A_2[H(v-\phi_3) - H(v-\phi_4)] \end{aligned} \tag{6.5.4}$$

where the ϕ's are functions of x and t, A_1 and A_2 are constants and H is the Heaviside function. The ϕ's are the boundaries of the water-bags in velocity space. Substitution of this form for $f(x, v, t)$ into the Vlasov equations gives the following equation for the ϕ's,

$$\frac{\partial \phi_i}{\partial t} + \phi_i \frac{\partial \phi_i}{\partial x} = -eE/m, \tag{6.5.5}$$

whilst Poisson's equation takes the form

$$\frac{\partial E}{\partial x} = -4\pi e\{A_1(\phi_1 - \phi_2) + A_2(\phi_3 - \phi_4) - n_o\}, \tag{6.5.6}$$

where n_o is the constant ion background density. We consider the special case of a warm stationary plasma ($\phi_4 = -\phi_3$) and a beam of relative density $\delta(= A_1(\phi_1 - \phi_2)/2A_3\phi_3)$ and velocity $v = (\phi_1 + \phi_2)/2$. A linear theory of perturbations about this state gives the following dispersion relation

$$D(\omega, k) = 1 - \frac{1}{\omega^2 - k^2} - \frac{\delta}{(\omega - kv)^2 - k^2 v_T^2} = 0 \tag{6.5.7}$$

where v_T is the effective thermal spread of the beam in units of the thermal spread of the stationary plasma $(=(\phi_1-\phi_2)/2\phi_3)$. We restrict attention to weakly unstable situations by defining a critical value of velocity v_c such that D and $\partial D/\partial\omega$ are zero for $v=v_c$. For values of v near v_c we have $\omega=\omega_c+\mathrm{i}\alpha$ where

$$\alpha^2=\frac{2(v-v_c)(\partial D/\partial v_c)}{\partial^2 D/\partial\omega^2}. \tag{6.5.8}$$

Thus for $v>v_c$ the system is unstable. For values of $v\simeq v_c$, the growth rate α is much smaller than ω_c so we have a situation of two distinct time scales. This is a situation where a nonlinear theory can be formulated using the multiple time perturbation method. Such a method is discussed in detail in Chapter 5 and was applied to the present problem by El-Labany and Rowlands (1986). In this way one finds

$$\Delta\phi_1=\phi_1-\phi_1=B_1\rho(t)\sin\psi-B_3\rho^3(t)\sin 3\psi+0(\rho^5), \tag{6.5.9}$$

where $\psi=\omega_p t-kx-\sigma(t)$. B_1 and B_3 are constants and depend on the plasma parameters v_c and δ. The quantities ρ and σ are related to the amplitude of the electric field by $E=\rho\exp(\mathrm{i}\sigma)$ and

$$\frac{\mathrm{d}^2E}{\mathrm{d}t^2}-\alpha^2E-b|E|^2E=0. \tag{6.5.10}$$

Here α is the linear growth rate given above whilst b is a function of v_c, δ and k. The above constitutes an approximate solution to the two stream problem, limited in amplitude because of neglect of terms of order ρ^5 and is the analogue of the solution (6.5.1) to the cold plasma problem.

Now we introduce a Langrangian type formulation by replacing the explicit solution (6.5.9) by the implicit one

$$\Delta\phi_1=A\sin\bar{\psi}+B\sin 2\bar{\psi}, \tag{6.5.11}$$

where

$$\psi=\bar{\psi}+\beta\sin\bar{\psi}. \tag{6.5.12}$$

The quantities A, B and β are now determined by expanding (6.5.12) in powers of β (essentially in powers of ρ) and equating powers of $\sin(n\psi)$ for $n=1$, 2 and 3. In this way we find in analogy with (6.5.2) that

$$\beta^2=8B_3\rho^2/B_1, A=(B_1-3B_3\rho^2)\rho$$

and

$$B=\rho^2\sqrt{2B_1B_3},$$

so that we may replace (6.5.11) and (6.5.12) by

$$\Delta\phi_1/B_1\rho=(1-3\lambda^2)\sin\bar{\psi}+\lambda\sqrt{2}\sin 2\bar{\psi}, \tag{6.5.13}$$

where

$$\psi=\bar{\psi}+8\lambda^2\sin\bar{\psi} \tag{6.5.14}$$

and $\lambda^2=B_3\rho^2/B_1$.

To $0(\lambda^4)$ equations (6.5.13) and (6.5.14) are identical to (6.5.9). However, in view of the success of the Lagrangian formulation for the cold plasma problem, we now assume a form of solution for $\Delta\phi_1$ as given by (6.5.13) and (6.5.14) for all values of λ, with λ related to ρ in some unknown manner, but to lowest order given by $\lambda=\sqrt{(B_3/B_1)}\rho$. Because we cannot solve the problem exactly we do not know the exact relationship, whereas for the cold plasma case we do, namely (6.5.2) is replaced by (6.5.3).

The explicit solution (6.5.13) can, by expansion in λ, be represented as an implicit solution as is (6.5.9) but with terms of all orders of ρ. However, only those to $0(\rho^4)$ are given correctly. Thus this formulation is equivalent to a selective resumming of all orders of perturbation theory.

A major difference between the explicit formulation as in (6.5.9) and the implicit one as in (6.5.13) is that in the latter $\Delta\phi_1$ can become a multivalued function of ψ for a sufficiently large value of $\lambda(8\lambda^2>1)$. In the analysis of the cold plasma case where the dependent variables were also physical ones, multi-valueness was not allowed as it led to physical nonsense such as three densities at one point. Here however, multi-valuedness in ϕ_i can be tolerated as long as it does not lead to multi-valuedness of the physical quantities such as density. The essential difference between the explicit formulation (6.5.9), which can be written as

$$\Delta\phi_1/B_1\rho=\sin\psi-\lambda^2\sin 3\psi \tag{6.5.15}$$

and the implicit form given by (6.5.13) and (6.5.14) is schematically illustrated in Fig. 6.11. Here the values of $\Delta\phi_1/B_1\rho$ as functions of ψ for a range of values of λ are shown. For small values of λ there is no significant difference between this and the explicit model but for larger values $(8\lambda^2>1)$ the implicit formulation allows multi-valuedness, an important qualitative change.

Berk and Roberts (1967) have solved the two-stream problem numerically using the water-bag model. However, they did not have to limit their work to the near stable situation. Their results are shown schematically in Fig. 6.12 where the velocity contours of the water-bags, that is the functions $\Phi_i(x,t)$, are shown as functions of x for various times. It is seen that initial small ripples on the ϕ lines grow and eventually become multi-valued. It is

Fig. 6.11. Two-stream water-bag plasma. Velocity contours follow Lagrangian calculation and multi-valuedness of $v(x)$ is seen in (c) and (d).

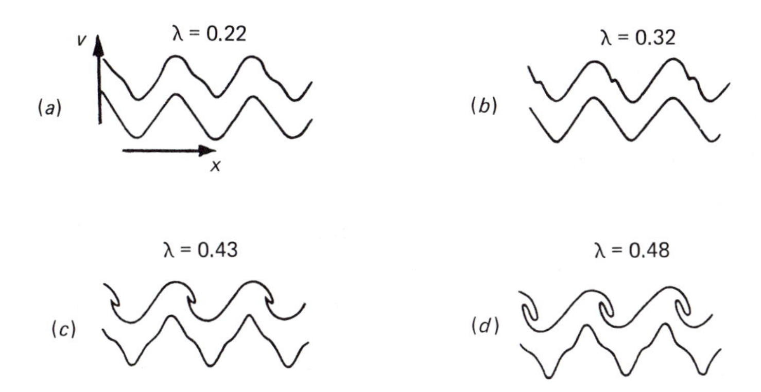

Fig. 6.12. Numerical results of Berk and Roberts. See caption of Fig. 6.11.

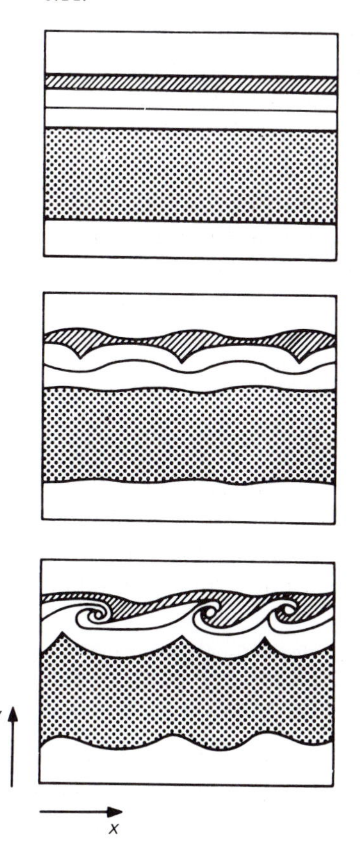

this behaviour which can be accounted for using the Lagrangian extrapolation formulation introduced above. Perturbation theory, no matter to what order, will always give a single valued form for the ϕ's.

The results presented in Figs. (6.11) and (6.12) are in qualitative agreement, as a function of time, if it is recalled that λ is in fact a function of time through $\rho(t)$.

In conclusion, we reiterate that the introduction of a Lagrangian extrapolation scheme allows one to reproduce a major qualitative feature of the exact solution: namely for sufficiently long times the contours of the water-bag become triple-valued over some region of space. The higher multi-valuedness found from the numerical computations is not a feature of the present scheme but could possibly be incorporated by making a further Lagrangian extrapolation over more limited regions of the ϕ contours. The multi-valuedness or transverse structure of the ϕ's seems to have much in common with the transverse structure of a strange attractor (see Chapter 10) and perhaps could be studied in terms of a simple one-dimensional recurrence relation of the type which leads to chaos (Chapter 10).

For the two-stream problem discussed in this Section, the Lagrangian transformation (6.5.12) which converted the explicit solution (6.5.9) to the implicit one (6.5.11) was suggested by analogy with the simpler problem of the cold plasma. However, the replacement of an explicit solution by an implicit one, which is at the heart of the whole analysis given in this Section, need not be related by a Lagrangian form such as (6.5.12). Other relations are possible. For example, ψ could be considered as a cubic function of $\bar{\psi}$. See also Infeld and Rowlands (1988).

Exercises on Chapter 6

Exercise 1

Derive equations similar to (6.2.4)–(6.2.6) but with three constants of integration. Give a transformation reducing these equations to (6.2.4)–(6.2.6). What are the physical implications of this transformation?

Exercise 2

Starting from (6.2.6), show that phase curves near the centre are almost ellipses (derive an equation analogous to (6.2.13)).

Exercise 3

Find the equations giving wave modes dependent on $\xi - v_0 t$ for the second collisional, ion acoustic plasma model of Chapter 5, (5.2.43b). How will the shock wave profile differ from that of Fig. 6.6?

Exercise 4

Find conditions on the constants in the trapped electron distribution (6.3.32) following from $f^{TR}(W_e) > 0$ and the vanishing of the total current. What additional condition is implied by demanding monotone W_e dependence of f^{TR} as in Fig. 6.2b? (Davidson (1972)).

Find the same conditions on the constants in (6.3.36). In particular show that $W_e/W_i = m_e/m_i$.

Exercise 5

Without performing a Lie generator calculation, show how (6.3.39) with non-zero a_4, a_5 only, follows from a rescaling. How does the subgroup with only a_1 non-zero lead to BGK solutions?

Exercise 6

Derive equation (6.4.16).

Exercise 7

Find the BGK wave solution to (6.4.16) such that $\psi = \psi(z - v_o\tau)$, in the form $\psi' = \psi'(\psi)$ for both the isothermal and adiabatic cases.

By using the identity

$$z = x + mE(x, 0)/4\pi\rho_o e$$

or otherwise, show that BGK waves in z, τ will also be such waves in x, t space (functions of $x - v_o t$).

7

Cartesian solitons in one and two space dimensions

7.1 Introduction

Most media are dispersive. That is, an initial very small amplitude disturbance in the form of a pulse will, as time evolves, tend to broaden and break up into its individual components. These are simple waves moving with different phase velocities. In the earlier chapters we looked at waves and instabilities propagating in various media. A linear analysis was seen to lead to a dispersion relation of the form $D(\omega, k)=0$, the solution of which gives $\omega=\omega(k)$. This ω may be complex. The phase velocity, ω_r/k (where we write $\omega=\omega_r+i\gamma$) is in general a function of k. In such a medium any measurable quantity, $\phi(x,t)$ say, takes the form

$$\phi(x,t)=\int \phi_o(k)e^{ik(x-\omega_r t/k)}e^{\gamma(k)t}dk. \tag{7.1.1}$$

Here $\phi_o(k)$ is related to the initial disturbance $\phi_o(x)$ at $t=0$. As we saw in Section 2, even for a stable medium in which $\gamma=0$, $\phi(x,t)$ evolves with time in a complicated manner depending on the form of $\omega_r(k)$. The exception is the special case $\omega_r/k=\text{constant}=c$. Then $\phi(x,t)=\phi_o(x-ct)$, that is the initial pulse $\phi_o(x)$ propagates with uniform velocity c. However, such behaviour is rare and in most media pulses will disperse.

The above considerations are based on two distinct assumptions. One is that the medium is dispersive, and the other that the disturbance is 'small' and hence a linear theory is applicable. If we are interested in a particular medium then the first assumption cannot be changed. The second can, however, be altered by considering large amplitude disturbances and then it is necessary to consider a nonlinear theory. In the past, with a few notable exceptions such as General Relativity and the theory of shock waves, which are inherently nonlinear, most physics was studied in its linear form. Linearization leads to such significant mathematical simplification that it almost came to be taken for granted that it was necessary if one wanted to make progress. A nonlinear theory was considered to be orders of magnitude more difficult and perhaps not worth the effort. It was thought to be easier to control the experiment such that one was always in the linear regime. There were exceptions, as we shall see below, but until the last twenty-five years nonlinear effects where considered to be somewhat of a taboo by the physics community. Strengthening of this attitude was due to the fact that one of the fundamental theories of physics, namely Quantum Mechanics, is in fact a linear theory.

This attitude, however, was not quite universal and with a little hindsight one can spotlight significant contributions to a study of effects which only exist because of the nonlinearity. One, very close to the content of the present book, was the observation of 'the Great Solitary Wave' by Scott-Russell in 1834. This was a 'well defined heap of water' which moved in a canal at a velocity of 8 to 9 miles per hour for a distance of a mile or so preserving its original shape (Section 8.1). The question that arises from this simple observation is why does such a 'heap' not disperse? The answer is that such disturbances cannot be described by a linear theory and somehow nonlinear effects can act such as to counterbalance the effects of dispersion. Scott-Russell appreciated this but otherwise it seems to have been treated in low key by the majority of the physics community.

To illustrate these effects we consider a weakly dispersive stable medium. It follows from the form of (7.1.1) that since $\phi(x,t)$ is a measurable and hence a real quantity, $\omega_r(k) = -\omega_r(-k)$. For long wavelength disturbances we write approximately $\omega_r = ck - \alpha k^3$, where c and α are constants. In this case it is readily seen that $\phi(x,t)$ satisfies the linear equation

$$\frac{\partial\phi}{\partial t} + c\frac{\partial\phi}{\partial x} + \alpha\frac{\partial^3\phi}{\partial x^3} = 0, \tag{7.1.2}$$

with the third term representing the dispersive qualities of the medium. To this we now add a nonlinear term to counterbalance the dispersion. There is no unique way of doing this. If we insist that $\phi(x,t)$ constant be a solution of the nonlinear equation and we choose the simplest nonlinear term we are led to the equation

$$\frac{\partial\phi}{\partial t} + c\frac{\partial\phi}{\partial x} + \alpha\frac{\partial^3\phi}{\partial x^3} + \beta\frac{\partial}{\partial x}\phi^2 = 0$$

(for a different argument see Subsection 1.2.1). This may be put into standard form by defining $u = 3^{\frac{1}{3}}(\beta\phi + c/2)$ and rescaling x to $\alpha^{\frac{1}{3}}x$. One then obtains the Korteweg–de Vries (KdV) equation in the form

$$\frac{\partial u}{\partial t} + 6u\frac{\partial u}{\partial x} + \frac{\partial^3 u}{\partial x^3} = 0. \tag{7.1.3}$$

Of course, the above is not a derivation of KdV but it illustrates its universality. It is the simplest equation which describes both dispersion and nonlinearity on an equal footing. This equation has been shown to describe many distinct physical phenomena and a more rigorous method of derivation was given in some detail in Section 5.2.

What of the solution? Solutions of the displacement type, that is where $u(x,t)$ is a function of $\xi(=x-Vt)$ only, are well known and were discussed in Chapter 6. In particular a soliton solution exists of the form

$$u=\frac{V}{2}\operatorname{sech}^2\left(\frac{V^{\frac{1}{2}}}{2}[x-Vt]\right). \tag{7.1.4}$$

This will be shown in Chapter 8 to be stable to small perturbations. Solutions of this form have been known to exist for over a hundred and fifty years and have been identified with the 'heap of water' found by Scott-Russell.

This in itself caused little excitement but then Zabusky and Kruskal (1965) discovered from numerical simulations the extraordinary structural stability of such solutions. Solitons could pass through one another and emerge unchanged. They behaved almost like particles and hence their name (Fig. 1.2).

Next Gardner, Greene, Kruskal and Miura (1967) showed how the KdV equation could be solved exactly as an initial value problem, not just as a final state solution such as (7.1.4). This was in itself interesting. However, it was not until Zakharov and Shabat (1971) showed how the method could be used to solve another nonlinear equation, namely the nonlinear Schrödinger equation, that it was realized that a powerful analytic method for the solution of nonlinear partial differential equations had been found. In subsequent years a number of distinct methods have been developed and applied to a range of nonlinear partial differential equations. An important restriction that applies to all these methods is that the solution must approach a constant value for $|x|\to\infty$ for all time. This, in fact, removes from consideration solutions in the form of an infinite train of nonlinear waves. Recently there has been some progress on the wave problem, see Novikov *et al.* (1984) or Novikov in the Bullough and Caudry book (1980).

It is the purpose of this Chapter to illustrate some of these methods and their applications (in the latter part in more space dimensions). For a more extensive and detailed coverage of the subject in one dimension, reference should be made to the books by Lonngren and Scott (1978), Bullough and Caudry (1980), and Newell (1985). Section 7.9 on two space dimensional solitons is somewhat less standard, whereas the material in Section 7.10 on a method of 'operator dressing' is expanded in the literature quoted there.

Presentation has been geared to acquainting the reader with as many methods as possible.

7.2 The direct method

The intrinsic stability of solitons can perhaps be best illustrated by applying the 'direct method' of Hirota to the KdV equation (7.1.3) with the boundary condition $u \to 0$ as $|x| \to \infty$. The method relies heavily on intuition, and the first thing to note is that we may write the soliton solution (7.1.4) in the form

$$u = 2\frac{\mathrm{d}^2}{\mathrm{d}x^2} \ln\left(\cosh\left[\frac{V^{\frac{1}{2}}}{2}(x - Vt)\right]\right). \tag{7.2.1}$$

Now solutions involving solitons will be well separated at some time and hence must be of the above form locally. This suggests we look for exact solutions of the form

$$u(x,t) = 2\frac{\mathrm{d}^2}{\mathrm{d}x^2} \ln F(x,t). \tag{7.2.2}$$

Because we have a solution involving a derivative we can integrate (7.1.3) once with respect to x. The integration constant is zero because of the boundary condition on u for $|x| \to \infty$. Now from (7.1.3) we obtain the following equation for F

$$F_{xt} - F_x F_t + F F_{xxxx} - 4F_{xxx} F_x F_x + 3F_{xx}^2 = 0, \tag{7.2.3}$$

where the subscripts denote partial differentiation with respect to x and t. This equation for F looks even more complicated than (7.1.3). However, it may be written in a much simpler form by introducing the idea of a bi-linear transformation. To this end we introduce the notation

$$D_t^m D_x^n A \cdot B = \operatorname*{Lt}_{t' \to t} \left(\frac{\partial}{\partial t} - \frac{\partial}{\partial t'}\right)^m \left(\frac{\partial}{\partial x} - \frac{\partial}{\partial x'}\right)^n A(x,t)B(x',t')_{x' \to x} \tag{7.2.4}$$

in which case (7.2.3) takes the form

$$D_x(D_t + D_x^3)F \cdot F = 0. \tag{7.2.5}$$

It will be noted that the operator is now linear and reflects the dispersive character of the problem only, but the equation itself is non-local (involves x and x').

The next stage in the analysis is to formally expand F and write

$$F = 1 + \varepsilon f_1 + \varepsilon^2 f_2 + \ldots \tag{7.2.6}$$

The parameter ε is introduced simply to keep track of the terms in the expansion. When a solution for the f_i is obtained, F is calculated from the above but with $\varepsilon=1$. Solutions such that the series terminates after a finite number of terms will now be looked for. If this search proves successful, we will have obtained exact solutions for u. Substitution of this form into (7.2.5) gives to first order ε

$$Lf_1=\frac{\partial}{\partial x}\left(\frac{\partial}{\partial t}+\frac{\partial^3}{\partial x^3}\right)f_1=0, \tag{7.2.7}$$

while to second order

$$Lf_2=-D_x(D_t+D_x^3)\mathbf{f}_1\cdot f_1=0. \tag{7.2.8}$$

Note that we now only need solve local differential equations since the right hand side of (7.2.8) can be evaluated, once f_1 is known, and the limit $x'\to x$ taken. A solution of (7.2.7) is

$$f_1=c\mathrm{e}^{\omega t-kx}, \tag{7.2.9}$$

where c and k are constants and $\omega=k^3$. Simple but tedious algebra then shows that for this case the right hand side of (7.2.8) is identically zero so that we can choose $f_2=0$. In this case we can also choose the higher f's to be zero and hence we have an exact solution of (7.2.5). Combining (7.2.6) and (7.2.9) gives

$$F=2\mathrm{e}^{(\omega t-kx+\eta)/2}\cosh[(\omega t-kx+\eta)/2] \tag{7.2.10}$$

where $c=e^{\eta}$. Then from (7.2.2) we finally have

$$u(x,t)=(k^2/2)\mathrm{sech}^2[(\omega t-kx+\eta)/2] \tag{7.2.11}$$

which is identical to (7.1.4) if we make the identification of k^2 with V and treat η as a simple phase shift.

The above is the one soliton solution. To consider more complicated solutions we must take a more general form for f_1. For example, if we write

$$f_1=\mathrm{e}^{\omega_1 t-k_1x+\eta_1}+\mathrm{e}^{\omega_2 t-k_2x+\eta_2} \tag{7.2.12}$$

with $\omega_1=k_1^3$ and $\omega_2=k_2^3$, then (7.2.7) is satisfied, and equation (7.2.8) reduces to

$$Lf_2=\tfrac{1}{2}(k_1-k_2)[(\omega_1-\omega_2)-(k_1-k_2)^3]\mathrm{e}^{(\omega_1+\omega_2)t-(k_1+k_2)x+\eta_1+\eta_2}, \tag{7.2.13}$$

so that

$$f_2=\frac{(k_1-k_2)^2}{(k_1+k_2)^2}\mathrm{e}^{(\omega_1+\omega_2)t-(k_1+k_2)x+\eta_1+\eta_2}. \tag{7.2.14}$$

Further algebra shows that the higher f's are identically zero so again we have an exact solution. A little later on we shall see that this solution corresponds to two solitons and hence it is not surprising that, if one keeps N distinct exponential solutions in the expression for f_1, the series terminates at f_N and the solution corresponds to N solitons.

The above method is due to Hirota and for further details one should consult his review article in the book by Bullough and Caudry (1980). He has applied the method to a number of different nonlinear partial differential equations.

Using equation (7.2.2) we obtain an exact solution of the KdV equation in the form

$$u(x,t)=\frac{2[k_1^2\phi_1+k_2^2\phi_2+A\phi_1\phi_2(k_1^2\phi_2+k_2^2\phi_1)+2(k_1+k_2)^2]}{(1+\phi_1+\phi_2+A\phi_1\phi_2)^2} \tag{7.2.15}$$

where $\phi_i=\exp(\omega_i t-k_i x+\eta_i)$ and $A=(k_1-k_2)^2/(k_1+k_2)^2$. If $k_1=k_2$ the above reduces to the single soliton solution (7.1.4). Note that for $|x|\to\pm\infty, u\to 0$.

To understand the nature of a soliton–soliton interaction it is best to 'sit' on one of them and watch the other pass by. This is accomplished mathematically by a change of coordinates. If we define the new spatial coordinate $\xi=x-k_2^2 t$ then

$$\phi_1=\exp(\eta_1-k_1[\xi-(k_1^2-k_2^2)t]), \phi_2=\exp(\eta_2-k_2\xi).$$

Without loss of generality we take $k_1>k_2$. We are interested in the limit $t\to\infty$. If we consider ξ finite then in this limit $\phi_1\to 0$ and

$$u(x,t)\simeq(k_2^2/2)\mathrm{sech}^2[\eta_2-k_2(x-k_2^2 t)/2], \tag{7.2.16}$$

a simple soliton. On the other hand, if we still consider $t\to-\infty$ but $\xi\simeq(k_1^2-k_2^2)t$ then ϕ_1 remains bounded and $\phi_2\to\infty$ in which case

$$u(x,t)\simeq(k_1^2/2)\mathrm{sech}^2[(\eta_1+\bar{\eta}-k_1(x-k_1^2 t))/2] \tag{7.2.17}$$

where $\bar{\eta}=\ln A$. Thus for $t\to-\infty$ the solution is composed of two distinct well-separated solitons, with the one with amplitude $k_1^2/2$ (the larger one since $k_1>k_2$) to the left of the soliton with amplitude $k_2^2/2$.

If we now consider the limit ξ finite but $t\to+\infty$ then ϕ_2 remains bounded but $\phi_1\to\infty$. This gives a solution

$$u(x,t)\simeq(k_2^2/2)\mathrm{sech}^2[(\eta_2+\bar{\eta}-k_2(x-k_2^2 t)/2], \tag{7.2.18}$$

whilst if we consider $\xi\simeq(k_1^2-k_2^2)t$ and take the limit $t\to+\infty$, then ϕ_1

remains bounded but $\phi_2 \to 0$ and

$$u(x,t) \simeq (k_1^2/2)\text{sech}^2[(\eta_1 - k_1(x - k_1^2 t))/2]. \tag{7.2.19}$$

Now for $t \to +\infty$ it will be noted that since $k_1 > k_2$ the larger amplitude soliton is to the right of the other. This phenomenon is illustrated in Fig. 1.2.

The solution discussed above illustrates the remarkable stability of solitons with respect to interactions. They seemingly pass through one another and the only change is in the phase factor. For the large soliton, the change of phase is from $\eta_1 + \bar{\eta}$ to η_1 whilst for the smaller it is from η_2 to $\eta_2 + \bar{\eta}$ as time progresses from $-\infty$ to $+\infty$. It is important to note that at other times $u(x,t)$ cannot be decomposed into two simple solitons, and indeed one would not expect this of a nonlinear theory. This stability of interacting solitons was first appreciated by Zabusky and Kruskal (1965) who solved the KdV equation numerically. When the solitons are separated, one can associate with each a trajectory, calculated by considering the motion of the maximum. From the assymptotic solution given above we see for the larger soliton that for $t \to -\infty$, the maximum follows the trajectory $x = (\eta_1 + \bar{\eta})/k_1 + k_1^2 t$ whilst for $t \to +\infty$, $x = \eta_1/k_1 + k_1^2 t$. Thus the only effect of the interaction of the solitons is a spatial displacement equal to $\bar{\eta}/k_1$.

One may conclude from the above discussion that solitons behave much like particles. They interact, but nevertheless emerge from the interaction essentially unchanged. This realization has led to some recent advances in statistical mechanics. For example when one wishes to discuss the statistical mechanics of a solid, the usual procedure is to consider the atomic oscillations in terms of small amplitude waves which when quantized are called phonons. At low temperature the atomic displacements are small and the phonons do not interact. The statistical mechanics model is then that of a gas of non-interacting phonons. This model has proved extremely useful (Kittel (1976) Chapter 5). As the temperature is increased, anharmonic terms in the interaction potential between atoms become important. This is equivalent to an interaction between phonons. The statistical mechanics then becomes very complicated. The alternative approach is to describe the atomic displacements, now nonlinear, in terms of solitons, not linear waves (phonons). This automatically takes into account the anharmonic effects. Now the solid is replaced by a gas of non-interacting solitons, to which conventional statistical mechanical methods may be applied. Such theories reveal effects unattainable from a phonon picture and in agreement with experiment (Krumhansl and Schrieffer (1975)). This picture has been justified directly from the fundamentals of statistical mechanics, the only difference being that the solitons induce a phase shift in the phonons and this must be taken into account.

7.3 Constants of motion

Why do KdV solitons have this remarkable structural stability? A short answer is that the KdV equation has associated with it an infinite number of constants of motion. Thus the time evolution must be of a very restricted nature. It is immediately seen by integrating (7.1.3) that

$$\frac{\mathrm{d}}{\mathrm{d}t}\int_{-\infty}^{\infty} u(x,t)\mathrm{d}x = 0, \tag{7.3.1}$$

subject only to the condition that u and its gradients tend to zero as $x \to \pm\infty$. Thus the first constant of motion is simply the spatial integral of u. Simple integration by parts shows that $\langle u^2 \rangle$ is also a constant. Miura and collaborators developed an ingenious method of generating a whole sequence of constants of motion, in fact an infinite sequence. (Miura *et al.* (1968)). First one introduces the function w defined such that

$$u = w + \varepsilon w_x + \varepsilon^2 w^2 \tag{7.3.2}$$

where ε is an arbitrary constant. Substitution of this into the KdV equation (7.1.3) shows that w must satisfy

$$\frac{\partial w}{\partial t} + \frac{\partial}{\partial x}\left[\frac{\partial^2 w}{\partial x^2} - 3w^2 - 2\varepsilon^2 w^3\right] = 0 \tag{7.3.3}$$

for all ε. This is in the form of a conservation equation with w the conserved density (and the term in brackets the flux). Integration over all x, assuming that w and derivatives vanish at $x = \pm\infty$, gives $\langle w \rangle = 0$, that is a constant of the motion. One may formally solve (7.3.2) by expanding in ε to give $w = w_o + \varepsilon w_1 + \varepsilon^2 w_2 +$ where $w_o = u, w_1 = -\partial u/\partial x, w_2 = \partial^2 u/\partial x^2 - u^2$. The important point is that u is independent of ε so that the condition $\mathrm{d}\langle w \rangle/\mathrm{d}t = 0$ leads to the *infinite* set of conditions $\mathrm{d}\langle w_n \rangle/\mathrm{d}t = 0$. Thus we have generated an infinite set of constants of the motion for the KdV equation, namely the integrated w_n. In particular $\langle w_o \rangle = \langle u \rangle$, $\langle w_2 \rangle = -\langle u^2 \rangle$, the two constants obtained above directly from the KdV equation. The next constant which derives from $\langle w_4 \rangle$ is $2\langle u^3 \rangle - \langle (\partial u/\partial x)^2 \rangle$. For another method of obtaining these constants, see Subsection 9.3.2.

The relevance of the existence of this infinite set of constants of motion is best seen by comparison with Hamiltonian systems. If for a system of N particles, described by a Hamiltonian with $6N$ degrees of freedom, there are $3N$ constants of the motion, the system is said to be integrable. This means that $3N$ angle-like variables can be defined with $3N$ corresponding action variables which are identified with the constants. If we label these constants of motion by J_i for $i = 1$ to $3N$ and the corresponding angle variables by θ_i, then the Hamiltonian equation of motion, $dJ_i/dt = -\partial H/\partial\theta_i$, implies that $H = H(J_i)$ only and $d\theta_i/dt = \partial H/\partial J_i$ so that

$$\theta_i(t) = \theta_{i,o} + \left(\frac{\partial H}{\partial J_i}\right)t. \tag{7.3.4}$$

We may interpret $\partial H/\partial J_i$ as a frequency. The dynamics of the system are then equivalent to a trajectory in the $6N$ dimensional phase space confined to a $3N$ dimensional torus and the time dependence is quasi-periodic. The fact that for KdV one has an infinite set of constants of motion makes one suspect that this equation is equivalent to an infinite order integrable Hamiltonian system, in which case relatively simple analytic solutions such as solitons should exist. Unfortunately, for infinite systems, the analogy is not strictly true and the existence of an infinite number of constants only gives a necessary condition for integrability but not a sufficient one (there are infinites and yet other infinities). From this we conclude that a necessary condition for a particular nonlinear partial differential equation to have N soliton type solutions is the existence of an infinite set of constants of motion. This is an important consideration when one is confronted with a new equation. (See Chapter 9 and especially Section 9.3, where this theme is elaborated somewhat in the context of an extension of KdV to more general geometries.)

The transformation (7.3.2) can also be viewed as one which relates nonlinear differential equations, namely the KdV equation (7.1.3) and equation (7.3.3). One particularly important case is where $w = q/\varepsilon$ and the limit $\varepsilon \to \infty$ is taken with q finite. This generates the modified KdV equation

$$\frac{\partial q}{\partial t} - 6q^2\frac{\partial q}{\partial x} + \frac{\partial^3 q}{\partial x^3} = 0 \tag{7.3.5}$$

with $u = \partial q/\partial x + q^2$. The modified KdV equation is also completely solvable. The above connection with the KdV equation was first appreciated by Miura (1968).

7.4 Inverse scattering method

Closely associated with the infinite number of constants of motion is another method that can be used to solve certain nonlinear partial differential equations. This is the so-called inverse scattering method. Somewhat out of the blue (though see Kaup (1980) for a rationale) let us consider the linear equation

$$\frac{\partial^2\psi}{\partial x^2} + (\lambda(t) + u(x,t))\psi = 0 \tag{7.4.1}$$

where $u(x,t)$ is the solution of the KdV equation. The above equation is identical to the quantum mechanical Schrödinger equation for a particle in a one-dimensional potential $u(x,t)$. Note that t appears only as a parameter. From (7.4.1) we may write $u = -\lambda - (\partial^2\psi/\partial x^2)/\psi$ which, when substituted into the KdV equation gives

$$\frac{\partial\lambda}{\partial t}\psi^2 = \frac{\partial}{\partial x}\left(\frac{\partial\psi}{\partial x}Q - \psi\frac{\partial Q}{\partial x}\right) \tag{7.4.2}$$

where

$$Q = \frac{\partial\psi}{\partial t} + \frac{\partial^3\psi}{\partial x^3} - 3(\lambda - u)\frac{\partial\psi}{\partial x}. \tag{7.4.3}$$

Integrating this equation over all x, and assuming that ψ and derivatives approach zero as $x \to +\infty$, gives

$$\frac{\mathrm{d}\lambda}{\mathrm{d}t}\int\psi^2(x,t)\mathrm{d}x = 0. \tag{7.4.4}$$

Equations of the form (7.4.1) are well-known to have two distinct types of solution: those identified as *bound* states and corresponding to ψ approaching zero exponentially as $x \to \pm\infty$, and the *scattering* states where ψ oscillates with x in this limit. For simplicity we initially assume that only bound states exist and these are discrete. Thus we associate with (7.4.1) N solutions, $\psi_n(x,t)$, which are bounded and such that the integral in (7.4.4) exists and is finite. This gives the remarkable result that if the potential $u(x,t)$ in (7.4.1) satisfies the KdV equation, then the eigenvalues λ_n are constants. We now impose a major constraint, namely, $u \to 0$ for $x \to \pm\infty$. This limits the discussion of solutions to the KdV equation to solitons and eliminates solutions corresponding to nonlinear waves. Since $u \to 0$ as $x \to \pm\infty$ we may solve for ψ in this asymptotic region and write for $x \to +\infty$, $\psi_n(x,t) = C_n(t)\mathrm{e}^{-k_n x}$ where $\lambda_n = k_n^2$. If this form is substituted into (7.4.2) one finds $C_n(t) = C_n(0)\mathrm{e}^{4k_n^3 t}$ where $C_n(0)$ is a constant of integration. This constant is obtained by solving (7.4.1) at $t=0$ and finding the asymptotic form. At $t=0$, the potential in (7.4.1) is just the initial condition $u(x,0)$ for the KdV equation. We have assumed that this is such as to give just a finite number of bound states and no scattering states (reflectionless potential). Using the above results we now have the asymptotic form for $\psi_n(x,t)$ for all time.

The inverse scattering problem in quantum mechanics and other wave problems is: given the asymptotic form for the solution, construct the potential. This is exactly the problem we need to solve, remembering that time t merely plays the role of a parameter. Thus the full solution of the KdV equation emerges from the solution of the inverse scattering problem. Fortunately, this latter problem has been solved and the solution of the KdV equation is finally expressed in the form

$$u(x,t)=2\frac{\partial}{\partial x}K(x,x) \tag{7.4.5}$$

where $K(x,y)$ satisfies the Gelfand–Levitan–Marchenko integral equation

$$K(x,y)+B(x+y)+\int_x^\infty B(y+z)K(x,z)\mathrm{d}z=0, \tag{7.4.6}$$

and

$$B(x)=\sum_{n=1}^{N}(C_n(0))^2\mathrm{e}^{8k_n^3t-k_nx}. \tag{7.4.7}$$

This equation is solved by assuming a solution of the form

$$K(x,y)=\sum_{n=1}^{N}K_n(x)\mathrm{e}^{-k_ny} \tag{7.4.8}$$

and substituting into (7.4.6). This leaves N linear algebraic equations for the K_n's which, when solved, gives the form for $K(x,y)$ and hence, by using (7.4.4), a unique solution for $u(x,t)$. For example, if there is just one bound state ($N=1$) we write $B(x)=B_o(t)\mathrm{e}^{-k_ox}$, $K(x,y)=K_o(x)\mathrm{e}^{-k_oy}$ and substitute into (7.4.6). This gives

$$K_o(x)=\frac{-2k_oB_o}{2k_o\mathrm{e}^{k_ox}+B_o\mathrm{e}^{-k_ox}} \tag{7.4.9}$$

so from (7.4.4)

$$u(x,t)=4k_oB_o\mathrm{e}^{-2k_ox}\Bigg/\left(1+\frac{B_o}{2k_o}\mathrm{e}^{-2k_ox}\right)^2. \tag{7.4.10}$$

The time dependence is in B_o. If we write $2k_o=u^{\frac{1}{2}}$ and introduce a phase η such that $C_o^2(0)/2k_o=\exp(-2\eta)$ then the above form for $u(x,t)$ reduces to that for a single soliton given by (7.1.4) (with the introduction of a phase factor). The two soliton solution obtained above by the direct method and given by (7.2.15) is similarly obtained.

It is important to note that the success of the inverse scattering method relies on the fact that the eigenvalues λ_n are constant and the time dependent equations one has to solve for the $C_n(t)$ reduce to trivial linear equations with coefficients independent of time. They have simple exponential type solutions. The method is somewhat similar to that of using Fourier transforms to solve linear partial differential equations. One takes the Fourier transform which reduces the problem to that of solving ordinary linear equations with constant coefficients. These are solved and the time dependent solution is then obtained by taking the inverse Fourier transform. The analogous process to the inverse transform is the solving of the Gelfand–Levitan–Marchenko integral equation (7.4.6).

In the above analysis a restriction was made for ease of presentation. It was assumed that the only solutions of the Schrödinger equation were those associated with bound states. This itself imposes restrictions on the initial conditions, that is on the form of $u(x,t=0)$. The generalization is to allow for the existence of scattering states which are such that $\psi(x,t)\to e^{-ikx}+b(k,t)e^{ikx}$ for $x\to+\infty$ and $\psi\to a(k,t)e^{ikx}$ for $x\to-\infty$. The solution of the KdV equation is still given by (7.4.5) and (7.4.6), but $B(x)$ now is of the form

$$B(x)=\sum_{n=1}^{N} C_n^2(0)e^{8k_n^3t-k_nx}+\frac{1}{2\pi}\int_{-\infty}^{+\infty} b(k)e^{ik(x-8k^2t)}dk. \tag{7.4.11}$$

The form for $C_n(0)$ and $b(k)$ being determined by the initial potential $u(x,t=0)$. Unfortunately, it is in general not possible to solve the basic integral equation (7.4.6) analytically except of course for the reflectionless potentials ($u(x,t=0)$ such that $b(k)=0$). We must resort to numerics at this stage. For the general case the long-time solution is in the form of N solitons travelling with different speeds to the right and noise-like behaviour travelling to the left.

It is of interest to note that the *nonlinearity* of the KdV equation has been replaced by the *non-locality* of the solution as expressed by (7.4.5) and (7.4.6). This interchange of nonlinearity for non-locality is also a feature of the direct method, (equation (7.2.5)). We will come back to this theme in Section 7.10.

Another point to note is that the number of solitons that eventually emerge is just the number of bound states. This number depends on the initial state $u(x,0)$, so that knowledge of this is sufficient to give the number of emerging solitons.

Finally we can be more explicit about the connection of the complete integrability of the KdV equation and integral Hamiltonian systems. We see from the above that the reflection amplitude $b(k,t)$ is of the form $b(k,t=0)e^{-i8k^3t}$. Thus we can associate with the system an infinite set of quantities $|b(k,t)|=|b(k,0)|$ which are time independent and a corresponding set $\arg(b(k,t))=\arg(b(k,0))-8k^3t$. Comparison with equation (7.3.4) allows us to identify $b(k,t)$ with the action variable J_i, and $\arg(b(k,t))$

with the angle variable θ_i. The infinite sequence of integers i is replaced by the continuous set represented by k and the frequencies $\partial H/\partial J_i$ by $-8k^3$.

The advantages of the inverse scattering method are: (i) the initial value problem for a nonlinear partial differential equation is reduced to a linear integral equation where the time appears only implicitly, (ii) solutions obtained are more general than those found by the direct method which are restricted to reflectionless initial conditions. A major restriction of the method is that $u(x,t)$ and various derivatives must approach zero as $x \to \pm\infty$. This means that spatially periodic solutions cannot be studied and consequently far less is known about such solutions. Certain restricted classes of nonlinear periodic solutions have been discussed in Chapter 6 and their stability with respect to small perturbations will be examined in Chapter 8. The latter property is intimately connected with their possible existence.

7.5 Bäcklund transformations

Yet another method of solving certain classes of nonlinear equations is associated with Bäcklund transformations. Such transformations either connect two distinct solutions of the same equation or else solutions of two distinct equations. They then allow one to progress from a simple solution, such as corresponds to a single soliton, to a many soliton solution of the same equation by repeated application of the transformation.

An example of the second type of use of these transformations has already been given, namely (7.3.2), a transformation between u and w which relates solutions of the KdV equation (7.1.3) to equation (7.3.3) and the modified KdV equation (7.3.5).

To illustrate the first of the above two uses of a Bäcklund transformation (sometimes called an auto-Bäcklund transformation), we consider two distinct solutions of the KdV equation (7.1.3) to be u and w and write

$$w_x = P(u, w, u_x, u_t, u_{xx}) \tag{7.5.1}$$

$$w_t = Q(u, w, u_x, u_t, u_{xx}) \tag{7.5.2}$$

where the suffixes denote differentiation and P and Q are functions of the variables indicated but not of the derivatives of w. If P, Q and $u(x,t)$ are known, then the above are just two uncoupled first-order equations for w. In solving (7.5.1), time is just a parameter whilst in solving (7.5.2), x may be taken as a constant parameter.

One must impose the integrability condition on w, $w_{xt} = w_{tx}$. This gives a relation between the derivatives of w essentially defining the equation w

satisfies. Unfortunately, the derivation of this equation or alternatively the forms of P and Q such that w and u satisfy the same equation is not at all straightforward. On the other hand, once a Bäcklund transformation has been discovered one has a relatively simple way of generating a hierarchy of solutions.

To illustrate the method we consider $P=-u_x-k^2+(w-u)^2$ and

$$Q=-u_t+4[k^4+k^2u_x-k^2(w-u)^2+u_x(w-u)^2+u_{xx}(w-u)], \tag{7.5.3}$$

where k is an arbitrary constant. The integrability condition gives, after a little algebra, that both w and u satisfy the equation

$$\phi_t-6\phi_x^2+\phi_{xxx}=0 \tag{7.5.4}$$

for all k. (This is an auto-Bäcklund transformation.) Differentiation of this equation with respect to x shows that $-2\phi_x$ satisfies (7.1.3). We now choose u to satisfy the above equation. One solution is simply $u=0$, in which case (7.5.1) and (7.5.2) reduce to

$$\frac{\partial w}{\partial x}=-k_o^2+w^2 \tag{7.5.5}$$

and

$$\frac{\partial w}{\partial t}=4(k_o-k_ow^2),$$

where k_o is the corresponding value of k. These equations are readily solved to give

$$w=-k_o\tanh\xi_o \tag{7.5.6}$$

where $\xi_o=k_o(x-4k_ot)$. This is simply related to the one soliton solution of the KdV equation. We now take this to be the value of u and substitute into P, with a different value of k, k_1 say, to obtain the next equation in the hierarchy. This is,

$$\frac{\partial \bar{w}}{\partial x}=k_o\text{sech}^2\xi_o-k_1^2+(\bar{w}+k_o\tanh\xi_o)^2. \tag{7.5.7}$$

Following Wahlquist and Estabrook (1973) we introduce a function ϕ defined such that

$$\bar{w}=(k_1^2-k_o^2)/(\phi-\bar{w}). \tag{7.5.8}$$

Using the above equations one finds that

$$\phi_x=-k_1^2+\phi^2 \tag{7.5.9}$$

which is identical in form to (7.5.5). If we use the solution analogous to (7.5.6) ($k_o \rightarrow k_1$), then the solution for $\bar{w}$ as given by (7.5.8) would become infinite for some x and this, of course, is not allowed physically. However, (7.5.9) has another solution, namely

$$\phi = -k_1 \coth \xi_1$$

where $\xi_1 = k_1(x - 4k_1^2 t)$. This is rejected as a solution of (7.5.4) because it is not bounded, but is admissable in (7.5.8) since it gives a bounded solution for $\bar{w}$. This solution corresponds to the two soliton solution found by the direct method.

We have shown in the above how the Bäcklund transformation can be used to transform the zero soliton solution to the one soliton solution (7.5.6) which itself can be transformed to the two soliton solution as expressed by (7.5.8). This process can be continued to give solutions with higher and higher numbers of solitons. At each stage, though, one has to solve equations (7.5.1) and (7.5.6). They can be reduced to the simpler form (7.5.9) by an algebraic transformation of the form of (7.5.8).

An alternate method of solving (7.5.7) is to define $z = \bar{w} + k_o \tanh \xi_o$ in which case (7.5.7) is reduced to a Riccati equation, which itself can be converted to a linear equation by writing $z = -\psi_x/\psi$. This procedure gives

$$\frac{d^2\psi}{d\xi_o^2} + (2\operatorname{sech}^2 \xi_o - k_1^2/k_o^2)\psi = 0.$$

This is of course a Schrödinger type equation and is in fact identical to (7.4.1) which forms the basis of the inverse scattering method. Now this equation defines the Bäcklund transformation between the one soliton solution defined by (7.5.6) and the two soliton solution given by (7.5.8).

7.6 Entr'acte

In the above sections three distinct methods: direct, inverse scattering, and Bäcklund transformations have been discussed and illustrated by application to the KdV equation. It has been shown how the N soliton solution can be obtained by these methods. Furthermore, the surprising structural stability of solitons has been demonstrated from these exact solutions. The methods are intimately related and some of these relationships have been discussed. (Other methods will be introduced to solve for two space dimensions later on. They could of course also be applied to the one-dimensional case treated above.)

What of course is of great importance is that these methods can be applied to a whole range of partial nonlinear differential and integral equations. These include the modified KdV equation (7.3.5), the sine-Gordon equation

$$\frac{\partial^2 v}{\partial x^2}-\frac{\partial^2 v}{\partial t^2}=\sin(v), \tag{7.6.1}$$

the non-linear Schrödinger equation

$$\mathrm{i}\frac{\partial \phi}{\partial t}+\frac{\partial^2 \phi}{\partial x^2}+a\phi+b|\phi|^2\phi=0 \tag{7.6.2}$$

where a and b are constants, and many others. A few more were indicated in Section 5.4. It would be out of place here to give an exhaustive list, even if it existed. Reference should be made to the review articles in Bullough and Caudry (1980), though the list is still expanding. What is important is that methods are available for the finding of exact solutions to the initial value problem for some nonlinear equations. What remains an open question is how does one recognize whether a particular equation can be solved by these methods? We will come back to this problem in Chapter 9. Finally it should be remembered that an important restriction that is applicable to all the solutions discussed above is that the solution u and derivatives must fall off to zero for $x\to\pm\infty$ (if u tends to one and the same constant at $+\infty$ and $-\infty$, say u_∞, we can just redefine $u\to u-u_\infty$).

7.7 Breathers and boundary effects

We have shown that N soliton solutions exist for the KdV equations and explicit solutions have been given for $N=1$ and 2. It was stated that a whole range of other nonlinear equations could be treated in a similar manner. However, it has been found that some of these other equations have solutions of a type not found in the KdV equation. An important example is the so-called breather solution. This is perhaps best illustrated by considering the sine-Gordon equation (7.6.1).

To this end we follow Lamb (1980) and assume solutions of (7.6.1) of the form

$$v(x,t)=4\tan^{-1}(\phi(x)/\psi(t)). \tag{7.7.1}$$

Note that, since $\tan^{-1}\theta=\pi/2-\tan^{-1}(1/\theta)$, the space and time variables are interchangeable in the above ansatz. Substitution of this form into (7.6.1) gives, after a little algebraic manipulation,

$$\frac{\psi^2}{\phi}\phi_{xx}+\frac{\phi^2}{\psi}\psi_{tt}=(\psi^2+2\psi_t-\psi\psi_{tt})+(-\phi^2+2\phi_x-\phi\phi_{xx}) \tag{7.7.2}$$

where the suffixes denote differentiation with respect to the argument.

Further differentiation with regard to both x and t eliminates the right hand side and allows the left hand side to be separated into functions of x and t to give eventually

$$\frac{(\phi_{xx}\phi)_x}{\phi\phi_x} = -\frac{(\psi_{tt}/\psi)_t}{\psi\psi_t} = -4k^2, \tag{7.7.3}$$

where k is the separation constant. These equations may in turn be integrated twice to give

$$\begin{aligned} \phi_{xx} &= -k^2\phi^4 + m^2\phi^2 + n^2 \\ \psi_{tt} &= k^2\psi^4 + (m^2-1)\psi^2 - n^2, \end{aligned} \tag{7.7.4}$$

where m and n are constants of integration. These equations may be solved in terms of elliptic functions and the solutions considered below are special cases. The single soliton solution is obtained by taking $k = n = 0$ and $m > 1$ in which case ϕ and ψ are simple exponential functions and

$$v = 4\tan^{-1}\left[\exp\frac{(\pm x - \beta t)}{\sqrt{(1-\beta^2)}}\right] \tag{7.7.5}$$

$$\beta = [\sqrt{(m^2-1)}]/m.$$

The two signs refer to the soliton, anti-soliton solutions respectively. The pulse-like nature of the solutions is best appreciated by considering $\partial v/\partial x$ or $\partial v/\partial t$ rather than v.

The two soliton solution is obtained by taking $k = 0, m > 1$ in which case

$$v = 4\tan^{-1}[\beta\sinh(mx)/\cosh(\beta mt)]. \tag{7.7.6}$$

This solution may be analysed in the same way as (7.2.15) to reveal the nature of the collision between two solitons. Then, just as for KdV, the solitons emerge from the interaction with a simple change of phase.

For $k \neq 0$: but $n = 0$ and $m^2 > 1$ one obtains a soliton–antisoliton solution of the form

$$v = -4\tan^{-1}\left[\frac{\sinh(\beta mx)}{\beta\cosh(mt)}\right]. \tag{7.7.7}$$

The soliton emerges unscathed from a collision.

The interesting new solution promised above and of a type not found for the KdV equation, is $(k \neq 0, n = 0, m^2 < 1)$

$$v = -4\tan^{-1}\left[\frac{m}{\sqrt{(1-m)^2}}\,\frac{\sin\sqrt{(1-m^2)}t}{\cosh(mx)}\right]. \tag{7.7.8}$$

This is the so-called breather solution and we see that for fixed x, v it is a periodic function of t with frequency $2\pi/\sqrt{(1-m^2)}$. The solution still has the pulse-like structure of a soliton but now one has the extra 'internal' time-like oscillations. The above solution is appropriate to a stationary breather, but more general propagating breather solutions exist and furthermore they interact with soliton type behaviour, emerging with only a phase change after interacting.

The 'breathing' may be given another interpretation. By analogy with linear waves we define the energy E of solitons, anti-solitons and breathers satisfying (7.6.1) to be

$$E = \int_{-\infty}^{\infty} \mathrm{d}x[\tfrac{1}{2}v_t^2 + \tfrac{1}{2}v_x^2 + 1 - \cos v].$$

For a single soliton or anti-soliton, equation (7.7.5), $E=8m$ whilst for the soliton–anti-soliton solution, equation (7.7.7), $E=16m$. This leads to an interpretation of the soliton–antisoliton solutions as corresponding to non-interacting solitons. For the breather again $E=16m$ but now $m<1$ (for all the other solutions $m>1$). Thus the energy is less than its components, suggesting an interpretation of the breather mode as a bound state between a soliton and an anti-soliton. Since the energy of a breather can approach zero whilst the other solutions approach a finite value ($m=1$, $E=8$ or 16), this also suggests that breather modes are more easily excitable and hence occur more frequently in systems away from absolute zero. Thus breathers must be taken into account in statistical mechanical treatments.

The above solutions may also be interpreted in terms of solitons interacting with a boundary. Consider the case of equation (7.6.1) being applicable for all $x<0$ and also that a boundary exists at $x=0$ and that v is held at zero value for all time at this boundary ($v(0,t)=0$). Examination of (7.7.4) shows immediately that this solution satisfies the differential equation (7.6.1) and the boundary condition. The solution (7.7.6) is a two soliton solution and the imposition of the boundary condition can be interpreted physically by saying that a soliton moves towards the boundary and is reflected as a soliton. The second soliton which exists in the space $x>0$ can be considered as an image soliton and the above result is an application of the method of images better known from electrostatics.

Following the above theme we may interpret the solution (7.7.7) as that appropriate to the boundary condition $\partial u/\partial x=0$ at $x=0$ for all t. Physically it corresponds to a soliton moving towards the boundary and being reflected as an anti-soliton.

More general solutions of the form (7.7.1) can be obtained by considering the most general form of solutions of (7.7.4), namely elliptic functions. This allows one to study solutions of (7.6.1) confined to finite regions of space by two boundaries, rather than the semi-infinite regions discussed above. It is found that the oscillations of the position of the soliton closely resemble those of a particle confined by an external potential. This reinforces the idea of treating solitons as particles. For details of this type of analysis, reference should be made to DeLeonardis and Trullinger (1980).

7.8 Experimental evidence

The story of solitons starts with Scott-Russell's observation of a 'heap of water' propagating down a canal. River bores had been known for countless centuries but these were just qualitative observations. Quantitative measurements came much later, mainly because solitons are inherently nonlinear whilst the consensus wisdom was based on linear ideas.

What characterizes a soliton? A KdV soliton is represented mathematically by (7.1.4) and it will be noted that the soliton is expressed by one arbitrary parameter V. It will also be noted that the amplitude scales with this parameter whilst the width is inversely proportional to the square root of V. These relations have been used to recognize solitons. The Scott-Russell type of experiment with water waves on canals is easily repeatable in laboratories and the above relationships checked (however, see Section 8.1 for limitations on these relations when V is large).

More detailed observations have been made for other physical situations, mainly because of their more immediate practical utilization. For example, soliton propagation in plasmas has received considerable attention because of a possible relevance to final state configurations in fusion devices. A plasma can support many different types of dispersive waves as is evident from the early chapters of this book. In the nonlinear regime there is usually a corresponding soliton type mode. Furthermore, many of these modes can be described in terms of a KdV equation and hence the above relations should hold. They do. For a more detailed account see, for example, the chapter by Ikezi in Lonngren and Scott (1978).

Transmission lines with nonlinear elements are also found to propagate in a soliton mode. Again to a good approximation they can be described by the KdV equation with of course an entirely different meaning to the quantity u in (7.1.3) than for plasma waves. However the relationship between amplitude, width and speed should still hold and this is verified experimentally (see for example the chapter by Lonngren in the same Lonngren and Scott book (1978)).

A more exotic soliton arises in a Josephson tunnel junction, which is a junction between two superconductors. The relevant nonlinear equation is no longer KdV but sine-Gordon. However, as we know, soliton solutions exist and they have been observed in these junctions and their characteristics studied in detail (see the chapter by Parmenter in Lonngren and Scott (1978)).

The conclusion one reaches from such studies is that soliton type solutions exist in real situations and that they have the scaling relations between amplitude and width as demanded for example by equations such as the KdV and sine-Gordon equations. What of their stability?

The clean passing of one soliton through another is of course the manifestation of a kind of soliton stability. This phenomenon is easily demonstrated in the laboratory with water waves in a trough. This aspect has also received considerable experimental attention in plasmas and transmission lines. The reference given above should be consulted for more details but the general conclusion is that solitons exist in real systems and behave in a manner that can be described by simple equations such as KdV or sine-Gordon. Stability with respect to small perturbations is a different problem and will be considered in Chapter 8.

7.9 Plane soliton interaction in two space dimensions

Fig. 1.4(*a*) pictures a beautiful natural instance of a two soliton interaction. As mentioned in Chapters 1 and 5, the simplest possible nonlinear, two space dimensional model equation known for water waves, the Kadomtsev–Petviashvili equation, describes this interaction very well. The present Section will look into this and related phenomena in some detail.

7.9.1 Introducing the trace method

There are by now several known methods for deriving N soliton solutions to nonlinear equations in x, y, t. These solutions exist (certainly if, and for $N \geqslant 2$ almost certainly only if, the equation is completely integrable, some of these equations were indicated in Section 5.4). One of the simplest approaches is the trace method, first applied to one space dimensional equations, KdV and modified KdV, by Wadati and Sawada (1980a, b).

The two space dimensional N soliton solution to KP was first seen to exist when this equation was solved by inverse scattering by Zakharov and Shabat (1974), and then found explicitly by the Hirota method of Section 7.2 two years later by Satsuma (1976). However, the trace method seems to be best for presentation. The few calculations of any length that are needed can be set as exercises in sums and matrix algebra, rather than being allowed to obscure the otherwise simple theory. Here calculations extend those of Okhuma and Wadati (1983) to both variants of KP (negative and positive dispersion). The relation of this method to inverse scattering is illustrated in Appendix 2, (which is best left until this Section has been read).

A similar and competitive method, based on a Wronskian formulation of N soliton solutions, was found by Nimmo and Freeman (1984). However, the trace method seems to be marginally simpler.

We will find it convenient to take KP in the form (each method has its own pet values of the numerical coefficients):

$$u_{tx}+12(uu_x)_x+u_{xxxx}+3u_{yy}=0. \tag{7.9.1}$$

(Later on we repeat the calculations for positive dispersion $-3u_{yy}$). The velocity potential, introduced by $u=\phi_x$, obeys, if $\phi\to 0$ as $x\to-\infty$,

$$\phi_t+6\phi_x^2+\phi_{xxx}+3\int_{-\infty}^{x}\mathrm{d}x\phi_{yy}=0. \tag{7.9.2}$$

Expanding ϕ formally (not worrying about small parameters as the series will not be truncated in this calculation):

$$\phi=\phi^{(1)}+\phi^{(2)}+\ldots \tag{7.9.3}$$

all $\phi^{(n)}\to 0$ as $x\to-\infty$, and substituting in (7.9.2), we obtain

$$\mathscr{L}\phi^{(1)}=(\partial_t+\partial_x^3+3\partial_x^{-1}\partial_y^2)\phi^{(1)}=0 \tag{7.9.4}$$

and this defines the linear operator $\mathscr{L}$. Also

$$\mathscr{L}\phi^{(2)}=-6(\phi_x^{(1)})^2 \tag{7.9.5}$$

$$\mathscr{L}\phi^{(3)}=-12(\phi_x^{(1)}\phi_x^{(2)}) \tag{7.9.6}$$

$$\mathscr{L}\phi^{(n)}=-6\sum_{l=1}^{n-1}\phi_x^{(l)}\phi_x^{(n-l)} \tag{7.9.7}$$

Now introduce

$$f_l(x)=\mathrm{e}^{q_l x} \tag{7.9.8}$$

$$g_l(x)=\mathrm{e}^{q_l x}C_l(y,t) \tag{7.9.9}$$

$$C_l(y,t)=C_l(0)\mathrm{e}^{-(p_l^2-q_l^2)y-4(p_l^3+q_l^3)t} \tag{7.9.10}$$

Substitution and simple algebra confirm that the functions

$$\phi^{(1)}=\sum_{l=1}^{N}f_l(x)g_l(x,y,t) \tag{7.9.11}$$

$$\phi^{(2)}=-\sum_{l=1}^{N}\sum_{m=1}^{N}\frac{1}{p_l+q_m}f_l(x)g_l(x,y,t)f_m(x)g_m(x,y,t) \tag{7.9.12}$$

satisfy (7.9.4) and (7.9.5) for any finite N. This number will be seen later on to be the number of solitons. We now introduce an $N\times N$ matrix **B** with elements

$$B_{lm} = \frac{1}{p_l + q_m} f_l(x) g_m(x, y, t) \tag{7.9.13}$$

to obtain immediately

$$\phi^{(1)} = \mathrm{Tr}\left(\frac{\partial \mathbf{B}}{\partial x}\right), \phi^{(2)} = -\mathrm{Tr}\left(\frac{\partial \mathbf{B}}{\partial x} \cdot \mathbf{B}\right). \tag{7.9.14}$$

More generally

$$\phi^{(n)} = (-1)^{n-1} \mathrm{Tr}\left(\frac{\partial B}{\partial x} \cdot \mathbf{B}^{n-1}\right) \tag{7.9.15}$$

(see Exercise 1). Thus, finally,

$$\phi = \mathrm{Tr}\left(\frac{\partial \mathbf{B}}{\partial x} - \frac{\partial \mathbf{B}}{\partial x} \cdot \mathbf{B} + \frac{\partial \mathbf{B}}{\partial x} \cdot \mathbf{B}^2 - \ldots\right) = \frac{\partial}{\partial x} \mathrm{Tr} \ln(\mathbf{I} + \mathbf{B}). \tag{7.9.16}$$

Since for any square matrix det $(\exp \mathbf{A}) = \exp(\mathrm{Tr}\mathbf{A})$, this will hold in particular for

$$\mathbf{A} = \ln(\mathbf{I} + \mathbf{B}) \tag{7.9.17}$$

yielding, together with (7.9.16)

$$u = \frac{\partial \phi}{\partial x} = \frac{\partial^2}{\partial x^2} \ln \det(\mathbf{I} + \mathbf{B}). \tag{7.9.18}$$

For positive dispersion, (7.9.10) is replaced by

$$C_l(y, t) = C_l(0) \mathrm{e}^{-(p_l^2 - q_l^2)y + 2(p_l + g_l)(p_l^2 + q_l^2 - 4p_l q_l)t} \tag{7.9.19}$$

(7.9.12) by

$$\phi^{(2)} = -\sum_{l=1}^{N} \sum_{m=1}^{N} \frac{(1+\mathrm{i}) f_l g_l f_m g_m}{p_l + q_m + \mathrm{i}(p_m + q_l)}, \tag{7.9.20}$$

and (7.9.13) by

$$B_{lm} = \frac{(1+i) f_l g_m}{(p_l + q_m) + i(p_m + q_l)} \tag{7.9.21}$$

and these values should be used in (7.9.18) (Exercise 2).

7.9.2 One and two soliton solutions

For $N = 1$, (7.9.18) yields for negative dispersion

$$u = \partial_x^2 \ln\left[1 + \frac{C_1(0)}{p_1 + q_1} \exp[(p_1 + q_1)x - (p_1^2 - q_1^2)y - 4(p_1^3 + q_1^3)t]\right] \tag{7.9.22}$$

$$=\tfrac{1}{4}(p_1+q_1)^2\operatorname{sech}^2[\tfrac{1}{2}\{(p_1+q_1)x-(p_1^2-q_1^2)y-4(p_1^3+q_1^3)t\}-\delta_1]. \tag{7.9.23}$$

Here

$$e^{-2\delta_1}=\frac{C_1(0)}{p_1+q_1}, C_1(0)>0. \tag{7.9.24}$$

Without loss of generality we will always assume $p_i+q_i>0$. If we took $C_1(0)$ negative, hyperbolic cosecant squared would replace (7.9.23). As this function is unbounded, we discard it as unphysical and demand $C_1(0)>0$. We also note that ω/k_x, minus the coefficient multiplying t divided by the coefficient multiplying x, is always positive. Soliton motion is thus from left to right.

For positive dispersion KP,

$$u=\tfrac{1}{4}(p_1+q_1)^2\operatorname{sech}^2[\tfrac{1}{2}\{(p_1+q_1)x-(p_1^2-q_1^2)y \\ -2(p_1+q_1)(4p_1q_1-p_1^2-q_1^2)t\}-\delta_1] \tag{7.9.25}$$

and, as expected, the two solitons (7.9.23) and (7.9.25) are identical for $p_1=q_1$, (no y dependence, for which case we just recover a KdV soliton). Horizontal motion from right to left is now possible. Static soliton solutions now exist for $p_1/q_1=2\pm\sqrt{3}$. Neither of the solitons (7.9.23) or (7.9.25) propagates straight up the y axis.

All in all at this level the two forms of KP seem to give very similar soliton solutions. However, the physics of two soliton *collisions* will be seen to be very different, on the whole yielding more variety for KP with negative dispersion. As if to compensate for this, KP with positive dispersion yields an additional class of dome-shaped solitons which will appear in Chapter 9.

It will sometimes prove convenient to use the notation

$$u=k_x^2\operatorname{sech}^2(k_xx+k_yy-\omega t-\delta_1) \tag{7.9.26}$$
$$k_x=\tfrac{1}{2}(p_1+q_1), k_y=\tfrac{1}{2}(q_1^2-p_1^2)$$

in which u is so expressed for both KP equations, but

$$\omega=2(p_1^3+q_1^3) \text{ for negative dispersion}$$
$$\omega=(p_1+q_1)(4p_1q_1-p_1^2-q_1^2) \text{ for positive dispersion.} \tag{7.9.27}$$

This ω dependence can be written jointly in terms of a soliton 'dispersion relation':

$$D(\mathbf{k},\omega)=k_x\omega-4k_x^4\mp 3k_y^2=0, \tag{7.9.28}$$

the upper sign being for KP with negative dispersion. For two soliton solutions we concentrate on the determinant that appears in (7.9.18). For negative dispersion

$$d=\det(\mathbf{I}+\mathbf{B})=\begin{vmatrix} 1+\dfrac{f_1g_1}{p_1+q_1}, & \dfrac{f_1g_2}{p_1+q_2} \\ \dfrac{f_2g_1}{p_2+q_1}, & 1+\dfrac{f_2g_2}{p_2+q_2} \end{vmatrix}=1+e^{2\eta_1}+e^{2\eta_2}+Ae^{2(\eta_1+\eta_2)} \tag{7.9.29}$$

$$\eta_i=[(p_i+q_i)x-(p_i^2-q_i^2)y-4(p_i^3+q_i^3)t]/2-\delta_i \tag{7.9.30}$$

$$e^{-2\delta_i}=\frac{C_i(0)}{p_1+q_i}>0 \tag{7.9.31}$$

$$A=\frac{(p_1-p_2)(q_1-q_2)}{(p_1+q_2)(p_2+q_1)}=-\frac{D(\mathbf{k}_1-\mathbf{k}_2,\omega_1-\omega_2)}{D(\mathbf{k}_1+\mathbf{k}_2,\omega_1+\omega_2)}. \tag{7.9.32}$$

For positive dispersion

$$d=\begin{vmatrix} 1+\dfrac{f_1g_1}{p_1+q_1}, & \dfrac{(1+i)f_1g_2}{(p_1+q_2)+i(q_1+p_2)} \\ \dfrac{(1+i)f_2g_1}{(p_2+q_1)+i(q_2+p_1)}, & 1+\dfrac{f_2g_2}{p_2+q_2} \end{vmatrix}=1+e^{2\eta_1}+e^{2\eta_2}+Ae^{2(\eta_1+\eta_2)} \tag{7.9.33}$$

$$\eta_i=[(p_i+q_i)x-(p_i^2-q_i^2)y-2(p_i+q_i)(4p_iq_i-p_i^2-q_i^2)t]/2-\delta_i \tag{7.9.34}$$

$$A=\frac{(p_1-p_2)^2+(q_1-q_2)^2}{(p_1+q_2)^2+(p_2+q_1)^2}=-\frac{D(\mathbf{k}_1-\mathbf{k}_2,\omega_1-\omega_2)}{D(\mathbf{k}_1+\mathbf{k}_2,\omega_1+\omega_2)}>0. \tag{7.9.35}$$

We now introduce Δ such that

$$\Delta=-\tfrac{1}{2}\ln A, \tag{7.9.36}$$

and demand that it be real. An important difference between the two cases now appears. Real Δ imposes a condition on A for negative dispersion ($A>0$) but not on positive dispersion KP solutions. In the first instance, not all p_i, q_i lead to finite solutions. However, if Δ *is* real, we can treat the two cases jointly for some considerations. Assuming this, we have for both cases

$$d=1+e^{2\eta_1}+e^{2\eta_2}+e^{2(\eta_1+\eta_2-\Delta)} \tag{7.9.37}$$

and we can forget for the moment which case we are considering.

We will now keep t fixed and look at asymptotic regions in space, where hyperbolic secant squared behaviour of u, similar to (7.9.26), is observed. To complete the investigation we should also prove that no other non-zero asymptotic behaviour is found, but we leave this to the reader (Exercise 2).

Table 7.1. *Four asymptotic arms of a two soliton solution assuming that* $\frac{k_{y2}}{k_{x2}} > \frac{k_{y1}}{k_{x1}}$

$d \simeq$	η_1	η_2	x	y	u
$1+e^{2\eta_1}$	$\simeq 0$	$\to -\infty$	$-\infty$ if $k_{y_1}<0$ $+\infty$ if $k_{y_1}>0$ x_o if $k_{y_1}=0$	$\to -\frac{xk_{x1}}{ky_1}$: $-\infty$	$k_{x_1}^2 \operatorname{sech}^2\eta_1$
$1+e^{2\eta_2}$	$\to -\infty$	$\simeq 0$	$+\infty$ if $k_{y_2}<0$ $-\infty$ if $k_{y_2}>0$ x_o if $k_{y_2}=0$	$\to -\frac{xk_{x_2}}{k_{y_2}}$: $+\infty$ $y \to +\infty$	$k_{x_2}^2 \operatorname{sech}^2\eta_2$
$e^{2\eta_2}[1+e^{2(\eta_1-\Delta)}]$	$\simeq \Delta$	$\to +\infty$	$+\infty$ if $k_{y_1}<0$ $-\infty$ if $k_{y_1}>0$ x_o if $k_{y_1}=0$	$\to -\frac{xk_{x_1}}{k_{y_1}}$: $+\infty$ $y \to +\infty$	$k_{x_1}^2 \operatorname{sech}^2(\eta_1-\Delta)$
$e^{2\eta_1}[1+e^{2(\eta_2-\Delta)}]$	$\to +\infty$	$\simeq \Delta$	$-\infty$ if $k_{y_2}<0$ $+\infty$ if $k_{y_2}>0$ x_o if $k_{y_2}=0$	$\to -\frac{xk_{x_2}}{k_{y_2}}$: $-\infty$ $y \to -\infty$	$k_{x_2}^2 \operatorname{sech}^2(\eta_2-\Delta)$

It follows from (7.9.18) and (7.9.37) that hyperbolic secant u follows when two of the four terms in (7.9.37) dominate. There are four possibilities of this happening and they are listed in Table 7.1 (when checking this Table it is useful to remember that $\partial_x^2 ln(\exp\Omega_i)=0$).
Asymptotic y values are obviously more convenient if we wish to write the results in compact form:

$$u = k_{x_1}^2 \operatorname{sech}^2\eta_1 + k_{x_2}^2 \operatorname{sech}^2(\eta_2-\Delta) \qquad y \to -\infty \tag{7.9.38}$$

$$= k_{x_1}^2 \operatorname{sech}^2(\eta_1-\Delta) + k_{x_2}^2 \operatorname{sech}^2\eta_2 \qquad y \to +\infty \tag{7.9.39}$$

The two soliton solution thus has arms displaced by Δ/k_{x_1} or Δ/k_{x_2} in x with respect to continuations of the opposite arms.

Behaviour in the intermediate region centred on $\eta_1 = \eta_2 = \Delta/2$ is especially simple for three special cases. They are

(1) $A \ll 1$, (2) $A \gg 1$, (3) $A \simeq 1$.

In the first case Δ is large and positive and the two simple exponentials in (7.9.37) dominate and

$$d \simeq e^{2\eta_1} + e^{2\eta_2}.$$

The discarded terms are both of order 1 and

$$u^{(1-2)} \simeq (k_{x_1} - k_{x_2})^2 \operatorname{sech}^2(\eta_1 - \eta_2). \tag{7.9.40}$$

For $A \gg 1, \Delta$ is negative and only the first and fourth terms in (7.9.37) survive

$$\begin{aligned} d &\simeq 1 + e^{2(\eta_1 + \eta_2 - \Delta)} \\ u^{(1+2)} &\simeq (k_{x_1} + k_{x_2})^2 \operatorname{sech}^2(\eta_1 + \eta_2 - \Delta) \end{aligned} \tag{7.9.41}$$

We call $u^{(1-2)}$ and $u^{(1+2)}$ *virtual solitons*, as their 'wavevectors' $\mathbf{k}_1 \pm \mathbf{k}_2$ and 'frequencies' $\omega_1 \pm \omega_2$ do not satisfy the one soliton 'dispersion' relation (7.9.28).

Finally, for the case $A \simeq 1$ we express this quantity in terms of $\mathbf{k}(k_x, k_y)$ defined in (7.9.26):

$$A = \frac{(k_{x_1} - k_{x_2})^2 \mp (\kappa_1 - \kappa_2)^2}{(k_{x_1} + k_{x_2})^2 \mp (\kappa_1 - \kappa_2)^2} \simeq 1 \tag{7.9.42}$$

$$\kappa_i = k_{y_i}/2k_{x_i}$$

(upper sign for KP with negative dispersion). Thus either one or both k_{x_i} must be very small, or one or both k_{y_i} very large. For all these cases and all η_1, η_2

$$d \simeq (1 + e^{2\eta_1})(1 + e^{2\eta_2}) \tag{7.9.43}$$

$$u \simeq u^{(1)} + u^{(2)} = k_{x_1}^2 \operatorname{sech}^2 \eta_1 + k_{x_2}^2 \operatorname{sech}^2 \eta_2 \tag{7.9.44}$$

and one can add solutions as if the equation were linear. When $k_{x_1} \gg k_{x_2}$, the second soliton will not be discernible during the interaction.

Fig. 7.1 gives a family of diagrams for one of the above cases of the interactions for KP with negative dispersion. Each little drawing pictures one case of two solitons interacting. The angle between interacting solitons ψ increases from left to right. The phase shift Δ is also drawn. In this picture we specify the bisector of the colliding solitons to be the y axis, and so we can take

$$k_{x_1} = k, k_{y_1} = -l \tag{7.9.45}$$

$$k_{x_2} = \alpha k, k_{y_2} = \alpha l, 0 < \alpha < 1, 0 < l < \infty \tag{7.9.46}$$

Thus the angle between the two solitons, $\psi = 2\tan^{-1}(\ell k^{-1})$, is the only parameter once α has been specified. We have

$$A = \frac{(1-\alpha)^2 k^2 - l^2/k^2}{(1+\alpha)^2 k^2 - l^2/k^2}. \tag{7.9.47}$$

Fig. 7.1. Diagram of two soliton collisions at various angles for fixed amplitudes (k^2 and $\alpha^2 k^2$) as found from KP with negative dispersion. Here $k_{y_1} = -1, k_{y_2} = +\alpha$. The phase shift $\Delta(\psi)$ is also indicated. Our simple theory fails to give bounded solutions between ψ_{c1} and ψ_{c2}, and the drawings in this region will be described later. Compare the third to last diagram with Fig. 1.4(a).

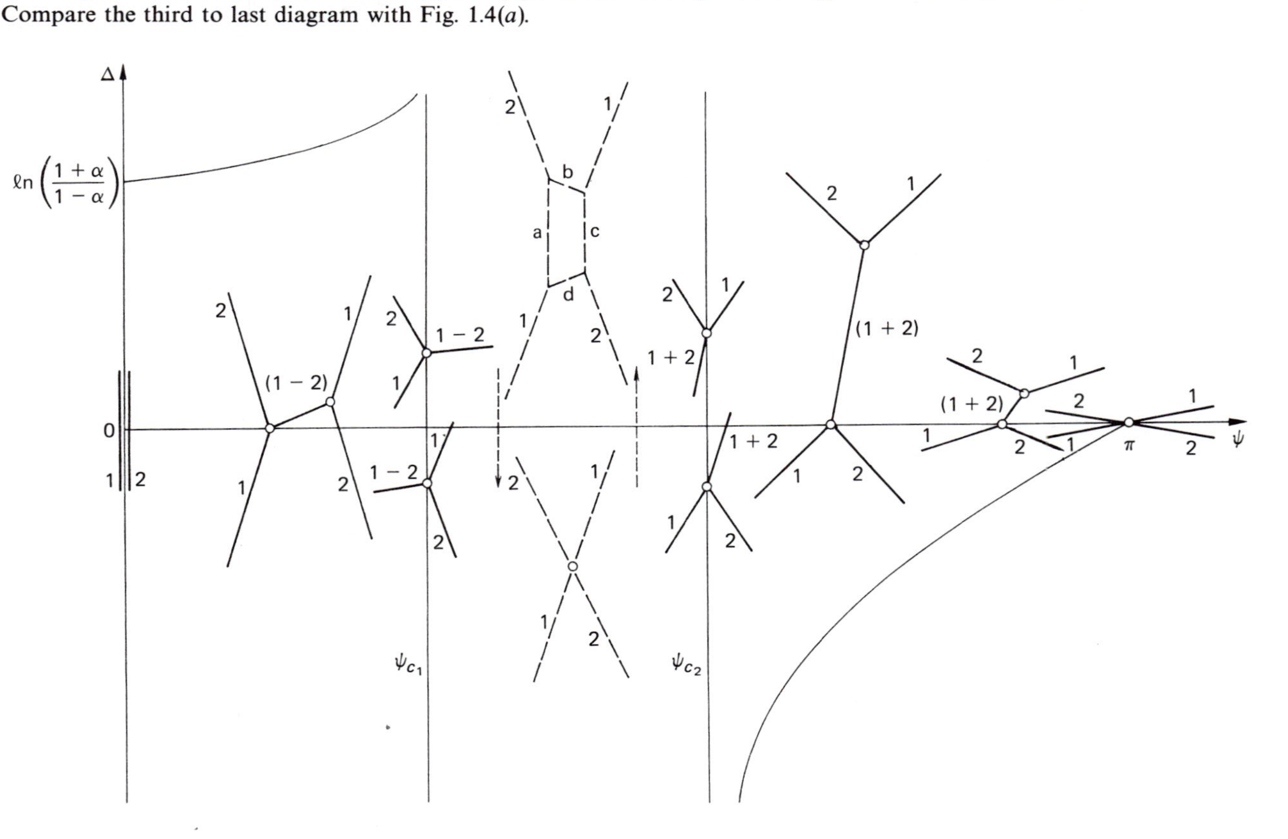

Thus A is negative between l_{c_1} and l_{c_2}, where

$$l_{c_1}=k^2(1-\alpha) \tag{7.9.48}$$

$$l_{c_2}=k^2(1+\alpha). \tag{7.9.49}$$

In this region our theory gives unphysical solutions and we must go beyond it to describe the collision. We defer this to the next Subsection. The $\alpha=1$ case (equal soliton amplitude) will now be considered separately. It is particularly popular with experimentalists as the simplest to realize. Now

$$l_{c_1}=0, l_{c_2}=2k^2, \tag{7.9.50}$$

$$\tan(\psi_{c_2}/2)=l/k=2\sqrt{a}, \tag{7.9.51}$$

where a is the common amplitude of the colliding solitons. Experimental confirmation of (7.9.51) is presented in Fig. 7.2. Fig. 7.3 gives a composite diagram similar to that of Fig. 7.1 for two soliton collisions as follow from KP with positive dispersion. The phase shift is always real and there is less variety.

Fig. 7.4 gives schematic diagrams for various three soliton interactions as follow from KP with negative dispersion.

Fig. 7.2. The resonance angle ψ_c v. the initial amplitude for two equal amplitude solitons, as following from two experiments: Nishida and Nagasawa (1980), and Lonngren *et al.* (1983), from which reference this figure was taken.

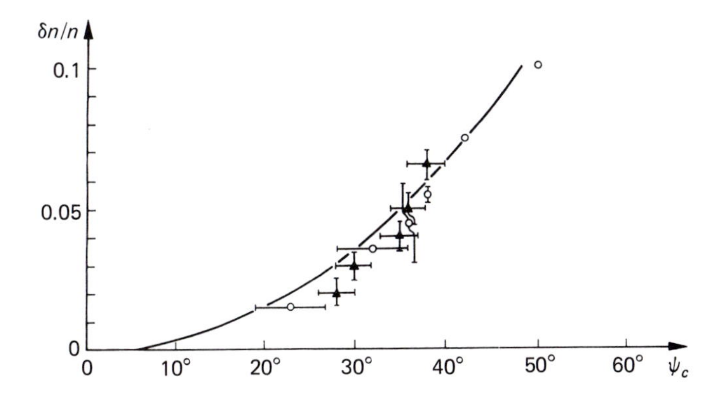

Fig. 7.3. A composite 'Feynman' diagram similar to Fig. 7.1, but for positive dispersion KP. Resonances cannot occur.

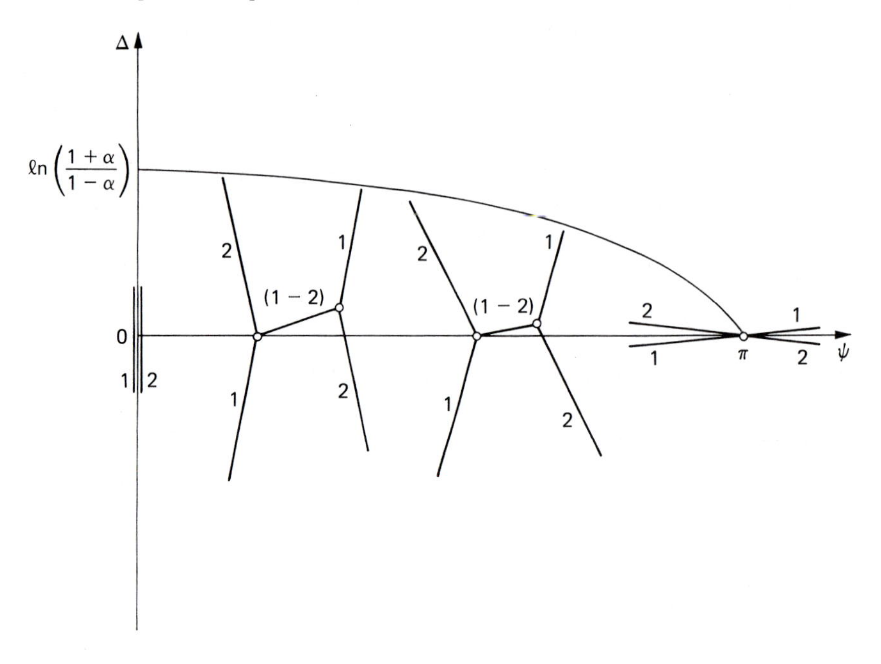

Fig. 7.4. A three infinite soliton solution with no resonances (a); and with production of various types of resonance solitons such that the total number of semi-infinite solitons is 5(b); or 4 (c & d). Based on KP with negative dispersion.

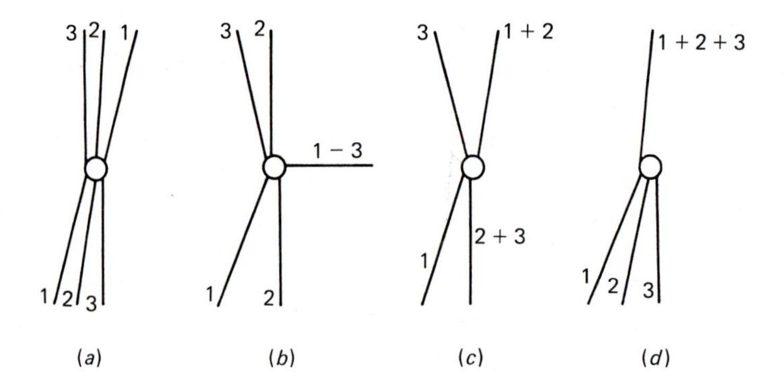

7.9.3 Some other developments and summary

As already indicated, a two soliton interaction as follows from KP with negative dispersion is not adequately described by our simple theory when

$$\psi_{c_1} < \psi < \psi_{c_2}. \tag{7.9.52}$$

Our solution (7.9.29)–(7.9.32), if applied mechanically, would involve hyperbolic cosecant squared functions and would become infinite along a curve in x, y space (Exercise 3). To obtain a more satisfactory theory, one would have to go back to (7.9.1), set up a two soliton situation at $t=0$:

$$u(x, y, 0) = k_{x_1}^2 \operatorname{sech}^2 \eta_1 + k_{x_2}^2 \operatorname{sech}^2 \eta_2 \tag{7.9.53}$$

such that $\mathbf{k}_1, \mathbf{k}_2$ form an angle ψ in the (7.9.52) interval, and then see what happens numerically. This does not seem to have been done for KP, but has for a version of the Boussinesq equation (Kako and Yajima (1980))

$$\partial_t^2 \phi - \nabla^2 \phi + \partial_t (\nabla \phi)^2 - \partial_{t^2} \nabla \phi = 0, \tag{7.9.54}$$

where

$$\nabla = (\partial_x, \partial_y), u = \partial \phi / \partial x. \tag{7.9.55}$$

A recurrent situation was found such that the configuration swings back and forth from one described by the Boussinesq equivalent of (7.9.53), containing no virtual solitons, to one in which a composite, four component virtual soliton is observed at the intersection. The little virtual solitons were found numerically to be 'almost' real in the sense that relations such as (Fig. 7.1):

$$\omega_c = \omega_1 + \omega_b$$
$$\mathbf{k}_c = \mathbf{k}_1 + k_b$$
$$D(\mathbf{k}_c, \omega_c) = 0$$

were found to be satisfied at the vertices within a few percent. The results are reproduced as Fig. 7.5 and were included schematically in Fig. 7.1 (two stages of the recurrence; this detail is drawn in a different convention from the rest as it is based on a numerical study of a different equation).

Interactions of plane ion acoustic solitons were investigated experimentally by Nagasawa and Nishida (1982), (1983). The theory of Kako and Yajima (1980) was confirmed, especially the quadrupole birtual soliton stage. Theoretically predicted phase shifts for $\psi > \psi_{c_2}$ have also been confirmed experimentally by Nagasawa *et al.* (1981). These experiments were performed in an apparatus called a Double-Plasma device that is described by Taylor *et al.* (1972).

The above investigations have been extended in several different directions. More complicated soliton interactions were considered on the basis of KP by Freeman (1979), (distributed solitons for KP with negative dispersion), Tajiri *et al.* (non-uniform background, (1982)) and van Dooren (three soliton interaction) (1985)). The work of Bryant extends some of these ideas to periodic waves (1982).

Soliton interactions in two space dimensions based on other more elaborate equations, have been considered by Yajima *et al.* (1978) and in the above mentioned Kako and Yajima paper (1980) (Boussinesq model for ion acoustic solitons), also by Anker and Freeman (1978) (Davey–Stewardson equation for water wave envelopes), and Johnson (1982), (fluid surface waves).

Resonant three soliton interaction has even been found for some very special one space dimensional equations (we know that KdV does not give these interactions as even KP with positive dispersion, which extends it, fails to). Two examples are: a variant of the Boussinesq equation, Tajiri and Nishitani (1982):

$$u_{tt}-u_{xx}+(u^2)_{xx}+u_{xxxx}=0, \tag{7.9.56}$$

Fig. 7.5. Solutions for initial value problem with initial condition (7.9.52) and (7.9.53) for the Boussinesq equation, which is more amenable to this type of calculation than KP. From Kako and Yajima (1980).

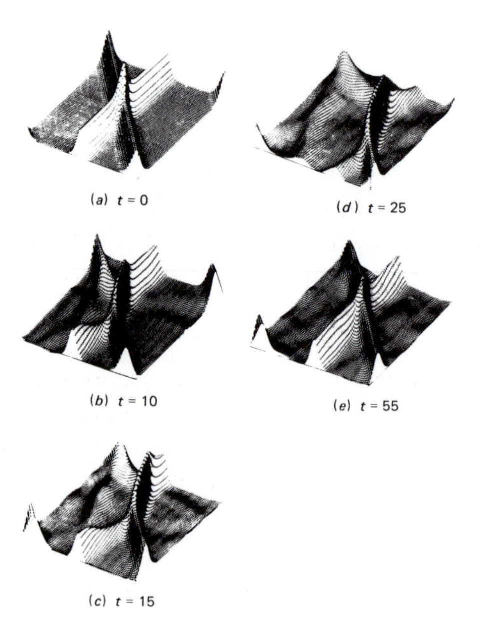

and a rather long-winded equation formulated by Sawada and Kotera (1974), see Hirota and Ito (1983). Even (7.9.56) is slightly exotic, as in most physical contexts such as water surface waves or ion acoustic waves in a plasma, a minus sign would insist upon preceding the last term. There is a fundamental difference between three soliton resonances in x, t and in x, y, t. In the former case two solitons fuse to produce a third (and are never seen again as separate entities). In the second all three exist for all t and the diagrams of Figs. 7.1 and 7.3 are in principle stationary. The *common* feature of these resonances is that

$$\mathbf{k}_3 = \mathbf{k}_1 + \mathbf{k}_2$$
$$\omega_3 = \omega_1 + \omega_2, \tag{7.9.57}$$

and all $\omega_i, \mathbf{k}_i$ satisfy the one soliton 'dispersion' relation $D(\mathbf{k}_i, \omega_i) = 0$. Composite diagrams in x, t space for the one-dimensional case resemble those in x, y space of Figs. 7.1 and 7.3.

In most physical situations, one must consider two space dimensions to observe a three soliton resonance. Indeed, this was initially thought to be mandatory, Newell and Redekopp (1977). The second point made in this pioneering work, namely that the resonance condition (7.9.57) is the same for solitons as for three linear dispersive waves to interact resonantly, is still true. However, $D(\mathbf{k}, \omega)$ for a soliton differs from that for a nonlinear wave. For KP it is (7.9.28):

$$D(\mathbf{k}, \omega) = k_x \omega - 4k_x^4 \mp 3k_y^2 = 0 \tag{7.9.58}$$

for a soliton and

$$D(\mathbf{k}, \omega) = k_x \omega + k_x^4 \mp 3k_y^2 = 0 \tag{7.9.59}$$

for a nonlinear wave (more generally $D_{\mathrm{SOL}} \leftrightarrow D_{\mathrm{LW}} ip\mathbf{k}, ip\omega$) if hyperbolic secant to the power 'p' appears in the one soliton solution). For solitons, KP with negative dispersion support resonances, whereas these will not occur for small amplitude or linear waves. On the other hand, KP with positive dispersion supports linear wave, but not soliton resonances! Thus, depending on the physics of a problem, nonlinearity can either suppress or give rise to a resonance as we proceed from the linear wave to the soliton limit.

The field of two and many soliton interactions is seen to be somewhat elusive. Simple formulae work most of the time, but there can be regions in parameter space where they do not, such as indicated in Fig. 7.1. When this happens we must resort to numerics, experiments as usual being the ultimate test of our answers.

So far, nothing has been said about stability with respect to small perturbations. This is the main theme of the next Chapter (8), where we will

Table 7.2. *Properties of linear waves and solitons as described by the two KP equations. All columns but the last relate to Cartesian waves and solitons.*

Equation	3 linear wave resonances	3 soliton resonances	N soliton solution bounded for all angles of collision	Stability of waves and solitons	Cylindrically shaped solitons exist
KP, negative dispersion	NO	YES	NO	YES	NO
KP, positive dispersion	YES	NO	YES	NO	YES

see that KP with negative dispersion yields stable waves and solitons, whereas with positive dispersion these entities are unstable. We summarize our findings (including a few yet to come) in Table 7.2.

Thus the two KP variants are certainly no Tweedledum and Tweedledee! On the contrary, this simple pair of equations is not only physically important, but also furnishes a complementary illustration of various types of wave and soliton behaviour. The tremendous difference of the physics described by the two variants derives from the difference in sign in the seemingly modest u_{yy} term only. We will return to this theme in Chapter 8.

7.10 Integrable equations in two space dimensions as treated by the Zakharov–Shabat method

As we saw in Section 7.4, a privileged group of partial differential equations (PDEs) exists. This is the family of completely solvable equations. These equations are believed to constitute a subset of measure zero in the set of all PDEs. However, as far as general interest is concerned, the situation is quite the opposite, most of the recent publications in fact treating only solvable PDEs.

There is still another method of dealing with solvable PDEs not so far mentioned in this book. It is based on finding a pair of linear operators associated with a given PDE, the Lax pair, usually denoted by L and A. However, finding this pair given a PDE is in general not an easy task. It is technically much simpler to go about things the other way round, postulating L and A and then finding what PDE they correspond to (and this equation will in principle be solvable). This is the procedure we propose to follow. We will then show how to generalize to two space dimensions.

It is a general trait of all solvable PDEs considered here (or perhaps of our limitations) that we are not up to solving them directly, but must always operate on some other entities that are associated with a given equation. Thus, as we saw in Section 7.4, in the ISM we pass on to the space of spectral parameters. In Hirota's method (Section 7.5) we use bilinear variables as an intermediate step. Thus it should come as no great surprise that we will now work with a pair of operators (L and A) rather than the PDE they represent. This will once more lead us to substitute non-locality for nonlinearity, as we did in Section 7.4. The reader desirous of pursuing these unifying considerations is referred to the article by Newell (1985).

7.10.1 Lax pairs and the PDEs they represent

Take the operator

$$L=\frac{\partial^2}{\partial x^2}+u(x,t) \tag{7.10.1}$$

with u vanishing at infinity but otherwise arbitrary, and look at the eigenproblem

$$L\psi+\lambda\psi=0 \tag{7.10.2}$$

treating t as a parameter. Thus our point of departure is the same as for the ISM (Section 7.4). The reader will no doubt notice further similarities in what follows. As time flows, $u(x,t)$ will change its form and so will L. This would seem to imply a time variance in ψ and λ. This time dependence should be tied to that of u. We will now find how this comes about.

Suppose the temporal evolution of ψ is given by

$$\psi_t=-A\psi \tag{7.10.3}$$

where A is some differential operator of the form

$$A=a_0\frac{\partial^n}{\partial x^n}+a_1\frac{\partial^{n-1}}{\partial x^{n-1}}+\ldots\ a_n. \tag{7.10.4}$$

where a_0 is constant and the other n coefficients a_i are functions of x and t. Differentiating (7.10.2) by time we obtain

$$L_t\psi+L\psi_t=-\lambda_t\psi-\lambda\psi_t. \tag{7.10.5}$$

However, in view of (7.10.3) we have

$$L\psi_t=-LA\psi \tag{7.10.6}$$

and, in view of (7.10.2) and (7.10.3)

$$\lambda\psi_t=AL\psi. \tag{7.10.7}$$

Thus when (7.10.2) is used, we obtain

$$[L, A] = LA - AL = L_t + \lambda_t. \tag{7.10.8}$$

We will now concentrate on time independent λ (or see the proof of Section 7.4), obtaining

$$[L, A] = L_t. \tag{7.10.9}$$

This will give us A. The pair of operators A, L are known as a Lax pair. In view of (7.10.4) this operator will be given by $n+1$ equations, as that is the number of coefficients in (7.10.4), whereas (7.10.9) is equivalent to $n+2$ equations. The remaining equation will give us a condition on $u(x, t)$. As is often the case, this is best seen by looking at an example. Look for A in the form

$$A = \frac{\partial^3}{\partial x^3} + a_1 \frac{\partial^2}{\partial x^2} + a_2 \frac{\partial}{\partial x} + a_3, \tag{7.10.10}$$

in particular

$$A = \frac{\partial^3}{\partial x^3} + \tfrac{3}{2} u \frac{\partial}{\partial x} + \tfrac{3}{4} \frac{\partial u}{\partial x}, \tag{7.10.11}$$

in which case (7.10.9) leads to the condition

$$u_t + \tfrac{1}{4}(6uu_x + u_{xxx}) = 0 \tag{7.10.12}$$

(Exercise 5). This is the Korteweg–de Vries equation.

7.10.2 Extension to x, y, t

A given Lax pair will generate an equation such as (7.10.12) in x, t space. One might wonder what solvable equation in x, y, t will generalize this equation naturally. The y dimension will, in the spirit of the above, be introduced on the level of the operators, therefore ensuring integrability of the new equation with

$$L \to L + \frac{\partial}{\partial y} = M. \tag{7.10.13}$$

Equation (7.10.9) becomes

$$\frac{\partial L}{\partial t} - \frac{\partial A}{\partial y} = [L, A]. \tag{7.10.14}$$

Again we compare coefficients multiplying the same powers of ∂_x. This will generate the new form of A and the PDE just as above. In our example we obtain

$$L=\frac{\partial^2}{\partial x^2}+u(x,y,t) \tag{7.10.15}$$

$$A=\frac{\partial^3}{\partial x^3}+\tfrac{3}{4}\left(2u\frac{\partial}{\partial x}+u_x\right)+w(x,y,t) \tag{7.10.16}$$

$$\tfrac{3}{4}u_y=-w_x \tag{7.10.17}$$

and, as an additional condition,

$$\frac{\partial}{\partial x}(u_t+\tfrac{3}{2}uu_x+\tfrac{1}{4}u_{xxx})+\tfrac{3}{4}\frac{\partial^2}{\partial y^2}\mathrm{u}=0 \tag{7.10.18}$$

the Kadomstev–Petviashvili equation once again!

7.10.3 How to proceed from the Lax pair to the general solution

Suppose we consider a solution to a given nonlinear differential equation (known to be solvable) that tends to a constant as $|x|\to\infty$. We then have

$$L(u)\underset{|x|\to\infty}{\to} L_0=\frac{\partial^2}{\partial x^2} \tag{7.10.19}$$

$$A\underset{|x|\to\infty}{\longrightarrow} A_0 = a_0\frac{\partial^{2n}}{\partial x^n}+a_1\frac{\partial^{n-1}}{\partial x^{n-1}}+\ldots a_n \tag{7.10.20}$$

$$M\underset{|x|\to\infty}{\longrightarrow} M_0=\frac{\partial}{\partial y}+L_0 \tag{7.10.21}$$

where the coefficients are now constants.

We will now bring a new operator in from the blue. It is known as the Fredholm operator F

$$F\psi=\int_{-\infty}^{\infty}\hat{F}(x,x')\psi(x')\mathrm{d}x', \tag{7.10.22}$$

the kernel of which, $\hat{F}(x,x')$, is assumed to be well-behaved. This operator can be factorized as follows

$$1+F=(1+K^+)^{-1}(1+K^-), \tag{7.10.23}$$

where K^+ and K^- are the Volterra operators

$$K^+\psi=\int_x^{\infty}\hat{K}^+(x,x')\psi(x')\mathrm{d}x' \tag{7.10.24}$$

$$K^-\psi=\int_{-\infty}^{x}\hat{K}^-(x,x')\psi(x')\mathrm{d}x' \tag{7.10.25}$$

(note the difference in limits of integration in F, K^+ and K^-). Suppose now that F has been chosen to commute with M_o:

$$[F, M_o]=0. \tag{7.10.26}$$

It is a fact of life, not to be proven here, that when (7.10.26) holds, the operator

$$M=(1+K^{\pm})M_o(1+K^{\pm})^{-1} \tag{7.10.27}$$

is purely differential, just as M_o was (however, the coefficients are now functions of x). Similarly, if

$$\left[F, \frac{\partial}{\partial t}+A_o\right]=0, \tag{7.10.28}$$

then

$$A=(1+K^{\pm})A_o(1+K^{\pm})^{-1} \tag{7.10.29}$$

is also a purely differential operator (the integral part is zero).

The operators M and A are a new Lax pair for the initial equation with a new $u(x,y,t)$. The explicit forms of M, A and the new equation for u are found. The process of finding the operators M and A with *variable* coefficients from the asymptotic operators M_o and A_o with *constant* coefficients is generally known as operator dressing and is due to Zakharov and Shabat (1974).

As is often the case an example is somewhat easier to follow than an outline of the general method.

7.10.4 An example: the Kadomtsev–Petviashvili equation

Take the operators L and A given by (7.10.15) and (7.10.16)

$$L=\frac{\partial^2}{\partial x^2}+u(x,y,t) \tag{7.10.30}$$

$$A=\frac{\partial^3}{\partial x^3}+\tfrac{3}{4}\left(2u\frac{\partial}{\partial x}+u_x\right)+w(x,y,t) \tag{7.10.31}$$

and the operators L_o and A_o corresponding to $u=0$ and $w=0$:

$$L_o=\frac{\partial^2}{\partial x^2},\ A_o=\frac{\partial^3}{\partial x^3}. \tag{7.10.32}$$

We must now find a Fredholm operator F that commutes with the operators

$$M_o=L_o+\frac{\partial}{\partial y} \text{ and } A_o+\frac{\partial}{\partial t}.$$

If we denote $\hat{F}(x,x,y,t)$ by $\hat{F}(x,x')$ and $\psi(x,y,t)$ by $\psi(x)$ for short, the commutability condition gives

$$\left[L_o+\frac{\partial}{\partial y}, F\right]\psi=\left(\frac{\partial^2}{\partial x^2}+\frac{\partial}{\partial y}\right)\int_{-\infty}^{\infty}\hat{F}(x,x')\psi(x')\mathrm{d}x'-\int_{-\infty}^{\infty}\hat{F}(x,x')$$

$$\left(\frac{\partial^2}{\partial x'^2}+\frac{\partial}{\partial y}\right)\psi(x')\mathrm{d}x'=\int_{-\infty}^{\infty}\left[\frac{\partial^2\hat{F}(x,x')}{\partial x^2}\right. \tag{7.10.33}$$

$$\left.-\frac{\partial^2\hat{F}(x,x')}{\partial x'^2}+\frac{\partial\hat{F}(x,x')}{\partial y}\right]\psi(x')\mathrm{d}x'=0.$$

This will be satisfied if

$$\frac{\partial\hat{F}}{\partial y}+\frac{\partial^2\hat{F}}{\partial x^2}-\frac{\partial^2\hat{F}}{\partial x'^2}=0. \tag{7.10.34}$$

Similarly we find that F commutes with $A_o+\dfrac{\partial}{\partial t}$ if

$$\frac{\partial\hat{F}}{\partial t}+\frac{\partial^3\hat{F}}{\partial x^3}+\frac{\partial^3\hat{F}}{\partial x'^3}=0. \tag{7.10.35}$$

We now take the simplest possible form of F that comes to mind

$$\hat{F}(x,x',y,t)=\hat{F}_o\mathrm{e}^{-\kappa x}\mathrm{e}^{-\sigma x'}\mathrm{e}^{Ry}\mathrm{e}^{St}. \tag{7.10.36}$$

Equations (7.10.34) and (7.10.35) give us R and S in terms of κ and σ. Finally

$$\hat{F}(x,x',y,t)=\hat{F}_o\,\mathrm{e}^{[(\sigma^2-\kappa^2)y+(\sigma^3+\kappa^3)t]}\mathrm{e}^{-\kappa x-\sigma x'}. \tag{7.10.37}$$

The next step is to find $\hat{K}^+$ once $\hat{F}$ is given. Multiplying (7.10.23) on the left by $1+K^+$ we obtain, for $x'>x$, in terms of the kernels $\hat{F}$ and $\hat{K}^+$,

$$\hat{F}(x,x')+\hat{K}^+(x,x')+\int_x^{\infty}\hat{K}(x,x'')\hat{F}(x'',x')\mathrm{d}x''=0, \tag{7.10.38}$$

the by now familiar Gelfand–Levitan–Marchenko equation.

When $\hat{F}(x,x')$ is separable, $\hat{K}^+(x,x')$ is particularly easy to find. In our example

$$\hat{K}^+(x,x')=-\frac{\hat{F}_o\mathrm{e}^{-\kappa x-\sigma x'}\mathrm{e}^{(\sigma^2-\kappa^2)y+(\sigma^3+x^3)t}}{1+\dfrac{\hat{F}_o}{\sigma+\kappa}\mathrm{e}^{-(\sigma+\kappa)x}\mathrm{e}^{(\sigma^2-K^2)y+(\sigma^3+K^3)t}}. \tag{7.10.39}$$

We must now find the explicit form of L given K^+ and L_o from (7.10.27)

$$M=(1+K^+)M_o(1+K^+)^{-1}, \tag{7.10.40}$$

which is equivalent to

$$(1+K^+)M_o=M(1+K^+). \tag{7.10.41}$$

As M is $\frac{\partial}{\partial y}+\frac{\partial^2}{\partial x^2}+u$, this can be written

$$\frac{\partial}{\partial y}+\frac{\partial^2}{\partial x^2}+K^+\frac{\partial}{\partial y}+K^+\frac{\partial^2}{\partial x^2}$$

$$=\frac{\partial}{\partial y}+\frac{\partial^2}{\partial x^2}+u+\frac{\partial}{\partial y}K^+ +\frac{\partial^2}{\partial x^2}K^+ +uK^+. \tag{7.10.42}$$

If we now use (7.10.24) and the identity

$$\frac{\partial}{\partial x}\int_x^\infty \hat{K}^+(x,x')\psi(x')\mathrm{d}x'$$

$$=-\hat{K}^+(x,x')\psi(x)+\int_x^\infty \frac{\partial \hat{K}(x,x')}{\partial x}\psi(x')\mathrm{d}x \tag{7.10.43}$$

in (7.10.42), we obtain

$$u=2\frac{\mathrm{d}}{\mathrm{d}x}\hat{K}^+(x,x'). \tag{7.10.44}$$

When we substitute our form of $\hat{K}^+$, Equation (7.10.39) in (7.10.44), we obtain

$$u(x,y,t)=\tfrac{1}{2}(\sigma+\kappa)^2\mathrm{sech}^2$$

$$\left((\sigma+\kappa)\left[x-(\sigma-\kappa)y+(\sigma^2-\kappa\sigma+\kappa^2)t-\frac{1}{\sigma+\kappa}\ln\frac{\hat{F}_o}{\sigma+\kappa}\right]\right), \tag{7.10.45}$$

the one soliton solution to (7.10.18), described in detail in Section 7.9. The different notation is used so as to avoid confusion due to different forms of KP used; (7.9.2) and (7.10.18).

Other forms of $\hat{F}(x,x')$ that satisfy (7.10.34) and (7.10.35) lead to other solutions of (7.10.18). As these equations are linear, a simple generalization of (7.10.36) will yield:

$$\hat{F}(x,x')=\sum_{n=1}^{N}\hat{F}_{on}\mathrm{e}^{-\kappa_n x-\sigma_n x'}\mathrm{e}^{[(\sigma_n-\kappa_n)y+(\sigma_n+\kappa_n)t]} \tag{7.10.46}$$

leading to an N soliton solution generalizing (7.10.45). Solutions can thus in principle be added just as for linear PDEs. However, one must know beforehand just what space to perform the exercise in.

The method presented in Subsection 7.10.2 gives us equations in x, y, t that extend known solvable equations in x, t. The so-called cylindrical KdV equation will play an important role in Chapter 9, where it will be shown to be solvable. It is

$$u_t + uu_x + \tfrac{1}{2}u_{xxx} + \tfrac{1}{2}\frac{u}{t} = 0. \tag{7.10.47}$$

(the coefficients appear naturally in the plasma context of Chapter 9). It would be interesting to see just what solvable equation in x, y, t corresponds to this cylindrical KdV equation by applying the above method. This was done by Dryuma (1983). The Lax pair for (7.10.47) is

$$L = t\frac{\partial^2}{\partial x^2} + \frac{tu}{3} - \frac{x}{6} \tag{7.10.48}$$

$$A = 2\frac{\partial^3}{\partial x^3} + u\frac{\partial}{\partial x} - \tfrac{1}{2}u_x$$

where $u = u(x, t)$. If we now extend u to be a function of x, y, t and apply (7.10.14), we obtain A and the equation we wanted (it is useful to multiply $\partial/\partial y$ by a constant α):

$$A = 2\frac{\partial^3}{\partial x^3} + u\frac{\partial}{\partial x} + \tfrac{1}{2}\left(u_x - \frac{\alpha}{t}\int u_y \mathrm{d}x\right) \tag{7.10.49}$$

$$\frac{\partial}{\partial x}\left(u_t + uu_x + \tfrac{1}{2}u_{xxx} + \frac{u}{2t}\right) + \frac{3\alpha^2}{2t^2}u_{yy} = 0. \tag{7.10.50}$$

Equation (7.10.50) is known as Johnson's equation and will appear in a physical context in Chapter 9 (water surface waves).

We have merely presented the main ideas of the Zakharov–Shabat method by treating an example. Those interested in a more thorough acquaintance with this powerful tool should see Zakharov and Shabat (1974), Zakharov *et al.* (1980), Zakharov in Bullough and Caudry (1980), Kuznetsov *et al.* (1984), and Novikov *et al.* (1984).

7.11 Summary

The subject matter of this Chapter reflects an important step in pushing back the boundaries of our ignorance about PDEs. Until recently, only linear equations could be dealt with comprehensively. This influenced the way people thought about the physical world. During the last twenty years, various exact methods for dealing with a restricted class of nonlinear PDEs have been developed. In particular, durable hump-like solutions called solitons have been investigated thoroughly. This in turn has led to interesting experiments.

Just as the mathematician's thorough understanding of linear PDEs cast a shadow on physics up to as recently as twenty years ago, we now observe the ISM strongly influencing present-day physics, and solvable equations receiving more than their share of attention. Indeed, one might be tempted to treat *linear and solvable nonlinear PDEs* as one joint class, as solutions can be added in some space in both cases.

There are many instances of two similar PDEs being known, the more physically important one currently receiving less attention just because the ISM cannot cope with it. However, as already mentioned, solvable PDEs seem to be a set of measure zero, though some physicists believe them to be much more important in the physical world than is immediately obvious (Newell (1985)). Just how limited this set is, was illustrated by the fact that a solvable equation in x, t generates just one new such equation in x, y, t in a natural way (Section 7.10); On the other hand, *several* solvable equations appeared in Figs. 5.9 and 5.10, where various models for fluids and plasma situations were considered.

The present authors feel that physicists should take a wider view and not loose sight of important non-solvable PDEs. Thus the next two Chapters will treat instances of both kinds of PDE, Chapter 8 for stability of solutions (wave and soliton) with respect to small perturbations, Chapter 9 for solitons with inherent geometry other than flat (cylindrical and spherical). We will try to treat the physics of these problems by discriminating as little as possible between the two fundamental kinds of PDE. Nevertheless, we hope that the reader has enjoyed his or her conducted tour of the various methods of dealing with solvable soliton equations. Our emphasis has been on the practical implications, some of the more mathematically inclined books on the subject having been mentioned at each stage.

Exercises on Chapter 7

Exercise 1

Show that

$$(-1)^n \mathrm{Tr}\left(\frac{\partial}{\partial x}\mathbf{B}\cdot\mathbf{B}^{n-1}\right) = (-)^{n-1}\sum_1\sum_2\cdots\sum_n \frac{\phi_1\psi_1\ldots\phi_n\psi_n}{(p_1+q_2)(p_2+q_3)\ldots(p_{n-1}+q_n)}$$

$$= \frac{(-)^{n-1}}{n}\sum_1\sum_2\cdots\sum_n \frac{(p_1+\ldots p_n+q_1+\ldots q_n)\phi_1\psi_1\ldots\phi_n\psi_n}{(p_1+q_2)\ldots(p_n+q_1)}$$

and that this quantity denoted by $\phi^{(n)}$ satisfies

$$\mathscr{L}\phi^{(n)} = -6\sum_{l=1}^{n-1} \phi_x^{(l)}\phi_x^{(n-l)}.$$

Here the notation of Section 7.9 is used. Hint: Use the identity

$$(p_1+\ldots p_n+q_1+\ldots q_n)^4 - 4(p_1+\ldots p_n+q_1+\ldots q_n)(p_1^3+\ldots p_n^3+q_1^3+\ldots q_n^3)$$

$$+3(p_1^2+\ldots p_n^2-q_1^2\ldots-q_n^2)^2 = 6\sum_{j=1}^{n}\sum_{j=1}^{n}(p_j+\ldots p_{j+k-1}+q_j+\ldots q_{j+k-1})$$

$$(p_{j+k-1}+q_{i+k})(p_{j+k}+\ldots p_{j+n-1}+q_{j+k}+\ldots q_{j+n-1})(p_{j+n-1}+q_i).$$

Exercise 2
By introducing

$$\tilde{p}_l = (p_l + iq_l)/(1+i)$$
$$\tilde{q}_l = (q_l + ip_l)/(1+i)$$

in (7.9.8)–(7.9.10) or otherwise, derive (7.9.19), (7.9.21) and prove (7.9.18) for KP with positive dispersion.

Show that u found from (7.9.29) and (7.9.35) is, for both KP equations,

$$u = 2[(k_{x1}^2 - k_{x2}^2)^2$$

$$\frac{\mp(k_{x1}^2+k_{x2}^2)(\kappa_1-\kappa_2)^2 + (k_{x1}^2\cosh[2_{\eta 2}-\Delta] + k_{x2}^2\cosh[2_{\eta 1}-\Delta])(n\cdot d)^{\frac{1}{2}}]}{[\sqrt{n}\cosh(\eta_1+\eta_2-\Delta)+d\cosh(\eta_1-\eta_2)]^2}$$

$$A = n/d > 0$$

where $\kappa_i = k_{yi}/2k_{xi}$. Repeat the analysis of Table 7.1 and show that it covers all non-zero asymptotic behaviour.

Exercise 3
By shifting x and y, find a new coordinate system in which, for $A\to 0$ in (7.9.37)

$$d \to e^{2\eta_1} + e^{2\eta_2} + e^{2(\eta_1+\eta_2)},$$

and show how this leads to a solution with three arms and $u^{(1-2)}$ on the left. What is the angle ψ between solitons 1 and 2?

Exercise 4
Repeat the calculation of the critical angle for interacting, equal amplitude, negative dispersion KP solitons, ψ_{c2}, but such that the bisector of the collision angle forms an angle β with the y axis. Show that (7.9.51) is generalized by

$$\tan(\psi_{c2}/2) = 2\sqrt{(a)}\cos^2\beta.$$

Is this physically feasible? Explain.

Exercise 5
Derive (7.10.11) by solving (7.10.9) and choosing the constants of integration properly. Derive (7.10.12). Perform the same exercise to obtain (7.10.49) and (7.10.50).

8

Evolution and stability of initially one-dimensional waves and solitons

8.1 A brief historical survey of large amplitude nonlinear wave studies

This Chapter singles out some nonlinear wave phenomena for detailed treatment. Although we believe these phenomena to be important, we also feel we owe the reader a broader view of various research areas that have developed over the years. Of course, the survey of this Section is still somewhat selective, but it does offer the opportunity of reaching much of the generic work on the subject.

During the 1960s the limitations of treating nonlinear waves by expanding in powers of their amplitude A became embarrassingly evident. Fluid dynamists were the first to realize the need for new methods, and they were closely followed by those working in plasma and condensed matter physics. Although, as we saw in Chapter 5, quite a lot of physics can be introduced through the back door of A^2 theory, some large amplitude effects cannot. A few examples, some of which have been known to navigators and fluid dynamists for quite a long time as observed phenomena, are:

1. The formation of sharp crests on steady profile waves
2. The formation of ordinary (single hump) and envelope solitons on the water surface and their dynamics
3. Wave breaking. (One way this can happen is when a large amplitude wave propagates into a region of shallow water. This is familiar from Summer holidays on the beach.)
4. Nonlinear waves evolving into soliton trains.

The first three effects are shown schematically in Fig. 8.1. The fourth will be described in Section 8.3 (though see Figs. 1.2 and 8.9(b) for two ways this can come about). The common denominator of all four effects is the negative one of not being amenable to linear or weakly nonlinear theory. In this introductory Section we will take a look at some of the most important progress in treating these and related nonlinear phenomena over, say, the last Century and a half. For the sake of completeness, some ground covered in Chapter 4 will be revisited briefly. Some other themes in this survey were indicated in Chapter 5, still others will be developed in greater detail in the rest of this Chapter. (The reader is asked to forgive the inevitable

Fig. 8.1. Examples of nonlinear water wave phenomena: (*a*) Formation of 120° cusp on a steady profile surface wave. (*b*) Two examples of a soliton (isolated, shape-preserving hump above the water surface propagating with a constant velocity which in principle is proportional to the amplitude). Left hand version for length of soliton $\gg$ height above bottom, right hand version for length $\sim$ height. (*c*) Envelope soliton. The wave envelope behaves like a soliton (*b*). (*d*) Wave breaking. Some water elements are faster than others and will overtake them.

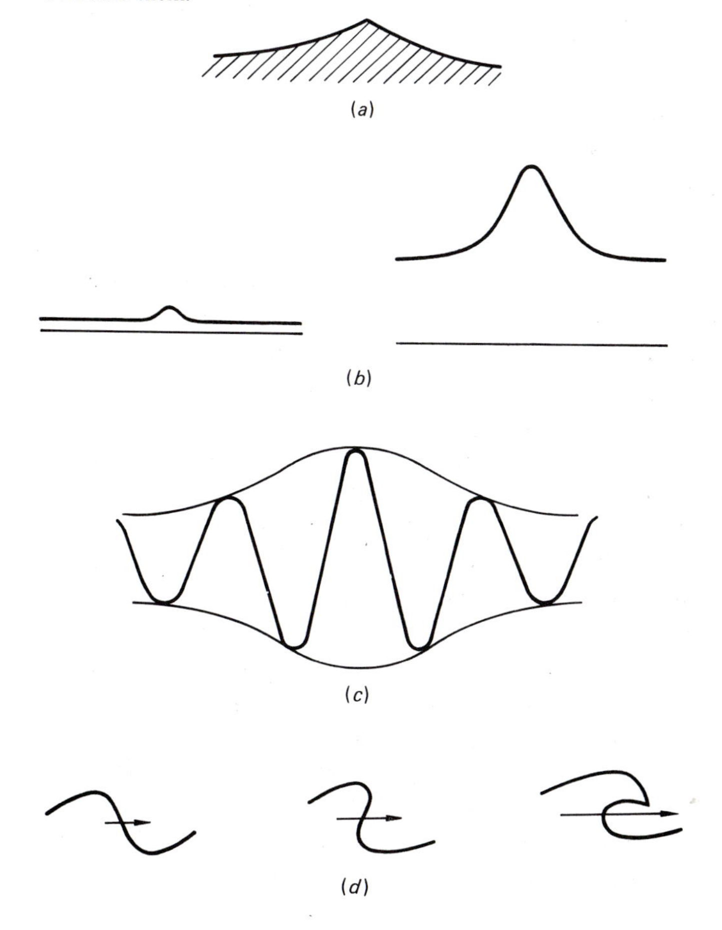

repetitions.) We will concentrate on fluid dynamics and also on plasma physics where methods used in hydrodynamics are often rediscovered, but sometimes also improved on.

Until fairly recently, *most* theoretical work on water surface waves was in fact based on an expansion in the wave amplitude, as in Chapter 5. Stokes initiated this method (1847). However, being a great man, he was by no means confined by it and also managed to prove an important result on the sharp crest mentioned above. Using ingenious complex plane arguments, he demonstrated that $2\pi/3$ is the *only* possible angle a stationary, cusped structure on the water surface can sustain. This, however, was a somewhat isolated result, as no calculations on waves of intermediate amplitudes seem to have been performed at the time (the cusp appears at $A = A_{MAX}$ for fixed wavelength). Half a century later, Michell found the profile of the cusped wave numerically (1893!), yielding $A_{max}/\lambda = 0.0706$, where A is defined as half the total height from crest to trough, λ is the wavelength, and infinite depth was assumed (finite water depth extensions are fairly recent, see Williams (1981), Bloor (1984) and papers quoted there).

8.1.1 Solitons

An even earlier result not based on small A expansion, concerned what we now call soliton propagation in canals. In August 1834, Scott-Russell, an outstanding engineer and scientist who somehow never succeeded in obtaining a university chair, happened upon an interesting wave phenomenon while investigating the Union Canal for its possible use by steamships (this was in present-day Edinburgh). He saw a boat, drawn along a narrow stretch of the canal by a pair of horses, suddenly stop. A single, hump-like mass of water detached itself from the boat and continued its course at eight to nine miles an hour. It was well over a foot high and thirty feet long (about 40 cm by 10 m). Fortunately for posterity, Scott-Russell was mounted on a horse and therefore in a position to comfortably follow this 'singular and beautiful phenomenon' for a mile or two. Without the undue haste characteristic of present day work, he reported his observations to the British Association in detail in 1844. Russell found that in principle the speed of a solitary wave increases with its amplitude. He initially called his wave the *Wave of Translation*, later the *Great Solitary Wave*, a term still in use, though usually shortened to 'solitary wave'. It is now often called a 'soliton' as already indicated in Chapter 1.

Scott-Russell once wrote that, that August day in 1834 was the happiest in his life. In hindsight, it certainly seems to have been the most significant, as witnessed by the avalanche of soliton papers in the current literature. Luck was also on his side that day. A recent attempt at repeating the experiment under similar conditions and at the same spot was a total failure (1984, Heriot-Watt University, Riccarton Campus, Edinburgh). An outline of Scott-Russell's career, together with a bibliography of his published work and some recent developments on soliton theory can be found in the book edited by Bullough and Caudry (1980). See also Emmerson's biography of Scott-Russell (1971).

How do Scott-Russell's findings tie up with water surface waves as found from the full equations (5.2.1)–(5.2.4), $T=0$, say? Existence of solitary wave solutions to these equations was first established for small amplitudes only; see Lavrientev (1943), (1960) and Friedrichs and Hyers (1954). Recently, however, existence has been established for amplitudes up to and including a maximum value when the water depth is finite, Amick and Toland (1981). This value is found to be roughly 0.83 times the depth (Williams (1981), Hunter and Vanden-Broeck (1983b)). Maximal velocity, however, is attained for a slightly smaller amplitude, 0.8 times the depth, see Fenton (1972) and Longuet-Higgins and Fenton (1974). The authors attribute this to the fact that the highest soliton lies beneath the slightly shorter soliton everywhere except in the immediate neighbourhood of the crest.

Shallow water waves of small amplitude were seen in Section 5.2 to be quite well described by an equation derived by two Dutchmen, Korteweg and de Vries. They assumed h/λ small, where h is the water depth (Chapter 5 gives a derivation and a critical analysis). We are thus dealing with long waves in shallow pools such that a wave 'feels' the bottom. We have now arrived at 1895, and again half a century separates the principal historical papers quoted, though both Boussinesq (1871) and Rayleigh (1876) worked on solitons in the intervening years (and the KdV equation was first formulated in de Vries' thesis a year earlier). Due to the simplifications of the Korteweg–de Vries model, one and the same differential equation in x, t space describes both the elevation of the water surface above the mean and the surface element velocity along the direction of propagation $v(x, t)$. This is indeed a drastic reduction from the full description of water-wave dynamics, involving complicated boundary conditions such as pressure balancing on the unknown water surface. As derived in Chapter 5, the KdV equation is essentially (leaving out the numerical coefficients)

$$\frac{\partial v}{\partial t}+v\frac{\partial v}{\partial x}+\frac{\partial^3 v}{\partial x^3}=0 \tag{8.1.1}$$

solved by a family of localized, simple humped entities;

$$v=3\eta\,\mathrm{sech}^2[\tfrac{1}{2}\sqrt{(\eta)}(x-\eta t)]. \tag{8.1.2}$$

This formula was in fact first suggested by Rayleigh (1876) long before KdV was formulated. Thus water surface solitons are faster if they are taller, slower if shorter and longer. Equation (8.1.2) is very successful in describing what Scott-Russell saw (Fig. 8.1(*b*), l.h.s.), but would cease to be for soliton height of the order of its width and the water depth (Fig. 8.1(*b*), r.h.s.). The Korteweg–deVries equation (8.1.1) embodies nonlinear (second term) and dispersive effects (third term). It has enjoyed an almost unparalleled career in classical physics. There is some justice in this success story. In spite of KdV's obvious simplicity, it describes nonlinear waves, solitons and their dynamics, including two soliton overtaking, and even the initial stages of wave breaking (as long as the third term can be neglected Fig. 1.2). However, the 120° cusp was lost in the derivation, though even this rather exotic feature could be restored by replacing the third term by a simple integral, the form of which is suggested by the dispersion properties of linear waves following from (5.1.11), $T=0$. A 110° cusp is obtained when this is done (Whitham (1974), p. 479). A feature more permanently lost was the possibility of describing a head-on soliton collision (as opposed to overtaking), as equation (8.1.1) is undirectional by its very nature.

As if the above mentioned assets of (8.1.1) were not sufficient, this equation was historically the first to be solved by the inverse scattering method, thus opening a whole new chapter in the saga of nonlinear partial differential equations, see Section 7.4 and Gardner *et al.* (1967).

Many model equations appearing in this book are extensions of (8.1.1) to include more physics and/or higher dimensional or different geometrical configurations. It will be interesting (and a bit bewildering) to see when these extensions deprive KdV of its solvability, and when not.

This is as good a place as any to give a few references for water surface solitons as well as solitons in other physical media. A plethora of contemporary books is available, many of them based on conference proceedings. Three from the seventies are exceptionally interesting: Kursunoglu (1977), Lonngren and Scott (1978), and parts 5 and 6 of volume **20** of *Physica Scripta* (1979).

As far as more recent volumes are concerned, three books published in 1980 can be recommended: Lamb (1980), Zakharov *et al.* (1980), and the much quoted here collection edited by Bullough and Caudry (1980). Eilenberger (1981), Drazin (1983), and Dodd *et al.* (1983), concentrate mainly, though not exclusively, on the inverse scattering method, the Dodd reference quoting from diverse branches of physics. Novikov *et al.* (1984) is entirely devoted to the ISM. Sym's book (1989) concentrates on the geometry of solitons, and Newell's (1985) on the role of the soliton concept in mathematical physics and on unifying ideas connecting various methods.

The book edited by Bishop and Schneider (1981) treats solitons in condensed matter physics, whereas Davydov's (1985) covers solitons in molecular systems. Rebbi and Soliani (1984) treat solitons and particles.

Some of the most recent developments were described in the transcript of a discussion organized by Atiyah *et al.* (1985). Another conference proceedings volume worth quoting is in Physica D **18**, nos. 1–3 (1986). Independent of conference proceedings, there are numerous single review articles, four of which we now list. Those of Scott *et al.* (1973), Makhankov (1980) and Miles (1980), (1981) have made some impact. Many interesting plasma soliton experiments are described by Lonngren (1983) and we will have occasion to quote from this reference again in Chapter 9. It would not be difficult to double or indeed triple the length of this list without lowering the level very much (for about a dozen more, see the Newell reference mentioned above).

8.1.2 Water waves are unstable

To finish the theme of small A expansion of periodic wave solutions, we should add that as late as 1967, when work was already being done on a systematic approach to fully nonlinear waves, Brooke-Benjamin and Feir (1967) obtained a fascinating result by amplitude expansion from the full equations: deep water, small A waves are unstable (Chapters 4 and 5, see also Zakharov (1966) and (1968)). Initially, they obtained instability for the deep water limit

$$h/\lambda \rightarrow \infty.$$

This is the opposite of the above mentioned KdV limit, for which all waves and solitons are stable (Section 8.3, see also Benjamin (1972)). Benjamin and Feir subsequently confirmed the existence of their instability in the National Physical Laboratory experimental pool at Feltham (Greater London), where it was in fact already known to the personnel. The result of this experiment is shown in Fig. 1.4(b). Benjamin then proceeded to calculate the critical value of kh, the wavenumber ($k = 2\pi/\lambda$) times the depth, above which small amplitude waves are unstable. It is 1.363, see Subsection 5.3.5 (and Benjamin (1967)). By then, however, the number of scientists no longer afraid of large amplitude waves had grown appreciably. The same issue of the Proceedings of the Royal Society that contained Benjamin's paper, edited by Lighthill (1967), includes a beautiful nonlinear paper also by Lighthill (1967). Among other ingenious results, it gives a simple nonlinear stability calculation. Lighthill wrote out the known averaged Lagrangian for small-amplitude, deep water waves, followed by the Lagrangian for the maximum-amplitude cusped wave found from Michell's calculation and, finally, the algebraically simplest model Lagrangian for

$$0 \leqslant A \leqslant A\text{max}$$

giving the previous two and the right values of the derivatives at the two A limits (this Lagrangian, when suitably rescaled, depends on just one dimensionless argument, ω^2/gk, which can be related to A). His result served as a basis for a general amplitude stability calculation. Lighthill found a critical value for the amplitude ($A/\lambda = 0.054$, 76% of maximum value) above which the Benjamin–Feir instability will not appear. Thus, all in all, wave amplitudes must be small enough and the water deep enough for the Benjamin–Feir instability. This rather cavalier approach of Lighthill's seems to have more appeal than many a pedantic and exhaustive calculation. Nevertheless, he subsequently performed a more exact analysis. The results do not seem to have been altered very much (Lighthill (1978)).

8.1.3 The geometrical optics limit

A full two years before the 'nonlinear wave' issue of the Proceedings of the Royal Society appeared, Lighthill's former research student, Whitham, formulated a theory of arbitrary-amplitude nonlinear waves, though limited to locally stationary forms (1965a). Due to the complicated boundary conditions that must be satisfied on a water surface, his theory cannot be immediately applied to arbitrary amplitude surface waves in hydrodynamics. To do this one must either assume small amplitude waves that are cosines in the lowest order A, or introduce a model that will somehow do away with the undesirable form of the boundary conditions. Examples of such models are KdV, Boussinesq, and the equation suggested by Davey and Stewartson for describing wave packet envelopes (1974). For a more complete list, see Section 5.4.

In plasmas, the physics of what happens inside the volume of the medium can often be more or less isolated and studied on its own, and then Whitham's approach is particularly suited. From now on, when discussing *plasmas* in this Chapter, we will always have volume effects in mind.

In Whitham's original theory (Whitham I) he introduced two assumptions:

1. The nonlinear wave is locally stationary and planar. Thus, the profile is locally a periodic function of one variable $\phi(\theta)$, where θ is the phase $\mathbf{k}\cdot\mathbf{x} - \omega t$. We take this function to be known. As we saw in Chapter 6, stationary plane waves of this kind are comparatively easy to find. In the linear wave limit, ϕ is sinusoidal.

Now we allow gradual changes in $\mathbf{k}, \omega, A$, assuming them to be noticeable only at distances much longer than k^{-1} and after times much exceeding ω^{-1}. This is known as the geometrical optics approximation. Although the parameters evolve, ϕ retains its general functional form in this approach.

2. The dynamics of the problem can be described in conservation form

$$\partial_t T_i + \nabla \cdot \mathbf{X}i = 0 \qquad i = 1, 2 \ldots n \tag{8.1.3}$$

(the original version envisaged only one space variable).

The next step is to average (8.1.3) over many wavelengths of ϕ to obtain

$$\partial_t \ll T_i \gg + \nabla \cdot \ll \mathbf{X}_i \gg = 0, \ll f \gg = \xi^{-1} \int_x^{x+\xi} f \mathrm{d}x, \ \xi \gg \lambda, \tag{8.1.4}$$

(Section 8.2). We further substitute averages over one wavelength $\langle f \rangle$ for $\ll f \gg$ and the errors involved in this will be discussed in Section 8.2. We thus obtain a set of equations governing the slow variations in $\mathbf{k}$, ω and A. This set of equations is the basis of the theory and in principle leads to a full description of the wave evolution in the geometrical optics approximation.

If equations (8.1.4) are hyperbolic, the nonlinear wave will be stable. Hyperbolic equations can lead to a set of surfaces on which given combinations of $\mathbf{k}, \omega, A$ are constant (Riemann invariants). These surfaces are known as characteristic surfaces and their velocities, not surprisingly, as characteristic velocities. More generally, if a set of variables r_n can be introduced such that

$$\frac{\mathrm{d}r_n}{\mathrm{d}t} + f_n(\mathbf{x}, t, \mathbf{r}) = 0 \text{ on } \frac{\mathrm{d}\mathbf{x}}{\mathrm{d}t} = \mathbf{C}_n(\mathbf{x}, t, \mathbf{r})$$

we say that r_n is a Riemann variable and $\mathbf{C}_n$ is still a characteristic velocity. Thus a Riemann *variable* becomes a Riemann *invariant* if $f_n = 0$. For example, for the small amplitude waves described by (5.3.6), $\mathrm{d}\omega/\mathrm{d}\mathbf{k}$ is the velocity of a double characteristic, k^2 is a Riemann invariant, and $|a|^2$ is a Riemann variable. The $f_n = 0$ case will play an important role in our subsequent analysis.

We will try in this Chapter to convince the reader that the characteristic velocities of nonlinear theory generalize group velocities of linear theory, though this view is not universally accepted (e.g. Exercise 8.1; Bhatnagar (1979)).

Characteristic surfaces (when they exist) coincide with the surfaces generated by an isotropic periodic point source. Thus, for example, they are circles on still water.

In the $A \to 0$ limit, characteristic velocities become the group velocities of linear theory $\partial\omega/\partial\mathbf{k}$ (there is no doubt about this). We have already seen in Chapter 5 that for finite A there are in general more characteristic velocities than branches of $\partial\omega/\partial\mathbf{k}$ in the linear limit. Thus, from the point of view of nonlinear wave theory, this limit is degenerate. Removal of this degeneracy leads to a splitting of upstream point signals in the presence of a nonlinear wave (Fig. 8.2).

It was a long time before this splitting was more than a theoretical effect. One of the first experimental confirmations came in the plasma physics context (Section 5.3 and Figs. 5.4 and 5.5). Generalization to more dimensions also seems to have taken some time (Infeld (1981a)).

Whitham (1965b) was quick to realize that his theory could be improved on. Rather than average n equations over a period, he suggested introducing a Lagrangian and averaging it over a wavelength (1965b). Of course this innovation involved proving that the following two procedures are equivalent:

Whitham I: dynamical equations in conservation form$\rightarrow$(averaging)$\rightarrow$equations for the dynamics of $A, \omega, \mathbf{k}$.

Whitham II: Lagrangian$\rightarrow$(averaging)$\rightarrow$averaged Lagrangian$\rightarrow$(Euler–Lagrange in $A, \omega, \mathbf{k}$ space)$\rightarrow$equations for dynamics of $A, \omega, \mathbf{k}$.

The best reference for this proof is Whitham (1974). Here we will simply take this to be the case.

Next in this line of thought came an elegant Hamiltonian formulation due to Hayes, this time for general three-dimensional problems (Section 5.5). Hayes applied his results to a three-dimensional analysis of water wave stability, assuming small amplitude (Fig. 5.6). Unfortunately, the Hamiltonian formulation as outlined in Section 5.5 seems to be of limited

Fig. 8.2. A periodic point signal at $x=0$ splits into two due to the presence of a nonlinear wave.

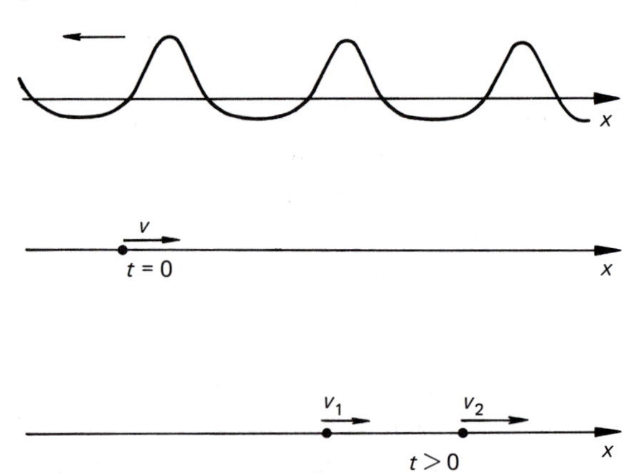

practical use whenever large amplitude waves are considered. Indeed, it is often easier to formulate a Hamiltonian theory than to apply it. We will come back to this theme with a specific example in Section 8.2.

Independent of the above trend, Rowlands initiated, and Infeld and Rowlands developed, a scheme based on the assumption that a nonlinear wave undergoes a long-wavelength perturbation (in addition to a λ periodic alteration). Thus, if the wave vector of the perturbation is $\mathbf{K}$, we assume $K \ll k$ (Fig. 8.3).

Physically speaking, we are treating the same situation as that considered by Whitham and Hayes. To complete the identity we should assume $\Omega/\omega \sim K/k$, where $\mathbf{K}$ and Ω are the wavevector and frequency of the perturbation, whereas $\mathbf{k}$ and ω describe the basic nonlinear wave structure (this convention is observed throughout this book).

We no longer need conservation laws, Lagrangians or Hamiltonians, and the class of tractable phenomena is somewhat extended. (Section 8.4 gives an example of a problem that can only be treated by this method.) The first step is to linearize the equations around a stationary nonlinear wave $\Phi(x)$, take perturbed quantities to be of the form $P(x)e^{i(\mathbf{K}\cdot\mathbf{x}-\Omega t)}$ where $P(x)$ is periodic with the same wavelength λ as Φ, and remove terms secular in x in each order in K/k. Integrations over $\lambda = 2\pi/k$ of the nonlinear wave are involved and equations of the general form

$$G(\mathbf{K}, \Omega, A) = 0, \tag{8.1.5}$$

are obtained as consistency conditions. This usually occurs in the second order (K^2) calculation. Equation (8.1.5) can be looked on as a nonlinear generalization of a linear dispersion relation. It contains the crucial information forthcoming from Whitham's calculation, as $\partial\Omega_i/\partial\mathbf{K}$ are the characteristic velocities when the original differential equations are hyperbolic. This approach thus naturally strengthens the interpretation of the characteristic velocities as nonlinear generalizations of group velocity. In Section 8.3 we will also see how to derive this interpretation from Whitham's approach.

Fig. 8.3. Wavelength of basic nonlinear structure (λ) and scale of perturbation (Λ) in a K expansion.

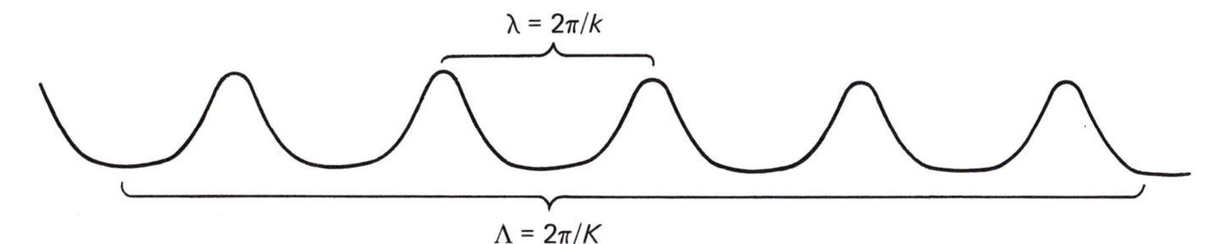

The K expansion method was given a mathematical basis by Rowlands in 1969 and applied to the problem of instabilities generated by two counterstreaming electron beams in a hydrogen plasma when a nonlinear electrostatic wave is present. This work was mentioned briefly in Section 6.3 (see also Infeld (1972) and Infeld and Rowlands (1975) and (1977)). The method was subsequently generalized to more space dimensions and a pictorial representation of $\Omega/K(\psi)$, where ψ is the angle spanned by $\mathbf{k}$ and $\mathbf{K}$, was introduced (Infeld, Rowlands and Hen (1978), Infeld and Rowlands (1979a), Infeld and Rowlands (1980), Infeld (1981b)). This pictorial representation will be presented in Sections 8.3–8.5. It generalizes the CMA diagrams of Section 2.3 to nonlinear wave theory.

The method has recently been extended to difference equations (Broomhead and Rowlands (1981)), and extension to differential-integral equations is at least possible in principle. All in all, this approach would appear to be the most versatile so far. We will have more to say about this versatility later on, once detailed examples of all four methods have been offered.

To summarize, we see that four methods for treating the dynamics (and hence stability) of arbitrary amplitude nonlinear plane waves have been developed, though variants and hybrids can also be found in the literature (Luke (1966), Dougherty (1970), Galloway and Kim (1971), Censor (1976)). We propose three lists (only surface space dimensions are counted in the hydrodynamic context; three-dimensional plasma problems can often be reduced to two dimensions locally);

1. *Chronology*

	Method	One dim.	Two dim.	Original context
(i)	Averaging conservation laws (Whitham I)	1965	None known to us, presumably due to loss of interest in view of (ii)	hydrodynamics plasma physics*
(ii)	Averaging Lagrangian (Whitham II)	1965	1974 (general) 1981 (example)	hydrodynamics plasma physics & hydrogen
(iii)	K/k expansion (Rowlands and Infeld)	1969	1978	plasma physics

* The plasma physics problem, that of nonlinear hydromagnetic waves propagating across a uniform magnetic field, was only *formulated* in Whitham (1965a). For the solution see Ziemkiewicz et al. (1981) where method (iii) was used.

(iv)	Averaged Hamiltonian (Hayes)	1973	1973 (small amp.)	hydrodynamics

2. *Beauty*: (iv), (ii), (i), (iii)
3. *Practicality*, predictably obtained by reading the second list backwards!

8.1.4 Recent results

The late 1970s and early 1980s saw quite a lot of progress on numerical and experimental work on nonlinear, large amplitude wave dynamics. This work was of course not limited by the above approximation. Some of it was briefly mentioned in Section 4.6.

Several computer calculations have been performed on large amplitude gravity waves on a deep water surface, telling us more about their shape, stability, and wave breaking than was previously known. (Schwartz (1974), Longuet-Higgins and Cockelet (1976), (1978), Longuet-Higgins (1978a, b)). Lighthill's result on the amplitude limitation of the Benjamin–Feir instability was confirmed. A second, larger growth-rate, instability was discovered for steep waves. In contradistinction to the Benjamin–Feir instability, it only appears when fully three-dimensional effects are taken into account unless the amplitude is at least 93% of its maximum value ($A/\lambda > 0.065$), when this second instability can also be two-dimensional (one surface dimension, along the direction of propagation, plus depth). Its wavelength is twice that of the basic wave as opposed to being many times longer for Benjamin–Feir. This new instability (Section 4.6), is reviewed by Saffman and Yuen, who also played an important role in researching it (1985). Some generalizations to water of finite depth can be found in McLean (1982), also reviewed briefly in Chapter 4. Interaction of the two instabilities (Benjamin–Feir and three-dimensional) has been investigated experimentally (Su and Green (1984)). Finally, Longuet-Higgins, whose work on gravity waves tends to concentrate on one surface dimensional waves, was able to develop an analytical description for overturning waves in this symmetry by using complex variable techniques (1980), (1981), (1983).

All that has been said so far about water waves assumed they were long enough to justify a neglect of surface tension ($\lambda > 10$ cm).* For gravity waves that are short enough for surface tension to be important, known as

* The effects of gravity and surface tension are equal for $\lambda = 1.7$ cm, for which value the phase and group velocities are equal and are both 23 cm/s. Gravity tries to achieve $V_{ph} > V_g$, surface tension strives for $V_g > V_{ph}$.

gravity-capillary waves, see Section 5.3, Kawahara (1975a), Chen and Saffman (1979), (1980), (1985), Vanden-Broeck (1980), and Hogan (1985).

The dynamics of very small-fry water surface waves ($\lambda < 0.5\,\mathrm{cm}$) are dominated by surface tension and for them *gravity* can be neglected in the restoring force. Faraday experimented on these waves and some of his experiments, performed in 1831, have been repeated and illustrated by Walker (1984). These waves were seen to form spoke-like patterns around vibrating solid objects (edge waves).

Travelling waves in which surface tension is the restoring force are known as capillary waves. They were encountered in Section 5.3, where one surface dimensional instabilities of these waves were found. In contradistinction to the Benjamin–Feir instability of gravity waves, the instabilities appear for all amplitudes (remember that the Benjamin–Feir instability is curbed by the Lighthill amplitude limitation). For the form of the finite amplitude capillary wave, see Crapper (1957). For the results of both an A^2 theoretical and arbitrary amplitude numerical stability analysis, see Section 5.3 and, for a fuller treatment, Ma (1984).

Returning to wave phenomena describable by the four methods on our list, we see that the averaged Lagrangian method was only recently applied to a two space dimensional problem. This was the Kadomtsev–Petviashvili equation (1970), derived in Section 5.2. It is, without the constants and for water depth well exceeding 0.47 cm:

$$\partial^2_{xt}\psi + (\partial_x\psi)(\partial^2_x\psi) + \partial^4_x\psi + \partial^2_y\psi = 0 \tag{8.1.6}$$

and in the water wave context ψ is the velocity potential:

$$\mathbf{v} = \nabla\psi$$

(Ablowitz and Segur (1979), Infeld and Rowlands (1979b), Santini (1981)). This equation generalizes (8.1.1) to a two-dimensional water surface context. It also describes ion acoustic waves in a plasma (Kako and Rowlands (1976)). The x axis has been chosen along the direction of propagation of the carrier wave, the very existence of which introduces anisotropy in the medium. The dynamics along x and y have received uneven coverage, as is evident from the very form of (8.1.6). This uneven coverage is an important idea in the construction of model equations in classical physics (Section 5.2: good physics is undemocratic!) Equation (8.1.6) has been seen to be useful when investigating the two-dimensional dynamics of locally planar waves and solitons. However, one should not forget how it was derived when applying it to fully two-dimensional situations such as solitons intersecting at an angle (for example, Exercise 4 of Chapter 7 illustrates a distortion introduced by (8.1.6)).

There is nothing wrong with looking for fully two-dimensional solutions to (8.1.6) as long as we take care when trying to fit them to phenomena found in the laboratory. This word of caution notwithstanding, we have seen in Chapter 7 that the seemingly modest coverage of y dynamics in (8.1.6) introduces some interesting physics (e.g. Section 7.9). We will return to this theme later on in the present Chapter.

Fig. 8.4 illustrates the characteristics of (8.1.6) in the presence of a stationary, planar nonlinear wave travelling from right to left. Along the x axis we just obtain a replica of Fig. 8.2. Two different characteristics meet at P and P'. This is the solution to the problem of generalizing Fig. 8.2 to the x, y plane in the simplified, Kadomtsev–Petviashvili model, Infeld (1981b). This problem will also receive more attention later on in this Chapter (Section 8.3).

8.1.5 What the remainder of Chapter 8 is about

In the remainder of this Chapter we will illustrate all four methods on our geometrical optics list by solving for an example in one space dimension (the nonlinear Klein–Gordon equation). Next, two of these methods will be illustrated by fully two space dimensional examples furnished by the Kadomtsev–Petviashvili and Zakharov–Kuznetsov models of Chapter 5. These calculations are instructive enough to merit detailed exposition (Section 8.3). The question of how much was lost when substituting models for full sets of equations is also of interest. We will compare the results obtained from the two models (KP and ZK) with those following from the fuller plasma physics descriptions that these equations were derived from. They are the ion acoustic wave equations without (KP) and with (KZ) an external magnetic field. Thus we will have a check on these models in one physical context. Finally, the nonlinear Schrödinger

Fig. 8.4. Shape of periodic signal emitted at $x=0, y=0$ after unit time in x, y space. Fig. 8.2 is the one-dimensional version of this drawing. From Infeld (1981b).

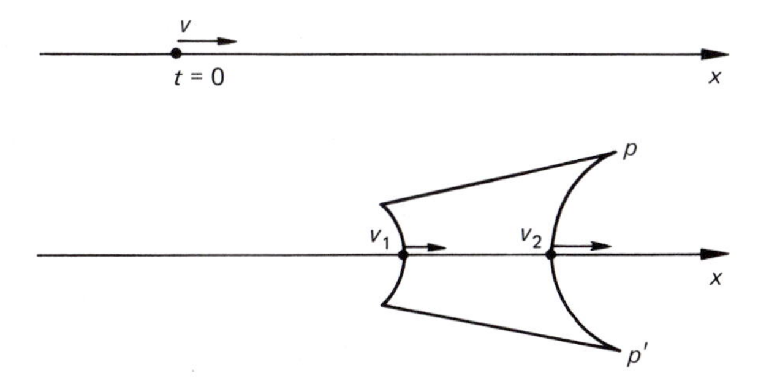

equation, seen in Chapter 5 to be important in small amplitude wave theory, will be investigated by one of the four methods.

General conclusions on large amplitude wave behaviour wind up the Chapter.

8.2 Four methods as illustrated by the nonlinear Klein-Gordon equation

As an illustration of the four methods outlined above, we propose to take the nonlinear Klein–Gordon equation

$$(\partial_t^2 - \partial_x^2)\phi + V'(\phi) = 0, \tag{8.2.1}$$

where V is a general nonlinear function of ϕ, but not its derivatives, and the prime denotes differentiation with respect to ϕ. This equation is the most natural nonlinear generalization of the wave equation and will serve to illustrate the salient points of the various approaches without involving excessively heavy calculations. It most probably first arose in a mathematical context with $V' = \exp(\phi)$ in the theory of surfaces of constant curvature, Liouville (1853). It is also a good physical equation in the sense that it appears in many fields of applications. For a cubic nonlinearity $V' = \phi^3 - \phi$, (8.2.1) can be used as a model field theory (Dashen *et al.* (1974)). A $\sin\phi$ term appears in the so-called sine-Gordon equation (rather unfairly to Klein). This latter version which was discussed in Sections 7.6 and 7.7 has been used to formulate a simple model of elementary particles (Perring and Skyrme (1962)), the motion of rigid pendula attached to a stretched wire (Scott (1970)), propagation of fluxons in Josephson junctions, dislocations in crystals, and other applications as described by Barone (1971), Gibbon *et al.* (1979), Bishop and Schneider (1981) and Davydov (1985). The sine-Gordon equation is solvable by inverse scattering (Chapter 7, Fadeev (1980) and references therein, Ichikawa (1985), Davydov (1985)).

More than one $\sin(\alpha_n\phi)$ term appears in some optical (propagation of resonant sharp line optical pulses through media with special symmetries), biological (one-dimensional organic polymer chains) and liquid helium (^{3}He below 2.6 mK) contexts. The equation is then known as the double, triple etc. sine-Gordon equation (Bullough and Caudry (1978), Rice (1981)). These more elaborate forms are no longer solvable by ISM (Eleonsky *et al.* 1985, Clarkson *et al.* 1986). See also Shockley (1987).

For numerical studies of all forms of (8.2.1) mentioned above, see the review by Eilbech (1981). Other possible nonlinearities and some solutions can be found in a survey by Bullough *et al.* in Bullough and Caudry (1980).

Here we will look at the dynamics of one stationary solution, allowing it to vary slowly in space and time. The exact form of the nonlinearity need not be specified until the very end of the day in three of our four methods (all but Hayes').

We now address ourselves to the problem of gradual modulating of periodic solutions.

8.2.1 Whitham I

The following conservation equations follow immediately from (8.2.1)

$$\partial_t(\tfrac{1}{2}\phi_t^2+\tfrac{1}{2}\phi_x^2+V)+\partial_x(-\phi_x\phi_t)=0 \tag{8.2.2}$$

$$\partial_t(-\phi_x\phi_t)+\partial_x(\tfrac{1}{2}\phi_t^2+\tfrac{1}{2}\phi_x^2-V)=0. \tag{8.2.3}$$

If we now look for a stationary solution to (8.2.1) depending on $X=x-Ut$ only, we obtain, upon integration,

$$\tfrac{1}{2}(U^2-1)\phi_X^2+V(\phi)=A. \tag{8.2.4}$$

By drawing phase diagrams of $\phi_X(\phi)$ as in Section 6.2, it is not difficult to see that periodic wave solutions exist for all $V(\phi)$ that have a minimum. Assume this to be the case and take $\phi(X, A, U)$ to be one of these waves. It will prove expedient to introduce (no need to further capitalize X):

$$2\pi W(U,A)=(U^2-1)\int\phi_x\mathrm{d}\phi=\sqrt{[2(U^2-1)]}\int\sqrt{[A-V(\phi)]}\mathrm{d}\phi$$
$$=\sqrt{[U^2-1]}G(A), \tag{8.2.5}$$

where the integral is taken over a wavelength λ of the nonlinear wave ϕ. This wavelength is given by

$$\lambda=\int_x^{x+\lambda}\mathrm{d}x=\int\mathrm{d}\phi/\frac{\mathrm{d}\phi}{\mathrm{d}x}=\sqrt{[\tfrac{1}{2}(U^2-1)]}\int\frac{\mathrm{d}\phi}{\sqrt{[A-V(\phi)]}}=2\pi\frac{\partial W}{\partial A} \tag{8.2.6}$$

We now derive (8.1.4) from (8.1.3). The first part

$$\ll\frac{\partial}{\partial t}T_i\gg=\frac{\partial}{\partial t}\ll T_i\gg,$$

is straightforward. The second follows once we use the definition of $\ll\ \gg$:

$$\ll\frac{\partial X_i}{\partial x}\gg=\frac{1}{\xi}\int_x^{x+\xi}\frac{\mathrm{d}X_i}{\mathrm{d}x'}\mathrm{d}x'=\frac{1}{\xi}[X_i(x+\xi)-X_i(x)]=\frac{\partial}{\partial x}\frac{1}{\xi}\int_x^{x+\xi}X_i(x')\mathrm{d}x'$$
$$=\frac{\partial}{\partial x}\ll X_i\gg.$$

Here ξ is fixed and contains a large number of wavelengths, not necessarily a whole number. Thus

$$\xi \gg \lambda.$$

We now treat X_i, in the integral, as a function of A and U, but not derivatives such as U_x, A_x. This introduces an error of the order ξ/L, where L is the scale of changes in U, assumed large:

$$L \gg \xi \gg \lambda.$$

We now treat X_i, in the integral, as a function of A and U, but not of

$$\langle f \rangle = \lambda^{-1} \int_x^{x+\lambda} f \mathrm{d}x.$$

Substitution of $\langle f \rangle$ for $\ll f \gg$ will involve new errors. That following from the fact that ξ may not be a multiple of λ is at most of order λ/ξ. The error introduced by omitting U_x, A_x in the average over one wavelength is of order λ/L.

The above reasoning shows that overall accuracy in the averaged quantities can, at the very best, be of the order λ/L, and indeed at this stage would seem likely to be worse than that. As the terms in

$$\frac{\partial}{\partial t}\langle T_i \rangle + \frac{\partial}{\partial x}\langle X_i \rangle = 0$$

themselves become of order λ/L upon differentiation, the very nature of the Whitham theory as formulated here precludes an $\mathrm{O}(\lambda/L)^2$ calculation. We will come back to this point in Section 8.2.

The next step is to evaluate the average values of all quantities over a period:

$$\langle \tfrac{1}{2}\phi_t^2 + \tfrac{1}{2}\phi_x^2 \rangle = \tfrac{1}{2}(U^2+1)\langle \phi_x^2 \rangle = \tfrac{1}{2}(U^2+1) \int_x^{x+\lambda} \phi_x^2 \mathrm{d}x/\lambda$$

$$= \tfrac{1}{2}(U^2+1)(U^2-1)^{-1} kW. \tag{8.2.7}$$

$$\langle -\phi_x \phi_t \rangle = kUW(U^2-1)^{-1}, \tag{8.2.8}$$

where

$$k = 2\pi/\lambda.$$

The first conservation equation, (8.2.2), takes the form

$$\partial_t[k(UW_U + AW_A - W)] + \partial_x[kU(UW_U + AW_A - W) - UA] = 0 \tag{8.2.9}$$

upon utilizing

$$W_U = UW(U^2-1)^{-1} \tag{8.2.10}$$

$$kW_A = 1. \tag{8.2.11}$$

Similarly, the second conservation law, (8.2.3) becomes

$$\partial_t(kW_U)+\partial_x(UkW_U-A)=0. \tag{8.2.12}$$

Now (8.2.9) can be expanded as

$$U[\partial_t(kW_U)+\partial_x(kUW_U-A)]+A[\partial_t(kW_A)+\partial_x(kUW_A-U)]$$
$$-W[\partial_t k+\partial_x(kU)]=0, \tag{8.2.13}$$

all other terms combining to give zero.

In view of (8.2.12) and $kW_A=1$, we finally obtain

$$k_t+(kU)_x=0 \tag{8.2.14}$$

(wave conservation). If we introduce

$$\omega = kU,$$

we can write this as

$$k_t+\omega_x=0, \tag{8.2.15}$$

and (5.3.47), obtained for weak amplitude waves, is seen to be exact:

$$k_t+\omega_k k_x+\omega_A A_x=0. \tag{8.2.16}$$

We can introduce $k=W_A^{-1}$ in (8.2.14) to obtain

$$\frac{\partial W_A}{\partial t}+U\frac{\partial W_A}{\partial x}-W_A\frac{\partial U}{\partial x}=0. \tag{8.2.17}$$

Simple manipulations on (8.2.12) yield

$$\frac{\partial W_U}{\partial t}+U\frac{\partial W_A}{\partial x}-W_A\frac{\partial A}{\partial x}=0. \tag{8.2.18}$$

If we now introduce

$$2\pi W=\surd(U^2-1)G(A),\ 2\pi W_A=\surd(U^2-1)G'$$

we can express our equations in a form that may, at first glance, seem less pleasing than (8.2.17) and (8.2.18), but will shortly lead to a characteristic form, thus rewarding our efforts.

$$G''A_t+UG''A_x+U(U^2-1)^{-1}G'U_t+G'(U^2-1)^{-1}U_x=0 \tag{8.2.19}$$

$$UG'A_t+G'A_x-G(U^2-1)^{-1}U_t-GU(U^2-1)^{-1}U_x=0. \tag{8.2.20}$$

Multiplying the first equation by $\sqrt{(-G/G'')}$ and adding it to the second gives a characteristic equation stating the constancy of R_+;

$$\frac{\partial R_+}{\partial t}+\frac{dX_+}{dt}\frac{\partial R_+}{\partial x}=0;\ R_+=\int_{U_0}^{U}(U^2-1)^{-1}dU-\int_{A_0}^{A}\sqrt{(-G''/G)}dA$$

along the characteristic C_+:

$$C_+:\frac{dX_+}{dt}=\frac{1+U\alpha}{U+\alpha};\ \alpha=\sqrt{(-GG''/G'^2)}.$$

Similarly, subtraction yields the constancy of R_-:

$$R_-=\int_{U_0}^{U}(U^2-1)dU+\int_{A_0}^{A}\sqrt{(-G''/G)}dA$$

along C_-:

$$C_-:\frac{dX_-}{dt}=\frac{1-U\alpha}{U-\alpha}.$$

The characteristic velocity splits, as we already saw it do in small amplitude theory, Section 5.3. One might expect one of the characteristic velocities to be the velocity at which energy is propagated. However, the reality is somewhat less straightforward. The velocity defined by the ratio of the energy flux to energy density, taken from (8.2.19), is

$$\frac{kU(UW_U+AW_A-W)-UA}{k(UW_U+AW_A-W)}=\frac{UG}{(U^2-1)AG'+G}. \tag{8.2.21}$$

This is definitely not one of the two characteristic velocities. However, the *increment* of energy, calculated on C_+, does flow with velocity X_- (and X_+ on C_-), as indicated in Section 5.5. Take

$$dU=(U^2-1)\sqrt{(-G''/G)}dA, \tag{8.2.22}$$

true on C_+, in the expression

$$\frac{d[kU(UW_U+AW_A-W)-uA]}{d[k(UW_U+AW_A-W)]}. \tag{8.2.23}$$

The velocity $(1-U\alpha)(U-\alpha)^{-1}$ is obtained. Once again, we see that the increment of energy propagates with the 'companion' characteristic velocity.

The velocity (8.2.21) does not seem to have a simple interpretation in fully nonlinear theory (see also Whitham (1965b), Lighthill (1965),

Bhatnagar (1979), Skorupski (1981)). Nevertheless see Exercise 1 for a not so simple interpretation.

We can now extend the conclusions of Section 5.3:
A small, $k_x \simeq A^{-1}A_x$. Group velocity $d\omega/dk$ is a double (degenerate) characteristic velocity, k constant on characteristics (Riemann invariant), A^2 in general not (Riemann variable). The velocity of energy flow is $d\omega/dk$.
A arbitrary, $k_x \simeq A_x$. Group velocity splits into two characteristic velocities, each with its own Riemann invariant. They are the velocities of energy increment flow when measured on the 'companion' characteristic (X_+ on C_- and X_- on C_+).

The splitting of characteristics will not occur if $G''=0$. Surprisingly enough, this singular behaviour is *not* an exclusive feature of the linear limit, $A \to 0$. One can easily check that for

$$V'(\phi) = \text{const}[(1+\phi)^{-\frac{1}{2}} - 1],$$

$$G'' = \frac{\partial \lambda}{\partial A} = 0$$

(Exercise 2). Thus, a fully nonlinear Klein–Gordon equation exists for which the characteristics will not split (Infeld and Rowlands (1979a)). Both characteristic velocities are equal to U^{-1}.

The fact that fully nonlinear field theories can, for suitably chosen Lagrangians, have single degenerate characteristics, just as linear theories do, is by no means widely appreciated. For example, this important fact was only realized for the nonlinear electrodynamic theory of Born and Infeld (1934) after thirty years had passed and much work had been done on other aspects of the theory (Boillat (1965) and (1970)). For interesting reviews of nonlinear electrodynamic theory, stressing this aspect, see Plebanski (1970) and Bialynicki-Birula (1983).

To return to the more general case of two characteristics, Figure 8.4 shows how the duality will lead to a separation of a modulation on a nonlinear wave into two single waves. Here, at $x=0$ we modulate a stationary nonlinear wave of amplitude A_0, between $t=0$ and t_0 only. Our initial disturbance at $x=0$ will separate into two simple waves at a time T_1. This will happen after a period of interaction $0<t<t_1$, as described in Chapter 5, where the velocities on C_+ and C_- were found in the A^2 approximation.

Opinions about whether each emerging, separate simple wave will spread or not seem to differ. Fig. 8.5 does not show spreading during the initial stages of the modulation splitting. However, before long, characteristics with slopes even slightly different from that of C_B^- and C_F^- in the first, and C_B^+ and C_F^+ in the second region will eventually emerge from the simple wave regions (broken lines). If superposition of two wavetrains is allowed at the intersections of characteristics of the same sign, then the region initially contained between C_B^- and C_F^- will eventually spread. We will not pursue this problem but will now consider a slightly different idealized set-up in which characteristics do not cross and the treatment of spreading is more straightforward.

Take at $x=0$:

(1) A uniform nonlinear wave, amplitude A_o for $t<0$

(2) Modulation between $t=0$ and t_o, such that $A(0)=A_o, A(t_o)=A_1$ (slightly different from A_o)

(3) A uniform nonlinear wave, amplitude A_1 for $t \geqslant t_o$.

Here A_o, $A(t)$ and A_1 are taken such that characteristics of the same sign do not intersect. Fig. 8.6 shows the two simple wave regions widening. Higher

Fig. 8.5. Initial value problem for modulation between $t=0$ and t_o at $x=0$ on an otherwise stationary nonlinear wave. The disturbance becomes disentangled after time t_1. B and F stand for back and front. Characteristics are straight lines outside the region of interaction.

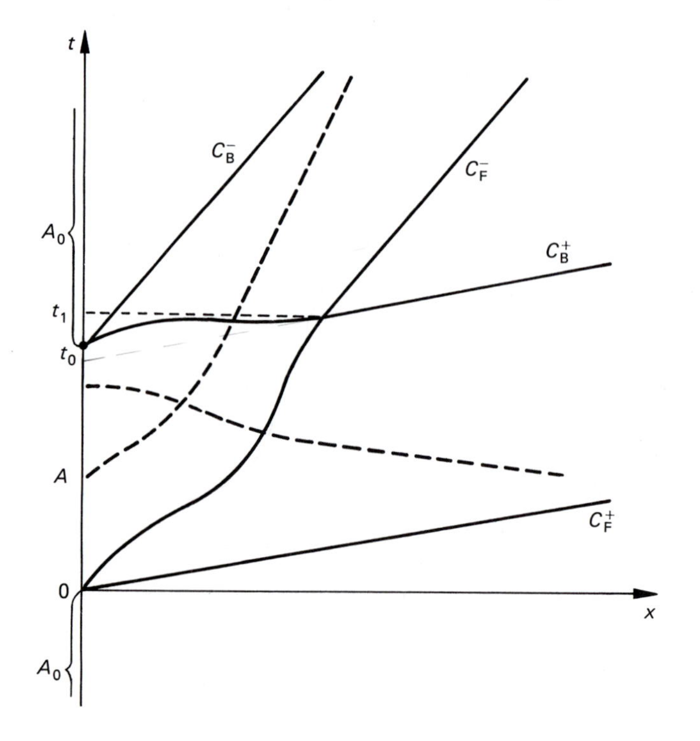

order terms in the modulation would have to be included if we wished to take the physics any further.

The experiments of Ikezi *et al.* (1978) seem to confirm this widening, though as was pointed out in Chapter 5, their wavepacket experiments only remotely correspond to the idealized situations treated here.

For the exceptional case $G''=0$, we would just have the simple modification from the beginning.

8.2.2 Whitham II

Equation (8.2.1) can be obtained from the Lagrangian density

$$L=\tfrac{1}{2}\phi_t^2-\tfrac{1}{2}\phi_x^2-V(\phi) \tag{8.2.24}$$

which, averaged over a wavelength, yields, see (8.2.7) and (8.2.8):

$$\begin{aligned}2\pi\mathscr{L}=2\pi\langle L\rangle&=\tfrac{1}{2}(U^2-1)^{\frac{1}{2}}kG(A)-2\pi A+\tfrac{1}{2}k(U^2-1)^{\frac{1}{2}}G(A) \quad (8.2.25)\\ &=(U^2-1)^{\frac{1}{2}}kG(A)-2\pi A.\end{aligned}$$

Fig. 8.6. As in Fig. 8.5, but now modulation between $t=0$ and t_o joins two stationary, constant amplitude waves of slightly different amplitudes A_o and A_1. Characteristics of the same sign never cross.

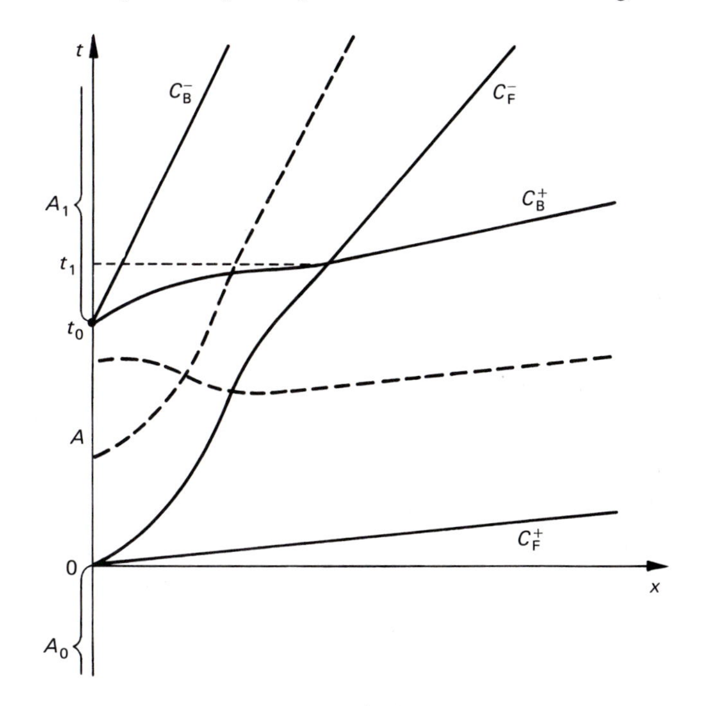

The A variational equation is just

$$\mathscr{L}_A = 0 \tag{8.2.26}$$

yielding

$$k = 2\pi/G'(U^2 - 1)^{\frac{1}{2}} \tag{8.2.27}$$

and, from the definition of ω,

$$\omega = kU = 2\pi U/G'(U^2 - 1)^{\frac{1}{2}} \tag{8.2.28}$$

The simplest way to obtain the second variational equation is to reintroduce phase θ such that

$$\omega = -\theta_t, k = \theta_x \tag{8.2.29}$$

so that

$$k_t + \omega_x = 0, \tag{8.2.30}$$

and write the Lagrangian as

$$2\pi \mathscr{L} = (\omega^2 - k^2)^{\frac{1}{2}} G(A) - 2\pi A = (\theta_t^2 - \theta_x^2)^{\frac{1}{2}} G - 2\pi A. \tag{8.2.31}$$

Now

$$\partial_t \mathscr{L}_{\theta_t} + \partial_x \mathscr{L}_{\theta_x} = 0 \tag{8.2.32}$$

yielding, reverting to $U = \omega/k$

$$\frac{\partial}{\partial t}\left(\frac{UG}{\sqrt{(U^2-1)}}\right) + \frac{\partial}{\partial x}\left(\frac{U_2}{\sqrt{(U-1)G'}}\right) = 0. \tag{8.2.33}$$

and (8.2.30) yields

$$\frac{\partial}{\partial t}\left(\frac{1}{\sqrt{(U^2-1)G'}}\right) + \frac{\partial}{\partial X}\left(\frac{U}{\sqrt{(U-1)G'}}\right) = 0. \tag{8.2.34}$$

These two equations are easily seen to be equivalent to (8.2.19) and (8.2.20) and the calculation proceeds as in the previous section to yield the same two characteristic velocities of the surfaces C_+ and C_-.

The above calculation was considerably simpler than the previous one, as only the one averaging procedure of (8.2.25) was involved. A Lagrangian is a useful thing to have when it is available.

8.2.3 K expansion

In this method we first transform the equation into a form in which the basic nonlinear structure is a function of one variable. Thus, if ϕ

depends on $X = x - Ut$, we take

$$X = \frac{x - Ut}{\sqrt{(U^2 - 1)}}, T = \frac{t - Ux}{\sqrt{(U^2 - 1)}}. \tag{8.2.35}$$

Equation (8.2.1) takes the form

$$(\partial_T^2 - \partial_X^2)\phi - V'(\phi) = 0. \tag{8.2.36}$$

We now perturb $\phi(X)$ and linearize in $\delta\phi$. It will be possible to Laplace and Fourier analyze in a generalized way. We take

$$\delta\phi \sim \delta\tilde{\phi} e^{i(KX - \Omega T)} \tag{8.2.37}$$

In this approach, A and U are fixed (they were slowly varying in both Whitham methods). Integrating (8.2.37) we find

$$\tfrac{1}{2}\phi_X^2 + V(\phi) = A \tag{8.2.38}$$

and $\phi(X, A)$ will be treated as a known, periodic structure. We find (again reverting to lower case x):

$$(\partial_T^2 - \partial_x^2)\delta\phi - V''\delta\phi = 0 \tag{8.2.39}$$

and, from (8.2.37)

$$(\Omega^2 + \partial_x^2)\delta\phi + V''\delta\phi = 0. \tag{8.2.40}$$

It follows from the periodicity of V in x and Floquet's theorem (Ince (1944)), that $\delta\phi$ will in general be a periodic function of x with the same period λ as $\phi(x,A)$, multiplied by e^{iKx}, K real. Thus $\delta\tilde{\phi}$ in (8.2.37) is a λ periodic function. It is seen from this treatment that K and $K + 2\pi n/\lambda$ will give the same Ω, as anticipated in Sections 4.6 and 5.3. We now assume both K and Ω to be small and of the same order, expanding all quantities in K:

$$\Omega = K\Omega_1 + K^2\Omega_2 + \ldots \tag{8.2.41}$$

$$\delta\tilde{\phi} = \delta\phi_o + K\delta\phi_1 + K^2\delta\phi_2 + \ldots \tag{8.2.42}$$

all $\delta\phi_n$ being periodic in x with wavelength λ.

Zero order We have:

$$L\delta\phi_o = (\partial_x^2 + V'')\delta\phi_o = 0. \tag{8.2.43}$$

One can easily see by differentiating (8.2.38) with respect to x, dividing though by ϕ_x and differentiating once again, that (8.2.43) is solved by

$$\delta\phi_o = \partial\phi/\partial x. \tag{8.2.44}$$

Similarly we obtain the second solution, $\partial\phi/\partial A$ which, however, will be seen to be secular. Thus the first order equation is

$$L\delta\phi_1 = (\partial_x^2 + V'')\delta\phi_1 = -2i\phi_{xx}. \tag{8.2.45}$$

Now we saw above that the corresponding homogeneous equation $L\delta\phi_1 = 0$ is solved by $\partial\phi/\partial x$ and $\partial\phi/\partial A$. Furthermore, a particular solution to (8.2.45) can be seen by inspection to be given by $-x\phi_x$. Thus we will look for a solution to (8.2.45) in the form

$$\delta\phi_1 = -ix\phi_x + \beta\phi_A \tag{8.2.46}$$

as the sum of two secular functions can be non-secular. This will involve a choice of β. (Any contribution from ϕ_x can be included in $\delta\phi_o$.) We now write ϕ as

$$\phi(x, A) = \psi(\xi, A), \xi = x\lambda(A)^{-1}, \tag{8.2.47}$$

and ψ is periodic in ξ with period one. Now

$$\frac{\partial\phi}{\partial x} = \lambda^{-1}\frac{\partial\psi}{\partial\xi} \tag{8.2.48}$$

$$\frac{\partial\phi}{\partial A} = \frac{\partial\psi}{\partial A} - x\lambda_A\lambda^{-2}\frac{\partial\psi}{\partial\xi} = \frac{\partial\psi}{\partial A} - x\lambda A\lambda^{-1}\frac{\partial\phi}{\partial x} \tag{8.2.49}$$

where $\partial\psi/\partial A$ is periodic. Thus secular terms will be avoided in $\delta\phi_1$, if $\beta = -i\lambda\lambda_A^{-1}$. Reverting to ϕ notation, we have

$$\delta\phi_1 = -i\lambda\lambda_A^{-1}\psi_A = -ix\phi_x - i\lambda\lambda_A^{-1}\phi_A, \tag{8.2.50}$$

and this function is non-secular.

The second order equation takes the form

$$L\delta\phi_2 + (\Omega_1^2 - 1)\phi_x - 2i\frac{\partial}{\partial x}(ix\phi_x + i\lambda\lambda_A^{-1}\phi_A) = 0. \tag{8.2.51}$$

Upon multiplication by ϕ_x and integration over a wavelength, (8.2.52) gives

$$\int \phi_x L\delta\phi_2 \mathrm{d}x + (\Omega_1^2 - 1)\int \phi_x^2 \mathrm{d}x - 2i\int \phi_x\partial_x(ix\phi_x + i\lambda\lambda_A^{-1}\phi_A)\mathrm{d}x = 0. \tag{8.2.52}$$

Using the fact that L is self-adjoint and ϕ_x is its eigenfunction, we readily obtain

$$\Omega_1 \int \phi_x^2 \mathrm{d}x + 2\lambda\lambda_A^{-1}\int \phi_x\phi_{Ax}\mathrm{d}x = 0, \tag{8.2.53}$$

(here calculations are simplest when the integral over a wavelength is taken between two zeros of ϕ_x).

In spite of L being second order, there is no other consistency condition, as the second eigenfunction ϕ_A has been seen to be non-periodic. Now introduce

$$g = G/\sqrt{(U^2-1)} = \int \phi_x^2 \mathrm{d}x = 2\int_{\phi_1}^{\phi_2} \phi_x \mathrm{d}\phi \tag{8.2.54}$$

$\phi_x = 0$ at ϕ_1 and ϕ_2.

It is easily seen that

$$g_A = 2\lambda. \tag{8.2.55}$$

Thus we have

$$\Omega_1^2 = -g_A/g g_{AA} = -(G')^2/GG'' \tag{8.2.56}$$

$$\Omega_1 = \lim_{K\to 0} \Omega/K = \partial\Omega/\partial K\,|_{\mathrm{o}} = \pm[(G')^2/GG'']^{\frac{1}{2}} = V_{\mathrm{CH}}(0). \tag{8.2.57}$$

The characteristic velocity is $\pm[(G')^2/GG'']^{\frac{1}{2}}$ when this quantity is real. If it is imaginary, the perturbation grows as $\mathrm{e}^{|\Omega_1|t}$ (the $\mathrm{e}^{-|\Omega_1|t}$ mode is clearly the loser in this competition) and the basic nonlinear wave $\phi(x, A, U)$ will be unstable. This approach is thus perhaps more amenable to a simple interpretation for Ω_1 imaginary than the previous two. The stability of shock and soliton solutions to (8.2.1) can be treated by performing the calculations for nonlinear waves and *then* taking $A\to A_{\mathrm{SOL}}$ or $A\to A_{\mathrm{SHOCK}}, \lambda\to\infty$. However, this method as formulated here requires ϕ periodic and should not be applied to soliton or shock solutions directly (see Appendix 3 for comments on this).

Our coordinate transformation (8.2.35) implies the following law of addition of velocities

$$V_{\mathrm{L}} = \frac{U+V_{\mathrm{CH}}}{1+UV_{\mathrm{CH}}} = \frac{U\pm\sqrt{(-G'^2/GG'')}}{1\pm U\sqrt{(-G'^2/GG'')}} = \frac{1\pm\alpha U}{U\pm\alpha}, \alpha = V_{\mathrm{CH}}^{-1} \tag{8.2.58}$$

where V_{L} is the characteristic velocity in the laboratory frame of reference. This agrees with the result following from (8.2.19)–(8.2.20).

We are now in a position to make some further observations on when K expansion can be applied, as compared with Whitham's methods. The above method (K expansion) can be used even when no Lagrangian exists and the number of conservation laws is insufficient to completely describe a system. Later on in this Chapter, an example of just such a problem will be given (Section 8.4). It might seem that this approach yields less information than the two Whitham methods, as we failed to find the Riemann invariants. However, in practical terms, Whitham's methods only yield these invariants for very special cases. In general the characteristic velocities (or growth rates of instabilities if they occur) are all one finds from any of these methods (Section 8.3).

The next order correction, Ω_2 in (8.2.41), can be found from K expansion by going to higher order and again removing secular terms (in fact Ω_2 is zero, and Ω_3 is real or pure imaginary in accord with Ω_1). No corresponding possibility follows as naturally in Whitham's method. The chance of doing this was lost when going from (8.1.3) to (8.1.4) and using quantities averaged over one wavelength (see Subsection 8.2.1); $K/k = \lambda/L$.

8.2.4 Hayes

Hayes' Hamiltonian method was introduced in Section 5.5 as the most natural for generalizing small A theory to arbitrary A. For our Lagrangian (8.2.31)

$$\mathscr{L}_\omega = I = \omega(\omega^2 - k^2)^{-\frac{1}{2}} G(A), \tag{8.2.59}$$

and the theory demands that I be promoted to being one of the two dependent variables (along with k). From (8.2.28)

$$\omega^2 = k^2 + 1/(G')^2. \tag{8.2.60}$$

Thus the equation to unravel, to obtain $A(I, k)$, is

$$I = \pm GG'\sqrt{[k^2 + 1/(G')^2]}. \tag{8.2.61}$$

Clearly we are heading for trouble, as the $A_\pm(I, k)$ will have to be tabularized, and even this will only be possible once $G(A)$, and therefore $V(\phi)$, have been specified. The first three methods are therefore preferable. However, very special choices of $V(\phi)$ can sometimes be painlessly treated by Hayes' method (Exercise 2).

8.3 Higher dimensional dynamics

For reasons that are apparent from Section 8.2, we will now concentrate on two methods: Whitham II and K expansion. A natural extension is to study fully three-dimensional problems in which a nonlinear plane wave is still locally a periodic function of one phase $\theta = \mathbf{k} \cdot \mathbf{x} - \omega t$, but $\mathbf{k}$ can now vary slowly in direction as well as length. Although a complete picture of our nonlinear wave as it evolves in time would in general demand a fully three space dimensional description, a local analysis, even on the large scale of $\mathbf{k}, \omega, A$ variation, can be performed in two dimensions. The x, y plane will then be defined by $\mathbf{k}$ and its increment which we denote by $\mathbf{K}$.

We will end the Section with a brief look at a third method, applicable for perpendicular perturbations on a soliton only, and so having no counterpart in x, t space.

8.3.1 Kadomtsev–Petviashvili as analysed by Whitham II

In Chapter 5 we found a Lagrangian density for the Kadomtsev–Petviashvili equation:

$$L = \psi_t \psi_x + \tfrac{1}{3}\psi_x^3 - \tfrac{1}{2}\psi_{xx}^2 + \tfrac{1}{2}\psi_y^2. \tag{8.3.1}$$

We now look for a solution to

$$\psi_{tx} + \psi_x \psi_{xx} + \tfrac{1}{2}\psi_{xxxx} + \tfrac{1}{2}\psi_{yy} = 0 \tag{8.3.2}$$

which is locally a stationary nonlinear wave. This implies a periodic function of some phase

$$\theta = \mathbf{k} \cdot \mathbf{x} - \omega t \tag{8.3.3}$$

the period in θ being 2π. However, only derivatives of ψ represent physical quantities, and ψ itself must contain terms linear in x and t if those physical quantities are to be allowed to have non-zero mean values (an example is the mean flow of a fluid). Thus

$$\psi = \psi_o + \phi(\theta), v = \nabla\psi = \nabla\psi_o + \nabla\phi(\theta), \tag{8.3.4}$$

$$\psi_o = \boldsymbol{\beta} \cdot \mathbf{x} - \gamma t, \nabla\psi_o = \boldsymbol{\beta}, \dot{\psi}_o = -\gamma, \tag{8.3.5}$$

$$\phi(2\pi) = \phi(0), \tag{8.3.6}$$

and ψ_o, $\boldsymbol{\beta}$ and γ are sometimes called the pseudo-phase, pseudo-wavenumber and pseudo-frequency. They have considerable nuisance value and were ignored in the general formulations of Section 8.1, as also previously in Section 5.5. Introduce

$$\eta = \psi_x = \beta_x + k_x \phi_\theta \tag{8.3.7}$$

$$\psi_y = \beta_y + k_y k_x^{-1}(\eta - \beta_x) \tag{8.3.8}$$

and then

$$\psi_{yy} = k_y^2 k_x^{-1} \eta_\theta \tag{8.3.9}$$

$$\dot{\psi} = -\gamma - \frac{\omega}{k_x}(\eta - \beta_x). \tag{8.3.10}$$

It follows from (8.3.2) that

$$\tfrac{1}{2}k_x^2 \eta_{\theta\theta\theta} + \eta\eta_\theta - \omega k_x^{-1}\eta_\theta + \tfrac{1}{2}k_y^2 k_x^{-2}\eta_\theta = 0, \tag{8.3.11}$$

and, upon integrating twice in θ, multiplying by η_θ and integrating once again, we obtain

$$k_x^2\eta_\theta^2+\tfrac{2}{3}\eta^3-2U\eta^2+4B\eta-4A=0 \tag{8.3.12}$$

$$U=\omega k_x^{-1}-\tfrac{1}{2}k_y^2k_x^{-2}. \tag{8.3.13}$$

This equation could be solved giving so-called cnoidal waves, just as in one-dimensional theory. A simple phase plane analysis as in Section 6.2 indicates the existence of nonlinear wave and soliton solutions.

For one of these waves L becomes, using (8.3.7)–(8.3.13):

$$\begin{aligned}L=&(\omega\beta_xk_x^{-1}-\gamma^{-1}-\omega k_x^{-1}\eta)\eta^{-1}-\tfrac{1}{2}k_x^2\eta_\theta^2+\tfrac{1}{3}\eta^3\\&+\tfrac{1}{2}(\beta_y+k_yk_x^{-1}\eta-k_yk_x^{-1}\beta_x)^2.\end{aligned} \tag{8.3.14}$$

The second bracket, when squared, will give $-2k_y^2k_x^{-2}\eta\beta_x$ and $k_y^2k_x^{-2}\eta^2$, among other terms. Using these in combination with ω to obtain U terms, we find

$$\begin{aligned}L=&(U\beta_x-\gamma-2B)\eta-k_x^2\eta_\theta^2\\&+2A+\tfrac{1}{2}[(2k_yk_x^{-1}\beta_y-k_y^2k_x^{-2}\beta_x)\eta-(\beta_y-k_yk_x^{-1}\beta_x)^2]\end{aligned} \tag{8.3.15}$$

What we really need is L averaged over the phase θ. This will follow from

$$\langle\eta\rangle=\frac{1}{2\pi}\int_0^{2\pi}\eta\,\mathrm{d}\theta=\beta_x \tag{8.3.16}$$

$$\langle\eta_\theta^2\rangle=\frac{1}{2\pi}\oint\eta_\theta\mathrm{d}_\eta=k_x^{-1}W \tag{8.3.17}$$

where

$$\begin{aligned}W(A,B,U)&=\frac{1}{2\pi}\oint\mathrm{d}\eta\sqrt{(4A-4B\eta+2U^2\eta-\tfrac{2}{3}\eta^3)}\\&=\frac{1}{\pi}\int_{\eta 2}^{\eta 3}\mathrm{d}\eta\sqrt{(4A-4B+2U^2\eta-\tfrac{2}{3}\eta^3)},\end{aligned} \tag{8.3.18}$$

and η_1,η_2,η_3 are the zeros of the cubic in W in increasing order. This cubic is positive between η_2 and η_3. The integral over a wavelength is just twice the integral from η_2 to η_3.

Finally, the averaged Lagrangian is

$$\mathscr{L}=\frac{1}{2\pi}\int_0^{2\pi}L\mathrm{d}\theta=-k_xW-2\beta_xB-\beta_x\gamma+U\beta_x^2+2A+\tfrac{1}{2}\beta_y^2, \tag{8.3.19}$$

and so the one-dimensional averaged Lagrangian is generalized to two dimensions by the somewhat surprising substitution

$$U_1 = \omega/k, \tag{8.3.20}$$

$$U_2 = \omega/k_x - \tfrac{1}{2}k_y^2/k_x^2 \tag{8.3.21}$$

$$\mathscr{L}(U_1) \rightarrow \mathscr{L}(U_2) + \tfrac{1}{2}\beta_y^2. \tag{8.3.22}$$

This is very different from Hayes' generalization for the deep water gravity wave problem, in which the two surface dimensional Hamiltonian for small amplitude waves was obtained from the x, t Hamiltonian virtually by underlining k and β, Hayes (1973)!

So far our varying parameters are A, B, ω, $\mathbf{k}$, γ, $\boldsymbol{\beta}$, though other combinations could be taken. Variation of (8.3.19) yields, as Euler–Lagrange equations

$$\delta B \mathscr{L}_B = 0 \rightarrow \beta_x = -\tfrac{1}{2}k_x W_B \tag{8.3.23}$$

$$\delta A \mathscr{L}_A = 0 \rightarrow k_x = 2W_A^{-1} \tag{8.3.24}$$

and it follows that

$$\beta_x = -W_B W_A^{-1} \tag{8.3.25}$$

and k_x, β_x can be removed. Our variables are now A, B, U, β_y, k_y. Since

$$\varphi = -\dot{\psi}_o, \beta = \nabla\psi_o, \tag{8.3.26}$$

the ψ_o Euler Lagrange equation is

$$(\mathscr{L}_\gamma)_t - (\mathscr{L}_{\beta x})_x - (\mathscr{L}_{\beta y})_y = 0 \tag{8.3.27}$$

and this will be used a little later. Wave conservation yields

$$(k_x)_t + \omega_x = 0$$

where

$$\omega = Uk_x + \tfrac{1}{2}k_y^2 k_x^{-2}.$$

In this calculation we will take $k_y = \beta_y = 0$ at the end of the day. Of course, these values can only be assigned at a point and after all differentiations are performed. The dependence of the wave dynamics on gradual changes in quantities like $\mathbf{k}$ and $\boldsymbol{\beta}$ is crucial in Whitham's methods.

With these simplifications, equations (8.3.23)–(8.3.27) yield

$$\partial_t W_A + U\partial_x W_A - W_A \partial_x U = 0. \tag{8.3.28}$$

Similar 'consistency conditions can be used to obtain three more straightforward equations. They are:

$$W_A^2\partial_x\beta_y + W_A\partial_y W_B - W_B\partial_y W_A = 0 \tag{8.3.29}$$

which follows from the consistency condition $\partial_x\beta_y - \partial_y\beta_x = 0$, and

$$\tfrac{1}{2}W_A^2\partial_x k_y + \partial_y W_A = 0 \tag{8.3.30}$$

which similarly follows from $\partial_x k_y - \partial_y k_x = 0$. Finally, $\partial_x\gamma + \partial_t\beta_x = 0$ in (8.3.27) yields

$$\partial_t W_B + U\partial_x W_B + W_A\partial_x B - \tfrac{1}{2}W_A\partial_x\beta_y = 0. \tag{8.3.31}$$

This was the last easy victory. The fifth and ultimate equation for slow variations in A, B, U, β_y and k_y is obtained from the equation for the x component of momentum, treating $\mathscr{L}$ as a function of θ and ψ_o and using the identities:

$$(\mathbf{k}, \omega) = (\nabla\theta, -\dot{\theta}) \text{ and } (\boldsymbol{\beta}, \gamma) = (\nabla\psi_o, -\dot{\psi}_o)$$

in

$$\partial_t(\theta_x\mathscr{L}_\theta + \psi_x\mathscr{L}_\psi) + \partial_x(\theta_x\mathscr{L}_{\theta x} + \psi_x\mathscr{L}\psi_x - \mathscr{L}) + \partial y(\theta_x\mathscr{L}_{\theta y} + \psi_x\mathscr{L}_{\psi y}) = 0. \tag{8.3.32}$$

The required form is

$$\partial_t(k_x\mathscr{L}_\omega + \beta_x\mathscr{L}_\gamma) + \partial_x(\mathscr{L} - k_x\mathscr{L}_{k_x} - \beta_x\mathscr{L}_{\beta_x}) + \partial_y(-k_x\mathscr{L}_{k_y} - \beta_x\mathscr{L}_{\beta y}) = 0.$$

We now write out the equation that follows from (8.3.32) above the previous four:

$$\partial_t W_U + U\partial_x W_U - W_A\partial_x A - \tfrac{1}{2}W_B\partial_y\beta_y + \tfrac{1}{2}(W_A W_U - \tfrac{1}{2}W_B^2)\partial_y k_y = 0 \tag{8.3.33}$$

$$\begin{aligned}
&\partial_t W_B + U\partial_x W_B + W_A\partial_x B - \tfrac{1}{2}W_A\partial_y\beta_y = 0\\
&\partial_t W_A + U\partial_x W_A - W_A\partial_x U = 0\\
&\partial_y W_A + \tfrac{1}{2}W_A^2\partial_x k_y = 0\\
&W_A\partial_y W_B + W_A^2\partial_y\beta_y - W_B\partial_y W_A = 0.
\end{aligned}$$

Changes in A, B, U, β_y, k_y are now assumed to be proportional to

$$e^{i(K_x x + K_y y - \Omega t)}$$

and (8.3.33) is linearized such that expressions quadratic in the changes are neglected. We introduce

$$\begin{aligned}
&\mathbf{K} = (K_x, K_y) = K(\cos\psi, \sin\psi)\\
&\dot{W}_U = W_{UA}\dot{A} + W_{UB}\dot{B} + W_{UU}\dot{U}.
\end{aligned}$$

The characteristic determinant for (8.3.33) must be zero and this leads to a cubic for Ω/k. It is (a change of sign in Ω and U will lead to interesting characteristics moving from the left to right as in Fig. 8.2).

$$
\begin{aligned}
&a_3(\Omega^*/K)^2(\Omega^* K^{-1}\cos\psi + \tfrac{1}{2}\sin^2\psi) \\
&+ a_2(\Omega^*/K)(\Omega^* K^{-1}\cos\psi + \tfrac{2}{3}\cos\psi \sin^2\psi) \\
&+ a_o\cos^4\psi + b_o\cos^2\psi\sin^2\psi + c_o\sin^4\psi = 0 \qquad (8.3.34) \\
&a_3 = W_A \,|\, \partial^2 W/\partial A_i \partial A_j \,|, (A_1, A_2, A_3) = (A, B, U) \\
&a_2 = W_A^2(2W_{AB}W_{UB} - 3W_{AU}^2 - W_{AA}W_{UU}) \\
&a_o = W_A^4 \\
&b_o = W_A^2(W_U W_{AA} + W_A W_{AU} - W_B W_{AB}), \\
&c_o = (W_U W_A - \tfrac{1}{2}W_B^2)(W_{UA}W_{AA} - \tfrac{1}{2}W_{AB}^2) \\
&\Omega^* = \Omega - KU.
\end{aligned}
$$

This cubic can be simplified. We see from (8.3.20) that

$$W_{AB} = 2W_{UA}$$

and

$$BW_A + UW_B + W_U = 0 \qquad (8.3.35)$$

$$AW_A + BW_B + UW_U - \tfrac{1}{2}W = \frac{1}{6\pi}\oint \eta^3(4A - 4B\eta + 2U\eta^2 - \tfrac{2}{3}\eta^3)^{-\frac{1}{2}} d\eta \qquad (8.3.36)$$

Differentiation of the last two identities by A, B, U gives five independent equations for all second derivatives of W (other than W_{BB}) in terms of first derivatives:

$$
\begin{pmatrix}
A, & B, & U, & 1/3, & 0 \\
0, & A, & 2B, & U, & 2/3 \\
B, & U, & 1, & 0, & 0 \\
0, & B, & 2U, & 1, & 0 \\
0, & 0, & B, & U, & 1
\end{pmatrix}
\begin{pmatrix}
W_{AA} \\ W_{AB} \\ W_{AU} \\ W_{BU} \\ W_{UU}
\end{pmatrix}
=
\begin{pmatrix}
-\tfrac{1}{2}W_A \\ -\tfrac{1}{2}W_B \\ 0 \\ -W_A \\ -W_B
\end{pmatrix}
\qquad (8.3.37)
$$

It is now possible to reduce the number of parameters without losing much generality. If we concentrate on a subset of all A, B, U (of course after all differentiations have been performed):

$$A = \tfrac{1}{6}\xi, B = -1/6, U = 0, \qquad (8.3.38)$$

the cubic in W becomes somewhat simpler. It is now proportional to

$$\eta - \eta^3 + \xi, \qquad (8.3.39)$$

and is non-negative for a finite range of η provided that

$$-2/\sqrt{27} \leqslant \xi \leqslant 2/\sqrt{27}. \qquad (8.3.40)$$

If we assume this and denote the three roots of (8.3.39) as $\eta_1(\xi), \eta_2(\xi), \eta_3(\xi)$ in increasing order, meaningful $\eta(x,\xi)$, leading to non-negative η_x^2, satisfy

$$\eta_2 \leqslant \eta \leqslant \eta_3.$$

We are now dealing with a subset of all parameters A, B, U that includes all interesting cases (linear limit for $\xi \to -2\sqrt{27}$, $\eta_2 \to \eta_3$; solitons for $\xi \to 2/\sqrt{27}, \eta \to \eta_2$). At the end of the day we would be able to regenerate the general A, B, U dispersion relation by the Galilean transformation (U) and by similarity (A, B). We will in fact only reinstate U. An example of the latter procedure will be given later in Section 8.4. However, much of the essential physics can be appreciated by just looking at all ξ. Introduce

$$Y = -2W_{B}/W_B = 2(\eta_1 + [\eta_3 - \eta_1]E(s)/K(s)) \qquad (8.3.41)$$

$$s^2 = \frac{\eta_3 - \eta_2}{\eta_3 - \eta_1} \leqslant 1.$$

Here E and K are the complete elliptic integrals

$$E(s) = \int_0^{\pi/2} \sqrt{(1 - s^2\sin^2\alpha)}\,\mathrm{d}\alpha, \quad K(s) = \int_0^{\pi/2} \mathrm{d}\alpha/\sqrt{(1 - s^2\sin^2\alpha)}.$$

Our dispersion relation (8.3.34) can now be simplified to (Exercise 5):

$$(\Omega^*/K)^2(\Omega^* K^{-1}\cos\psi + \tfrac{1}{2}\sin^2\psi) + \tfrac{2}{3}\Omega^* K^{-1}\frac{3Y(Y+6\xi)+4}{Y^3 - 4Y - 8\xi}(\Omega^* K^{-1}\cos\psi$$
$$+ \tfrac{2}{3}\cos\psi\sin^2\psi)$$

$$+ \frac{(8/27)(27\xi^2 - 4)}{Y^3 - 4Y - 8\xi}\cos^4\psi - \frac{(2/9)(8Y + 9\xi Y^2 + 12\xi)}{Y^3 - 4Y - 8\xi}\sin^2\psi\cos^2\psi \qquad (8.3.42)$$

$$+ \frac{2(1/3 - Y^2/4)^2}{Y^3 - 4Y - 8\xi}\sin^4\omega = 0,$$

where $\Omega^* = \Omega - KU$.

We have reinstated U via the Galilean transformation. Polar plots of $\Omega^*/K(\psi)$ are given in Fig. 8.7 for four different values of ξ: the linear wave limit, two nonlinear waves, and the soliton limit.

If we re-do the calculation for KP with positive dispersion (this variant appears in hydrodynamics and solid state physics Section 5.2 and Kunin (1975)):

$$\dot{\psi}_x + \psi_x\psi_{xx} + \tfrac{1}{2}\psi_{xxxx} - \tfrac{1}{2}\psi_{yy} = 0$$

the only alteration will be a minus sign in front of all $\sin^2\psi$ terms in (8.3.42). Fig. 8.8 gives polar plots for this case.

All waves and solitons of the KP equation with negative dispersion (8.3.7) are stable. Positive dispersion KP, on the other hand, leads to nonlinear waves that are unstable for a range of ψ around $\pi/2$. This unstable angle spreads as the amplitude of the basic nonlinear wave is increased up to the soliton amplitude and solitons are unstable for all ψ other than zero and π. Note how the instability develops from a double root of the linear limit seen in Fig. 8.8(a). This was forseen in Section 4.5.

For $\psi = 0$, the characteristic velocities of KdV are obtained (small K, Ω^*

Fig. 8.7. Polar plots of Ω^*/K for four ξ values: (a) linear limit, (b) & (c) fully nonlinear waves, (d) soliton. KP with negative dispersion.

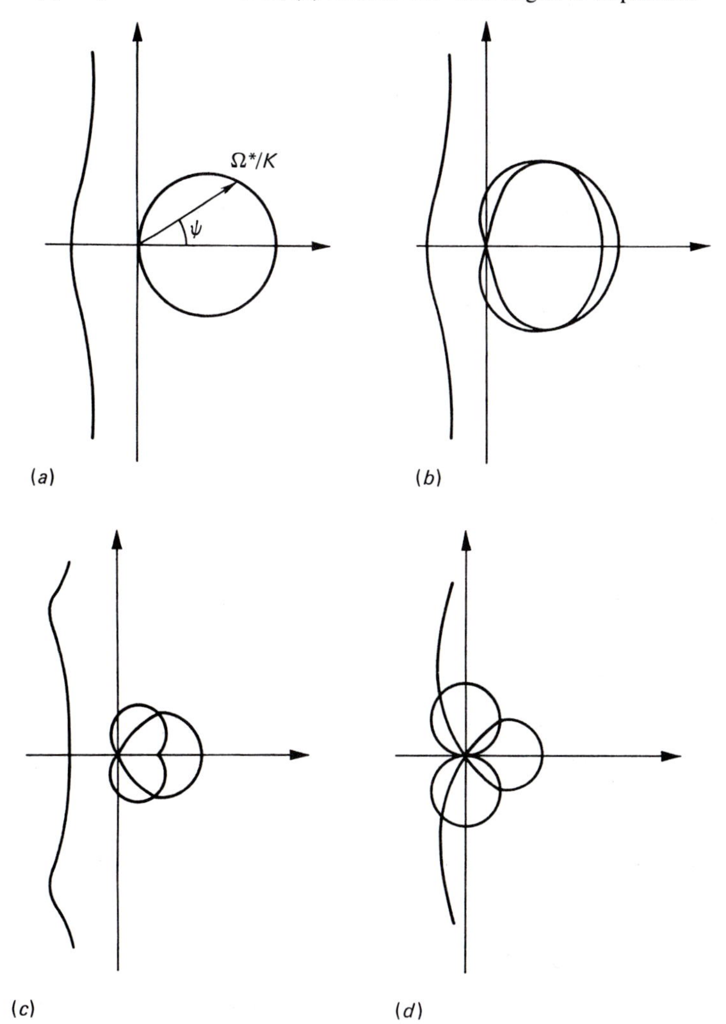

phase and group velocities are identical in one dimension since $\partial\Omega^*/\partial K \sim \Omega^*/K$). They are given in Fig. 8.9 for all ξ.

The above work on both KP equations, based on Infeld *et al.* (1978) and Infeld (1981b), was redone by Kuznetsov *et al.* (1984). These authors developed a new method, extendable in principle to all equations in (x, y, t)

Fig. 8.8. As in Fig. 8.7 but for KP with positive dispersion. Real Ω^*/K indicated by solid lines, pure imaginary by broken lines. Fully complex roots not shown, but unstable angles are indicated. KP indicates the original Kadomtsev–Petviashvili result (1970).

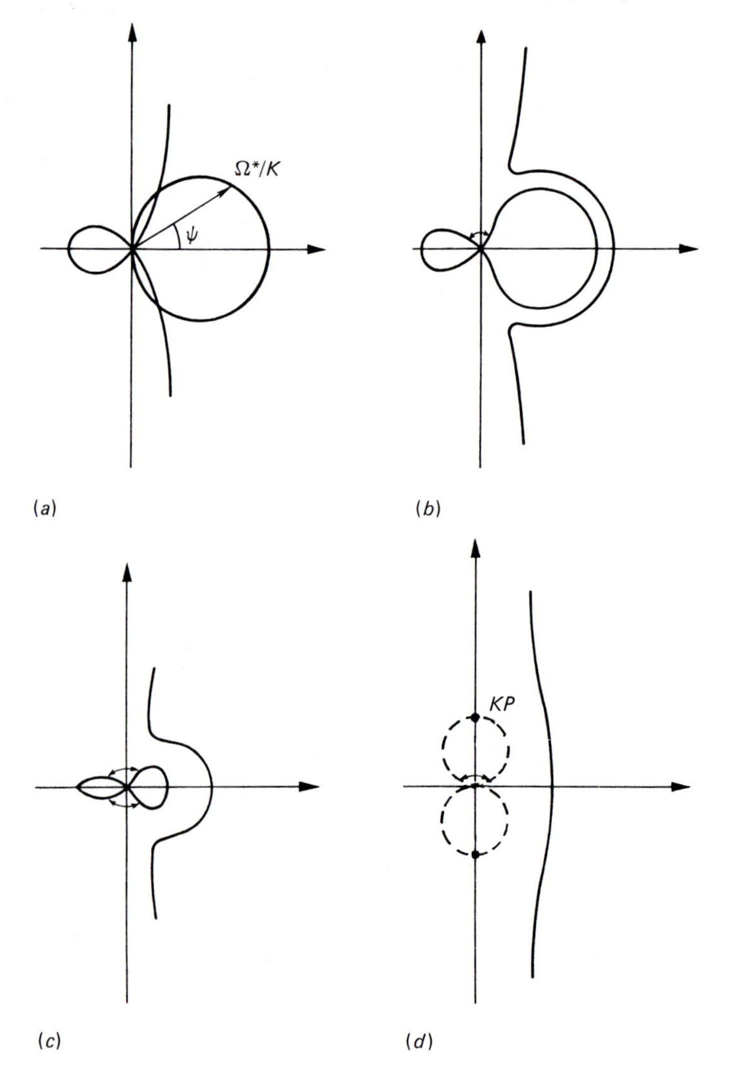

space that admit a Zakharov–Shabat representation (Section 7.10), as KP does:

$$\partial_t \mathrm{L} - \partial_y \mathrm{A} - [L, A] = 0$$

where the operators appearing in A and L are in x space only. Our results (stability for negative, instability for positive dispersion KP waves and solitons) were confirmed. Figs. 8.7 and 8.8 could of course be reproduced from their calculation.

Fig. 8.9. One-dimensional wave and soliton dynamics for KdV: (*a*) Characteristic velocities of KdV, obtainable from KP with $\psi = 0$, for all ξ. Here $\mu = (\xi_{\mathrm{LIN}} - \xi)/(\xi_{\mathrm{LIN}} - \xi_{\mathrm{SOL}})$. Near the linear limit, the separation of the two positive branches is proportional to $\sqrt{\mu}$, or A. (*b*) A slight amplitude increase in the direction of motion in a soliton train will cause the spacing to grow linearly in time. This is the opposite of the spreading of a weak amplitude wavetrain, in which a slight gradient in the spacing between modes will cause the amplitude to diminish while the spacing between crests remains constant, Fig. 5.2(*a*).

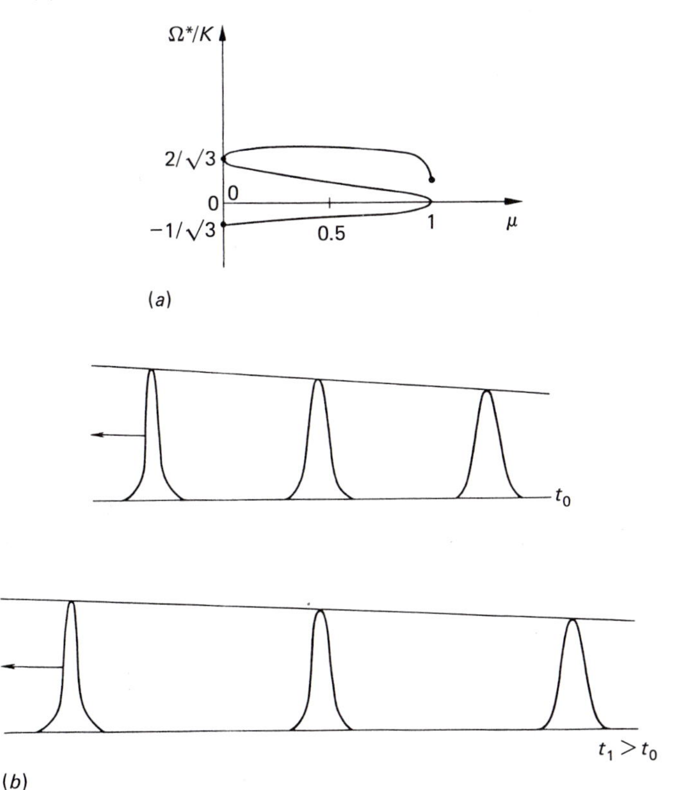

8.3.2 Various limits

To obtain the linear limit we take $\xi \to -2/\sqrt{27}, Y \to 2/\sqrt{3}$ yielding

$$(\Omega^*/K - 2\cos\psi/\sqrt{3})^2(\Omega^*/K + \cos\psi/\sqrt{3} + \tfrac{1}{2}\sin^2\psi/\cos\psi) = 0 \quad (8.3.43)$$

for KP with negative dispersion, and a change of sign in front of $\sin^2\psi$ for positive dispersion (Figs. 8.7(a) and 8.8(a)). In the soliton limit ($\xi \to 2/\sqrt{27}$, $Y \to -2/\sqrt{3}$) we obtain

$$(\Omega^{*2}/K^2 \pm (2/\sqrt{27})\sin^2\psi)(\Omega^*/K - \cos\psi/\sqrt{3} \pm \tfrac{1}{2}\sin^2\psi/\cos\psi) = 0 \quad (8.3.44)$$

illustrated by Figs. 8.7(d) and 8.8(d).

8.3.3 Common features of the weak amplitude and soliton limits for $\psi = 0$

For one-dimensional perturbations, $\psi = 0$, the cubic (8.3.42) can be solved to give three roots for Ω^*/K. They are, in increasing order,

$$\tfrac{1}{3}(\eta_2 - \eta_3)/(1 - E/K), \tfrac{1}{3}(\eta_3 - \eta_2)(1 - s^2)/(E/K - 1 + s^2), \tfrac{1}{3}(\eta_2 - \eta_1)E/K. \quad (8.3.45)$$

(Fig. 8.9(a)). Here in the figure, we introduced $\mu = (\xi - \xi_{\mathrm{LIN}})/(\xi_{\mathrm{SOL}} - \xi_{\mathrm{LIN}})$. (Alternatively, we could have used s^2 as our parameter, leading to a very similar drawing.) The first root in (8.3.45) is non-positive, the second and third non-negative. Notice how different roots merge in the linear and soliton limits in Fig. 8.9(a). This illustrates how very interrelated the individual characteristics are in nonlinear wave dynamics. The small amplitude splitting of a characteristic, seen on the left of the figure, was described in some detail in Chapter 5 in the framework of small A theory.

As mentioned above, two characteristics merge in the soliton limit ($\mu \to 1$), and thus a train of well-separated solitons might be expected to have *some* features in common with a linear wave limit. Indeed, in the small $1-\mu$ approximation, we can show that there is once more a degenerate characteristic velocity (U) and A^2 and k are again Riemann variables, just as in the small $A, \mu \to 0$ limit described in Chapter 5 (in the soliton train context, k is 2π times the reciprocal of the soliton spacing). However, it transpires from an analysis similar to that of Section 5.3, that for the soliton train, A^2 is constant and k decreases as t^{-1} on a degenerate characteristic. One way to visualize this without repeating the calculation is to take a sequence of solitons, given by (8.1.2), well-separated but equally spaced, and with η gradually but uniformly increasing in the direction of motion by some fixed value $\delta\eta$ from soliton to soliton. This corresponds to a small uniform value of η_x or A_x. If the solitons are far enough apart there will be no change of shape of each soliton as the soliton train evolves, but separation will increase proportionally to $\delta\eta t$ while remaining uniform in space (Fig. 8.9(*b*)). Thus, roughly speaking, A^2 and k have changed roles as compared with the small amplitude limit (Fig. 5.2(*a*)). This is not really so surprising in view of the fact that the small amplitude limit is $A^2 \to 0$, whereas the soliton limit corresponds to $k \to 0$. In the latter limit, each individual soliton has constant amplitude and propagates along a characteristic, just as a constant k point did in the former (linear) limit.

To next order in $1-\mu$, the degenerate characteristic bifurcates (breaks into two, each of which propagates a true Riemann invariant).

The *single* characteristic modes in the linear and soliton limits can be interpreted individually. They correspond to sound modes propagating from right to left slightly faster (small amplitude limit, $\mu \to 0$), or slightly slower (soliton, $\mu \to 1$) than the basic nonlinear structure, also taken to be moving from right to left. Sound modes in the opposite direction were scaled out of our model due to the way coordinates were stretched (the velocity of sound is now of order ε^{-1} as compared with those analyzed in Fig. 8.9(*a*)).

In Section 8.4 we intend to investigate the full plasma model from which both KP and KdV were derived in Section 5.2. (This problem will be used to fall back on for elucidation whenever a feature of the KdV family of equations seems obscure, at least in the remainder of this Chapter.) A fourth root, corresponding to sound modes propagating from left to right, will reappear (Fig. 8.15). The difference in velocity between the two sound modes will be seen to be exactly 2 in both the linear and soliton limits, but only in these limits. In both extreme cases, the basic nonlinear wave energy per unit length tends to zero and the overall alteration of the properties of the medium vanishes. Infinitesimal sound modes test the overall properties of the medium and their relative velocity will of course be 2 (in units of sound velocity) in a virtually unaltered medium. *Les extremes se touchent*!

8.3.4 Group velocity

Equations (8.3.33) differ from (8.2.15) and (8.2.20) in that they are not easily separated into equations for the three Riemann invariants of the form

$$\partial_t R_j + V_j \partial_x R_j + W_j \partial_y R_j = 0, j = 1, 2, 3 \qquad (8.3.46)$$

though Whitham has done this for x, t space, obtaining $R_j = \eta_1 + \eta_2 + \eta_3 - \eta_j$ and the velocities given by (8.3.45) for the characteristics. The characteristic equations so obtained led to a quantitative formulation of some of the physics quoted immediately above (Whitham (1974)).

Suppose, for the moment, that the (almost) impossible has been done and the characteristic (8.3.46) found. Take changes in R_j, $R_j = R_{jo} + \delta R_j$ in the form

$$\delta R_j \sim e^{i(K_x x + K_y y - \Omega t)}$$

equation (8.3.46) gives upon linearization,

$$\Omega_j = V_j K_x + W_j K_y \tag{8.3.47}$$

$$\frac{\partial \Omega_j}{\partial \mathbf{K}} = (V_j, W_j) \tag{8.3.48}$$

where $\partial\Omega/\partial\mathbf{K}$ is a characteristic velocity. This strengthens our interpretation of characteristic velocities as natural generalizations of linear group velocity which was of the form $\partial\omega/\partial\mathbf{k}$ and was a double characteristic (for k and a^2).

Write (8.3.42) as

$$G(\mathbf{K}, \Omega, \xi) = 0. \tag{8.3.49}$$

Then, in the coordinate system of the nonlinear wave $U = 0$ and

$$\frac{\partial \Omega_j}{\partial \mathbf{K}} = -(G_{\mathrm{K}}/G_{\Omega})_{\Omega = \Omega j}. \tag{8.3.50}$$

The surfaces defined by $\mathbf{X} = \partial\Omega i/\partial\mathbf{K}$ corresponding to unit time are shown in Fig. 8.10 for KP with negative dispersion.

For small Ω, Ω_{K} is independent of the modulus K and thus these surfaces will not depend on Ω. Indeed, had (8.3.42) been the correct dispersion relation for all Ω, we would have been in a position to present Fig. 8.10 as representing all group curves generated by an instantaneous impulse at the origin, and there would have been no spreading for given angle ψ. However, an instantaneous impulse $\sim\delta(t)$ will generate all Ω, small and large (with nonlinear K dependence), taking us beyond the present theory and leading to a spreading of group curves. Thus this spreading is due to dependence of Ω/K on the *modulus* of K.

Figures such as 8.10 (polar plots of $\partial\Omega/\partial\mathbf{K}$ in the presence of a nonlinear wave) can be considered to be nonlinear extensions of the Friedrichs diagrams of linear theory, just as the $\Omega/K(\psi)$ plots introduced earlier extend the polar plots of Clemmow, Mullaly and Allis.

The linear and soliton limits in Fig. 8.10, (*a*) and (*c*), are comparatively easy to generate, as the $\Omega_j(\mathbf{K})$ are given by (8.3.43) and (8.3.44). In the intermediate, fully nonlinear case, it is best to use (8.3.50) rather than unravel (8.3.42).

Fig. 8.10. Characteristics produced by an isotropic point pulse at the origin after unit time for three ξ values $(-2/\sqrt{27}; 0; 2/\sqrt{27})$. The little diagrams on the right of (*a*) and (*c*) depict the corresponding characteristics for the full plasma problem. The scale of (*b*) differs from that of (*a*) and (*c*). From Infeld (1981a).

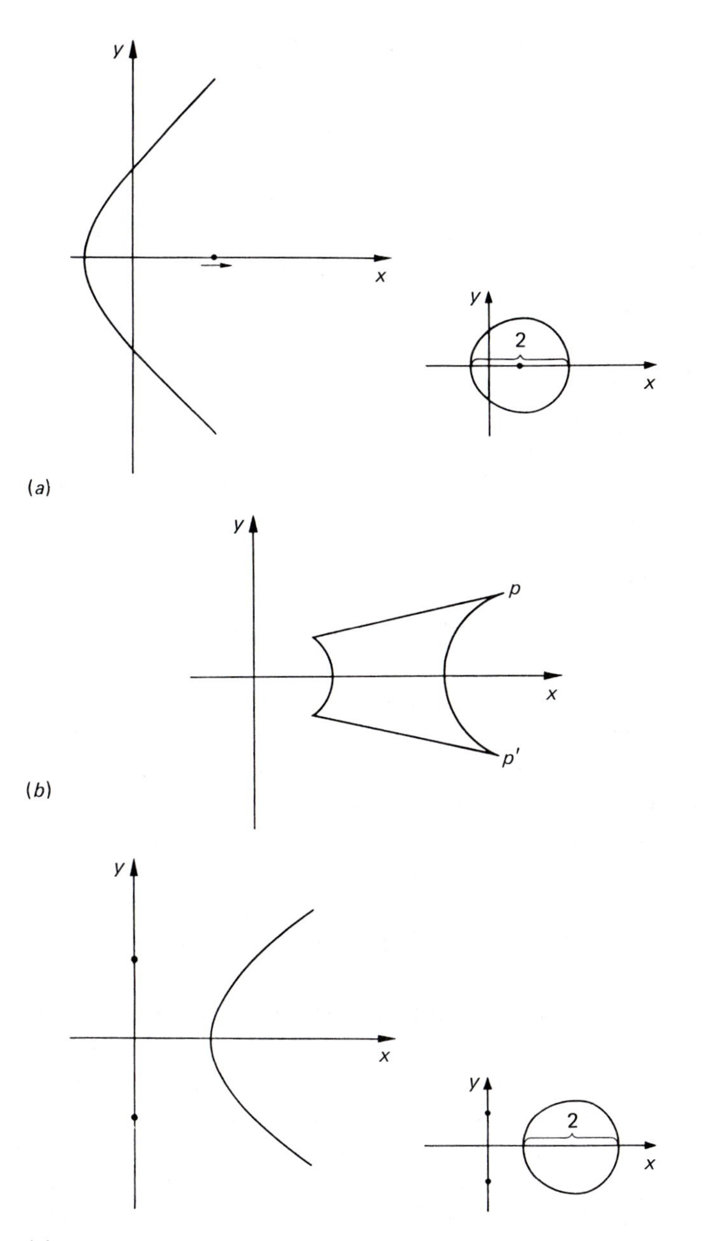

Figs. 8.7 and 8.10 give the same velocities on the $\psi=0$ axis, where phase and group velocities coincide for small Ω, K. All this is saying is that $\Omega/K \simeq \partial\Omega/\partial K$ for small Ω, K and $\psi=0$.

All in all, we have been able to extend some of the concepts of linear wave theory (phase and group velocities, energy transport) to the nonlinear problem, though with varying credibility. The presence of the nonlinear wave or soliton can, in Figs. 8.7 and 8.10, be thought of as a material property of the medium. This enables us to look on Ω and $\mathbf{K}$ as we do in linear theory, though with more branches $\Omega(\mathbf{K})$ present due to the alteration of the medium. Figs. 8.7 and 8.8 are then just polar plots for phase velocities $\Omega/K(\psi)$ of plane waves propagating in this altered medium and drawn in the coordinate system of the altering agent (nonlinear wave or soliton, not shown but understood in the drawings). Fig. 8.10 gives characteristic surfaces in this medium. They tell us how a periodic point signal propagates out from the origin.

It might seem surprising that characteristics produced at the origin should extend out to infinity after unit time (Fig. 8.10(*a*) and (*c*)). This is of course unphysical and, as already mentioned, is a consequence of the uneven way coordinates have been stretched. If we were to go back to, say, the full plasma wave problem (Chapter 5 and Section 8.4), this would not occur. In the lower right hand corner of the same diagram, characteristics of the full problem, as described by (5.1.7)–(5.1.9) with $\Omega_c=0$ and solved in Section 8.4, are given for comparison. Now characteristics are of finite extent, as they should be. Case (*b*), however, becomes unstable in the full description and real characteristics no longer exist (Section 8.4).

It takes a little experience to be able to separate the real physics from undesirable effects introduced by model equations.

8.3.5 Zakharov–Kuznetsov as analyzed by K expansion

In Chapter 5, we found a Lagrangian for the ZK equation and so the dynamics of the nonlinear wave and soliton solutions could be investigated by Whitham II as above (Exercise 7). However, we will use this example to illustrate how K expansion can be extended to x, y, t space. This will be presented in some detail, as the method is not as widely known as Whitham's. Presentation follows Infeld (1985).

We take (5.2.66) in the slightly more convenient form

$$\partial_t n + n\partial_x n + \tfrac{1}{3}\partial_x^3 n + \tfrac{2}{3}\partial_{xyy} n = 0 \qquad (8.3.51)$$

obtained by rescaling and taking $\Omega_c = 1$. We will regenerate the general Ω_c equations after all calculations have been performed.

We now look for stationary nonlinear solutions of the form

$$n_o'(x') = n_o + U, x' = x - Ut \tag{8.3.52}$$

thus adding a constant to n_o which we simply include in zero order (Section 5.3). Thus, in the new variables, dropping the primes and integrating twice,

$$\begin{aligned} &1/3\ \partial_x^2 n_o + \tfrac{1}{2} n_o^2 = C \\ &1/6\ \partial_x^2 n_{ox} = C n_o - \tfrac{1}{6} n_o^3 + D. \end{aligned} \tag{8.3.53}$$

We now concentrate on

$$C = 1/6, D(C = 1/6) = \xi/6 \tag{8.3.54}$$

to obtain

$$n_{ox}^2 = n_o - n_o^3 + \xi, \tag{8.3.55}$$

and n_{ox} non-negative in an interval yields the inequalities

$$-2/\sqrt{27} \leqslant \xi \leqslant 2/\sqrt{27}. \tag{8.3.56}$$

Suppose a periodic wave solution to (5.3.55) is perturbed such that the wave vector $\mathbf{K}$ of the perturbation forms an angle ψ with $\mathbf{x}$, the common direction of $\mathbf{B}_o$ and of the nonlinear wave. In the coordinate system of the basic wave we have

$$n = n_o(x) + \delta n = n_o(x) + \delta n(x) e^{i[(K\cos\psi)x + (K\sin\psi)y - \Omega t]} \tag{8.3.57}$$

$$\mathbf{K} = K(\cos\psi, \sin\psi) = (K_x, K_y) \tag{8.3.58}$$

$$B_o = (B_o, 0),$$

and $\tilde{\delta n}(x)$ is λ periodic. We now assume K small and expand:

$$\begin{aligned} &\Omega = \Omega_1(\psi)K + \Omega_2(\psi)K^2 + \ldots \\ &\tilde{\delta n} = \delta n_o + K\delta n_1 + K^2 \delta n_2 + \ldots \end{aligned} \tag{8.3.59}$$

Consistency in second order will be seen to yield a relationship of the form

$$G(\Omega, \mathbf{K}, \xi) = 0, \tag{8.3.60}$$

again generalizing a dispersion relation of linear theory. As we will be working within the framework of (8.3.51) and will find it necessary to proceed to second order in K, we must assume

$$1 \gg K^2 \gg \varepsilon. \tag{8.3.61}$$

We will now neglect terms quadratic in δn. Therefore, in view of the fact that linearization was performed after the expansion that led to ZK we have the hierarchy of small parameters:

$$1 \gg K^2 \gg \delta n / n_o \gg \varepsilon. \tag{8.3.62}$$

Up to now, we have tried to observe the convention that the basic nonlinear wave is moving from right to left (Figs. 8.2 and 8.4), leading to interesting characteristics moving from left to right, a situation most of us find more natural. To continue in this tradition, we must take $\partial_t \to -\partial_t$ in (8.3.51), and a change of sign in Ω occurs. Fortunately, this nuisance only occurs for unidirectional equations such as (8.3.7) and (8.3.51).

Introducing

$$L = \tfrac{1}{3}\partial_x^2 + n_o \tag{8.3.63}$$

we find, upon changing the sign of Ω,

$$\partial_x L\delta n = \tfrac{1}{3}\partial_x^3 \delta n + \partial_x n_o \delta_n = -i\Omega\delta n - iK\cos\psi n_o \delta n - iK\cos\psi \delta n_{xx} + K^2\cos^2\psi\delta n_x + \tfrac{2}{3}K^2\sin^2\psi\delta n_x + 0(K^3). \tag{8.3.64}$$

Zero order

In zero order, the right hand side of (8.3.64) vanishes and δn_o satisfies

$$L\delta n_o = M \tag{8.3.65}$$

where M is a constant, solved by

$$\delta n_o = \partial_x n_o + A(\partial_x n_o)\int^x n_{ox}^{-2}\mathrm{d}x + 3M(\partial_x n_o)\int^x n_o n_{ox}^{-2}\mathrm{d}x. \tag{8.3.66}$$

The notations $\partial_x n_o$ and n_{ox} are used interchangeably. The second and third terms are both secular. This suggests the following notation:

$$\begin{aligned} \rho &= n_{ox}\int n_{ox}^{-2}\mathrm{d}x = \beta x n_{ox} + Q_o(x) \\ \kappa &= n_{ox}\int n_o n_{ox}^{-2}\mathrm{d}x = \gamma x n_{ox} + Q_1(x), \end{aligned} \tag{8.3.67}$$

where Q_o and Q_1, are periodic with wavelength λ. Integrals are principal values as understood for second order poles (see Gelfand and Shilov (1964)). Removal of secular terms in (8.3.66) determines A:

$$A = -3M\gamma/\beta$$

thus

$$\delta n_o = n_{ox} + 3M(Q_1 - \gamma Q_o/\beta). \tag{8.3.68}$$

First order

We now have

$$\partial_x L\delta n_1 = -i(\Omega_1 + n_o\cos\psi)\delta n_o - i\cos\psi\delta n_{o_{xx}}. \tag{8.3.69}$$

Integration of (8.3.69) over a period gives a condition on Ω_1 and M. If we first multiply on the left by n_o and then integrate over a period, we obtain a second condition. The two conditions are only both satisfied for $M=0$. Thus $M=A=0$. Introduce

$$\Omega_1=\hat{\Omega}_1\cos\psi,\ \delta n_1=\delta\hat{n}_1\cos\psi \tag{8.3.70}$$

and integrate (8.3.69) to obtain

$$L\delta\hat{n}_1=-i\hat{\Omega}_1 n_o-\frac{i}{2}n_o^2-in_{oxx}+\bar{E}. \tag{8.3.71}$$

We now make use of the following relations:

$$\begin{aligned} &L\cdot\kappa=1/3 \\ &L\cdot\rho=0 \\ &L\cdot 1=n_o \\ &L\cdot(x\partial_x^2 n_o)=\tfrac{2}{3}\partial_x^2\ n_o. \end{aligned} \tag{8.3.72}$$

Using (8.3.53) with $C=1/6$ in the above equations, (8.3.71) and (8.3.72), we obtain

$$\delta\hat{n}_1=-i\hat{\Omega}_1-ixn_{ox}+3E\kappa+F\rho \tag{8.3.73}$$

$$\delta\hat{n}_{1\,\mathrm{SEC}}=(3E\gamma-i+F\beta)xn_{ox}=0 \tag{8.3.74}$$

where $E=\bar{E}-1/6$.

Thus δn_1 will be periodic if

$$F=(i-3E\gamma)/\beta. \tag{8.3.75}$$

Finally,

$$\delta\hat{n}_1=-i\hat{\Omega}_1+3E\bar{Q}+iQ_o/\beta \tag{8.3.76}$$

$$\bar{Q}=Q_1-\gamma Q_o/\beta. \tag{8.3.77}$$

The following straightforward identities will also be required in the next order:

$$\langle n_o\delta\hat{n}_1\rangle=-i\hat{\Omega}\langle n_o\rangle+E \tag{8.3.78}$$

$$\tfrac{1}{2}\langle n_o^2\delta\hat{n}_1\rangle=-\langle\delta\hat{n}_1\rangle/6-i\hat{\Omega}_1\langle n_o^2\rangle+\frac{2i}{3}\langle n_{ox}^2\rangle+E\langle n_o\rangle$$

where

$$\langle f\rangle=\lambda^{-1}\int_o^\lambda f\mathrm{d}x.$$

Second order expansions and consistency

We will mercifully not need δn_2, but will merely adjust the two constants that have already appeared to ensure non-secularity to second order. Equation (8.3.64) yields, in this order,

$$\partial_x L \cdot \delta n_2 = -i\Omega_2 \delta n_o - i\Omega_1 \delta n_1 - i n_o \delta n_1 \cos\psi - i\delta n_{oxx} \cos\psi + \delta n_{ox} \cos^2\psi + \tfrac{2}{3}\delta n_{ox} \sin^2\psi. \tag{8.3.79}$$

This equation integrates over a period to give

$$-i\hat{\Omega}_1 \langle \delta \hat{n}_1 \rangle - i\langle n_o \delta \hat{n}_1 \rangle = 0. \tag{8.3.80}$$

Thus

$$\hat{\Omega}_1[-i\hat{\Omega}_1 + \langle 3E\bar{Q} + iQ_o/\beta \rangle] + E - i\Omega_1 \langle n_o \rangle = 0, \tag{8.3.81}$$

and so

$$E = \frac{i(\hat{\Omega}_1^2 + \hat{\Omega}_1[\langle p_o \rangle - \langle Q_o \rangle/\beta]}{1 + 3\hat{\Omega}_1 \langle \bar{Q} \rangle}. \tag{8.3.82}$$

We note in passing that a constant generated in the lth order is usually determined in the $l+1$st order.

Finally, we multiply (8.3.70) by n_o on the left, integrate over a period, use the self-adjoint property of L, and then integrate by parts:

$$\langle n_o \partial_x L \cdot \delta n_2 \rangle = -\langle \delta n_2 L \cdot \partial_x n_o \rangle = 0. \tag{8.3.83}$$

Thus

$$i\hat{\Omega}_1 \langle n_o \delta \hat{n}_1 \rangle + i\langle n_o^2 \delta \hat{n}_1 \rangle + i\langle n_{oxx} \delta \hat{n}_1 \rangle + \langle n_{ox}^2 \rangle + \tfrac{2}{3}\langle n_{ox}^2 \rangle \tan^2\psi = 0. \tag{8.3.84}$$

This, together with (8.3.82), is in principle a dispersion relation of the required form (8.3.60).

The dispersion relation in terms of complete elliptic integrals

Using (8.3.53) with $C = 1/6$ and $D = \xi/6$ we obtain

$$n_{ox}^2 = \tfrac{2}{3}n_o + \tfrac{2}{3}n_o n_{oxx} + \xi. \tag{8.3.85}$$

If we divide through by n_{ox}^2 and integrate over a period, we obtain

$$1/3 = \tfrac{2}{3}\gamma + \xi\beta. \tag{8.3.86}$$

Similar manipulations yield

$$\begin{aligned} \langle Q_o \rangle &= -\gamma + \beta \langle n_o \rangle \\ \langle Q_1 \rangle &= -\beta/3 + \gamma \langle n_o \rangle \\ \tfrac{5}{3}\langle n_{ox} \rangle &= \xi + \tfrac{2}{3}\langle n_o \rangle \\ -\tfrac{1}{3}\langle n_o \rangle &= \xi\gamma + \tfrac{2}{9}\beta. \end{aligned} \tag{8.3.87}$$

These identities, together with the definition of Q (8.3.77), give enough equations to determine $\langle\bar{Q}\rangle, \langle Q_o\rangle, \beta, \gamma$ in terms of $\langle n_{ox}^2\rangle$ and ξ. A useful combination of these two quantities, suggested by the form of $\langle n_o\rangle$, is

$$Y = 2\langle n_o\rangle = 5\langle n_{ox}^2\rangle - 3\xi = 2(n_1 + [n_3 - n_1]E(s)/K(s)), \qquad (8.3.88)$$

$$s^2 = \frac{n_3 - n_2}{n_3 - n_1} \leqslant 1,$$

and $E(s)$ and $K(s)$ are the complete elliptic integrals defined immediately after equation (8.3.41). Simple algebra yields

$$\langle\bar{Q}\rangle = (Y^2 - 4/3)/(4Y + 12\xi) \qquad (8.3.89)$$
$$\langle Q_o\rangle/\beta = (Y^2 + 6\xi Y + 4/3)/(2Y + 6\xi).$$

It is now a matter of straightforward, if somewhat lengthy, algebra to obtain a cubic in Ω/K from (8.3.82) and (8.3.84). Dropping the subscript on Ω_1, reinstating K, U and $\cos\psi$, we arrive at

$$(\Omega^*/K)^3 + \tfrac{2}{3}\frac{(\Omega^*/K)^2[4 + 3Y(Y + 6\xi)]}{Y^3 - 4Y - 8\xi}\cos\psi + \tfrac{4}{15}(\Omega^*/K)$$

$$\frac{(Y + 3\xi)(Y^2 - 4/3)}{Y^3 - 4Y - 8\xi}\sin^2\psi + \tfrac{8}{27}\frac{27\xi^2 - 4}{Y^3 - 4Y - 8\xi}\cos^3\psi \qquad (8.3.90)$$

$$+\tfrac{16}{45}\frac{(Y + 3\xi)^2}{Y^3 - 4Y - 8\xi}\sin^2\psi\cos\psi = 0.$$

Note that for $\psi = 0$, (8.3.42) and (8.3.90) coincide; the one-dimensional dynamics of a wave are not altered by a magnetic field along the direction of propagation. Once again, the equivalence of Whitham II and K expansion is demonstrated (to complete this demonstration for all ψ, see Exercises 5 and 7).

The solution for general ψ_c is recovered from (8.3.90) by multiplying the third and fifth terms by $\frac{1}{2}(1 + \Omega_c^{-2})$.

Phase diagrams (phase velocities Ω/K for plane wave perturbations as functions of ψ) are given for $U = 0, \Omega_c = 1$ and three ξ values in Fig. 8.11. An instability appears for $\psi \simeq \pi/2$ when the wave amplitude is small. Unstable ψ spread out as ξ is increased (Fig. 8.11(*b*)). In the soliton limit $\xi \to 2/\sqrt{27}$, $\lambda \to \infty$, all ψ other than 0 and π are unstable. There is a certain similarity between this behaviour and that of a positive dispersion KP wave (Fig. 8.8). The external magnetic field has a destabilizing influence which at first glance seems somewhat similar to that of a sign change in the perpendicular dispersion properties of the medium.

For $\psi = \pi/2$, purely growing modes are obtained with growth rates given by

$$\Gamma/K = \left[\tfrac{2}{15} \frac{(1+\Omega_c^{-2})(Y+3\xi)(Y^2-4/3)}{Y^3-4Y-3\xi} \right]^{\frac{1}{2}} \tag{8.3.91}$$

For general waves, $\pi/2$ is the only angle for which purely growing modes appear (such modes are known as soft instabilities, Fig. 8.12(*a*)).
The linear limit is $(\xi \to -2/\sqrt{27})$

$$\left(\Omega/K + \frac{1}{\sqrt{3}}\cos\psi\right)\left(\Omega/K - \frac{2}{\sqrt{3}}\cos\psi\right)^2 = 0 \tag{8.3.92}$$

Fig. 8.11. Polar plots of Ω/K for the three ξ values $(-2/27, 0, 2/27)$ and $\Omega_c = 1$. The basic nonlinear wave or soliton is propagating along $\mathbf{B}_o$ from right to left. An unstable region appears at $\psi = \pi/2$ and spreads as ξ is increased, finally covering all non-zero angles in the soliton limit. LS = Laedke Spatschek. Note similarity to Fig. 8.8.

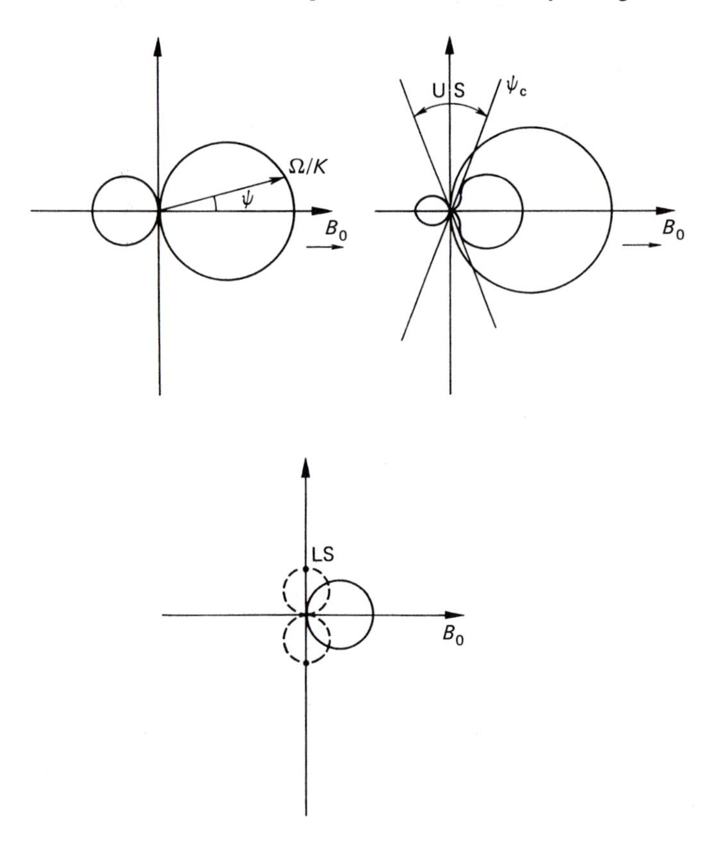

Fig. 8.12. Dynamics of nonlinear waves and solitons as described by the Zakharov–Kuznetsov equation. (*a*) Small K growth rate Γ for $\psi=\pi/2$, $\Omega_c=1$, all μ. Soliton limit on the right. (*b*) Finite K growth rates for solitons, $\psi=\pi/2(K=K_y), \Omega_c=1$; as found by Laedke and Spatschek (1982b), broken lines, and from a numerical calculation by Infeld and Frycz (1987). Theoretical values for instability cut-off: Katyshev and Makhankov (1976) (x); Infeld and Frycz (1987) (·). (When checking the slope at zero with (8.3.97), 32/15 should replace 4/15). (*c*) Destabilization of a nonlinear wave. (*d*) Overtaking of an unperturbed soliton by an initially perturbed soliton. From Infeld (1985) (*a*); Infeld and Frycz (1987) (*b*–*d*).

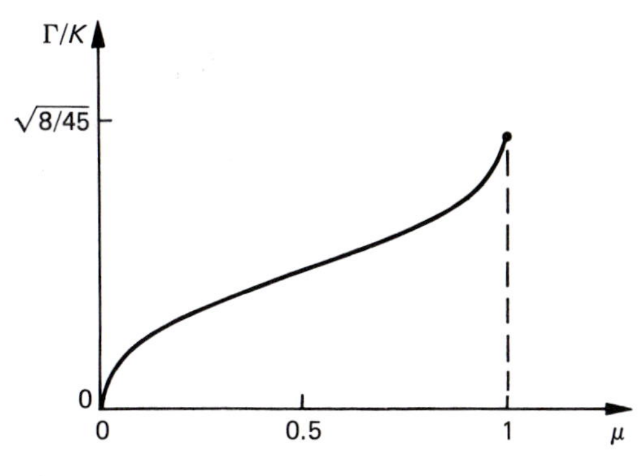

(*a*)

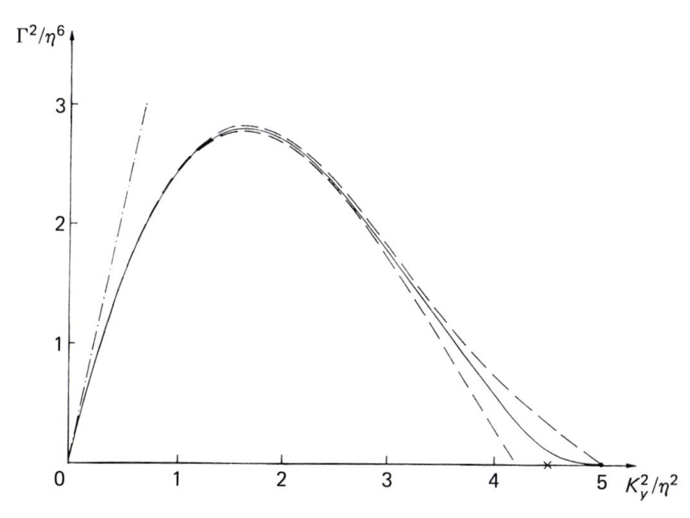

(*b*)

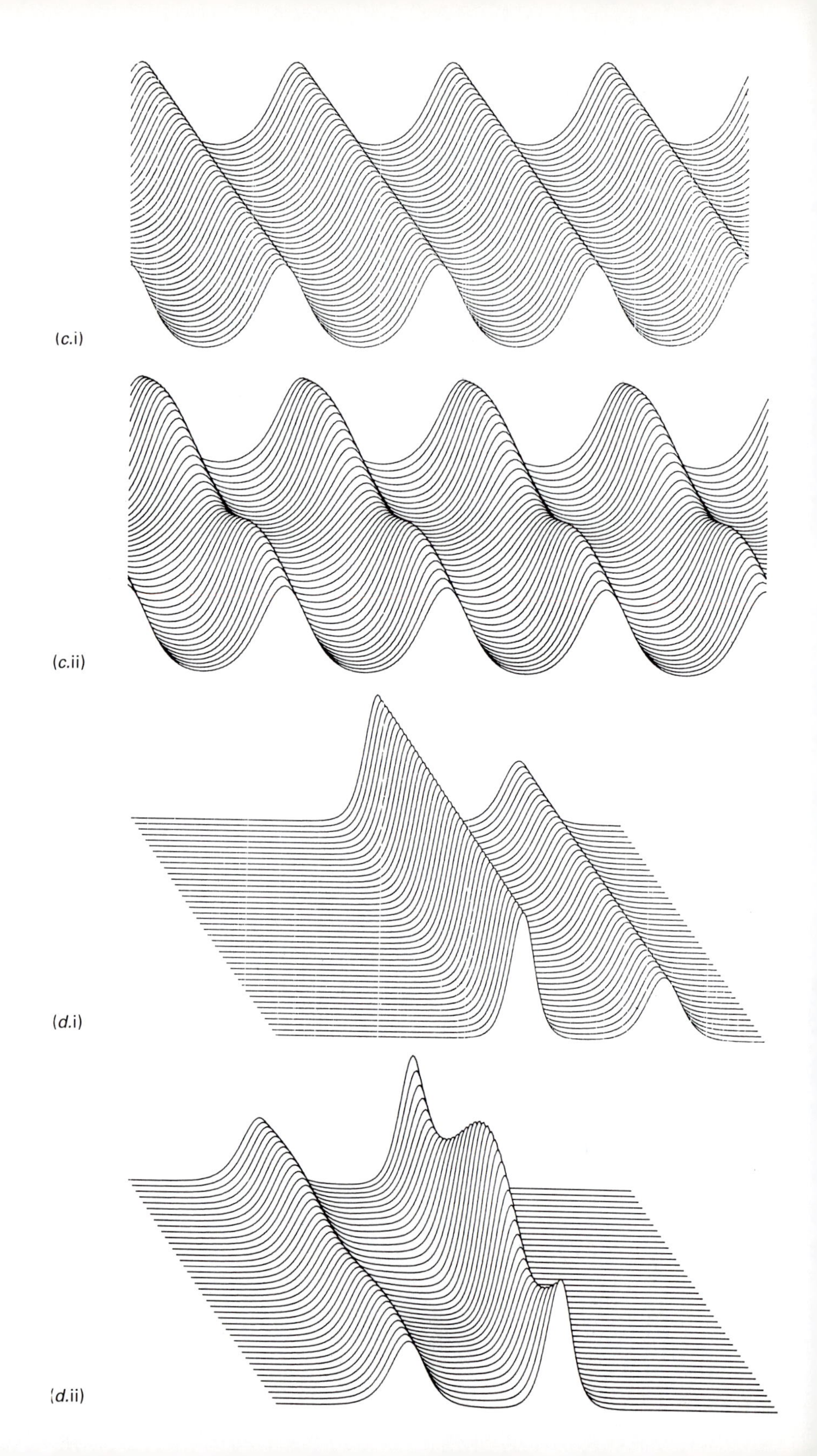

(c.i)
(c.ii)
(d.i)
(d.ii)

and the soliton limit $\xi \to 2/\sqrt{27}$

$$\left(\Omega/K - \frac{1}{\sqrt{3}}\cos\psi\right)(\Omega^2/K^2 + \tfrac{8}{45}\sin^2\psi) = 0. \tag{8.3.93}$$

The roots following from the condition that the second bracket vanishes have been obtained by Laedke and Spatschek (1982b) for the special case $\psi = \pi/2$. We will now write our result in more general notation, obtaining consistency with Laedke and Spatschek and, at the same time, illustrating the generation of more general dispersion relations by similarity. In the soliton limit we have, from (8.3.55) when $\xi = +2/\sqrt{27}$

$$n_2 = -1/\sqrt{3}, n_3 = 2/\sqrt{3}. \tag{8.3.94}$$

Thus the soliton amplitude is $n_3 - n_2 = \sqrt{3}$. If we wish to recover more general amplitudes, say $3\eta^2$, we simply use the similarity transformations

$$n_o(x, y, t) \to \alpha n_o(\alpha^{-\frac{1}{2}}x, \alpha^{-\frac{1}{2}}y, \alpha^{-3/2}t) \tag{8.3.95}$$

with $\alpha = \sqrt{3\eta^2}$. Now Ω^2/K^2 should be multiplied by α^2:

$$\Omega^2/K^2 = -\Gamma^2/K^2 = -\tfrac{8}{15}\alpha^2\sin^2\psi = -\tfrac{8}{15}\eta^4\sin^2\psi. \tag{8.3.96}$$

Thus the growth rate divided by K is proportional to η^2. If we also reintroduce Ω_c we have

$$\Gamma^2/K^2 = \tfrac{4}{15}(1 + \Omega_c^{-2})\eta^4\sin^2\psi \tag{8.3.97}$$

which is essentially the result obtained by Laedke and Spatschek but now extended to all ψ.

Laedke and Spatschek considered the problem of extending the soliton, $\psi = \pi/2$ result (8.3.91) to finite K. A single maximum in Γ and a cut-off $K_c \simeq \eta$ for unstable K were found, though the full $\Gamma(K)$ curve was not found. This same behaviour was observed for nonlinear waves, $\psi = \pi/2$, extending (8.3.91) (Infeld and Frycz (1987)). However, Laedke and Spatschek's method (complementary variational principles) is not extendable to non-soliton solutions (or even $\psi \neq \pi/2$). We give the result in Fig. 8.12(*b*), together with a numerically found dispersion curve, seen to fit in the Laedke–Spatschek region. Fig. 8.12(*c*) shows how a wave destabilizes due to this perpendicular instability, and 8.12(*d*) shows how instability can be passed on to a second soliton during a collision.

8.3.6 The variational method

Another method for checking the very limited case of $\psi = \pi/2$, that is for perturbations propagating across a soliton solution is the variational method (Lavrienter (1943), Morse and Feshbach (1953), Makhankov

(1978), Anderson *et al.* (1979)). This method is at times referred to as variation of constants, at others as variation of action.

Consider a two-dimensional, nonlinear evolution equation

$$P(u, u_t, u_x, u_y, u_{xx}, \ldots) = 0 \tag{8.3.98}$$

admitting soliton solutions

$$u = u_s(t, x, A_1, A_2 \ldots). \tag{8.3.99}$$

We further assume (8.3.92) can be derived from a Langrangian density $L(u, u_t, u_x, u_y \ldots)$

$$\delta \int\int\int L \mathrm{d}x \mathrm{d}y \mathrm{d}t = 0. \tag{8.3.100}$$

We now integrate the x dependence out, but still allow the A_j to vary slowly with y and t:

$$\int L(u, u_t, u_x \ldots) \mathrm{d}x = \langle L \rangle. \tag{8.3.101}$$

The dynamics of the perturbed soliton are then obtained from the reduced variational principle

$$\delta \int\int \langle \mathrm{L} \rangle \mathrm{d}t \mathrm{d}y = 0, \tag{8.3.102}$$

leading to the Euler–Lagrange equations

$$\delta \langle L \rangle / \delta A_j = 0 \qquad j = 1, 2. \tag{8.3.103}$$

After linearization around the unperturbed values of the parameters, (8.3.103) yields an approximation for the dispersion relation $\Omega(K_y)$ around a soliton. In the traditional form of the calculation, the variations are linked by the same relations as the quantities varied. Thus, for example, the variation of the amplitude is proportional to that of the square of the width as in (8.1.2). This gives a good estimate for Ω/K_y or Γ/K_y when K_y is small; when amplitudes and widths alter *very* gradually in y. However, Anderson *et al.* (1979) found that, for *finite* K_y, a superior model for $\Omega(K_y, A_i)$ is obtained by allowing these quantities to vary independently. Thus, for example, a correct estimate of the cut-off K_{yc} for instability of the ZK soliton can be obtained (Infeld and Frycz (1987)).

The simpler version of the above method for investigating solitons for perpendicular stability is covered in more detail by Makhankov (1978).

8.4 A more physical approach leading to an assessment of models

We will now see how the above results, obtained from model equations, stand up to results taken from a more complete physical analysis.

8.4.1 Form of the waves considered

In Chapter 5 we derived the two models of Section 8.3 (KP and ZK). The initial set of equations for a hot electron, cold ion, magnetized plasma in which the electron inertia is neglected was

$$\partial_t n + \nabla(n\mathbf{v}) = 0 \tag{8.4.1}$$

$$\partial_t \mathbf{v} + (\mathbf{v}\cdot\nabla)\mathbf{v} + \nabla\phi + \Omega_c \hat{\mathbf{x}} \wedge \mathbf{v} = 0 \tag{8.4.2}$$

$$\nabla^2\phi = e^{\phi} - n. \tag{8.4.3}$$

For details see Section 5.2. We now look for stationary solutions along **x**, travelling from right to left, and take them to be functions of x only. Upon integration:

$$nv = M \tag{8.4.4}$$

$$\tfrac{1}{2}v^2 + \phi = \tfrac{1}{2}M^2 \tag{8.4.5}$$

$$\tfrac{1}{2}\phi_x^2 = e^{\phi} + M\surd(M^2 - 2\phi) - C, \tag{8.4.6}$$

where M and C are positive constants of integration. We take $M > 1$. When $n \to 1, \phi \to 0$ at $|x| \to \infty, C = 1 + M^2$ and M can be interpreted as the soliton Mach number. In general there would be three constants of integration, one of which can always be removed by rescaling variables (and shifting ϕ) to obtain (8.4.4)–(8.4.6).

The phase analysis performed as in Section 6.2 leads to nonlinear wave solutions between the linear wave limit, given in parametric form by ($s = \phi_{\mathrm{LW}}$):

$$M_{\mathrm{LW}} = e^s\surd[2s/(e^{2s} - 1)] \tag{8.4.7}$$

$$C_{\mathrm{LW}} = e^s(e^{2s} + 2s - 1)/(e^{2s} - 1), \tag{8.4.8}$$

and the soliton value

$$C_S = 1 + M_S;\ M \leqslant 1.5852 \tag{8.4.9}$$

(Fig. 8.13).

For greater M, the large amplitude boundary curve is given by the condition that the square root in (8.4.6) be non-negative and so

$$C_{\mathrm{MW}} = e^{M^2}/2. \tag{8.4.10}$$

for the maximum amplitude wave (square root is zero).

The quantity $C_{\mathrm{LIN}} - C$ measures the nonlinearity (its departure from a cosine), and for interesting parameters the soliton is the wave of maximum amplitude, as well as being the 'most nonlinear wave' of the M family.

In the small $M-1$ limit, we can take $C=1+M^2$ and expand (8.4.4)–(8.4.6) in powers of $M-1$ to obtain

$$\phi \simeq 3(M-1)\mathrm{sech}^2[\{\tfrac{1}{2}(M-1)\}^{\frac{1}{2}}x], \qquad (8.4.11)$$

as the approximate soliton solution (Exercise 4 of Chapter 5). This tells us how to rescale KP or ZK results so that amplitudes coincide, as in Section 8.3 (Exercise 8).

8.4.2 Unmagnetized plasmas, $\Omega_c=0$

Formulas for Ω/K for this case are given in Appendix 3.

Fig. 8.14 gives phase diagrams $\Omega/K(\psi)$ for plane wave perturbations to the solutions (8.4.4)–(8.4.6) propagating out at an angle ψ between the wave vector of the perturbation **K** and the direction at which the basic nonlinear wave or soliton propagates, that of **x**. It is convenient to introduce

$$\mu=\frac{C_{\mathrm{LIN}}-C}{C_{\mathrm{LIN}}-C_{\mathrm{S}}}, 0\leqslant\mu\leqslant 1, M\leqslant 1.5852 \qquad (8.4.12)$$

$$=\frac{C_{\mathrm{LIN}}-C}{C_{\mathrm{LIN}}-C_{\mathrm{MW}}}, 0\leqslant\mu\leqslant 1, M\geqslant 1.5852.$$

Thus the existence region of nonlinear wave solutions in Fig. 8.13 is

Fig. 8.13. Existence of nonlinear ion acoustic waves and solitons in parameter space. Region limited by linear waves (LW), solitons (S), and nonlinear waves (W).

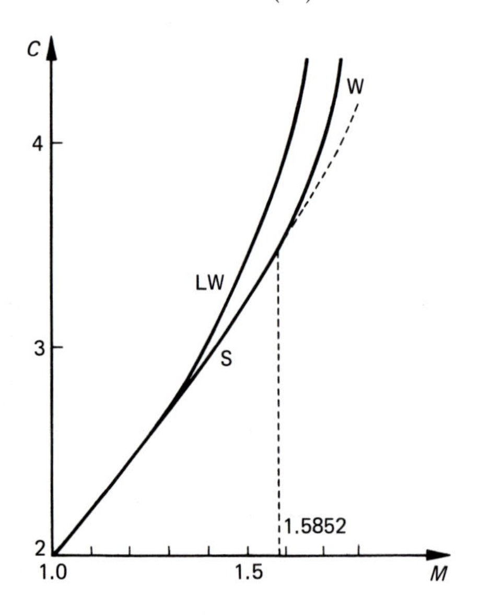

transformed into a semi-infinite rectangular slab, which is preferable for presentation.

As mentioned in Section 8.3, each little polar plot corresponds to one nonlinear wave or soliton (one μ, M pair). It is possible to think of the whole diagram (Fig. 8.14) as representing a huge plasma pond in which, at any given μ, M point, a nonlinear wave or soliton is propagating from right to left. This wave is not indicated, but is understood. The polar plot at this point then illustrates how a linear plane wave modulation will propagate out locally in the coordinate system of the basic wave structure. By slanting a ruler at an angle ψ to the horizontal at the origin of the polar plot, we find

Fig. 8.14. CMA type diagram showing polar plots of phase velocities for different values of μ, M. The origin of each polar plot is indicated by a dot. The M axis corresponds to the linear wave limit. The soliton limit is on the right below $M = 1.5852$. Only real Ω/K are shown, missing angular segments corresponding to instabilities. The KP plots, drawn underneath the main diagram, are seen to be very similar to the plots for small $M - 1$. From Infeld and Rowlands (1979a).

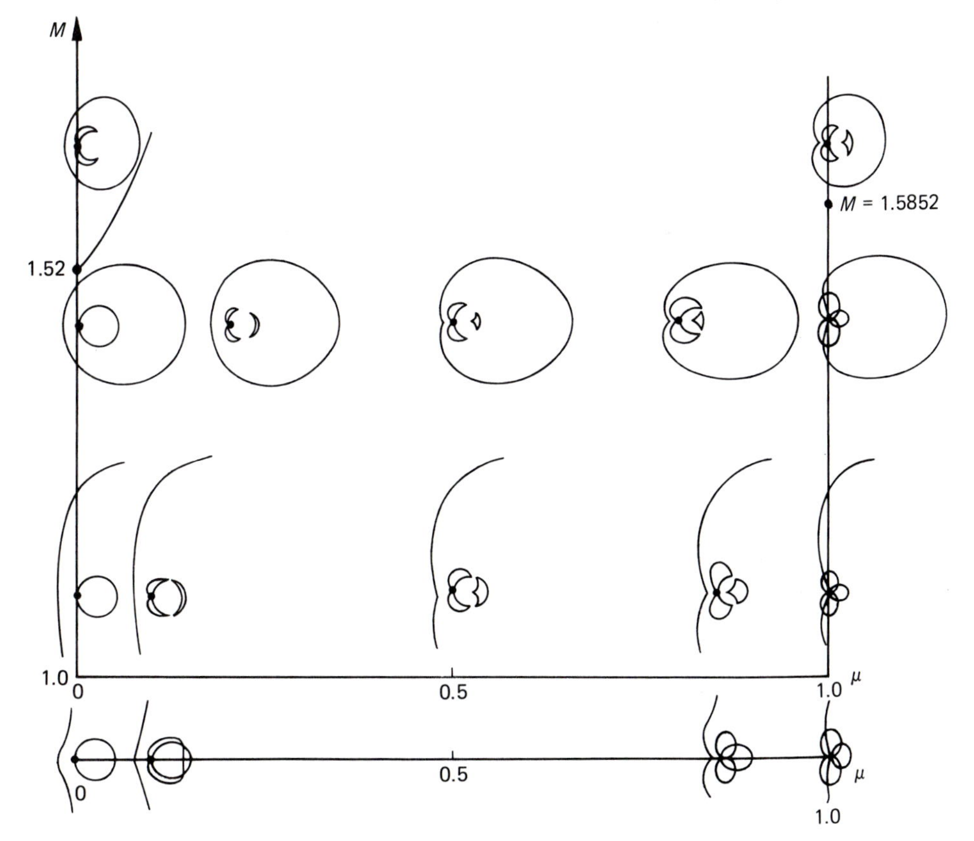

the phase velocity $\Omega/K(\psi)$ of a plane wave propagating out at angle ψ to the basic nonlinear structure (wave or soliton). The missing angular segments correspond to unstable small amplitude perturbations. Plots are not drawn to scale, but the outer root at $\pi/2$ is always 1. This diagram is a nonlinear generalization of a so called CMA diagram of linear theory (from the names Clemmow, Mullaly and Allis, see Chapter 2 and Clemmow and Mullaly (1955), Stix (1962)). By a CMA diagram we will designate the whole collection of polar plots, thus Fig. 8.14 is one CMA diagram. The main difference between our Fig. 8.14 and a classical CMA diagram, apart from the presence of the nonlinear wave, is that each little polar plot is moving on its own conveyor belt at a uniform velocity in the negative x direction. If not for this, the outer phase curves would be more circular.

Instabilities are never found for $\psi = \pi/2$. Solitons ($\mu = 1$) are stable for all

Fig. 8.15. Schematic diagrams of real characteristic velocities for ion acoustic wave problem, $\psi = 0$. This diagram shows the nature of the variation of the four values of Ω/K for fixed $M \leqslant 1.52$ as a function of the amplitude. The linear root, which splits in two, varies initially as $\sqrt{\mu}$. For $M > 1.52$ drawings are similar, but roots merge to the right of the M axis (this instability seems of very little physical importance, as it is limited to very small amplitude waves indeed). This figure was drawn for $M = 1.3$. It is instructive to compare it with Fig. 8.9(a). From Infeld and Rowlands (1979a).

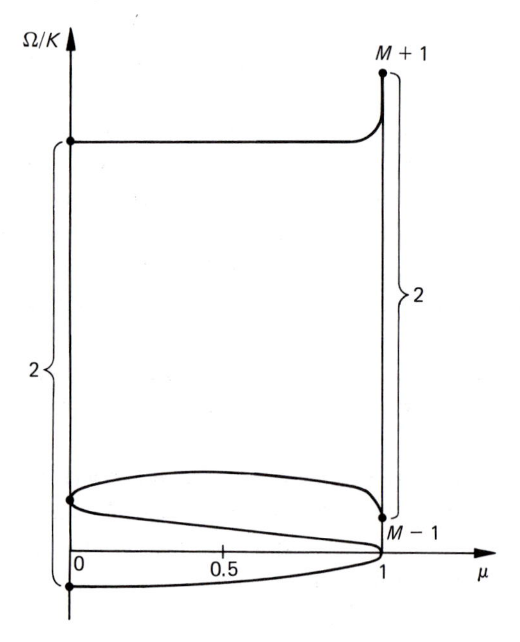

ψ. Their phase curves are given by

$$(\Omega^2/K^2 - f^2(M)\sin^2\psi)([\Omega/K - M\cos\psi]^2 - 1) = 0, \quad (8.4.13)$$

corresponding to a sound mode in the coordinate system of the soliton, plus two stable modes, leading to group signals propagating up and down the y axis (Fig. 8.10(c)).

For general μ, M and all interesting cases other than solitons, instabilities appear between two angles, both acute. As there are two real roots in this region and the coefficients of the quartic are real, the remaining two roots must be complex conjugate and so one represents an instability. These acute angle instabilities are comparatively weak and their growth rates are less than 1% of the characteristic frequency of the problem, $\sqrt{(4\pi ne^2/m_i)}$ (one in our units).

Underneath the main CMA diagrams, polar plots of $\Omega/K(\psi)$, following from KP, are shown. Here, in the notation of Section 8.3,

$$\mu = \frac{\xi_{LIN} - \xi}{\xi_{LIN} - \xi_{SOL}}. \quad (8.4.14)$$

The plots are seen to be very similar to those for small $M-1$ on the main diagram, but they miss the instability found for all non-zero $M-1$, non-soliton solutions. Instead, the KP plots show a crossing of the curves at a given angle. One could say that in the full problem a bifurcation about this angle has occurred, leading to an instability. A small amplitude A^2 calculation, as in Chapter 5, confirms this bifurcation (Infeld and Rowlands (1979a), and, without the interpretation, Kako and Hasegawa (1976)). In the language of Section 4.5, this is a class I, $M=2$ stability.

When the amplitude of the nonlinear wave or soliton in the KP model is adjusted to that of the full model (this is equivalent to specifying ε in the derivation of Section 5.2), the angle at which roots cross will move up or down as this amplitude is increased or decreased, due to the uneven stretching of x and y (Section 5.2). In Fig. 8.14, KP amplitudes roughly correspond to those of the waves and soliton of the bottom row drawn in the μ, M diagram.

In the soliton limit, unstable angular segments disappear and the KP model is virtually flawless. The same is true of the small amplitude limit $M \to 1$.

Apart from furnishing a description of the propagation properties of a plasma medium sustaining a nonlinear wave structure, Fig. 8.14 gives some insight into the applicability of KP in plasma physics. This model passes the test quite well, even though only one term represents two-dimensional effects. Quite a lot is thus accomplished on very little.

Fig. 8.15 shows characteristic velocities obtained for one-dimensional propagation of perturbations, $\psi=0$. It should be compared with Fig. 8.9 to gain an assessment of the KdV model in the one space dimensional plasma physics context. This model is seen to be very successful when viewed in this light.

Recently the above calculation was generalized to non-zero T_i, but amplitude small (left hand strip, Fig. 8.14). The unstable angular section is shifted down towards the x axis (Infeld and Rowlands (1981)). This effect was pronounced even for $T_i/T_e=10^{-1}$. Thus one should make sure that $T_i/T_e<10^{-1}$ in a given situation before applying (8.4.1)–(8.4.3) or KP.

8.4.3 Magnetized plasma, $\Omega_c>0$

For this case no Lagrangian is known and K expansion is the only available method. When it is applied, the dispersion relation found from (8.4.1)–(8.4.3) for non-zero external magnetic fields is still a quartic in Ω/K and is given in Appendix 3. Of course, both the basic nonlinear (wave or soliton) profiles and the $\psi=0$ dynamics are unaffected by the magnetic field.

A full CMA diagram would have to be three-dimensional in parameter space, as a third Ω_c axis should now be added to Fig. 8.14. Here we will just analyse one nonlinear wave (μ, M) in some detail, admitting all Ω_c in an interval. Salient features of general wave dynamics will be covered, though the soliton case $(\mu=1)$ must be treated separately.

In practical, physical terms, $\Omega_c=1$ corresponds to a very strong external magnetic field. In other words, physical instances of Ω_c much larger than one, either in the solar atmosphere or laboratory context, are rare. (It is several orders of magnitude smaller in most cosmic contexts.) Since Ω_c scales as $B_o n_i^{-\frac{1}{2}}$ (Section 5.2) this expression can be of the same order in very different situations.

Fig. 8.16 shows how the dynamics of one nonlinear wave vary as Ω_c is increased from zero. As Ω^2/K^2 is always real for $\psi=\pi/2$ (Appendix 3 and Exercise 10), we have singled this angle out for presentation. The first part of Fig. 8.16(a) shows minus this value, Γ^2/K^2, as a function of Ω_c. Whenever Γ^2/K^2 is positive, the nonlinear wave is unstable. Resonances for $\Omega_c=\Omega_{cn}$:

$$\Omega_{cn}=2\pi n\lambda^{-1}\langle v^{-1}\rangle^{-1}, n=1.2\ldots \tag{8.4.15}$$

complicate the overall picture. These are introduced by the K expansion and the true dispersion curve would not include them for finite K (it would thus resemble one of the curves of Fig. 8.17).

The KP model is seen to give a good estimate for very small Ω_c. The large Ω_c model, ZK, is also confirmed. Paradoxically, it is 'better' than the equations it was derived from in the sense that a K expansion introduces no resonances for ZK. Following the $\Gamma^2/K^2(\psi=\pi/2, \Omega_c)$ chart, polar plots for a few chosen Ω_c values are given in Fig. 8.16(*b*).

The polar plots of Fig. 8.14 for $\Omega_c=0$ showed how very particular and limited the $\psi=\pi/2$ analysis can be. Even when perpendicular perturbations were stable, acute angle instabilities tended to appear. Now similar plots for Ω_c non-zero confirm this. Even when stability is indicated by Fig. 8.16(*a*), acute unstable angles usually appear (Fig. 8.16(*b*)).

Fig. 8.16. (*a*) Γ^2/K^2 as a function of Ω_c for $\psi=\pi/2$. Here $\mu=0.939$ and $M-1=5\times10^{-3}$. Both the KP and ZK results are indicated by broken lines. The resonances are introduced by the expansion and so Γ^2/K^2 should be visualized without them and resembles one of the curves of Fig. 8.17. (*b*) Polar plots go through cycles as Ω_c is increased. Compare with Fig. 8.11. From Infeld *et al.* (1985).

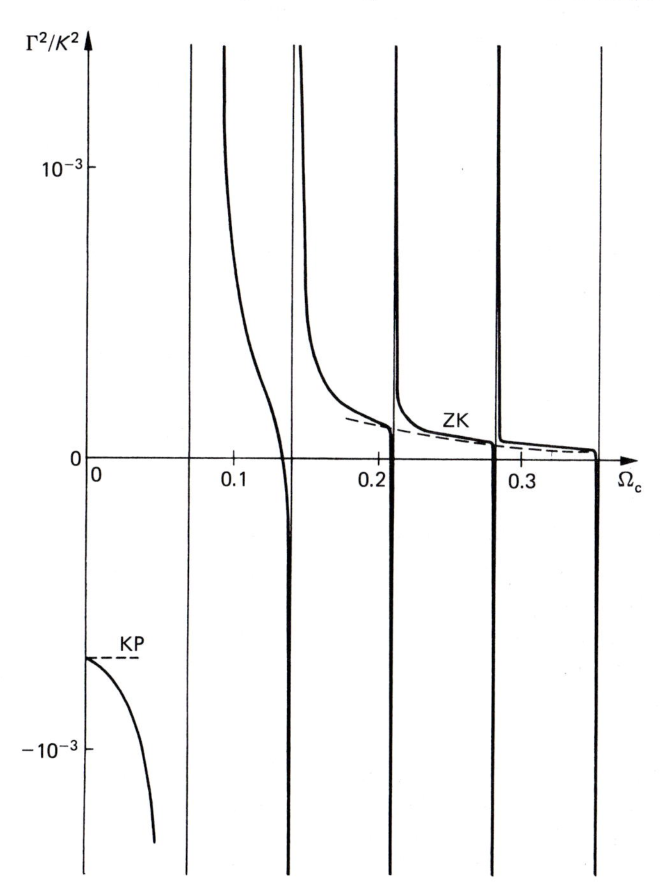

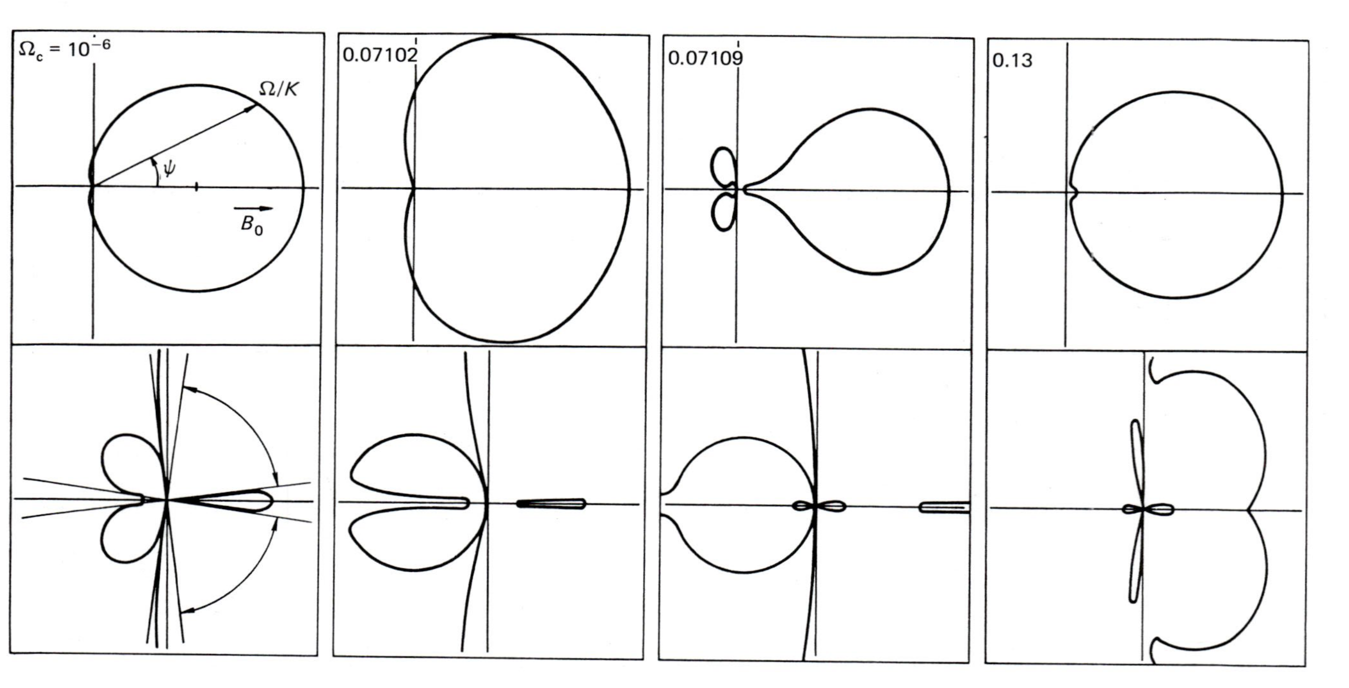

(*b*.i)

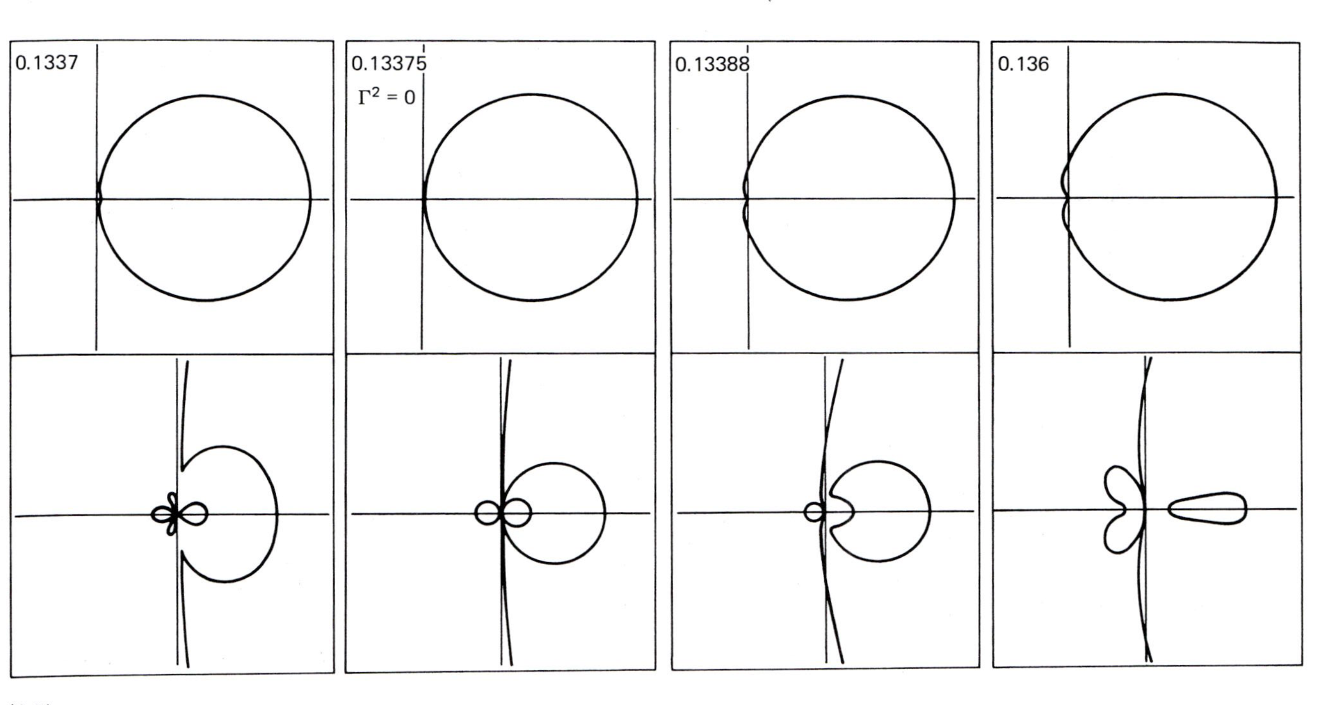

(*b*.ii)

This figure should be compared with Fig. 8.11 to assess ZK for all ψ.

In the soliton limit $\lambda \to \infty$, resonances disappear. Growth rates for $\psi = \pi/2, \mu = 1$ and three M values are given as functions of Ω_c in Fig. 8.17. Again ZK is seen to be a good model.

The formulae backing these figures and their discussion can be found in Appendix 3, but derivation would take up too much space and would involve little that is conceptually new. Procedure follows that of the ZK calculation (Section 8.3) though considerably more work is involved.

Fig. 8.17. Γ/K as a function of Ω_c for $\psi = \pi/2$ for three solitons propagating along a magnetic field. Thus $\mu = 1$. The ZK values are again indicated by broken lines and are good fits for large Ω_c and $M - 1$ small. All but broken lines and M values from Laedke and Spatschek (1981).

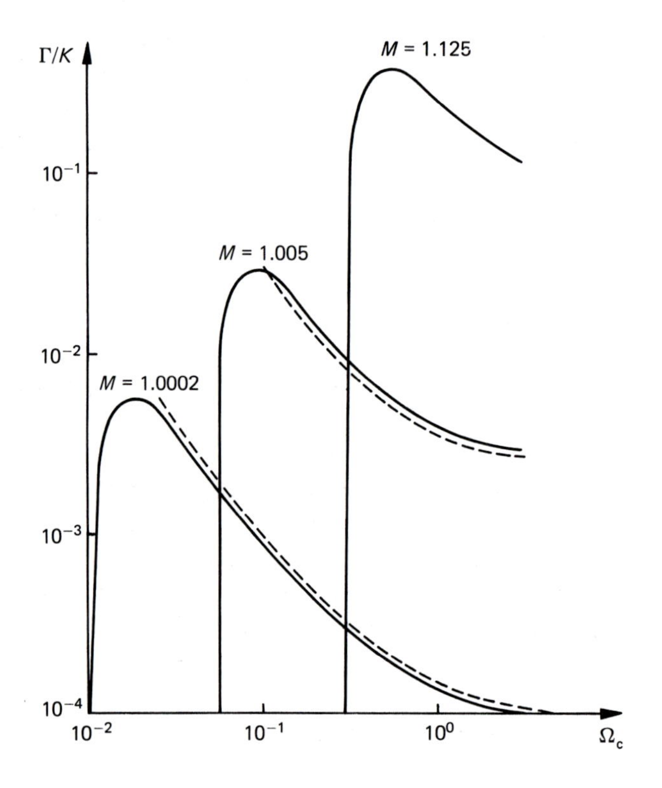

8.5 Dynamics of nonlinear wave, shock and soliton solutions to the cubic nonlinear Schrödinger equation

The cubic nonlinear Schrödinger equation figured prominently in Chapter 5 (Sections 5.3, 5.4, and 5.6), where it played an important role in weakly nonlinear wave theory. It appears in classical physics in both elliptic:

$$i\partial_t\phi + c(\partial_x^2 + \partial_y^2)\phi + b\phi + a|\phi|^2 = 0 \tag{8.5.1}$$

and hyperbolic

$$i\partial_t\phi + c(\partial_x^2 - m^2\partial y^2)\phi + b\phi + a|\phi|^2\phi = 0 \tag{8.5.2}$$

form. Here all constants a, b, c, m are taken to be real, though complex constants have appeared in a solid state theory context, see Stewartson and Stuart (1971), and Pawlik and Rowlands (1975). The elliptic form (8.5.1) describes the envelope of electron Langmuir waves in a plasma (Chapter 9). It also describes at least two other plasma wave modes that occur in the ionosphere (whistlers and ion cyclotron waves), and makes an appearance in nonlinear optics (Shimizu and Ichikawa (1972), Kako (1972) and (1974), Taniuti and Washimi (1968), Hasegawa (1972)).

The hyperbolic form (8.5.2), with $m^2 = 1$, models the envelope of electromagnetic (lower hybrid) waves in a plasma (Sen *et al.* (1978)). As we saw in Section 5.3, the small amplitude wave envelope on deep water is described by $m^2 = 2$ (Davey and Stewartson (1974), Dysthe (1979)).

Without loss of generality, we can assume $a = 1, c = \pm 1$, all obtainable by rescaling ϕ and $\mathbf{x}$, and look for common solutions to (8.5.1) and (8.5.2) that are functions of x only. Any one such solution $\phi_o(x)$ will generate a whole class via the transformation

$$\phi = \phi_o(x - V_D t)e^{i[V_D(x - V_D t)/2 + \omega_o t]} \tag{8.5.3}$$

just be redefining b. The function ϕ_o can be complex even when a, b, c are real and for this case a reasonably complete geometrical optics stability analysis was performed by Infeld and Ziemkiewicz (1981a). However, we will now concentrate on real ϕ_o, a case which is found to cover most of the interesting behaviour (Infeld and Rowlands (1980)). The constant, nonlinear wave, shock and soliton solutions were found and the corresponding curves drawn in (ϕ_{ox}, ϕ_o) phase space in Section 6.2 (Fig. 6.1).

As (8.5.1) and (8.5.2) are complex, so will the perturbations be. Thus we take for perturbations around $\phi_o(x)$ and one chosen Ω:

$$\phi(\mathbf{x}, t) = \phi_o(x) + \delta\phi_1(\mathbf{x})e^{i(\mathbf{K}\cdot\mathbf{x} - \Omega t)} + \delta\phi_2(\mathbf{x})e^{i(-\mathbf{K}\cdot\mathbf{x} + \Omega^* t)} \tag{8.5.4}$$

We insert this in (8.5.1) and linearize, using both (8.5.1) and its complex conjugate. An alternative method is to perturb the amplitude and phase of $\phi(\mathbf{x}, t)$ separately and *then* linearize. In both methods we have two unknown, perturbed quantities and as many equations. The first method is more usual (Rowlands (1974), Infeld and Rowlands (1980)), but the second was used by Infeld and Ziemkiewicz (1981a).

8.5.1 Results of a general stability calculation

Either method and a K expansion as in Section 8.3 yield the following biquadratic in Ω/K:

$$\begin{aligned}
&(\Omega/K)^4 + D(\psi)(\Omega/K)^2\cos^2\psi + A(\psi)\cos^4\psi = 0 \\
&D(\psi) = a_1(\psi) + b_1(\psi) + a_2 b_2, \\
&A(\psi) = a_1(\psi) b_1(\psi), \\
&K_x = K\cos\psi.
\end{aligned} \tag{8.5.5}$$

For the moment we will concentrate on (8.5.1). If, as in the analysis of Section 6.2, ϕ_o satisfies

$$\frac{c}{2}\phi_{ox}^2 = B - \frac{b}{2}\phi_o^2 - \tfrac{1}{4}\phi_o^4, \tag{8.5.6}$$

three separate formulae for the a_i and b_i functions result, depending on the signs of b, c and B as follows:

Case 1. ($c=1$, any b, $B>0$):

$$\begin{aligned}
&p^2 = 2\sqrt{(b^2+4B)},\, s^2 = (\sqrt{(b^2+4B)} - b/p^2) \leqslant 1,\, x = E(s)/K(s) \leqslant 1, \\
&H(x) = 1 - 2x + x^2/(1-s^2),\, A_1 = 1 - (1-2s^2)x/(1-s^2),\, A_2 = 1 - 2x \\
&A_3 = 1 - x/(1-s^2) \\
&a_1 = 2s^2p^2A_3[1 + A_1^2(1-s^2)\tan^2\psi/3s^2]/HA_1 \\
&b_1 = 2s^2p^2A_1[1 - A_3^2(1-s^2)\tan^2\psi/s^2]/HA_3 \\
&a_2b_2 = -2s^4p^2[A_2A_3 - A_1]^2/H^2A_1A_3,
\end{aligned} \tag{8.5.7}$$

Case 2. ($c=1, b\leqslant 0, -b^2/4 \leqslant B \leqslant 0$)

$$\begin{aligned}
&q^2 = 2\sqrt{(b^2+4B)}/(\sqrt{(b^2+4B)} - b) \leqslant 1,\, x = E(q)/K(q) < 1, \\
&B_1 = 1 - (2-q^2)x/2(1-q^2) \\
&H(x) = (1 - q^2 - x^2)(2 - q^2) \\
&a_1 = 2|b|q^4x[1 - 4B_1^2\tan^2(\psi)(1-q^2)/3q^4]/B_1H \\
&b_1 = 8|b|(1-q^2)^2B_1[1 + x^2\tan^2\psi/(1-q^2)]/xH \\
&a_2b_2 = -8|b|(1-x)^2(1-q^2-x)^2(1-q^2)(2-q^2)/xB_1H^2
\end{aligned} \tag{8.5.8}$$

Case 3. $(c=-1, b\leqslant 0, -b_2/4\leqslant B\leqslant 0)$

$$s^2=[1-\surd(1+4Bb^{-2})]/[1+\surd(1+4Bb^{-2})]\leqslant 1, \mathrm{x}=E(s)/K(s) \tag{8.5.9}$$

$$A_1=1-x, A_2=1-(1+s^2)x/(1-s^2), A_3=1-2x/(1-s^2)$$
$$H=1-(2x-x^2)/(1-s^2)$$
$$a_1=-4|b|s^2A_1[1-A_2^2\tan^2\psi/3s^2]/(1+s^2)HA_2$$
$$b_1=-4|b|s^2A_2[1+A_1^2\tan^2\psi/s^2]/(1+s^2)HA_1$$
$$a_2b_2=-4|b|s^4(A_2-A_1A_3)^2/(1-s^4)A_1A_2H^2.$$

Fig. 8.18. Parameter space (b, B) as divided into various types of solution, and CMA diagrams. For basic nonlinear structures, omitted here, see Fig. 6.2. (*a*) $c=1$, elliptic form of NLS; (*b*)$c=-1$, elliptic form. For detailed behaviour when $\psi=0$ see Fig. 8.20.

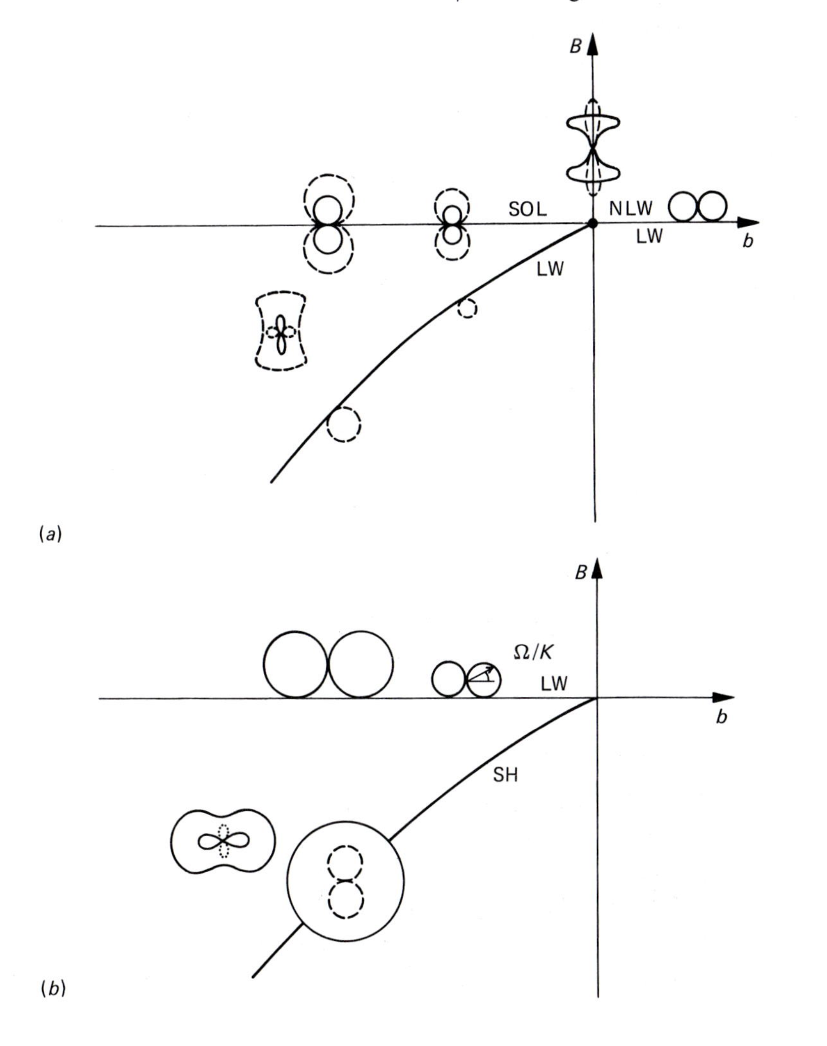

Fig. 8.19. (*a*) Schematic of changes in some polar plots when solving (8.5.2) as compared with (8.5.1). (*b*) $c=1$, hyperbolic cubic NLS (8.5.2). The basic nonlinear structures and their stability properties are the same as in Fig. 8.18(*a*) for $\psi=0$.

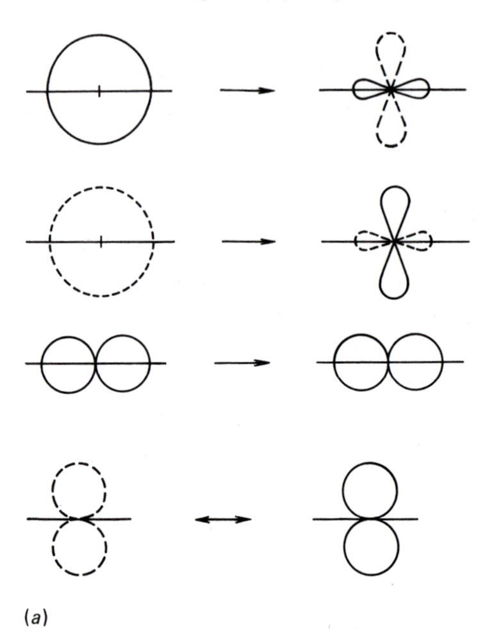

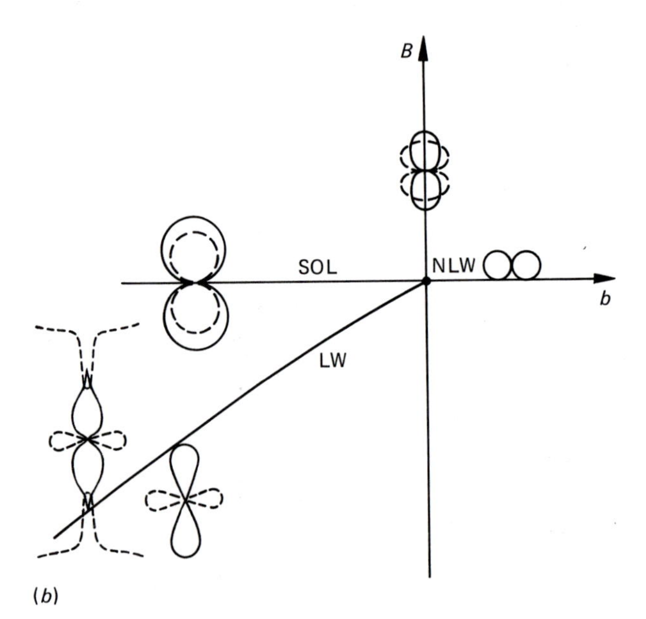

These formulae yield the linear wave limits found in Chapter 5 when s or q tend to zero. For (8.5.2) we simply change the coefficient of $\tan^2\psi$:

$$\tan^2\psi \to -m^2\tan^2\psi. \tag{8.5.10}$$

CMA diagrams for the elliptic case are given in Fig. 8.18. Fig. 8.19 shows how some simple polar phase plots (usually corresponding to linear or soliton limits) are affected by (8.5.10). Fig. 8.19(*b*) gives the CMA diagram for the hyperbolic case.

8.5.2 One-dimensional dynamics: $\psi=0$

Fig. 8.20 shows the $\psi=0$ values of Ω/K, which as we know are also characteristics in one dimension, for Case 3. All roots are real and different characteristics are seen to merge in the linear and shock wave limits* (left and right hand borders). Near the shock limit, two characteristics coalesce, just as they do for the soliton KdV limit (Fig. 8.9). Indeed, for small $1-s^2, |b|-|\phi_o|^2$ is the shape of a soliton train and Walstead (1980) has actually derived the KdV equation from the x, t version of (8.5.1) for this

* Shocks and solitons are really shock and soliton shaped *envelopes* of near-cosine shaped waves in this Section.

Fig. 8.20. Ω/K for $c=-1$, (8.5.1) as a function of the argument of the complete elliptic integrals s^2 defined in text. Linear waves to the left, shock limit on the right. This is a cross section of Fig. 8.18(*b*) from top to bottom, but limited to $\psi=0$. Note the similarity of the right hand limit to that of Figure 8.9(*a*).

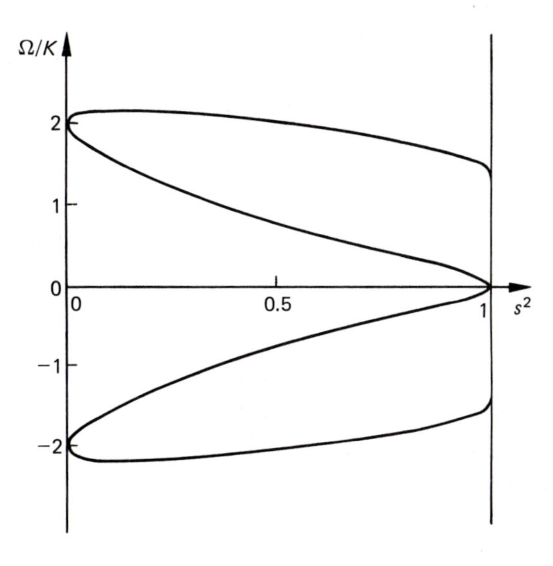

limit. Thus the dynamics of our 'soliton train' can be described by KdV and the conclusions of Section 8.3, namely constant amplitude and linearly increasing spacing of solitons for large times ($k \sim t^{-1}$) on characteristics, will be valid (Fig. 8.9(*b*)). This might have been expected (or at least suspected) from a comparison of the right hand limits of Figs. 8.9 and 8.20. For general s, Riemann variables for $\psi = 0$ have not been found. One dimensional stability for $c = -1$ is a particularly important result in the biological context, as the cubic NLS equation furnishes an approximate description of the propagation of impulses in proteins (Davydov (1981), (1985)). Numerical calculations confirm stability (Hyman *et al.* (1981)).

8.5.3 Oblique and perpendicular propagation of perturbations

Three-dimensional effects are seen to be dangerous, as they can destabilize all nonlinear stationary structures, including solitons. We see from Figs. 8.18(*a*) and 8.19(*b*) that both electron plasma and deep water wave envelope solitons are most unstable with respect to perpendicular perturbations, though instability is found for all $\psi \neq 0$. Here solitons have been treated as limits of nonlinear waves as $\lambda \to \infty$. As seen from Figs. 8.18(*a*) and 8.19(*b*), two different classes of nonlinear waves give the same soliton phase diagrams as $B \to 0$ limits ($B > 0$ and $B < 0$).

Fig. 8.18(*a*) is seen to give instabilities at all angles. It corresponds, among other physical problems, to that of nonlinear wave and soliton shaped envelopes of a plasma wave (Langmuir wave, Chapter 9). It is understood in this context that the changes in plasma density involved are small compared to the background density. Thus the nonlinear structures should differ strongly from linear wavetrains in the *shape* of the envelope, but the *amplitude* should remain small for NLS with a cubic nonlinearity to apply. In most soliton experiments, however, $(n_{SOL} - n_o)/n_o$ is of order 15% (Kim *et al.* (1974), Wong and Quon (1975), Ikezi *et al.* (1976)). To cope with this inconsistency, Wilcox and Wilcox (1975) and D'Evelyn and Morales (1978) derived a more realistic nonlinearity involving an exponential of $-|\phi|^2$, therefore saturating for large amplitudes. A three-dimensional analysis of planar waves and solitons satisfying this improved model equation indicates a decrease in growth rates of instabilities (Anderson *et al.* (1979) for envelope solitons, Infeld and Ziemkiewicz (1981b) for general nonlinear wave envelopes). Conversely, nonlinear Schrödinger equations with steeper (as compared with cubic) nonlinearities tend to lead to enhanced instability (Spatschek and Laedke (1982)). We will come back to the influence of the steepness of various nonlinearities in the NLS equation on stability in Sections 8.6 and 9.5.

Recent numerical investigations on the instability of nonlinear Langmuir wave envelopes (interior of Fig. 8.18(*a*)) confirm our result; instabilities appear at all angles ψ (Tajima *et al.* (1981)). After a while, recurrent behaviour was observed, as found in Section 5.6 from a theoretical calculation.

The nonlinear and shock waves of Fig. 8.18(*b*) are one dimensionally stable, but destabilize when more general geometry is considered. This drawing, apart from covering several specific problems mentioned above, should serve as a warning against relying too strongly on one-dimensional analysis, when describing waves in the real world.

The stability problem for the envelopes of deep water waves is formally identical to that for plasma waves when $\psi=0$ (Fig. 8.19(*b*)). However, for non-zero ψ, the hyperbolic form of the differential operator generates the substitution (8.5.10), leading to important differences. In the linear limit (LW) we see the Benjamin–Feir instability mentioned in Section 8.1 and spanning $\psi=\pm\tan^{-1}(1/\sqrt{2})$. Nonlinear wavetrains are seen to exhibit a new instability (Infeld (1980b)) for all ψ. (Infeld, Ziemkiewicz and Rowlands (1987) give an explanation of how this comes about in terms of branch pairing when the carrier wave amplitude is increased from zero.) This is an instability of a wave *envelope* and should not be confused with the large amplitude instabilities of nonlinear waves proper mentioned in Sections 8.1 and 8.6. However, they do have one feature in common as far as theory is concerned, as they only appear when a finite amplitude calculation is performed. Envelope solitons (SOL) are one-dimensionally stable, but unstable with respect to all non-zero ψ, small K perturbations. The perpendicular instabilities are the strongest and will cause initially planar envelope solitons to self-focus, eventually leading to a two-dimensional structure. (An interesting interpretation of the physics of this phenomenon is given in Kadomtsev's book, (1979).) This instability of envelope solitons for $\psi=\pi/2$ but general K has been investigated by Saffman and Yuen (1978). Instabilities were found for all K between zero and a critical value. In this context see also Cohen *et al.* (1976).

For some further results concerning the $\psi=\pi/2$ problem for *finite* K_y and both wave and soliton structures, see Martin *et al.* (1980).

8.6 Some general conclusions and possible future lines of investigation

Once a Fourier–Laplace analysis of a linearized physical problem tells us that a wave of the form $a\cos(\mathbf{k}\cdot\mathbf{x}-\omega t)$ is unstable and perturbations grow as $e^{\gamma t}$, one could raise two basic groups of questions; one 'geometrical' and one time-behavioural:

1. How would the growth rate Γ be affected by a more realistic description of the wave, that is by using the exact form $\phi(A, \mathbf{k}\cdot\mathbf{x}-\omega t)$ in place of a cosine, but keeping the perturbation small? This is especially important when the wave amplitude is large. How does the geometry considered affect the small perturbation stability? Would a strong external magnetic field improve the stability in the plasma wave case?
2. What would a fuller description of the time development of the perturbation $\delta\phi$ lead to? Obviously exponential growth must level off sooner or later. To answer this question we must relax the condition of infinitely small perturbations, as linearization invariably leads to exponential time dependence.

We will now try to summarize our answers to these two questions. Conclusions will be based on insight gained both in Chapter 5 and the present chapter.

1. Nonlinearity has an important effect on the stability of a wave, though this effect cannot be summarized by any simple rule or prescription. For example, there is no general rule that nonlinear waves will become more or less unstable as they steepen, and examples of both kind of behaviour are known. Introducing more than one dimension, on the other hand, almost invariably increases the family of instabilities. More often than not, a nonlinear wave or soliton that is one-dimensionally stable will become unstable in three dimensions, and therefore in the real physical world. This effect could, in some laboratory investigations, be limited by conducting experiments in very long, thin tubes, thus eliminating small K_y instabilities, which tend to be the most dangerous. Although Nature, or the conditions of an experiment, occasionally decree one-dimensionality, as in proteins or optical fibres, we are usually dealing with fully three-dimensional dynamics in the physical world. This is so in most hydrodynamic, ionospheric solar, and cosmic plasma applications.

   Strong external magnetic fields will by no means always stabilize nonlinear plasma waves, and indeed a destabilizing influence has been found.
2. (Our answer to this question will be based on an analysis of the cubic NLS equation, Section 5.6.) For spatially periodic systems

and times longer than that of one e folding, the perturbation either cyclicly exchanges its energy with the system, leading to a see-saw effect (Fermi–Ulam–Pasta recurrence), or a temporarily non-periodic recurrence is obtained, the system being represented by ergodic motion in phase space. Again, including more than one space dimension can change the picture fundamentally, adding the possibility of a slow thermalization. For other scenarios, see Chapter 10.

Thus, all in all, a one-dimensional analysis is seen to be too restricted, even though the basic nonlinear structure was planar. In fact, as mentioned in Section 4.6, observations of gravity waves on a deep water surface even furnish examples of one surface dimensional, nonlinear structures (surface profile depending on $x-ct$) developing into ordered, two surface dimensional nonlinear structures with both $x-ct$ and y periodicity! The agent is a two surface dimensional instability (Section 4.6 and Su *et al.* (1982)).

At present it would seem impossible to say beforehand how a nonlinear, stationary structure satisfying a given nonlinear differential equation will evolve, depending on its amplitude, steepness etc., and the answer for long-wave perturbations (geometrical optics) is best obtained by Whitham II or K expansion though there are other methods of less universal applicability (Hayes, complementary variational principles, Kuznetsov *et al.* (1984)). When there is no Lagrangian and the number of conservation laws is insufficient to completely describe the problem, K expansion may well be the only possible tool. It is also the only method that would enable one to find the K^n corrections ($n>1$).

When instability is obtained, the answer as supplied by any of these methods describes the initial stages of evolution only.

On the other hand, the correlation between the structure of a nonlinear equation and the dynamics of its stationary solutions might just possibly be simpler to grasp. A possible example of this is the enhanced instability of NLS waves as the steepness of the nonlinearity is increased (see also Section 9.5). The same effect has been observed for generalized KdV ($v^n v_x$ instead of vv_x). Perhaps a proper theory could be built around these observed correlations?

Thus, all in all, some general stability rules may emerge from future research. They seem to be lacking at the moment.

Exercises on Chapter 8

Exercise 1

Show that $\Omega_2 = 0$ and find Ω_3 for (8.2.40).

Find $\partial\omega/\partial k$ for the nonlinear Klein–Gordon wave keeping $\mathscr{L}_\omega$ fixed. Show that you have obtained the velocity of energy propagation (8.2.21). (This is a general result, Lighthill (1965).)

Exercise 2

Show that $\partial\lambda/\partial A = 0$ for the following choice of $V'(\phi)$:

$$V'(\phi) = \text{const.}[\sqrt{(1+\phi^{-1})} - 1].$$

(If you get stuck, see Section 6.3.)

Complete the Hayes calculation of Section 8.2 for this case, using

$$\frac{dx_\pm}{dt} = \mathscr{H}_{\text{IK}} \pm \sqrt{\mathscr{H}_{\text{kk}}\mathscr{H}_{\text{II}}}.$$

Show that both characteristic velocities are equal to U^{-1}.

Exercise 3

Find a stationary solution to the Sine–Gordon equation

$$(\partial_t^2 - \partial_x^2)\phi + \sin\phi = 0.$$

Investigate its stability using any of the four methods outlined in Section 8.2.

Exercise 4

Solve the stability problem of the Kadomtsev–Petviashvili equation to obtain (8.3.42) by K expansion.

Exercise 5

Derive (8.3.42) from (8.3.34) for two special cases:

(a) $\xi = 0$

(b) $Y = 0$.

Obtain (8.3.44) for the soliton limit by applying de l'Hospital's theorem to (8.3.42).

Exercise 6

(a) Find all three $\partial\Omega_j/\partial\mathbf{K}$ for KP in the

(i) linear limit

(ii) soliton limit

for both positive and negative dispersion. Can all branches for both dispersion signs be interpreted as characteristic surfaces? Describe how Fig. 8.10(*b*) could be drawn.

(b) Find $\Omega(K_y, \eta)$ for perpendicular perturbations in the soliton limit by variation of one constant. Is the result identical to that following from (a)(ii) for small K_y?

Exercise 7

Starting from

$$L = \dot{\psi}\psi_x + \tfrac{1}{3}\psi_x^3 - \tfrac{1}{2}\psi_{xx}^2 - \psi_{xy}^2,$$

for the Zakharov–Kuznetsov equation, show that

$$\mathscr{L} = -k_x W\sqrt{(1+2k_y^2/k_x^2)} - 2\beta_x B - \beta_x \gamma + U\beta_x + 2A$$

and re-obtain (8.3.90) by the Whitham II method by analogy with the KP derivation of Section 8.3.

Exercise 8
Show how to reduce the number of constants from three to two as in (8.4.4)–(8.4.6). Derive (8.4.7) and (8.4.8).

Assume $M-1$ small in (8.4.1)–(8.4.3) and expand. Show that for $C=1+M^2$,

$$\phi \simeq 3(M-1)\mathrm{sech}^2[\{\tfrac{1}{2}(M-1)\}x^{\frac{1}{2}}].$$

How should the results of Section 8.3 for KP be rescaled to model a given M value?

Exercise 9
Derive

$$\tfrac{1}{2}\phi_x^2 = \phi_{xx} + M(v+v^{-1}) - C$$

from (8.4.3) and (8.4.6). Use this equation to derive (A.3.5)–(A.3.7).

Exercise 10
Read Appendix 3.

Derive $\overline{\delta v_y}$. Try to derive equation (A.3.4) by introducing

$$i\Omega\delta\phi = n\delta v_x + v\delta n,$$

and reducing (8.4.1), (8.4.3) and x component of (8.4.2) to two equations for δn and $\delta\phi$ (δv_{y_1} is known). (For $\mathbf{K}\cdot\mathbf{B}_0 = 0$ we expand in K_y and the calculation simplifies.)

Exercise 11
Show that the equation

$$\partial_t n + \tfrac{1}{2}\partial_x^3 n + n\partial_x n + \tfrac{1}{2}(1+\Omega_c^2)(\partial_x^2 + \Omega_c^2)^{-1}\partial_{xyy} n = 0$$

gives KP and ZK in the appropriate limits. Show that a K expansion will give resonances as obtained in Section 8.4. Find the formula that replaces (8.4.15). (This equation is an improvement on one obtained by Laedke and Spatschek (1982a) in that the $\dfrac{1+\Omega_c^2}{2}$ factor extends validity to large Ω_c.)

Exercise 12
Investigate complex $\phi_0(x)$ solutions to (8.5.1). If

$$\phi_0 = \psi(x)e^{i\sigma(x)}$$

where ψ, σ are real, show how the $c=-1$ solutions, for which $\psi(x)$ tends to constant values at $x\to\pm\infty$, yield the shock solutions of Chapter 6 when $\sigma\to 0$.

Show that ϕ given by (8.5.3) solves (8.5.1) with b redefined. Find the new value of b. Derive as many CMA plots of Fig. 8.18 as you can by the small A methods of Chapter 5.

Read Martin *et al.* (1980) and try to relate their findings to those of Section 8.5.

9

Cylindrical and spherical solitons in plasmas and other media

9.1 Interest in higher dimensional plasma solitons

As we have seen, quite a lot is now known about one-dimensional plasma waves and solitons. Possibly as a continuation of this effort, or again possibly as a result of most scientists' impatience with a field once it has disclosed some of its secrets, a new discipline of cylindrical and spherical plasma soliton waves has come into being. Investigations began fairly recently (Maxon and Viecelli (1974a, b)), but progress is rapid. Similar phenomena in other fields of classical physics are mentioned briefly at the end in Section 9.7.

In mathematical terms, the cylindrical and spherical plasma (and hydrodynamic) solitons treated here are often described quite well by variants of the model equations that have already appeared in this book. Examples are the Korteweg–de Vries (KdV) cum Boussinesq family (ion acoustic solitons), and the nonlinear Schrödinger family in various geometries and with diverse nonlinearities, the simplest being a cubic term (Langmuir envelope solitons). Both are of course classes of idealized equations. Here we will see how they can be obtained in the higher dimensional, non-Cartesian plasma physics context, what their properties are and also some of the effects they do not describe. Finally, we will see how some of their predictions stand up to laboratory and numerical experiments. A few extensions of the above models will be suggested.

Comparatively little theoretical work has been done on the stability of higher dimensional solitons, existing analyses being either very restricted or incomplete. One could, of course, argue that the very existence and perseverence of many of these solitons in both laboratory and numerical experiments suggest that at least *some* specimens possess a healthy degree of stability. We therefore present some recent progress and collect the results to date.

Finally, semantics. Some people use the word 'soliton' to describe an object only if it is rigorously shape-preserving and stable even after a collision with or overtaking of a second soliton. Others tend to include collapsing and expanding objects such as will be encountered here. (The more orthodox, shape-preserving class are sometimes rather pretentiously called 'aristocratic solitons'.) We will simply use the term in its broader sense as denoting a consistently localized and basically single humped wave.

9.2 Unidirectional cylindrical and spherical ion acoustic solitons

We now consider a hydrogen plasma in which the electrons are hot, isothermal and weightless, and the ions cold and heavy (see Chapters 1 and 5). Thus, in our model

$$m_e/m_i \to 0, T_i/T_e \to 0.$$

As in Chapter 5, we work with the equations in dimensionless form, velocities being normalized to $\sqrt{K_B T_e/m_i}$, the ion sound velocity, ion density to the far field value n_{io}, and electric potential ϕ to $K_B T_e/e$. The electron momentum equation becomes

$$\nabla\phi - n_e^{-1}\nabla n_e = 0,$$

solved by $n_e = \exp(\phi)$. It is now possible to describe ion dynamics by a set of equations in which electron quantities do not appear at all (though they have their say through ϕ):

$$\partial_t n + \nabla(n\mathbf{v}) = 0 \tag{9.2.1}$$

$$\partial_t \mathbf{v} + (\mathbf{v}\cdot\nabla)\mathbf{v} + \nabla\phi = 0 \tag{9.2.2}$$

$$\nabla^2\phi = e^{\phi} - n. \tag{9.2.3}$$

These are the equations of continuity, momentum balance, and electric field divergence (Poisson). The ion subscript has been omitted. All dissipative effects are neglected in this description.

9.2.1 Model equations in non-Cartesian geometry

Starting from (9.2.1)–(9.2.3) we will now derive some simplified model equations for small amplitude cylindrical and spherical solitons or groups of solitons. This will be an extension of the procedure introduced in Chapter 5. Each equation will describe one class of solitons, all moving in the same manner with collapse, expansion, or longitudinal propagation velocities always slightly in excess of the ion sound velocity. Just from symmetry, there could be two basic soliton modes for which a cylindrical system would be optional: one tube shaped and collapsing (or expanding) on the axis (away from the axis); the other of constant shape and propagating along the axis of symmetry. We will denote these two basically different types by CI and CII respectively. Only CII in fact satisfies the purist's definition of a soliton (see Section 9.1). Each type will be described by a different model equation. Spherical solitons arising from (9.2.1)–(9.2.3) and treated here, *all* either collapse or expand.

9.2.2 Cylindrical soliton equations CI and CII

Equations (9.2.1)–(9.2.3) are, when written in cylindrical variables r, θ, z:

$$\eta_{1t}+r^{-1}(rnv_r)_{1r}+r^{-1}(nv_\theta)_{1\theta}+(nv_z)_{1z}=0 \tag{9.2.4}$$

$$v_{r1t}+v_r v_{r1r}+r^{-1}v_\theta v_{r1\theta}-r^{-1}v_\theta^2+v_z v_{r1z}+\phi_{1r}=0 \tag{9.2.5}$$

$$v_{\theta 1t}+v_r v_{\theta 1r}+r^{-1}v_\theta v_{\theta 1\theta}+r^{-1}v_r v_\theta+v_z v_{\theta 1z}+r^{-1}\phi_{1\theta}=0 \tag{9.2.6}$$

$$v_{z1t}+v_r v_{z1r}+r^{-1}v_\theta v_{z1\theta}+v_z v_{z1z}+\phi_{1z}=0 \tag{9.2.7}$$

$$r^{-1}(r\phi_{1r})_{1r}+r^{-2}\phi_{1\theta\theta}+\phi_{1zz}=e^\phi-n \tag{9.2.8}$$

The undesirable, explicit r dependence renders these equations much more difficult to solve as compared to the Cartesian versions. Our next step is to introduce coordinate systems co-moving with a soliton or group of unidirectional, small amplitude solitons. Thus cases CI and CII will require separate treatment.

For CI type imploding solitons, $r+t$ is a natural choice for one of the coordinates. We assume small amplitudes and weak coordinate dependence of the solitons, leading to the following choice of stretched variables (θ is unaltered)

$$\begin{array}{lll} \rho=-\varepsilon^{\frac{1}{2}}(r+t) & v_r=-\varepsilon v^{(1)}+\ldots & n=1+\varepsilon n^{(1)}+\ldots \\ z'=\varepsilon z & v_\theta=\varepsilon^2 u^{(1)}+\ldots & \phi=\varepsilon\phi^{(1)}+\ldots \\ \tau=\varepsilon^{3/2}t & v_z={}^{3/2}w^{(1)}+\ldots & \end{array} \tag{9.2.9}$$

We will also need

$$r=-\varepsilon^{-3/2}\tau-\varepsilon^{-\frac{1}{2}}\rho.$$

Signs are chosen such that $v^{(1)}$ is positive and ρ increases in the direction of motion. One might again wonder why the various powers of ε are chosen as they are. Some of the answers can be found in Section 5.2. The v_r expansion can be justified by finding a small amplitude solution to (9.2.4)–(9.2.8) as in Appendix 4. It satisfies

$$v_r x(\text{width})^2 \simeq \text{constant}, r\to\infty,$$

suggesting that the stretching of v_r should be the square of that for the principal space dependence.

In any case, we will presently see that (9.2.9) does lead to a sensible model. To lowest order we obtain, dropping the prime in z and the strokes for differentiation,

$$\begin{aligned}&\phi^{(1)}=n^{(1)}=v^{(1)},\ w_\rho^{(1)}=\phi_z^{(1)}\\&v_\rho^{(1)}=-\tau^{-1}\phi^{(1)}.\end{aligned} \tag{9.2.10}$$

The next order quantities satisfy

$$\begin{bmatrix}\partial_\rho, & 0, & 0, & 0, & -\partial_\rho\\ 0, & \partial^\rho, & 0, & 0, & -\partial_\rho\\ 0, & 0, & \partial_\rho, & 0, & \tau^{-1}-_\theta\\ 0, & 0, & 0, & \partial_\rho & -\partial_z\\ \partial_\rho, & -\partial_\rho, & 0, & 0, & 0\end{bmatrix}\begin{bmatrix}n^{(2)}\\ v^{(2)}\\ u^{(2)}\\ w^{(2)}\\ \phi^{(2)}\end{bmatrix}=\begin{bmatrix}v^{(1)}v_\rho^{(1)}-v_{\rho\rho\rho}^{(1)}\\ v_\tau^{(1)}+v^{(1)}v_\rho^{(1)}\\ v_\tau^{(1)}+v^{(1)}v_\rho^{(1)}+\rho\tau^{-2}v_\theta^{(1)}\\ w_\tau^{(1)}+v^{(1)}w_\rho^{(1)}\\ v_\tau^{(1)}+2v_\rho^{(1)}v^{(1)}+w_z^{(1)}+\tau^{-1}v^{(1)}\end{bmatrix} \tag{9.2.11}$$

The first, second and fifth rows are linearly dependent:

$$\mathbf{R}_1-\mathbf{R}_2-\mathbf{R}_5=0, \tag{9.2.12}$$

leading to a consistency condition involving first order quantities only. No other linear combinations of rows vanish, thus consistency will be ensured if

$$v_\tau^{(1)}+v^{(1)}v_\rho^{(1)}+\tfrac{1}{2}v_{\rho\rho\rho}^{(1)}+\frac{1}{2\tau}v^{(1)}+\tfrac{1}{2}w_z^{(1)}=0 \tag{9.2.13}$$

and, from (9.2.10),

$$w_\rho^{(1)}=v_z^{(1)}. \tag{9.2.14}$$

Explicit r dependence has merely been exchanged for a τ dependence, but we have otherwise simplified our description somewhat. As $\mathbf{v}^{(1)}$ is curl-free, a velocity potential ψ can be introduced

$$\mathbf{v}^{(1)}=\nabla\psi=(\partial_\rho,-\tau^{-1}\partial_\theta,\partial_z)\psi,$$

yielding

$$\mathrm{CI}\psi_{\rho\tau}+\psi_\rho\psi_{\rho\rho}+\tfrac{1}{2}\psi_{\rho\rho\rho\rho}+\frac{1}{2\tau}\psi_\rho+\tfrac{1}{2}\psi_{zz}=0 \tag{9.2.15}$$

(Infeld 1983). The ρ,τ version of this equation has been known ever since 1969. It also describes ring solitons on a shallow water surface (Lugovtzov and Lugovtzov (1969)), and was derived five years later in the plasma physics context (Maxon and Viecelli (1974a)). Here τ is taken in the range $(-\infty,0)$, and the equation breaks down in the vicinity of $\tau=0$.

Calculations for *expanding* cylindrical solitons are very similar, the basis of ρ being $r-t$.

For CII, introduce

$$\begin{array}{lll} z'=\varepsilon^{\frac{1}{2}}(z-t) & v_r=\varepsilon^{3/2}v^{(1)}+\dots & n=1+\varepsilon u^{(1)}+\dots \\ \rho=\varepsilon r & v_\theta=\varepsilon^{3/2}u^{(1)}+\dots & \phi=\varepsilon\phi^{(1)}+\dots \\ \tau=\varepsilon^{3/2}t & v_z=\varepsilon w^{(1)}+\dots & \end{array} \tag{9.2.16}$$

At first, this choice of expansion may seem hard to swallow, as it differs considerably from (9.2.10). The idea behind both (9.2.9) and (9.2.6), apart from introducing coordinates that follow the solitons, is that coordinate dependence is stronger along the principal direction of motion than perpendicular to it, just as for Cartesian structures in Section 5.2. Once the r, z, t stretching is accepted, the powers of ε in front of the three components of $\mathbf{v}$ in both CI and CII follow if we assume that they are derivable from a velocity potential $\psi \sim \varepsilon^{\frac{1}{2}}$. This however, is no more than hindsight.

Calculations similar to those for CI yield $\phi^{(1)}=n^{(1)}=w^{(1)}$, $\nabla \wedge \mathbf{v}^{(1)}=0$. Again we can introduce a velocity potential $\mathbf{v}^{(1)}=\nabla\psi$ and obtain CII

$$\psi_{z\tau}+\psi_z\psi_{zz}+\tfrac{1}{2}\psi_{zzzz}+\frac{1}{2\rho}(\rho\psi_\tau)_\rho+\frac{1}{2\rho^2}\psi_{\theta\theta}=0 \tag{9.2.17}$$

this equation was proposed (though not derived by expansion) by Petviashvili (1981).

9.2.3 Spherical solitons

The equation for spherical solitons is derived in Appendix 5 by the Lagrangian method of Section 5.2. The result is, in ρ, θ, v, τ variables,

$$\text{S} \quad \psi_{\rho\tau}+\psi_\rho\psi_{\rho\rho}+\tfrac{1}{2}\psi_{\rho\rho\rho\rho}+\frac{1}{\tau}\psi_\rho=0 \tag{9.2.18}$$

the ρ, τ dependent version of (9.2.18) is known and is identical to (9.2.18) (Maxon and Viecelli (1974b)). However, (9.2.18) as an equation in three space dimensions does contain additional information; that the angular dependence is of higher order.

9.2.4 Summary

Collecting all three equations we have $\mathbf{v}=\nabla\psi, n-1=\phi$ and

$$\text{CI} \quad \psi_{\rho\tau}+\psi_\rho\psi_{\rho\rho}+\tfrac{1}{2}\psi_{\rho\rho\rho\rho}+\frac{1}{2\tau}\psi_\rho+\tfrac{1}{2}\psi_{zz}=0, \ \phi=\psi_\rho$$

$$\text{CII} \quad \psi_{z\tau}+\psi_z\psi_{zz}+\tfrac{1}{2}\psi_{zzzz}+\frac{1}{2\rho}(\rho\psi_\rho)_\rho+\frac{1}{2\rho^2}\psi_{\theta\theta}=0, \ \phi=\psi_z \tag{9.2.19}$$

$$\text{S} \quad \psi_{\rho\tau}+\psi_\rho\psi_{\rho\rho}+\tfrac{1}{2}\psi_{\rho\rho\rho\rho}+\frac{1}{\tau}\psi_\rho=0, \quad \phi=\psi_\rho.$$

We see that the first three terms are essentially the same. All these models can be extended to include effects of non-zero ion temperature, two or more ion components, and/or collisions, as has already been done for the ρ, τ version of CI in the literature. The first two effects merely change coefficients, whereas the third adds a $\partial_\rho^3\psi$ term (in CI and S) or a $\partial_z^3\psi$ term (CII) when the right ordering is assumed (Section 5.2). This is called the Burgers term (Maxon (1976), Panat (1976), Tagare and Shukla (1977)).

Equation CI is particularly interesting in that its very form suggests that small amplitude, imploding or expanding cylindrical plasma solitons can deform in the z direction but should be azimuthally smooth. This result is in full agreement with experiments on plasma solitons (Hershkovitz and Romesser (1974), Nishida *et al.* (1978)).

Our three equations are relatively simple and represent a certain degree of mathematical rigour, having been obtained as consistency conditions on the second order equations in an expansion scheme (alternatively, when derived from a Lagrangian as in Appendix 5, they are the second non-trivial Euler–Lagrange equations). The main limitation of these equations is their complete inability to describe solitons in head-on collisions. A less rigorous set, spanning two expansion orders but including the possibility of soliton collisions, is derived in Appendix 4.

9.3 Properties of unidirectional soliton equations

The equations derived in Section 9.2 have some remarkable properties in common, some leading to negative conclusions (such as not being completely solvable).

9.3.1 Integrability by inverse scattering

Oddly enough, all three unidirectional equations CI, CII, and S have self-similar solutions that satisfy the same ordinary differential equation:

$$F'''+(F')^2-\tfrac{2}{3}\xi F'+\tfrac{4}{3}F=0 \tag{9.3.1}$$

where

$$\psi=\tau^{-1/3}F(\xi), \tag{9.3.2}$$

and

$$\begin{aligned}
\xi&=\tau^{-1/3}\rho+\tfrac{1}{2}\tau^{-4/3}z^2, \text{CI};\\
\xi&=\tau^{-1/3}z+\tfrac{1}{2}\tau^{-4/3}\rho^2, \text{ CII};\\
\xi&=\tau^{-1/3}\rho, \qquad \text{S}.
\end{aligned}$$

This extends the self-similar solution of KdV (Rosales (1978), Tajiri and Kawamoto (1982)).

Equation (9.3.1) is not of Painlevé type. In plain language this means that the critical points of the solutions depend on the constants of integration. It is generally believed, and has been demonstrated for very restricted classes of equations, that any ordinary differential equation obtained from a completely *integrable* partial differential equation by a similarity transformation such as (9.3.2) must have the Painlevé property (see Ablowitz *et al.* (1980a, b), Ablowitz (1981), Lakshmanan and Kaliappan (1983), Weiss *et al.* (1983), Weiss (1983)). Thus we do not expect any of our three equations to be solvable by inverse scattering. However, the ρ, τ version of CI($\partial_z = 0$) *does* lead to a Painlevé type equation and is solvable by ISM (Calogero (1978a, b)).

Painlevé classification apart, there are several more or less accepted tests for whether a differential equation is in principle solvable by inverse scattering or not. For the ρ, τ version of CI, the answer is simple, as a transformation from this equation to the ordinary KdV equation exists (Lugovtzov and Lugovtzov (1969), rediscovered in the English language literature by Hirota (1979a)). The KdV equation was the first ever to be solved by the ISM (Gardner *et al.* (1967)). Equally painless reduction to an equation known to be solvable is rare, however. Another test is the existence of a Bäcklund transformation with an arbitrary parameter (Chapter 7; Kaup (1980), Alberty *et al.* (1982)).

9.3.2 Conservation laws

We will now concentrate on yet another test for existence of the ISM. It is the existence of a suitable set of conserved densities. The specific form of those quantities will prove useful when constructing models.

There seems to be a feeling among mathematicians that if both:

1. A differential equation admits an infinite sequence of conserved quantities.
2. Not all of those quantities are generated by the Lie group,

then the nonlinear PDE should be solvable by inverse scattering.

The Lie group is the group of all transformations involving both dependent and independent variables that leave the Lagrangian unaltered. In the more general, Lie-Bäcklund group, which will play a role in what follows, *derivatives* of the dependent variables join the club. Readers unfamiliar with these concepts should not worry too much. However, those who do will find both Blumam and Cole (1974) and Kumei (1977) useful. For more advanced theory, see Barut and Raczka (1980).

The reader may be somewhat dissatisfied with statements like 'it is generally believed', 'there is a feeling among mathematicians' and such like. The sad fact is that our knowledge of just how absolute various criteria of solvability of nonlinear partial differential equations are, is far from complete. Indeed, it is still an open field and to be recommended to mathematically minded theoretical physicists.

The existence of an infinite Lie–Bäcklund sequence of conservation laws for CI with $\partial_z = 0$ has been demonstrated by two different methods so far (Calogero and Degasperis (1978c), Nakamura (1981), but will be derived here by yet another, considerably simpler method (Infeld (1981a), Infeld and Frycz (1983), Frycz (1988)). Each step will yield a new conservation law whereas the other methods only yield one every other step.

Equation CI, $\partial_z = 0$, can be written in terms of $v^{(1)}$ only (9.2.13). The form

$$u_t + 6uu_x + u_{xxx} + \frac{1}{2t}u = 0 \tag{9.3.3}$$

in which notation has been altered and suffixes denote partial differentiation, will prove more convenient. This form can be expressed as a conservation law upon multiplication by $t^{\frac{1}{2}}$:

$$(t^{\frac{1}{2}}u)_t + (t^{\frac{1}{2}}[3u^2 + u_{xx}])_x = 0 \quad [1,0] \tag{9.3.4}$$

whereas tu times (9.3.3) yields

$$(tu^2)_t + (t[4u^3 + 2uu_x - u_x^2])_x = 0 \quad [2,0] \tag{9.3.5}$$

We have introduced a label in which the highest power of u in the conserved density is followed by the highest power of x.

We will now derive a sequence of conserved densities from (9.3.3) and (9.3.4). The method is as follows; take a conservation law in which x does not appear explicitly, such as (9.3.4),

$$T_t + X_x = 0, \tag{9.3.6}$$

multiply it by x and write the result as

$$(xT)_t + (xX)_x - X = 0. \tag{9.3.7}$$

This will lead to a new conservation law if X can be written as a sum of derivatives

$$X = (X_1)_t + (X_2)_x, \tag{9.3.8}$$

(9.3.7) taking the form of a new conservation equation

$$(xT - X_1)_t + (xX - X_2)_x = 0. \tag{9.3.9}$$

The next law in the sequence will in turn follow if X_2 can be written as a sum of derivatives. This process can continue *ad infinitum* unless, at some stage, a decomposition similar to (9.3.8) is no longer possible.

The first four conserved densities so obtained are $[n, n-1]$, $n=1$ to 4:

$$\begin{aligned}
&t^{\frac{1}{2}}u,\\
&xt^{\frac{1}{2}}u-6t^{3/2}u^2,\\
&\tfrac{1}{2}x^2t^{\frac{1}{2}}u-6xt^{3/2}u^2+24t^{5/2}u^3-12t^{5/2}u_x2,\\
&\tfrac{1}{6}x^3t^{\frac{1}{2}}u-3x^2t^{3/2}u^2+24xt^{5/2}u^3-12xt^{5/2}u_x2\\
&-t^{3/2}u-72t^{7/2}u^4-\tfrac{72}{5}t^{7/2}u_{xx}^2+144t^{7/2}uu_x2.
\end{aligned}$$

A standard Lie calculation yields integrals of [1,0], [2,0], [2,1] and [3,2] as the complete set of Lie invariants. Thus we have found some non-Lie, or Lie–Bäcklund invariants and both 'conditions' for solvability by inverse scattering are satisfied.

For large modulus, negative t, (9.3.3) is well approximated by the regular KdV equation. Thus for $t\to-\infty$, (9.3.10) should yield the regular KdV conserved densities. Indeed, the coefficients of the highest powers of t are, when suitably normalized

$$\begin{aligned}
&u\\
&\tfrac{1}{2}u^2\\
&\tfrac{1}{3}u^3-\tfrac{1}{6}u_x^2\\
&\tfrac{1}{4}u^4+\tfrac{1}{20}u_{xx}^2-\tfrac{1}{2}uu_x^2,
\end{aligned}\tag{9.3.11}$$

the first four densities for KdV $[n,0]$ (Section 7.3, Miura (1968), Miura *et al.* (1968)). Fig. 9.1 sums up our derivation of conserved densities by 'x integration' for (9.3.3) as generalized to

$$u_t+6uu_x+u_{xxx}+\gamma u/t=0.\tag{9.3.12}$$

We see that for γ other than 0, $\frac{1}{2}$, the ISM cannot be expected to work. This is bad news for the spherical case, $\gamma=1$, for which the only conserved densities known at the time of writing are

$$\begin{aligned}
&tu[1,0]\\
&t^2u^2[2,0]\\
&xtu-3t^2(\ln t)u^2[2,1].
\end{aligned}\tag{9.3.13}$$

Puny lot as they are, they can be combined to give interesting information. For example, the centre of mass of an N soliton configuration will propagate according to

$$S\langle x\rangle=\int_{-\infty}^{\infty}xu\mathrm{d}x\Big/\int_{-\infty}^{\infty}u\mathrm{d}x=A+B\ln(t/t_{\mathrm{o}}),\tag{9.3.14}$$

where A and B are constants. A similar calculation for cylindrical solitons gives

$$C\langle x\rangle = C + (t/t_0)^{\frac{1}{2}} \tag{9.3.15}$$

(Nakamura (1980)). These results will be useful when modelling one soliton solutions in Section 9.4.

Fig. 9.1. Conservation laws as generated by the 'x-integration' method (skew arrow). The $y=0$ laws are obtained from those for $\gamma=1/2$ in the $t\to\infty$ limit. There are only three conservation laws for γ other than 0, 1/2, 1/3. L = Lie group.

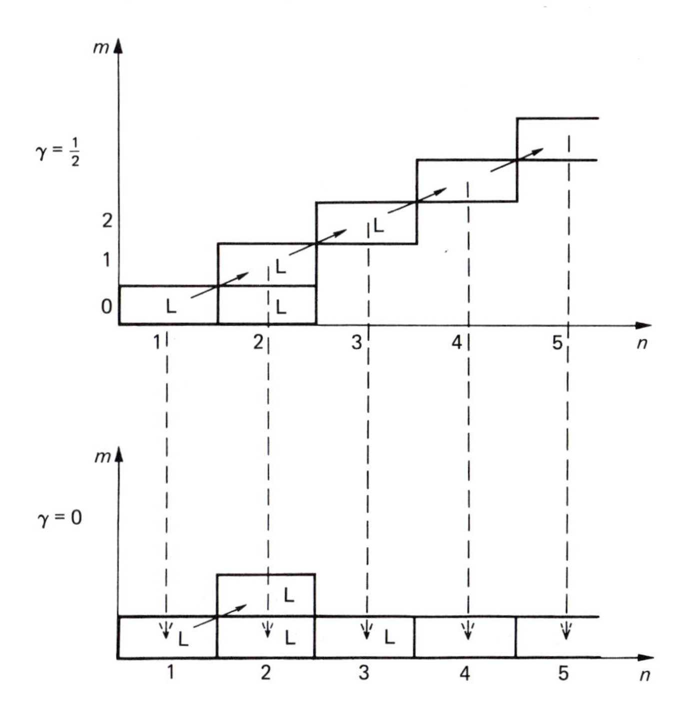

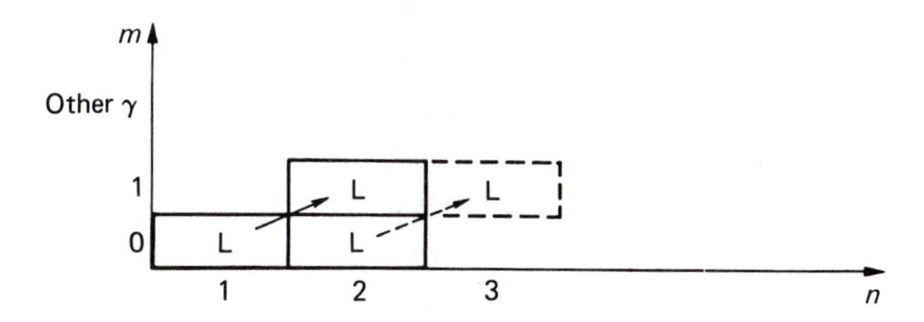

9.4 Soliton solutions as compared with numerics and experiments

In this Section we will concentrate on collapsing and expanding ion acoustic solitons, the more so as no fully two space dimensional, stationary and localized solutions to CII are known to the authors.

We note in passing that when the sign of the third term of CII is changed to a minus, a numerical solution of the form

$$\psi(z-ct,\rho),\ c>0,$$

is known. It describes a shape preserving, axially symmetric soliton (Petviashvili (1981)). This is formally reminiscent of an *analytical* solution to the KP equation, also with changed sign (positive dispersion):

$$\psi_{z\tau}+\psi_z\psi_{zz}-\tfrac{1}{2}\psi_{zzzz}+\tfrac{1}{2}\psi_{\rho\rho}=0.$$

(Exercise 1 and Manakov *et al.* (1977)). For this solution, as the radial distance tends to infinity, the physical quantities are rational functions that only fall off as ρ^{-2} locally, but such as to give finite integrals over all space.

Unfortunately, the corresponding analytical solution to KP with negative dispersion (plus in front of all terms) is no longer bounded. Likewise, no finite solution to CII of this type seems to be known.

9.4.1 Exact solutions to CI

As we have seen, the cylindrical and spherical equations introduced here are inconsiderate enough to exhibit an explicit dependence on one of the independent variables (first on r and then on τ after variable stretching). There is therefore no hope of finding stationary solutions such as were found for the Cartesian versions (Chapter 6).

However, as already mentioned, the $\partial_z=0$ version of CI can be solved exactly by inverse scattering. There are in fact much simpler methods for finding exact, N soliton solutions (Chapter 7 and, in this context, Hirota (1979b), Nakamura (1980), Nakamura and Chen (1981)). The first step is to substitute

$$u=2(\ln f)_{xx} \tag{9.4.1}$$

in (9.3.3) (Hirota (1971)). Solutions to the resulting bilinear f equation that correspond to N solitons can be found either via a sequence of Bäcklund transformations, or by an approach known as the Hirota bilinear method in which binary operators are introduced. The resulting equation for f is expanded in a small parameter, $f=1+\varepsilon f^{(1)}$, and then solved by a Laplace contour calculation. The solution obtained is exact, all higher order terms in ε being zero! (see Exercise 2).

Fig. 9.2. Five stages in a two imploding cylindrical soliton overtaking. The larger and faster soliton transfers some of its mass to the slower one and velocities are exchanged (relay type overtaking); $\rho_1/\rho_2 = 10^3$.

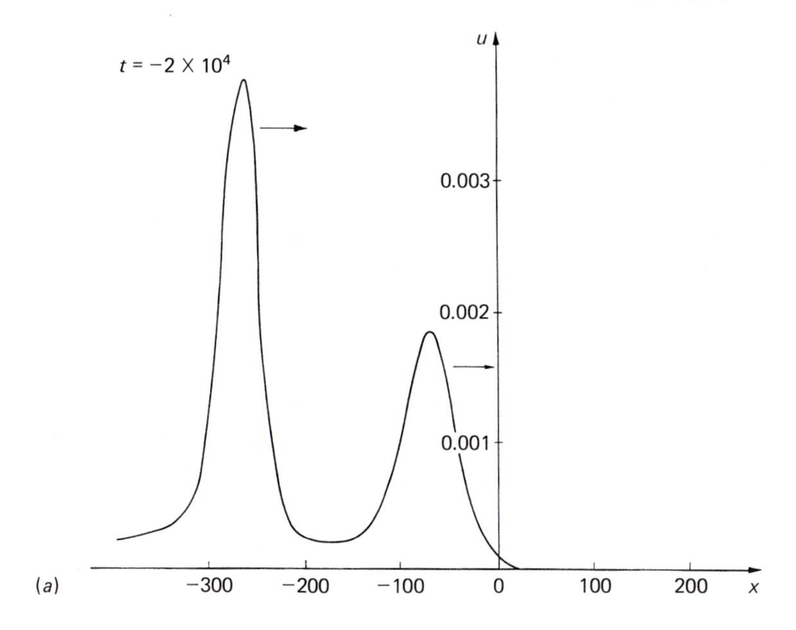

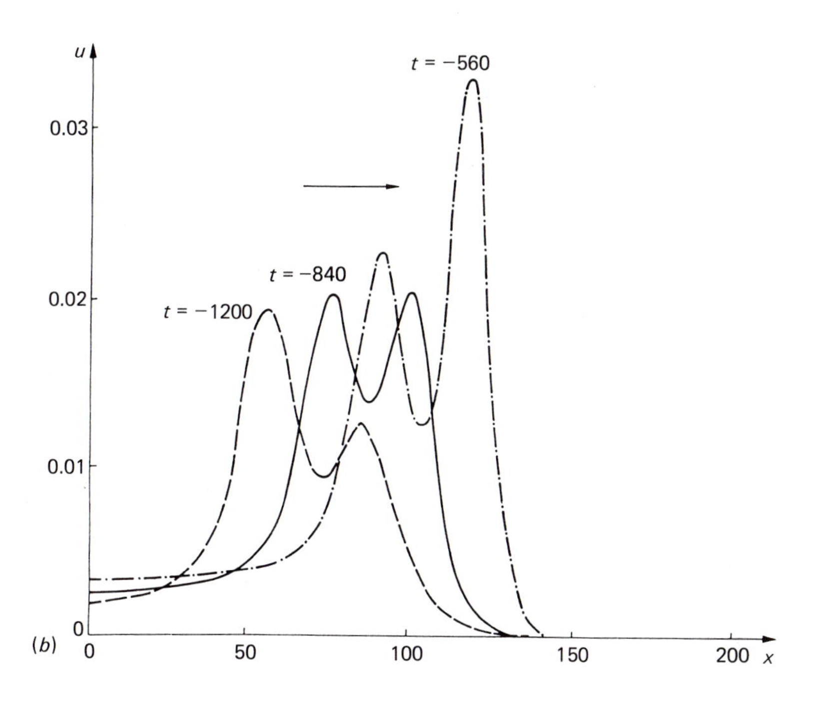

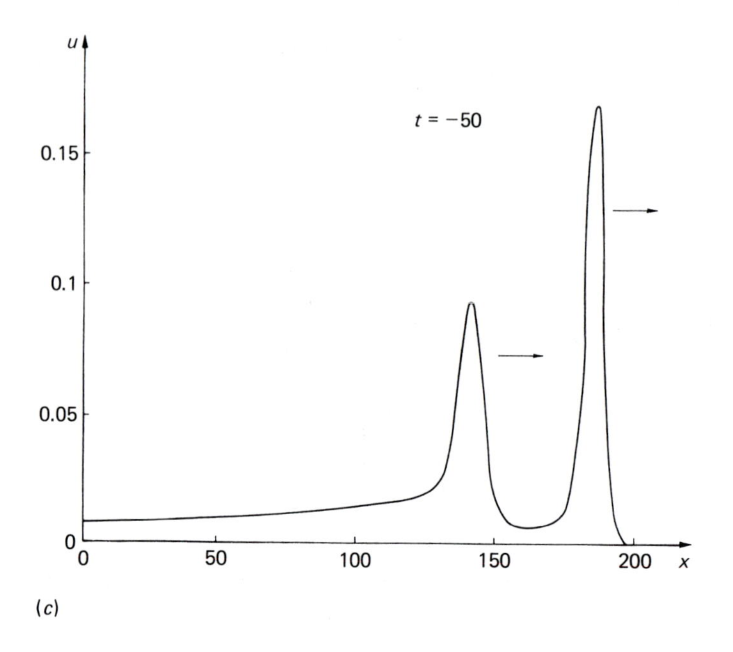

Fig. 9.3. Identity preserving overtaking, $\rho_1/\rho_2 = 10^6$.

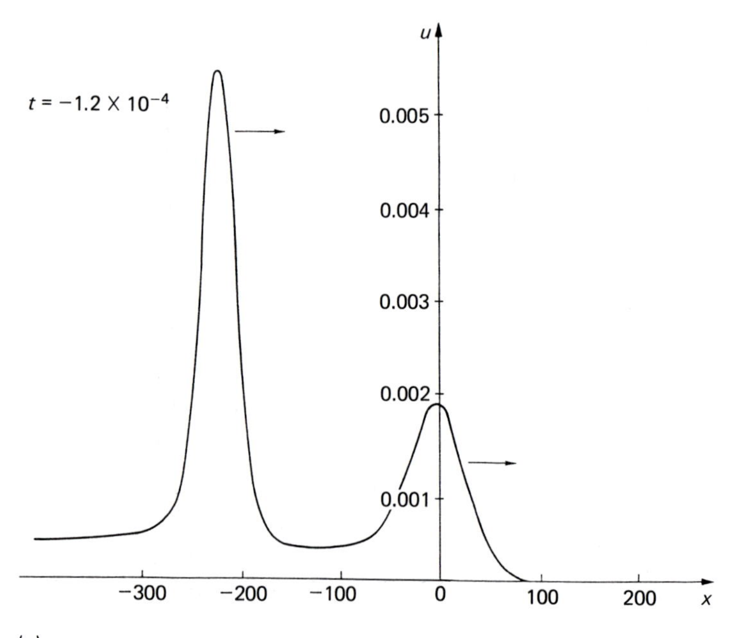

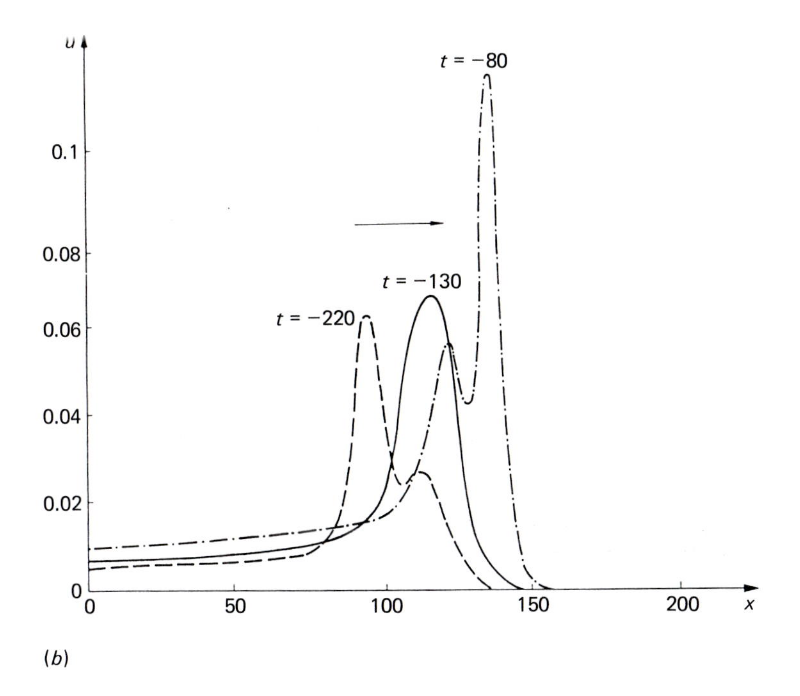
u
t = −80
0.1
0.08
t = −130
t = −220
0.06
0.04
0.02
0
0
50
100
150
200
x

(b)

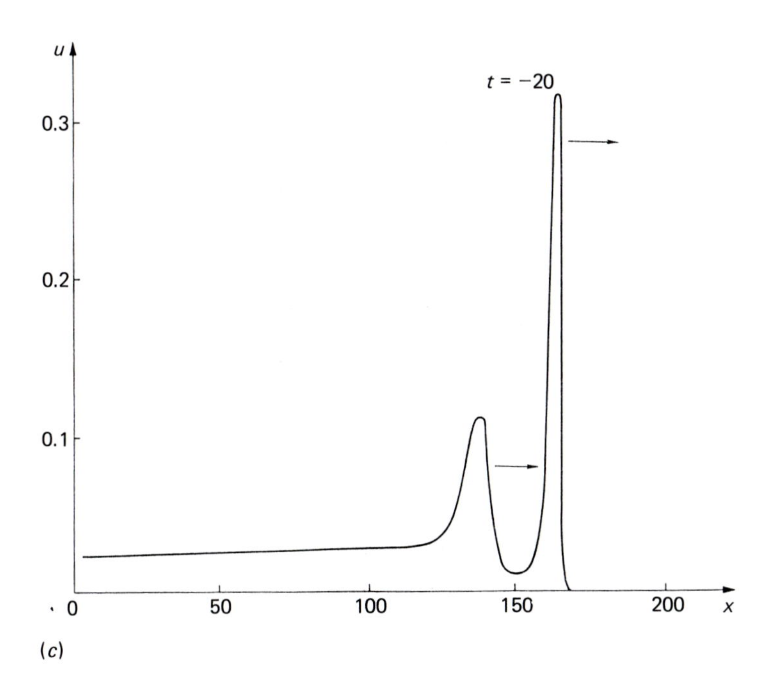
u
t = −20
0.3
0.2
0.1
0
50
100
150
200
x

(c)

The one soliton solution is found by either method to be

$$f=1+\rho_1^2\int_x^\infty [(12t)^{-1/3}Bi\{(12t)^{-1/3}(x-x_1)\}]^2 dx$$
$$\equiv 1+\rho_1^2\int B_1^2. \tag{9.4.2}$$

Here ρ_1 and x_1 are constants and an abbreviated notation is introduced for further convenience (Hirota (1979b), Nakamura (1980), Nakamura and Chen (1981)). *Bi* is the second Airy function.

The two soliton f is given by the determinant

$$f=|\delta_{ij}+\rho_i\rho_j\int B_iB_j| \tag{9.4.3}$$
$$i=1,2.$$

Here the abbreviated notation introduced in (9.4.2) is used. Thus for $t\to-\infty$, we just recover the sum of two solitons. Figs. 9.2 and 9.3 show five consecutive stages of two soliton overtakings for two different values of ρ_1/ρ_2. These figures were obtained from the exact solutions (9.4.1) and (9.4.3). Notice how different the scenarios of overtaking are in the two figures.

9.4.2 Initial value problem and experiments

Maxon and Viecelli (1974a, b) investigated the initial value problem numerically for the ρ,τ (flat) version of CI and S. If not for the $\tau^{-1}\psi_\rho$ term, both equations would be exactly solved by a hyperbolic secant square soliton. Thus they took

$$v=3\,\mathrm{sech}^2\left(\frac{\rho-\rho_o}{\sqrt{2}}\right), \tau_o<0, \tag{9.4.4}$$

as their initial value and investigated the time development for both geometries. They found that the soliton grew taller and thinner as time elapsed, and it propagated inwards. The amplitude times the square of the width was approximately constant. A positive value tail developed behind the main body of the soliton. As τ increased, the cylindrical soliton thus became more and more like the exact solution to CI described above, the main difference being the finite extension of the tail. The amplitude of the *spherical* soliton developed more rapidly, but essential features were similar. Diverging solitons, on the other hand, were found by Chen and Schott to develop oscillatory tails (1977). In fact, KdV type solitons seem to develop tails at the slightest provocation, corrections for cylindrical and spherical geometries being only two of many instances. Others are: dissipative effects (Watanabe (1978), Tran (1979), Bona *et al.* (1980)), higher order nonlinear effects (Konno *et al.* (1977)), and permanent perturbations represented by forcing terms (Karpman (1979)).

If not for the ψ_{ρ}/τ term, an exact solution to CI, extending (9.4.4) in time, would be

$$v = 3\,\mathrm{sech}^2[(1/\sqrt{2})(\rho - \rho_o - \tau + \tau_o)]. \tag{9.4.5}$$

We might therefore expect that far from the axis, cylindrical solitons might have the value of

$$= \text{amplitude} \times (\text{width square})$$

seen to be almost constant in numerical experiments, to be approximately 6. Similarly

$$\beta = \text{peak velocity/amplitude}$$

Fig. 9.4. Quantities α and β defined in text for outgoing cylindrical solitons as functions of the 'distance' from the cylinder at which they were excited, $\tau - \tau_o$. Here the radius of excitation is treated as a parameter. After Chen and Schott (1977).

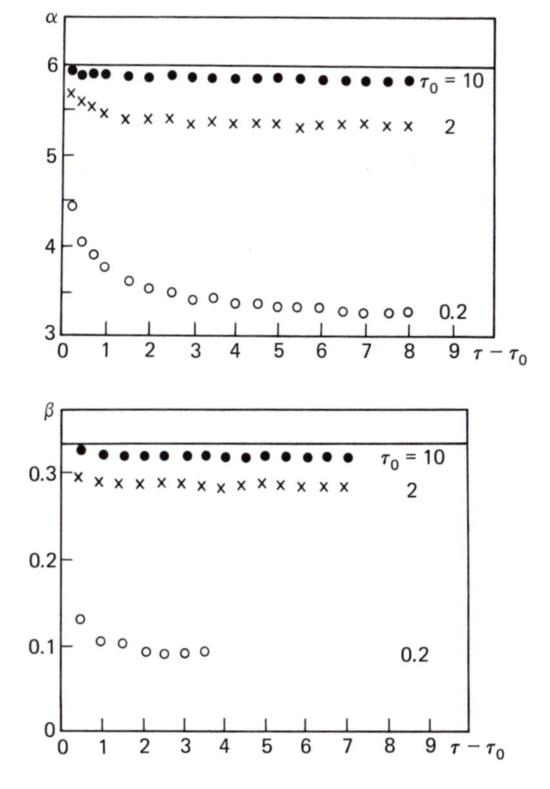

Fig. 9.5 Schematic of the dependence of asymptotic β on the radius of excitation for both incoming (I) and outgoing (O) cylindrical ion acoustic solitons. This diagram was suggested by Chen and Schott (1977) and several experimental papers, all quoted in text.

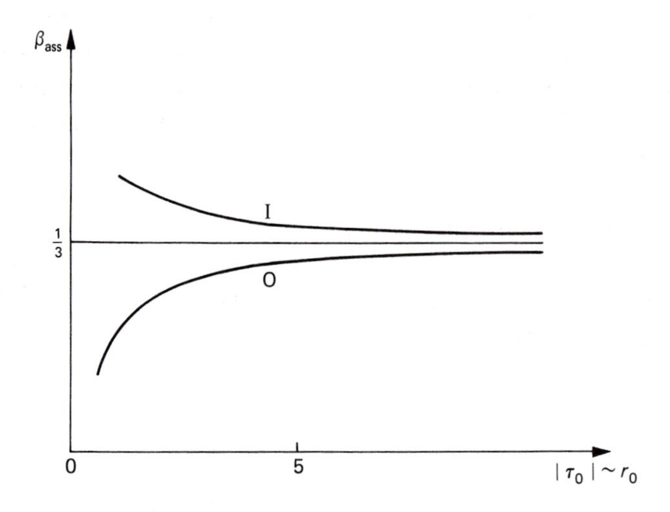

Fig. 9.6. Observed density perturbations for an initially small, negative, spherical pulse. From Nakamura and Ogino (1982).

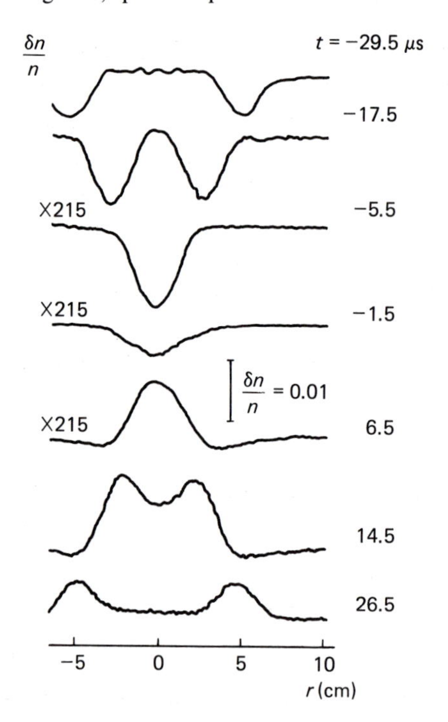

might be expected to be somewhere near $\frac{1}{3}$. Those quantities were investigated numerically by Chen and Schott (1977) for *outgoing* cylindrical solitons. They found that, when starting from (9.4.4), and thus from $\alpha=6, \beta=\frac{1}{3}$, both these quantities fell rapidly below 6 and $\frac{1}{3}$ and then more or less settled at decreased values, as seen in Fig. 9.4. As the initial value of τ was decreased, the asymptotic values of α and β fell further and further below 6 and $\frac{1}{3}$. Since τ_o is a measure of the initial distance from the axis r_o, this is hardly surprising; deviations from Cartesian behaviour will increase if we start our solitons off nearer the axis. See Fig. 9.5.

9.4.3 Reflection from the axis (centre)

Although linearized equations can certainly not govern soliton behaviour, they can sometimes give us a clue to reflection properties. When (9.2.1)–(9.2.3) are linearized around $n=1, \mathbf{v}=0, \phi=0$, we obtain

$$\begin{aligned} &\delta n_t + \nabla\cdot\delta\mathbf{v} = 0 \\ &\delta\mathbf{v}_t + \nabla\delta\phi = 0 \\ &\nabla^2\delta\phi = \delta\phi - \delta n, \end{aligned} \tag{9.4.6}$$

invariant with respect to two time reflection transformations:

$$\begin{aligned} &\text{I} \quad t\to -t, \delta n\to -\delta n, \delta\phi\to -\delta\phi, \delta\mathbf{v}\to\delta\mathbf{v}; \\ &\text{II} \quad t\to -t, \delta n\to\delta n, \delta\phi\to\delta\phi, \delta\mathbf{v}\to -\delta\mathbf{v}. \end{aligned}$$

The full nonlinear set, on the other hand (9.2.1)–(9.2.3), is only invariant with respect to II$(t\to -t, \mathbf{v}\to -\mathbf{v})$. Therefore symmetry arguments would lead to the following ramifications, expressed in terms of density:

1. Large amplitude solitons must reflect from the origin as solitons,
2. Very small amplitude solitons would so far seem to have the choice of reflecting as depressions or as solitons ($1+\delta n\to 1-\delta n$ as indicated by I, or $1+\delta n\to 1+\delta n$ following II).

We are thus in need of a further argument to determine what will happen in an experiment in which δn is very small.

Equation (9.4.6) is easily seen to give, for r symmetry (see also Appendix 4)

$$\delta\phi_{tt} - \Delta_r\delta\phi - \Delta_r\delta\phi_{tt} = 0, \Delta_r = r^{-m}\partial_r r^m\partial_r.$$

If the last term can be considered to be small near $r=0$, the above equation is solved by

$$\begin{aligned} \delta n \simeq \delta\phi &\simeq [f_c(t+r) + g_c(t-r)]\ln r, m=1 \\ &\simeq [f_s(t+r) + g_s(t-r)]r^{-1}, m=2, \end{aligned}$$

and so $f=-g$ if these quantities are to remain finite at $r=0$. A phase reversal thus occurs at the origin and the system opts for symmetry I.

Numerical and laboratory Double Plasma experiments confirm this reflection duality between small and large amplitude soliton reflection properties. For $\delta n/n \simeq 10^{-2}$, solitons are reflected as depressions (negative pulses and vice versa, (see Fig. 9.6)), whereas for $\delta n/n \simeq 10^{-1}$ positive solitons are reflected as positive solitons. It is not clear exactly what the critical value of $\delta n/n$ is but it seems to depend on whether we collapse positive or negative pulses, being considerably larger for the latter if indeed it exists. Work on cylindrical soliton reflection is reported by Tsukabayashi *et al.* (1981), on spherical ion acoustic solitons by Nagasawa *et al.* (1980) and Nakamura and Ogino (1982) (the last mentioned reference reports a numerical confirmation of the r^{-1} behaviour near the centre, as predicted above, to within a few per cent). Water ring convergences are reported and illustrated by a photograph by Tsukabayashi and Yagishita (1979). For a theory of water surface rings, see Johnson (1983), where a model equation is derived. This equation is mathematically similar to our (9.2.15), though the last term is divided by τ^2:

$$\psi_{\rho\tau}+\psi_\rho\psi_{\rho\rho}+\tfrac{1}{2}\psi_{\rho\rho\rho\rho}+\frac{1}{2\tau}\psi_\rho+\frac{1}{2\tau^2}\psi_{\theta\theta}=0.$$

Fig. 9.7. Relative density perturbation against r for expanding spherical ion acoustic solitons as found experimentally. From Ze *et al.* (1979b). The solid line is $\sim r^{-4/3}$.

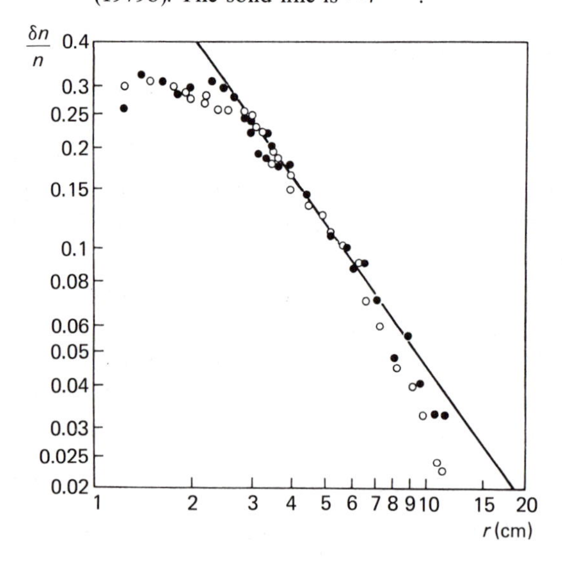

Unlike (9.2.15), it is solvable by ISM (Oevel Wand Steeb (1984)). However, photographs of water rings tend to be very smooth in the azimuthal direction, casting the physical necessity for the last term in some doubt.

9.4.4 Models

It would seem desirable to have a simple model for the symmetric part of a cylindrical or spherical soliton far away from the axis or centre, assuming it begins life as a simple hump at some negative τ_o. For $\tau \to -\infty$, the model equations, CI with $\partial_z = 0$ and S, are locally just KdV, and so we will try a generalization of the KdV soliton:

$$v = a\,\text{sech}^2\left[\left(\frac{a}{6}\right)^{\frac{1}{2}}\left(\rho - \rho_o - \tfrac{1}{3}\int_{\tau_0}^{\tau} a\mathrm{d}\tau\right)\right] \tag{9.4.7}$$

(for $a = 3$ and $\rho_o = 0$, this is just (9.4.5)). Numerical and experimental results point at α and β being 6 and $\frac{1}{3}$ respectively in the far field, so (9.4.7) seems to be consistent with all we know. We will, however, allow 'a' to be a slowly varying function of τ. In early attempts the τ dependence of 'a' was (inadvertently) chosen such that the peak velocity of the symmetric part was equal to the centre of mass velocity of the whole soliton (symmetric part plus tail (9.3.14), (9.3.15); Maxon and Viecelli (1974a, b), Ogino and Takeda (1976)). Thus $a_{\text{cyl}} \sim t^{-\frac{1}{2}}, a_{\text{sph}} \sim t^{-1}$ in those calculations. However, this proved inadequate, roughly speaking because more mass then energy is lost to the tail in the initial stages of motion. Energy conservation should thus be the basis of our calculation (Cumberbatch (1978)). For both geometries, from the constancy of α and energy conservation;

(a) α = amplitude × (width square) = constant

(b) energy ~ (amplitude square) × width ~ r^{-m}, $m = 1$, for cylindrical, $m = 2$ for spherical solitons.

Combining these two relations we obtain $a \sim r^{-2m/3}$, or $\tau^{-2m/3}$. This result is obtained by a different argument in Appendix 4. As if this was not enough evidence, it also follows from the constancy of the [2.0] invariant of Section 9.3. For experimental confirmation see Fig. 9.7.

Our model, (9.4.7) with $a \sim \tau^{-2m/3}$, if accepted, can even supply information about the tail. Consider imploding solitons, τ ranging from a negative τ_o to 0. Since the total mass is conserved, ([1,0] of Section 9.3), the mass of the tail is given by the difference between the constant M and the decreasing mass of the symmetric part as found from (9.4.7). Thus

$$M_{\text{tail}} = M[1 - (\tau/\tau_o)^{m/6}],\ M = M_{\text{sol}} + M_{\text{tail}}. \tag{9.4.8}$$

All one soliton predictions of our model (amplitude, position of peak, fraction of mass in symmetric part M_s) are compared with the numerical results of Maxon and Viecelli (1974a, b) in Fig. 9.8. Even the times and positions of two soliton collisions (but not the resulting phase shifts) can be found from (9.4.7), but this is not shown in Fig. 9.8.

The simple model (9.4.7) is seen to be very successful. It has also been derived (including next order corrections) by a local, two time scale expansion of the differential equations (Ko and Kuehl (1978), (1979), Grimshaw (1979)). Unfortunately, the higher order corrections obtained are secular in the far field and the expansion is only valid in the vicinity of the symmetric part of the soliton.

Our model, (9.7.7) with $a \sim \tau^{-2m/3}$, can also be derived from an action principle by starting from a Lagrangian density (Appendix 5), taking v to be

$$v = a(\tau) f\left(\frac{\rho - b(\tau)}{c(\tau)}\right) \tag{9.4.9}$$

Fig. 9.8. Position of peak, amplitude, and fraction of total mass in the symmetric part of the soliton as functions of τ_0/τ for models of imploding cylindrical and spherical solitons (9.4.7), as compared with numerical results of Maxon and Viecelli (1974a, b). Values for which half the mass has leaked to the tail are indicated and are taken to limit applicability.

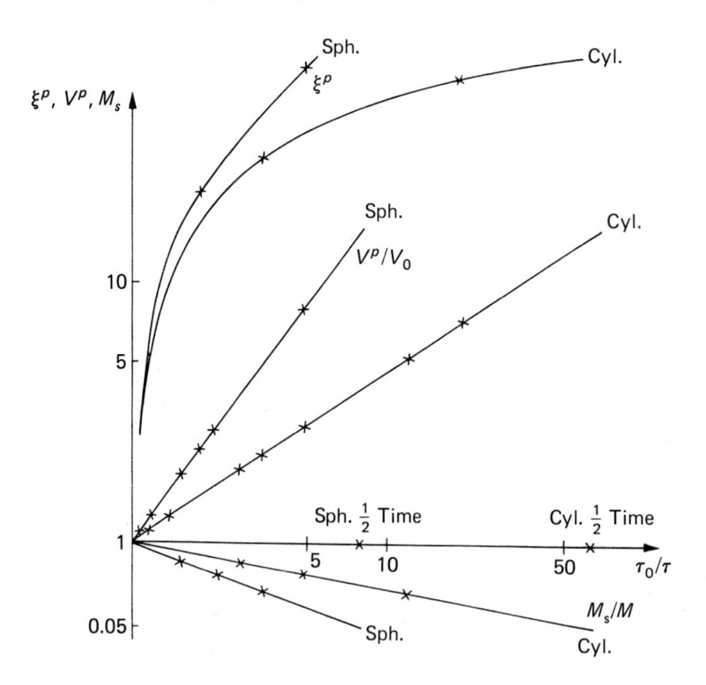

where f is any symmetric function prescribed by the initial value. The Euler–Lagrange a, b, and c equations yield, after some manipulation,

$$ac^2=\alpha, \quad a^{-1}b_\tau=\beta, \quad a\tau^{2m/3}=\sigma \tag{9.4.10}$$

where α, β, σ are constants determined by the choice of f (Exercise 3). The Euler–Lagrange equation for f gives,

$$f(x)=\text{sech}^2(x) \tag{9.4.11}$$

proving that this function is the best possible fit to the symmetric part of both the cylindrical and spherical one soliton solutions over the entire τ range, the whole diversity of the class of solitons being unloaded on $x(\tau)$ and $a(\tau)$. Equation (9.4.11) is particularly useful for the spherical case, exact solutions not being known even for one soliton. For hyperbolic secant square f, we find $\alpha=6$ and $\beta=\frac{1}{3}$, a welcome result.

Still another approach, based on the ISM, is outlined by Kaup and Newell (1978).

9.4.5 Stability of cylindrical solitons

Both numerical and laboratory experiments described above imply a certain degree of stability. We now outline a simple stability analysis.

Solution to CI will be sought in the form

$$\phi=\phi_s(\rho,\tau)+\delta\phi(\rho,\tau)e^{i(K_z z-\Omega t)}$$

where ϕ_s is given in our model by

$$\phi_{s/\rho}=v=v_o(\tau_o/\tau)^{2/3}\text{sech}^2(v_o/6)^{\frac{1}{2}}(\tau_o/\tau)^{1/3}(\rho-\rho_o-v_o\tau_o+v_o\tau_o^{2/3}\tau^{1/3}),$$

neglecting the $\phi_\rho/2\tau$ term when linearizing and using the results of Section 8.3 we obtain

$$\Omega^2\sim K_z^2(\tau_o/\tau)^{2/3}>0,$$

and so cylindrical solitons are stable against long wave z perturbations *in the far field.* Nearer the axis the possibility of 'sausage' (azimuthally symmetric) type instabilities cannot be ruled out at this stage.

9.5 Langmuir solitons

Plasma oscillations with frequency ω_{pe} are the basic modes in a cold plasma. When an electron temperature *is* present, plasma oscillations are modified by a coupling with the ion acoustic mode. The resulting electric field $\mathscr{E}$ can be decomposed into a fast time-varying component and a slow modulation:

$$\mathscr{E} = \mathbf{E}(\mathbf{r}, t)e^{i\omega_{pe}t} \tag{9.5.1}$$

The modulation **E** can take the shape of a Langmuir soliton, an intense, localized electric field in a plasma density cavity. For **E** not *too* large, a possible pair of model equations in dimensionless units was given by Zakharov (1972)

$$\begin{aligned} &i\mathbf{E}_t + \nabla(\nabla \cdot \mathbf{E}) - \alpha\nabla \wedge \nabla \wedge \mathbf{E} = n\mathbf{E} \\ &n_{tt} - \nabla^2 n = \nabla^2 |\mathbf{E}|^2 \\ &n = n_i - 1, \alpha = m_e c^2/3K_B T_e. \end{aligned} \tag{9.5.2}$$

The simplest case, $\alpha = 1$, is often considered, simply because of the vector identity

$$\nabla \wedge \nabla \wedge \mathbf{E} = \nabla(\nabla \cdot \mathbf{E}) - \nabla^2 \mathbf{E},$$

which leads to

$$\begin{aligned} &i\mathbf{E}_t + \nabla^2 \mathbf{E} = n\mathbf{E} \\ &n_{tt} - \nabla^2 n = \nabla^2 |\mathbf{E}|^2. \end{aligned} \tag{9.5.3}$$

However, α being the square of the velocity of light divided by three times the square of the electron thermal velocity, tends in the real world to be a large number. So, in general, a good deal of physics is being sacrificed to mathematical tractability. Numerical work on the full set with $\alpha = 10$ has been performed on two space dimensional problems (Wardrop and ter Haar (1979)).

If we further assume the time variation of n to be slow we can take $n \simeq |E|^2$ to solve the second member of (9.5.3). We then obtain

$$i\mathbf{E}_t + \nabla^2 \mathbf{E} + f(|E|^2)\mathbf{E} = 0, f(x) = x \tag{9.5.4}$$

the cubic nonlinear Schrödinger equation. For large amplitude **E**, Wilcox and Wilcox (1975) and, independently D'Evelyn and Morales (1978), found (9.5.4) with saturating nonlinearity $f(x) = 1 - \exp(-x)$ to be a better model. (This suggests that there is room for improvement in many of the model equations of this book.) The saturating nonlinearity of course takes us beyond the Zakharov picture.

Equation (9.5.4) with still another f, namely $\ln x$ has been investigated (Bialynicki-Birula and Mycielski (1975), (1979), more about which in Section 9.7). Although this model has not been found to apply to any plasma physics problem, results on stability will be included here when we try to understand general f behaviour of (9.5.4).

9.5.1 Integrability

There is an essential difference between the cubic NLS equation and (9.5.3) in one space dimension x, t. The cubic NLS equation is solvable by the ISM, and exact N soliton solutions were found in 1971 (Zakharov and Shabat). For this model an infinite set of conserved densities, generated by the Lie–Bäcklund (as distinct from Lie) group is known. Furthermore, if two solitons are collided in a numerical experiment, they emerge unscathed (Abdulloev *et al.* (1974)). This ceases to be true when the cubic nonlinearity is exchanged for $f = 1 - e^{-x}$, though collisional changes in momentum are very small (of the order 10^{-2}. D'Evelyn and Morales (1978)).

Only three conservation laws are known for the one-dimensional Zakharov equations (9.5.3). Numerical collisions of two pulses can even lead to merging and/or emission of finite bursts of radiation (Pereira (1977), Payne *et al.* (1983)). Thus (9.5.3) seems not to be solvable by the ISM. Further proof of this was furnished by Goldstein and Infeld (1984), who demonstrated that the Zakharov equations in x, t are not of Painlevé type, as understood for partial differential equations (Weiss *et al.* (1983)). Interestingly enough, there is a half-way house between the cubic NLS equation and the Zakharov equations, all in x, t:

$$iE_t + E_{xx} = nE$$
$$n_t - n_x = -\tfrac{1}{2}(|E|^2)_x$$

the second equation being unidirectional. This set is solvable (Yajima and Oikawa (1976); Painlevé analysis Goldstein and Infeld (1984)).

Higher dimensional versions, even of the cubic NLS equation, do not seem to be solvable (Zakharov and Shulmann (1980)). We expect their soliton collisions to be inelastic.

9.5.2 Stability of Langmuir solitons

Returning to cylindrical and spherical solitons and using (9.5.4), one can easily show that radial, localized solutions exist in the form $E_r \exp(i\omega t)$ for both geometries (collapsing and expanding r, t dependent Langmuir solitons also exist, but will not be considered here). By inflating E_r such that the mass invariant is unaltered:

$$\begin{aligned} &r \to \lambda r \\ &E_r \to \lambda^{p/2} E_r \\ &\int |E_r|^2 \mathrm{d}_r^p = \text{const}, p = \text{dimension} \end{aligned} \tag{9.5.5}$$

one can prove instability when (Exercise 4, Ladyzhenskaya (1961),

Table 9.1. *Indications of an r dependent, inflationary stability analysis for various NLS equations.*

'Stable' indicates stability with respect to mass preserving inflations. Additional information, based on other calculations, is indicated in brackets.

$f(x)$	Flat Symm $p=1$	Cyl $p=2$	Sph $p=3$	Plasma context
x	'Stable' (stable, all $\parallel$ perturb., unstable, $\perp$ perturbations).	Inconclusive (For numerical experiments and θ perts see text.)	Unstable	Small amplitude Langmuir solitons.
Saturating (a) $1-e^{-x}$ (b) $x(1+x)$	'Stable' (stable, all $\parallel$ perturbations, unstable, $\perp$ perturbations).	'Stable'	'Stable' (Unstable, $\omega_c > \omega > 0$, see text).	(a) Arbitrary amplitude Langmuir solitons (b) Model for *a*.
lnx	'Stable'	'Stable'	'Stable' (also stable to θ perts)	None

Gibbons *et al.* (1977));

$$\operatorname*{Lt}_{x\to\infty} f(x) \sim x^q, q > 2/p \tag{9.5.6}$$

(*stability* cannot be established by such a restricted analysis). Conclusions for particular modes are displayed in Table 9.1. In a more realistic stability calculation (as compared to inflations) neighbouring off-tune, radial soliton solutions in ω parameter space are compared. Instabilities appear when

$$\frac{\partial}{\partial\omega}\int |E_r|^2 d^p r < 0. \tag{9.5.7}$$

(Vakhitov and Kolokolov (1973)). For the cubic NLS equation, conclusions are the same as for the inflationary calculation (Exercise 4). For saturating nonlinearities, $p=1$ and $p=2$ solitons are still stable but for $p=3$ (spherical solitons), an unstable band of w appears. This band was found by Vakhitov and Kolokolov for $f = x(1+x)^{-1}$ to be

$$0.08 > \omega > 0$$

for nodeless solitons, higher ω solitons being radially stable (solitons exist for $1 > \omega \geqslant 0$ for both saturating nonlinearities considered). For

$f = 1 - \exp(-x)$, the unstable band is

$$0.101 > \omega > 0,$$

(Laedke and Spatschek (1984)). Comparison of results for the three forms of f would seem to add strength to our conjecture, put forward in Chapter 8, that steeper nonlinearities are the more unstable. See also Kusmartsev (1987).

The effects of azymuthal perturbations on cylindrical Zakharov and both cylindrical and spherical NLS solitons have been studied by several authors (Denavit *et al.* (1974), Walstead (1980), Bialynicki-Birula and Mycielski (1976), Oficjalski and Bialynicki (1978)). Cylindrical Zakharov solitons are strongly destabilized by these perturbations.

As we saw in Chapter 8, one-dimensional solitons of the cubic NLS equation are strongly destabilized by perpendicular perturbations. However, this nuisance does not necessarily carry over to higher dimensions! A beautiful illustration of the difference can be seen in Fig. 9.9, based on

Fig. 9.9. Collision of a one-dimensional solitary wave and a two-dimensional, nodeless soliton solution to the scalar NLS (Exercise 4). Here $|E|^2$ is shown at $t=0$; 10; 20; 40; 50; 100. From Walstead (1980).

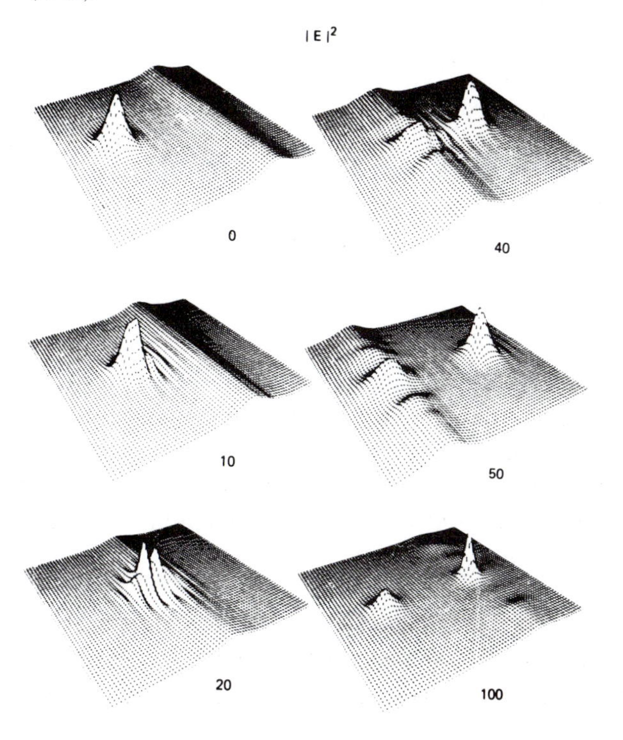

Walstead (1980). A one-dimensional cubic NLS soliton is made to collide with a stationary cylindrical soliton. The former breaks up completely, the latter being relatively unaffected by either its own motion or the collision. Calculations were performed numerically for the scalar cubic NLS equation in x, y, t. Presumably they could be re-done for (9.5.4), the difference only being important during the collision.

9.6 Interacting solitons and some conclusions

In the early 1980s, the behaviour of interacting, non-concentric cylindrical and spherical ion acoustic solitons received considerable attention.

Before we present some results a reminder of Cartesian soliton interactions considered in Section 7.9 might be useful:

When the resonance condition

$$\omega(\mathbf{k}_1+\mathbf{k}_2)=\omega(\mathbf{k}_1)+\omega(\mathbf{k}_2) \tag{9.6.1}$$

is satisfied by two overlapping, Cartesian solitons, in which the argument is treated as a phase, only one soliton will emerge. When $|\mathbf{k}_1| = |\mathbf{k}_2| = k$, this condition defines a resonance angle ψ_r between the two vectors $\mathbf{k}_1, \mathbf{k}_2$. However, if $\psi_r > \psi > 0$ for the same k, two solitons will again emerge from the overlapping. A composite soliton is then generated at the intersection.

The Cartesian soliton intersection work has been extended to interactions of two non-concentric, outgoing cylindrical solitons (Fig. 9.10; here (*b*) is based on a cylindrical Boussinesq equation solved numerically). Experiments were performed in cylindrical Double Plasma devices (Nakamura *et al.* (1982); Kako and Yajima (1982)). When two non-concentric, outgoing *spherical* solitons intersect, a toroidal, and hence fully three-dimensional soliton can form (Ze *et al.* (1979b): Lonngren *et al.* (1983)). Four hemispheres mounted on a plane have also been used to generate outgoing, spherical intersection solitons (Khazei *et al.* (1982), Lonngren *et al.* (1982)).

When analyzing these resonant interactions of cylindrical and spherical outgoing solitons, Cartesian soliton theory is often used, based on KP or Boussinesq. Diagrams like Fig. 9.10(*a*) differ from Cartesian versions in that arcs replace straight lines, but otherwise the reasoning is similar. Three wave relations such as the Manley–Rowe relations (9.6.1) should be satisfied locally at each intersection. However, in this geometry time-dependence is no longer periodic.

Although interest in cylindrical and spherical ion acoustic soliton experiments, both laboratory and numerical, seems largely to have shifted to the above mentioned non-concentric interactions, much further work on concentric interactions is needed.

For example, a comprehensive, three-dimensional stability theory, even for radial cylindrical and spherical $m=0$ Langmuir solitons and based on simple models suggested in Section 9.5, seems far in the future.

In particular, it would be interesting to see if cylindrical ion acoustic solitons are stable with respect to 'sausage' (cylindrically symmetric) instabilities throughout the latter stages of motion. We know that they are stable in the far field (Section 9.4). Recent calculations of ours indicate stability.

In spite of these and many other questions and grey areas, one cannot fail to be impressed by the ground covered lately. This is true of both ion acoustic and higher dimensional Langmuir solitons. One of the healthiest aspects of the field is the way experiments march side by side with theory and numerics. This is an interesting subject to watch in the near future.

Fig. 9.10. (*a*) Schematic of a two cylindrical soliton interaction such that a virtual soliton is created. (*b*) Virtual state of interacting ion acoustic cylindrical soliton configuration. Numerical experiment reported in Tsukabayashi *et al.* (1983).

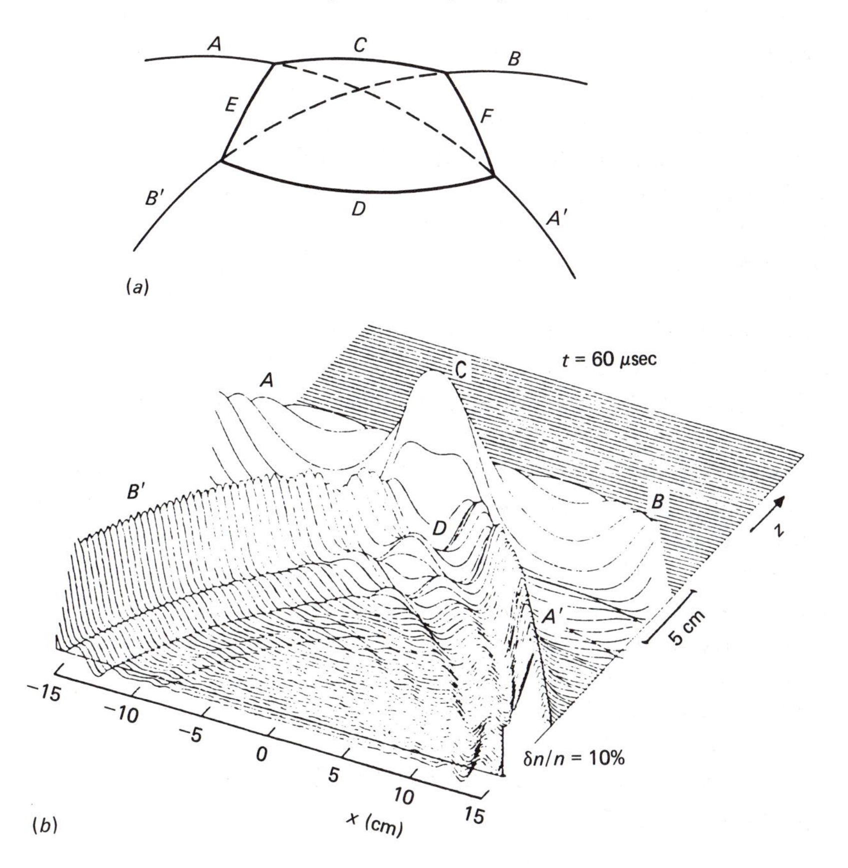

9.7 Epilogue. Some other examples of spherical and cylindrical solitons

Quite a lot of work has been done on other two- and three-dimensional solitons in various fields of contemporary physics. Back in the 1950s, a spherical vortex solution to the equations of magnetohydrodynamics was found. The solution was in fact inspired by a hydrodynamic vortex solution known in the nineteenth century and known as Hill's vortex (Lamb (1932)). The magnetohydrodynamic vortex so obtained has subsequently been generalized somewhat (Whipple (1960), Infeld (1971)).

Many simple model equations of classical physics admit imploding and expanding cylindrical and spherical solutions. They are often functions of simple combinations of ρ and t or r and t, apart from a phase when complex equations such as the nonlinear Schrödinger equation are considered. These solutions are called *self-similar* and examples have been encountered in this Chapter (Section 9.3). Apart from a multitude of these self-similar solutions, the cubic NLS equation in x, y, t admits a structure preserving class of solitons, one example of which appears in Fig. 9.9 and Exercise 4. Solitons with rotating phase of the form $m\theta-\omega t, m=1,2\ldots$ are also known, that of Fig. 9.9 and Exercise 4 being the $m=0$ mode. The $m=1$ soliton is often used to represent a vortex line in superfluid helium ^{4}HeII (Pitayevski (1961), Fetter (1972)). Direct photographs of a system of vortex lines in a cylindrical vessel containing superfluid helium seem to indicate stability (Gordon *et al.* (1978)). An analytical analysis of the NLS model, performed by Rowlands (1973), also gave stability.

As mentioned in Section 9.4, the KP equation with a minus sign in front of the last term admits cylindrical, structure preserving soliton solutions (Exercise 1, Manakov *et al.* (1977)). These odd structures are completely unaltered by interactions, as follows from an exact N soliton solution. Even the phase shifts of one-dimensional overtakings are absent! A restricted stability calculation, based on inflation, implies that these solitons are two-dimensionally stable (see Kuznetsov and Turitsyn (1982)).

A group of Japanese physicists investigated a cylindrical vortex in a low pressure plasma permeated by a strong magnetic field B_o. The model used was the so-called Hasegawa–Mima equation in x, y, t, the magnetic field being along z. The vortices move along x. The same equation describes Rossby waves in the atmosphere (Hasegawa and Mima (1977), Makino *et al.* (1981a)). In the plasma problem, this equation for the electric field potential $\psi(x, y, t)$ is, in dimensionless form,

$$\partial t(\nabla^2\phi-\phi)+v\frac{\partial\phi}{\partial x}-(\partial_y\phi\partial_x-\partial_x\phi\partial_y)\nabla^2\phi=0, \tag{9.7.1}$$

where ∇^2 is in x, y space only and v is a drift velocity due to a density gradient. The wave structures are in uniform motion along the x axis when well separated.

When (9.7.1) represents Rossby waves, ϕ is the vertical displacement of the surface of the atmosphere and v is still a drift velocity. (Mima and Hasegawa (1978), Hasegawa *et al.* (1979), and Makino *et al.* (1981b)).

Fig. 9.11 presents a numerical experiment in which two solitary vortices are collided according to (9.7.1) and, following a tidier collision than would seem possible, emerge virtually unscathed. Thus collisions are apparently elastic, at least for these (and several other trial) initial conditions.

When the density gradient is inhomogeneous, collisions of two vortices become very inelastic, Mikhailovskaya and Erokhin (1987). For an analysis of the discontinuities involved see Nycander (1987), and for further extensions, are Isichenko and Marnachev (1987). For beautiful water flow simulations of further Rossby solitons, purported to model the Great Red Spot of Jupiter, see Antipov *et al.* (1986) and Nezlin *et al.* (1987).

When the nonlinearity in the cubic NLS equation is exchanged for a logarithmic term: $\phi \ln \phi$, three-dimensional stationary solutions can be found. The original family were spherically symmetric and stable. This was demonstrated by both a numerical and a theoretical investigation.

Fig. 9.11. Numerical collision of two vortex solutions of the Hasegawa–Mima equation. (Makino M and Kamimura T, *Ann. Rev. Inst.) Plasma Phys.*, Nagoya 1979–80, 165.)

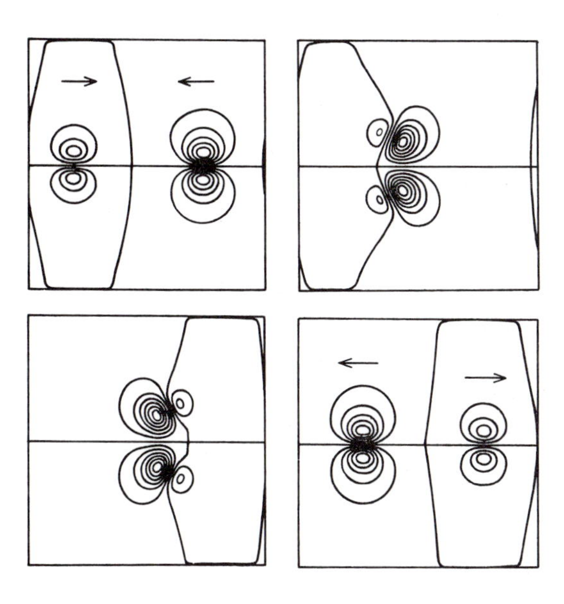

Depending on the parameters, both apparently elastic and inelastic collisions were found (one can never be sure that a numerical collision is one hundred per cent elastic). These structures were still stable, at least against a restricted class of perturbations.

Of course, the physical relevance of this interesting form of the nonlinear Schrödinger equation is an open question. However, it does have a unique and attractive property; the factorization of wave functions for composed systems as demanded by quantum mechanics (Bialynicki-Birula (1978), Bialynicki-Birula and Mycielski (1979)).

Cylindrical vortex solutions of the sine-Gordon equation have also been found (Borisov *et al.* (1985)).

Exercises on Chapter 9

Exercise 1

Show that the KP equation with positive dispersion (9.3.15) can be put in the form

$$\psi_{z\tau}+\psi_z\psi_{zz}-\tfrac{1}{2}\psi_{zzzz}+\tfrac{1}{2}\psi_{\rho\rho}=0.$$

Show that this is solved by $\psi(z-c\tau,\rho)$ rational, describing a soliton in uniform motion along z. Show that the invariant

$$\int\psi_\rho\rho\mathrm{d}\rho\mathrm{d}z$$

is finite in spite of the ρ^{-2} local dependence in the far field. For solution see Manakov *et al.* (1977).

Exercise 2

Could the method of Nakamura and Chen (1981) be applied to the spherical equation (9.3.12) with $\gamma=1$? Explain.

Exercise 3

Derive (9.4.10) and (9.4.11) from the spherical Lagrangian density given in Appendix 5. Take v to be given by (9.4.9), f symmetric.

Exercise 4

Show that $\mathbf{E}=\mathbf{i}_\rho E(\rho)e^{i\omega t}$, $\rho^2=x^2+y^2$ is a solution of (9.5.4) if $E(\rho)$ satisfies an ordinary differential equation. Find this equation and, choosing $E(0)=1$, $E_\rho(0)=0$, solve it numerically for the eigenvalue ω such that $E(\rho)$ has one maximum only (this is the cylindrical soliton that appears in the initial stage of Fig. 9.9). A personal computer is sufficient. Assuming (9.5.5), prove (9.5.6). Show that the same stability criterion follows from (9.5.7).

10

Non-coherent phenomena

10.1 Introduction

In the earlier chapters of this book the emphasis has been on a study of the existence and stability of nonlinear waves and solitons, that is of coherent structures. Such structures are found in Nature and thus certainly deserve our attention. However, a much more universal type of behaviour is described under the umbrella of turbulence. One envisages turbulence as a phenomenon where some measurable quantity has a rapid space and/or time dependence. For example, in the case of water passing over a weir, the complicated behaviour is apparent in the local velocity of the water. One sees eddies (or vortices) of a range of sizes. They not only move with some background velocity but also interact with one another to produce a continually changing picture.

For another example, consider turbulence in the wake of a cylinder if the water flow is very high Reynolds number (see Fig. 1.5(*c*), (*d*)).

The problem of trying to understand turbulence has been with us for centuries but it still remains a basic unresolved problem. (The beauty and complexity of turbulence was well appreciated by Leonardo da Vinci as is evidenced in his drawings of vortices in water, Fig. 10.1).

Somewhat ironically, the study of turbulence in plasmas, which themselves are much more complicated media, is more tractable than in water and considerable progress has been made in the last twenty years. However, most theories to date are restricted in that they assume that the energy in the fluctuations is small compared to the kinetic energy of the particles. For plasmas near thermodynamic equilibrium the ratio of these energies is of order $g = 1/n\lambda_D^3$. (Here n is the plasma density, and λ_D the Debye length (Section 1.3).) As we saw in Section 1.3, for many plasmas of interest, ranging from thermonuclear to interstellar, this quantity is less than unity (see Chapter 1). In such plasmas one expects the amplitude of fluctuations in electric fields for example, resulting from an intrinsic instability, to be small and hence amenable to study by some form of perturbation theory. Such theories may be elaborate (and often are) but at least there exists a basic expansion parameter and this allows one to understand the essential aspects in terms of a simple model.

Consider some measurable quantity such as the electrostatic potential, ϕ, which is a function of space and time. A linear analysis, as described in Chapter 2, gives $\phi(\mathbf{r},\mathbf{t})=A\mathrm{e}^{\mathrm{i}(\mathbf{k}\cdot\mathbf{r}-\omega t)}$ where A is a constant and $\omega=\omega(\mathbf{k})$ satisfies a dispersion relation $D(\omega,\mathbf{k})=0$. Quite generally we can consider D to depend on an external parameter (sometimes called a control parameter), μ say. For example in the two-stream instability this could be the relative velocity of the streams. One can envisage the situation where a critical value μ_c exists such that for $\mu>\mu_c$, the system is linearly unstable. This may be expressed by writing the solution of the dispersion relation for $\mu\simeq\mu_c$ in the form $\omega=\omega_R+i\lambda^2(\mu-\mu_c)$. Thus one may imagine controlling the system by changing the parameter such that $\mu>\mu_c$ causes the system to go linearly unstable, and ϕ grows exponentially with time.

Of course there must exist some saturation mechanism which limits this growth. The identification of such a mechanism is rarely a trivial problem,

Fig. 10.1. A page from Leonardo's notebook with a self-portrait and some studies of water flow. From the Royal Library at Windsor Castle, 12579.

but assuming this can be done one argues, following Landau, that one may write $\phi = A\exp(\mathrm{i}\cdot\mathbf{k}\cdot\mathbf{r} - \mathrm{i}\omega_{\mathrm{R}}t)$ *plus harmonics*, but where A is time dependent and satisfies an equation of the form

$$\frac{\mathrm{d}A}{\mathrm{d}t} = \lambda^2(\mu - \mu_c)A - \beta A^3. \tag{10.1.1}$$

Here β is a parameter, assumed positive, which depends on the plasma parameters but is finite at $\mu = \mu_c$ and thus its dependence on μ may be neglected. For $\mu > \mu_c$ and small values of A the above equation reproduces the exponential growth of the linear theory, but as time evolves the second term must be taken into account. The above equation is readily solved and one finds that for large time, A approaches a saturated state given by $A_s^2 = \lambda^2(\mu - \mu_c)/\beta$. This new state is readily found to be linearly stable. See Exercise 1 in this context.

Rather than consider the full time dependence of A it is convenient to concentrate on the asymptotic state, A_s, and consider this as a function of μ. This behaviour is shown schematically in Fig. 10.2. The dotted line denotes the unstable equilibrium $A = 0$ for $\mu > \mu_c$.

The above is a simple example of a pitchfork bifurcation. Equations of the form of (10.1.1) or simple generalizations have been obtained for a large number of distinct physical systems. Equation (10.1.1) describes a single mode laser with A being a measure of the intensity of the laser and the coefficient multiplying A representing the difference between the destabilizing effect of the pump field and simple radiation losses (Haken (1983)). The Lorenz model of turbulence, of which more will be said later, can be solved approximately and the equation for the amplitude of a particular fluid pattern reduced to (10.1.1). In this case μ is essentially the Reynolds number (Rowlands (1983)).

Fig. 10.2. Bifurcation diagram showing the variation of amplitude A with control parameter μ. The broken line corresponds to an unstable state.

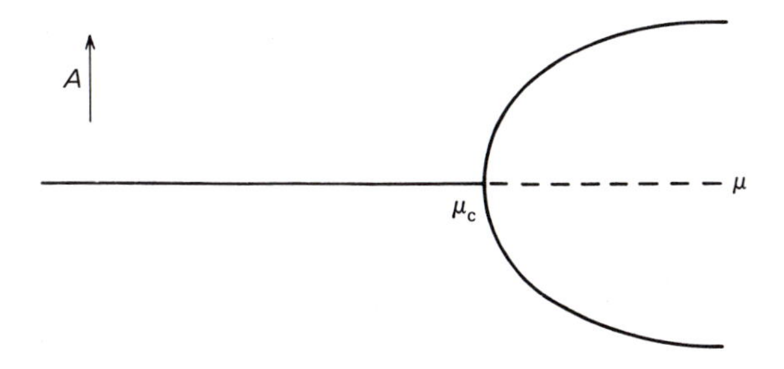

The above simple picture of a saturation effect is of course limited such that $\mu-\mu_c$ is assumed small. However, one imagines that as $\mu-\mu_c$ increases, even though λ and β are now functions of μ, and the nonlinear terms become more complicated, the basic concept of a saturated amplitude survives. This then leads to a non-trivial disturbance $\phi(\mathbf{r}, t)$ which has a fundamental frequency ω_R (which is a function of μ) and a whole host of harmonics. Such a disturbance will be called a 'quasi-mode' in analogy with situations in other branches of physics. For an example we could take a polaron in solid-state physics which is a quasi-particle associated with an electron together with its polarization cloud, (Kittel (1976), Chapter 10). Thus one imagines the existence of a quasi-mode (a stable entity) even for values of $\mu-\mu_c$ no longer small. The qualitative features of a quasi-mode remain the same as μ increases, but quantitative features such as the basic frequency ω_R and the harmonic content change. The concept of a quasi-mode will be basic to the theories of turbulence discussed later in this Chapter. Of course, one mode by itself will not be able to describe turbulence, as it is basically time-periodic.

The so-called quasi-linear theory of plasma turbulence is based on a kinetic description of the plasma. The fluctuating electric or magnetic fields are described by equations of the form of the Landau equation, but the parameter λ is now related to the kinetic aspects of the plasma. In particular it depends on the equilibrium velocity distribution function. The distribution function itself satisfies an equation which depends on ϕ. Thus an equation of the form of (10.1.1) has to be solved together with an infinite set of equations describing the kinetics. It is this interaction with the infinite number of degrees of freedom that gives rise to the complicated time and space dependence identified with plasma turbulence in this model. This model is extremely useful and has been used to explain many nonlinear plasma phenomena. It is a successful theory of plasma turbulence but is only applicable when a kinetic description of the plasma is appropriate. For detailed accounts of quasi-linear theories see, for example, Davidson (1972).

The alternative description of a plasma in terms of fluid-like quantities, namely magnetohydrodynamics, is, as the name suggests, closely related to ordinary fluid mechanics and one would expect that a common theory of turbulence might exist for both cases. (Both these descriptions were mentioned in Section 1.3 and regions of applicability in parameter space were indicated.) So far no such theory is known.

The major effort in studying fluid turbulence has in the past been associated with the problem of homogeneous turbulence. This is the study of an infinite medium where in some sense the turbulent properties are the same throughout. Unfortunately in this situation there does not seem to be a small parameter which could be used as the basis of a perturbation theory. All modes that can be identified in a linear theory are equally important. This has led people to discuss the turbulent state in terms of the statistical mechanics of a large number of interacting modes, the turbulence arising precisely because of the large number of modes. Such theories have their place but are not applicable to the turbulent phenomena to be discussed in this Chapter. For a thorough discussion of this approach see, for example, the book by Leslie (1973).

However, there are many situations in Nature where turbulent type solutions exist and are controllable by an external parameter, μ, such that $\mu-\mu_c$ is small. A classic example is the Bénard cell problem. Here a homogeneous liquid is heated from below. For small values of the temperature gradient (essentially μ) the flow of heat is homogeneous. For larger values of this gradient, convective fluid cells are formed which serve to transport the heat through the system. At this stage the liquid behaves in a coherent fashion and the analysis discussed earlier in the book is applicable. However, further increase of the temperature gradient causes the fluid to move in a random sort of way, (the cells start a wavy motion along their axis). The fluid is now turbulent. The cell stage is illustrated in Fig. 1.6(*b*). Many more examples exist, ranging from fluid dynamics through chemical systems, where the concentrations of various chemicals can change in a turbulent manner, to biological systems where the population of a species is the relevant quantity showing turbulent behaviour. The realization of the common features of these seemingly diverse systems has led to them being studied collectively as 'synergetic' systems. For further discussion of this amalgamation see for example the books on synergetics by Haken (1978), (1983) and later books in the same series edited by this author.

The earliest theory of turbulence incorporating these ideas is due to Landau. Basically one envisages that as the parameter μ increases above a critical value μ_c, a quasi-mode with its associated frequency ω, is excited. Further increase of μ above a second critical value μ_2 causes a second quasi-mode to be excited with frequency ω_2, whilst the first mode retains its basic identity. This generation of quasi-modes continues as μ is increased further but the important assumption is that they do not interact or at least only weakly. They retain their individuality. Then for large μ, where a large number of modes have been excited, each with its characteristic frequency, the total time and space dependence is extremely complicated and non-periodic. This is identified with the turbulent solution. The Landau scenario of turbulence is illustrated schematically in Fig. 10.3.

A radically different theory was proposed by Ruelle and Takens (1971). In this treatment is is only necessary to have three modes, but they must be strongly coupled. This is sufficient to produce complicated time dependence which may be associated with a turbulent type solution. The solution takes the form of a 'strange attractor'. Several years earlier Salzman (1962) and Lorenz (1963) had proposed the following set of coupled equations

$$\left.\begin{aligned}\frac{\mathrm{d}x}{\mathrm{d}t}&=\sigma y-\sigma x\\ \frac{\mathrm{d}y}{\mathrm{d}t}&=-xz+rx-y\\ \frac{\mathrm{d}z}{\mathrm{d}t}&=xy-bz,\end{aligned}\right\}\qquad(10.1.2)$$

where σ, b and r are parameters, with the third of these parameters, a reduced Reynolds number, playing the important role of control parameter μ. Lorenz solved (10.1.2) numerically and found, surprisingly, that these simple deterministic equations gave rise to seemingly random (pseudo-random) behaviour. Such is the behaviour of a 'strange' attractor and as such the Lorenz equations have been extensively studied as a model of strange attractors in general (Sparrow 1982).

The conclusion from these early studies is that the complicated

Fig. 10.3. Power spectra and corresponding bifurcation diagrams as functions of the control parameter for the Landau scenario. (For simplicity, the harmonics are not shown.)

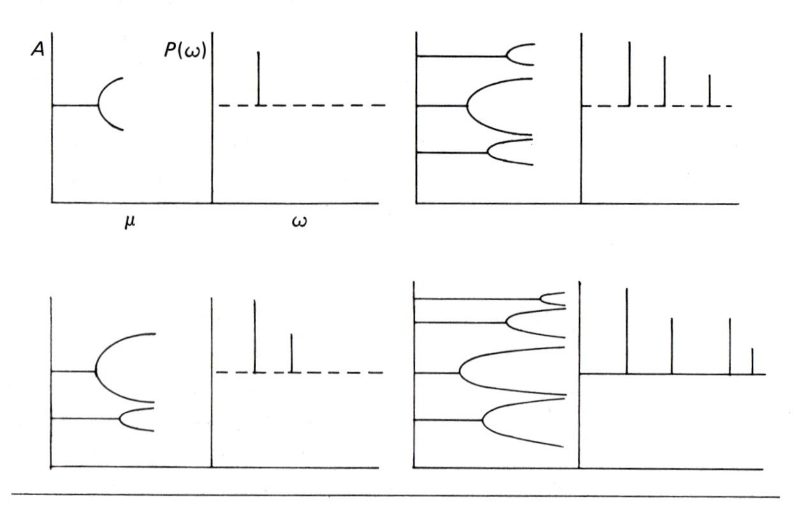

behaviour characteristic of turbulence can arise from a few simple coupled ordinary, but necessarily nonlinear, differential equations. Ruelle and Takens showed that the type of behaviour found by Lorenz is generic (that is, it is the most likely type of behaviour to occur under a wide range of conditions).

Lorenz derived the above equations from the Navier–Stokes equations mentioned in Section 1.4 by assuming that the solution was the sum of the three most unstable modes according to linear theory. This solution is substituted into the full equations and, by neglecting all harmonics, a closed set of three equations for the three amplitudes is obtained. In dimensionless form these are given by (10.1.2) for the amplitudes x, y and z. Unfortunately the neglect of the harmonics can only be justified for $r \simeq 1$ whilst the interesting time behaviour occurs for $r \simeq 26$. This method has been used in other problems in fluid mechanics (Benney (1962), Bretherton (1964), Fornberg and Whitham (1978)) with apparent success and by Infeld (1981c) to study Fermi–Ulam–Pasta recurrence phenomena (see Section 5.6). In the latter paper the agreement between theory and numerical computations has a wider range of validity than what one would normally expect. This also seems to be the case for the Lorenz equations.

A possible justification of these equations could be that x, y, and z are in reality the amplitudes of quasi-modes and that the Lorenz equations represent the simplest model incorporating interaction of such modes consistent with overall symmetry considerations. The parameter r is then a renormalized parameter not necessarily limited to values around unity. A possible method of justification could be based on the idea of 'enslaving'. According to this argument the higher order harmonics are enslaved, that is they follow the lower order modes in time. This is a method that has been particularly successful in the study of nonlinear lasers. It has been used by Rowlands (1980) to study the Fermi–Ulam–Pasta problem mentioned above.

Irrespective of their mode of derivation the Lorenz equations are certainly a good model of a strange attractor and their solution will be considered in more detail later. The Ruelle–Takens scenario of turbulence is illustrated in Fig. 10.4.

An alternative theory of turbulence has more recently been proposed and is now associated with the name of Feigenbaum. In its simplest form one considers the system described by a single parameter μ. When μ exceeds the critical value μ_c a single quasi-mode is excited which is described by an equation such as (10.1.1). That is the steady state ($A=0$), which is the stable state for $\mu<\mu_c$, bifurcates for $\mu>\mu_c$ into the states described by $A_s^2=\lambda^2(\mu-\mu_c)/\beta$. Now further increase of μ causes these two states ($\pm A_s$)

to bifurcate, in contrast to the Landau theory where it is assumed that a new quasi-mode is excited independently of the first one. This process of bifurcations continues as μ increases, until a further critical value μ_{ch} is reached where the bifurcation sequence reaches an accumulation point. For $\mu > \mu_{ch}$ the possible values of A_s are essentially random between certain limits and are said to be chaotic. Most importantly, Feigenbaum (1980) was able to show that for $\mu < \mu_{ch}$ the sequence of bifurcations had a universal character. If we associate A_n with the amplitude of the nth quasi-mode then

$$\operatorname*{Lt}_{n\to\infty} \frac{A_{n+1} - A_n}{A_n - A_{n-1}} = \alpha \tag{10.1.3}$$

where $1/\alpha$ is a universal constant equal to 2.5029.

For $\mu < \mu_{ch}$ one has a finite number of quasi-modes (each one associated with a distinct branch of the bifurcation process) namely 2^n, where n is the number of bifurcations that have taken place. The time dependence is then the result of the addition of these uncoupled quasi-modes and in some ways resembles the Landau theory. However, much more importantly, the sequence of bifurcations has a universal character as described by equation (10.1.3). For $\mu > \mu_{ch}$ the underlying behaviour is chaotic and the time dependence is complicated and of a form that could represent turbulence. This behaviour is illustrated in Fig. 10.5.

The main feature of the three distinct scenarios of turbulence discussed above are illustrated in Figs. 10.3 to 10.5 where a schematic representation of the quasi-modes is related to the frequency dependence of the power

Fig. 10.4. Power spectra and bifurcation diagrams for the Ruelle–Takens scenario.

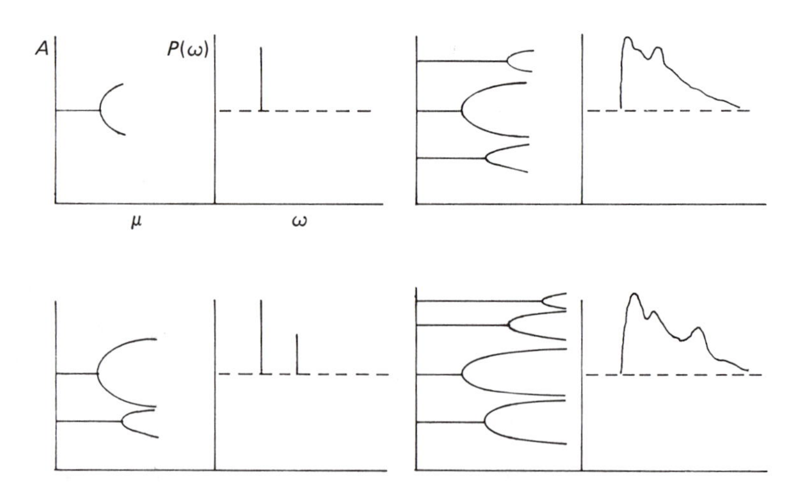

spectrum. It is immediately apparent that one should expect other scenarios, for example one where there is non-trivial interaction between the modes in the Feigenbaum sequence. However, we shall not elaborate further on such possibilities.

The reader may by now be wondering about the experimental situation. Swinney *et al.* (1977) studied the power spectrum of the velocity of a fluid produced when the fluid is constrained to flow between two concentric cylinders, one of which rotates with an angular velocity ω relative to the other (Fig. 1.8). Identifying ω with the control parameter μ, these experiments show that as ω is increased, first one, then two, and next three peaks appear in the power spectrum. This one tentatively associates with quasi-modes. Further increase of ω leads to broad band noise with little structure, presumably reflecting the presence of a strange attractor. These results strongly suggest that the Landau picture of turbulence is inadequate but that at least the qualitative features of the Ruelle–Takens theory are correct. There is no evidence for a Feigenbaum sequence of modes. However, power spectrum measurements of heat transport in liquid helium (Libchaber and Maurer (1980)) do show a sequence of period bifurcations in qualitative agreement with the Feigenbaum scenario.

The experimental evidence for the existence of strange attractors is reviewed in Section 10.6.

In conclusion, it is fair to say that there is good experimental evidence for both the Ruelle–Takens and Feigenbaum scenarios of turbulence, but little support for Landau theory. Some systems exhibit both Feigenbaum and

Fig. 10.5. Power spectra and bifurcation diagrams for the Feigenbaum scenario.

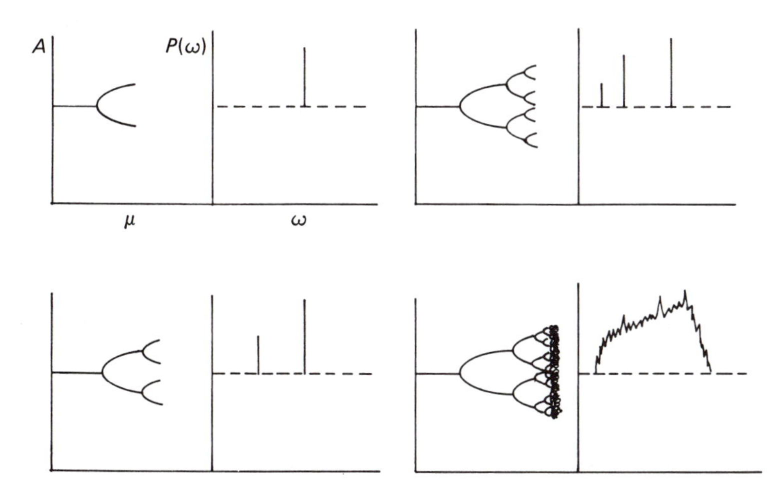

Ruelle–Takens type behaviour. A number of intriguing questions still remain. For example, why do some systems seem to follow one scenario whilst others follow another? Can the same system show both types of behaviour under different spatial conditions? Whatever the answers may be, it is plain that the phenomena of bifurcation sequences, chaotic motion, and strange attractors warrant further study.

In the above it has been stressed that turbulent-like behaviour can be described by as few as three coupled nonlinear equations. This does not mean to say that all such equations show such behaviour. The outstanding counter-example, relative to the subject of this book is in the study of three wave interactions. This has been treated in Chapter 5 and more extensively by Weiland and Wilhelmsson (1977). The usual type of behaviour found in these studies is a periodic time variation, not chaos. In the relevant phase space this is motion on a simple torus (an attractor but *not* a strange one). Though Ruelle and Takens argue that a strange attractor is the generic behaviour (to be expected most of the time), these arguments can be invalidated if some symmetry property must be satisfied. Satisfying such properties imposes conditions on the coefficients and form of the nonlinear terms and as a consequence one may obtain coherent solutions, the antithesis of chaos.

The rest of this Chapter is devoted to a study of simple mathematical models that illustrate the above phenomena. Two points need stressing. The first is that the models are simplistic so that their relationship to the real world of fluid dynamic or plasma turbulence is somewhat tenuous. Secondly, the concepts discussed often have a much wider field of application than just to the theory of turbulence.

10.2 Bifurcation sequences and chaos

Most of the features of bifurcation sequences and chaos can be understood by studying the simple one-dimensional difference equation

$$x_{n+1} = \mu x_n(1 - x_n) \tag{10.2.1}$$

where μ is a control parameter and x_n the value of x after n iterates of the equation. Difference equations of this form are commonly called mappings, as the right hand side of the equation maps x_n to the value x_{n+1}. In the following we shall only be interested in the case where x_n is bounded and so the following restrictions will be imposed: $0 \leqslant \mu \leqslant 4, 0 \leqslant x_0 \leqslant 1$. This equation, commonly called the logistic equation, has an immediate application in population dynamics where x_n is the population of some species after n breeding cycles. It was in this context that some of the intriguing properties of this equation were first revealed (May (1976 a and b)). However, there does not seem to be any immediate interpretation in the physical sciences and the equation is best considered just as a model which illustrates some relevant mathematical phenomena.

The classical physical sciences deal mainly with continuous systems described by differential equations (space and time being continuous) and so an obvious question is whether the above equation can be replaced by some appropriate differential one. If one assumes that $x_n = x(n)$ is a slowly varying function of its argument and write

$$x_{n+1} = x(n+1) \simeq x(n) + \frac{\mathrm{d}x}{\mathrm{d}n}$$

then equation (10.2.1) reduces to

$$\frac{\mathrm{d}y}{\mathrm{d}\tau} = y(1-y) \tag{10.2.2}$$

where $\tau = (\mu - 1)n$ and $y = \mu x/(\mu - 1)$. This is readily solved to give

$$y = A\mathrm{e}^{\tau}/(1 + A\mathrm{e}^{\tau}) \tag{10.2.3}$$

where A is a constant of integration. For small but positive τ one obtains exponential growth which for large τ saturates such that $y \to 1$, and so exhibits similar qualitative features to the solution of (10.1.1).

Equation (10.2.1) is extremely simple to solve on a small calculator and it is readily found that the difference equation (10.2.1) and the differential equation have many of the same qualitative features, namely, an exponential growth for small τ, (small n) and an asymptotic state for large τ (large n). In both cases the exponential growth factor is the same and so is the value of the fluid saturated state $\left(\frac{\mathrm{d}y}{\mathrm{d}\tau} = 0, x_{n+1} = x_n\right)$, namely

$$x_s = (\mu - 1)/\mu. \tag{10.2.4}$$

This final asymptotic state is termed a fixed point of the map, defined by $x_{n+1} = x_n = x_s$ and independent of n.

Whether the solution of the differential or difference equations will evolve to this asymptotic state depends on the stability of this state. If it is unstable it cannot be reached. Conversely, if it is stable then it may be reached. A conventional linearized stability analysis based on the differential equation (10.2.2) gives

$$x(n) = x_s + \alpha\mathrm{e}^{-(\mu-1)n}. \tag{10.2.5}$$

Using the difference equation (10.2.1), writing $x_n = x_s + \delta x_n$ and linearizing, one obtains

$$\delta x_{n+1} = (2-\mu)\delta x_n, \tag{10.2.6}$$

which is readily solved to give

$$x_n = x_s + \alpha(2-\mu)^n. \tag{10.2.7}$$

For $1 < \mu < 3$ both equations show that the final asymptote is stable (note that for the difference solution the condition is $2-\mu < 1$), from which we conclude that for μ in this range the differential and difference equations have the same qualitative features. This is shown in Fig. 10.6 where x_n and $x(n)$ are plotted as functions of n.

The evolution of the solution, that is the variation of x_n with n, is illustrated pictorially in Fig. 10.7. The diagonal line represents $x_{n+1} = x_n$, the vertical line the solution and the dashed horizontal lines the construction which shows how x evolves with n. The figure illustrates the solution going to the fixed point $x_s (\mu < 3)$.

For $\mu > 3$ we expect differences to appear since the asymptotic state of the difference equation is unstable. However, even for $\mu < 3$ there is a fundamental difference between the discrete and the continuous equations. Namely in the continuous state the relevant equation (10.2.2) can be solved

Fig. 10.6. Variation of the solution of the differential equation (10.22) and the solution x_n of the difference equation (10.2.1) for $\mu = 2.5$.

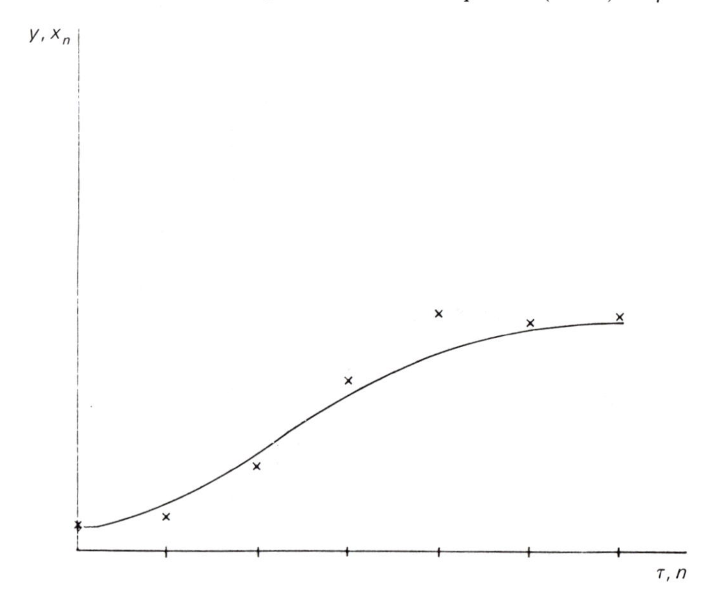

for negative as well as positive values of τ, that is forward or backward in time. However, if we attempt a similar calculation based on the discrete case we find from (10.2.1) that

$$x_n = [1 \pm (1 - 4x_{n+1}/\mu)^{\frac{1}{2}}]/2 \tag{10.2.8}$$

that is, the mapping is double valued and a procedure based on such a map would not lead to a unique value. The discrete case does not allow one to go backward in 'time'. Such maps are called non-invertable.

It is readily confirmed numerically that, for $\mu > 3$, there is no constant stable asymptotic state. However, what is found is that for $3.499 > \mu > 3$ *successive* iterates of the difference equation settle down asymptotically to constant values, that is, for large m, $x_m = x_{m+2} = x_{m+4} = x_a$, and $x_{m+1} = x_{m+3} = x_{m+5} = x_b$. Assuming that this is so, one has from (10.2.1) that

$$x_b = \mu x_a(1 - x_a) \tag{10.2.9}$$

and

$$x_a = \mu x_b(1 - x_b). \tag{10.2.10}$$

These equations are readily solved by realizing that they must contain the solutions corresponding to $x_a = x_b = x_s = (\mu - 1)/\mu$. In this way one finds that x_a, x_b are the solutions of

$$\mu^2 x^2 - \mu(1 + \mu)x + (\mu + 1) = 0 \tag{10.2.11}$$

Fig. 10.7. Geometric construction of the solution of the difference equation (10.2.1) for $\mu < 3$.

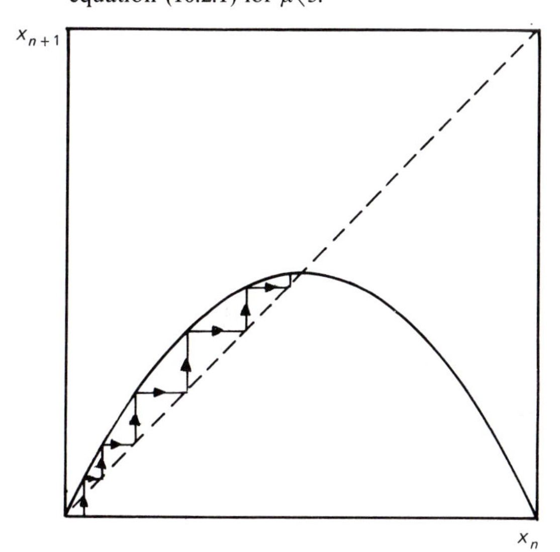

Real solutions exist for $\mu > 3$. A linear analysis about this asymptotic state gives

$$\delta x_{n+2} = (4 + 2\mu - \mu^2)\delta x_n. \tag{10.2.12}$$

For $\mu > 1 + \sqrt{6}(\simeq 3.449)$ this gives rise to instability. We conclude that for $3 < \mu < 1 + \sqrt{6}$ a simple bifurcated asymptotic state exists which is stable. This again is easily confirmed numerically.

Fig. 10.8. First and second fold iterate, $f(x)$ and $f^2(x)$, of the logistic map for $\mu \langle 3(a)$ and $3 \langle \mu \langle 3.4(b)$.

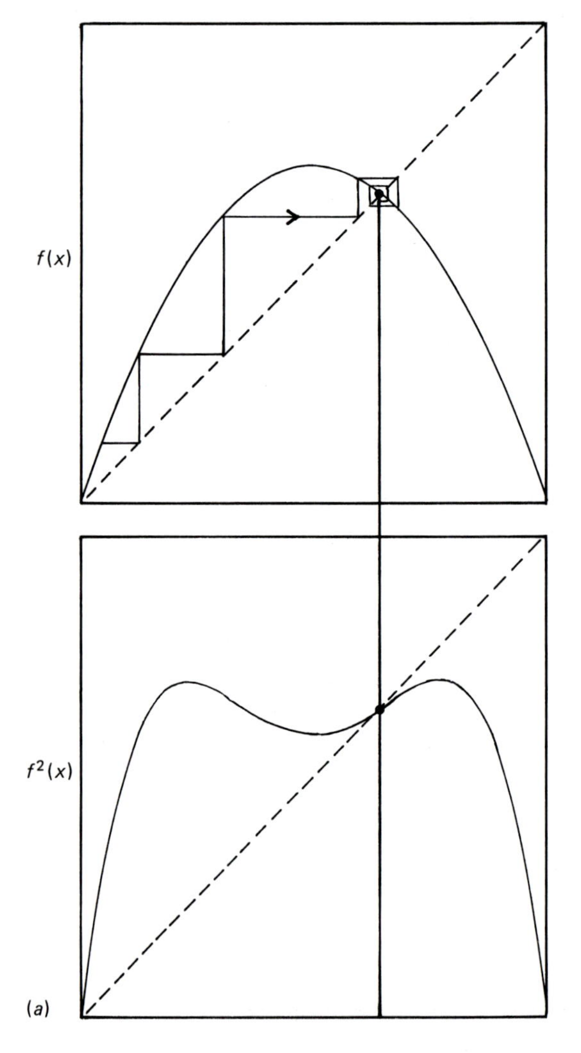

If we write the logistic equation in the general form

$$x_{n+1} = f(x_n, \mu), \tag{10.2.13}$$

with $f(x) = \mu x(1-x)$ in particular, then whereas the asymptotic value for $\mu < 3$ was interpreted as the fixed point of the map $x_s = f(x_s)$, the bifurcated asymptotic state which satisfies $x_b = f(x_a)$ and $x_a = f(x_b)$ can be interpreted as the fixed point of the map $f(f(x))$. This is usually, somewhat misleadingly, written as $f^2(x)$, so that $x_a = f^2(x_a)$ and of course $x_b = f^2(x_b)$. The stability of these fixed points can be discussed as follows. Writing $x_n = x_s + \delta x_n$ gives, in a linear approximation, $\delta x_{n+1} = f'(x_s)\delta x_n$, where the dash denotes differentiation. Then the stability criterion is simply $|f'(x_s)| < 1$. Similarly for

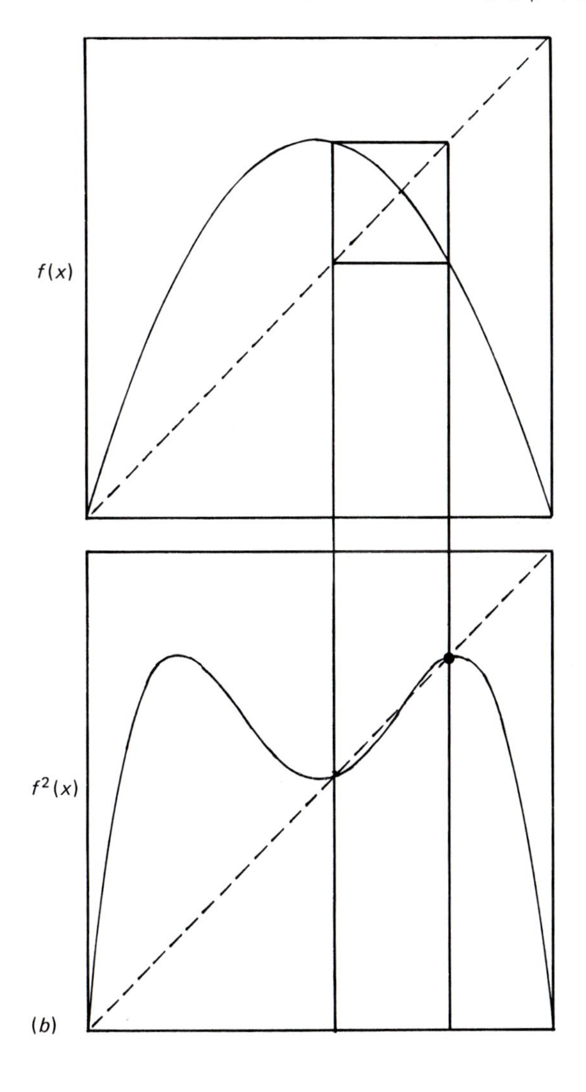

the bifurcation state one considers the map $f^2(x)$ and then the condition is $|(f^2(x_a))'|<|$. This last condition is readily seen to be equivalent to $|f'(x_a)f'(x_b)|<1$ (that is a condition on an average slope). These results are illustrated graphically in Fig. 10.8 where $f(x)$ and $f^2(x)$ are shown as functions of x for $\mu=2.5(<3)$ and $\mu=3.2(3<\mu<3.449)$. The fixed points are defined by the intersection of $f(x)$ and $f^2(x)$ with the diagonal. It should be noted that the slope of the tangent at the fixed points (greater or less than unity) reflects the stability of these points. If $1+\sqrt{6}<\mu$ it is found that an asymptotic state exists but now every fourth value of x_n is a constant so the fixed point is defined by $x=f^4(x)$. The state has bifurcated again. This process of bifurcation of the asymptotic state continues as μ is further increased and is illustrated in Fig. 10.9, where the possible asymptotic states are shown as functions of μ. The solid curves correspond to stable states, the dotted lines to unstable ones.

Two important points are immediately obvious from Fig. 10.9. If μ_n represents the critical value of μ for the onset of a bifurcation then the values of the 'window' $\Delta\mu_n=(\mu_n-\mu_{n-1})$ decrease as n increases. The analytic results above give $\Delta\mu_1=2, \Delta\mu_2=0.449$ whilst simple numerics give decreasing values for $\Delta\mu_3, \Delta\mu_n$. This is recognized as a decaying sequence and one finds that the following limit exists

$$\underset{n\to\infty}{\mathrm{Lt}}\ \frac{\Delta\mu_{n+1}}{\Delta\mu_n}=1/4.669201. \tag{10.2.14}$$

A critical value μ_c must exist which is the termination point of this sequence. Numerically it is found that $\mu_c=3.568$.

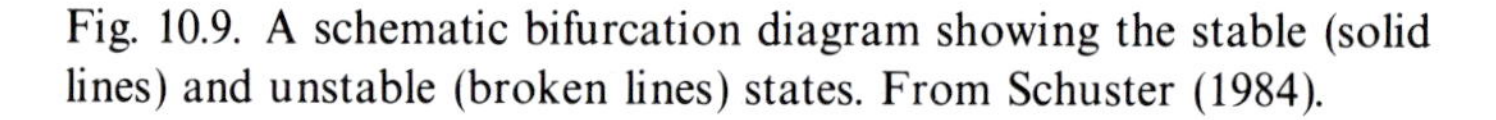
Fig. 10.9. A schematic bifurcation diagram showing the stable (solid lines) and unstable (broken lines) states. From Schuster (1984).

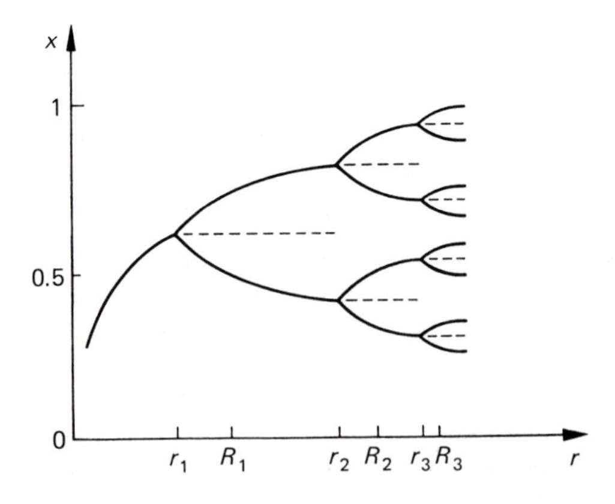

The other feature of the bifurcation diagram is that the difference between the bifurcation solutions decreases as μ increases. If A denotes the amplitude of a state just before it bifurcates, so that $A_o = 1 - 1/\mu$ with $\mu = 3$ and $A_1 = [(1+\mu)(\mu-3)]^{\frac{1}{2}}/\mu$, with $\mu = 1+\sqrt{6}$, then $A_o = 2/3$, $A_1 = 0.35$. Then one finds from the numerical results that

$$\operatorname*{Lt}_{n\to\infty} \frac{A_n}{A_{n+1}} = 2.5029.$$

The behaviour of x_n as a function of n is perhaps best illustrated graphically. This is done in Fig. 10.10 for values of μ which lead to a unique asymptotic state ($\mu < 3$), a simple two point state ($3 < \mu < 3.449$) and a four point state ($3.449 < \mu < 3.544$).

In summary, we have that for $\mu < \mu_c$ the asymptotic state (fixed point) of the solution of (10.2.1) is in general not constant, but consists of 2^m distinct values with the value of m depending on μ. ($m = 0, \mu < 3, m = 1, 3 < \mu < 3.449$). This means that the asymptotic state is a periodic function of n with period 2^m. Most importantly, this final state is reached by iterating (10.2.1) irrespective of initial conditions, that is, independent of the value of initial value x_o.

The main features of the above may be illustrated analytically by applying a simple form of renormalization group theory (RNG). First the logistic map is recast in a more appropriate form by considering deviations from the asymptotic state. By writing $x_n = x_s + (2-\mu)\phi_n/\mu$, equation (10.2.1) takes the form

$$\phi_{n+1} = \eta\phi_n(1-\phi_n), \tag{10.2.15}$$

$\eta = 2 - \mu$. We immediately see that for $\mu > 3 (\eta > 1)$ the state is unstable and as we have seen above, the asymptotic state bifurcates. The period doubled map, $f^2(x)$ has an asymptotic state given by (10.2.11). Near to this state we may expand in a Taylor series $x_n = \bar{x} + y_n$ to give ($f^2(\bar{x}) = \bar{x}$)

$$y_{n+1} = (f^{2\prime})y_n + (f^{2\prime\prime})y_n^2/2. \tag{10.2.16}$$

This can be put in the form of (10.2.15) with $\eta = (f^2(\bar{x}))'$, which after a little algebra reduces to $\eta = 4 + 2\mu - \mu^2$. This whole process can be repeated since at each stage we get the same map, with a simple change in parameter value. Then in the spirit of RNG theory we concentrate on the change of the parameter each time the above process is carried out. If we write μ_n as the parameter after n bifurcations then

$$\mu_n = F(\mu_{n+1}) = \mu_{n+1}^2 - 2\mu_{n+1} - 2, \tag{10.2.17}$$

since $n+1$ bifurcations have occurred for the old map. The critical point of

Fig. 10.10. Solutions of the logistic equations (10.2.1) for different values of the parameter μ: 2.5(a); 3.3(b); 3.5(c). From Haken (1983).

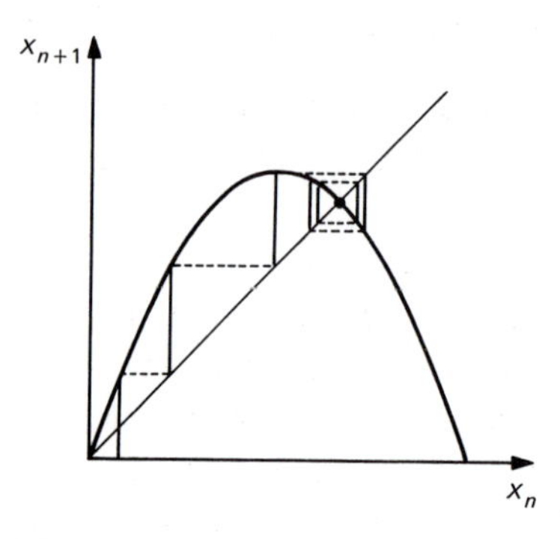

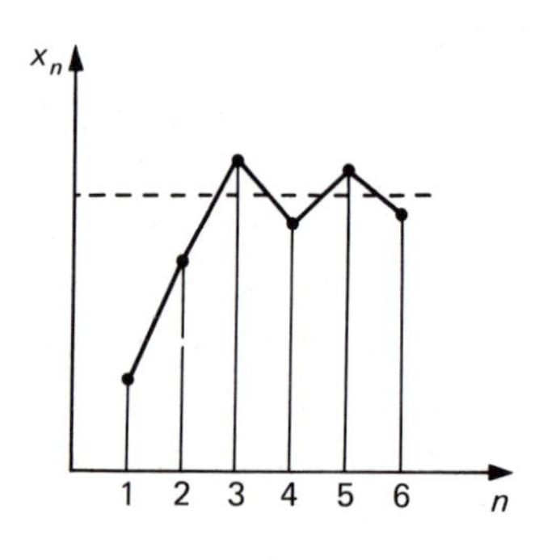

(a)

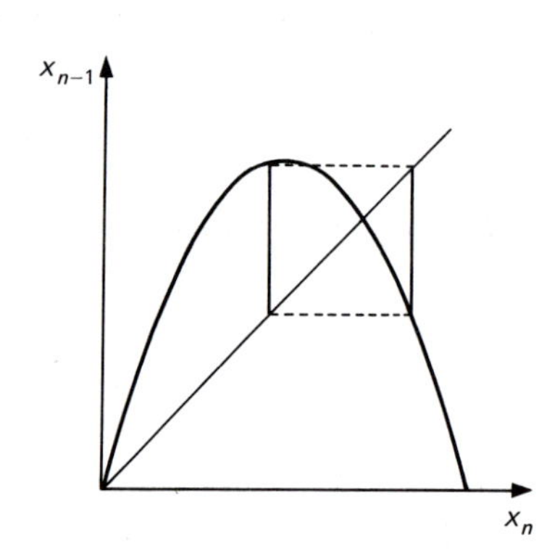

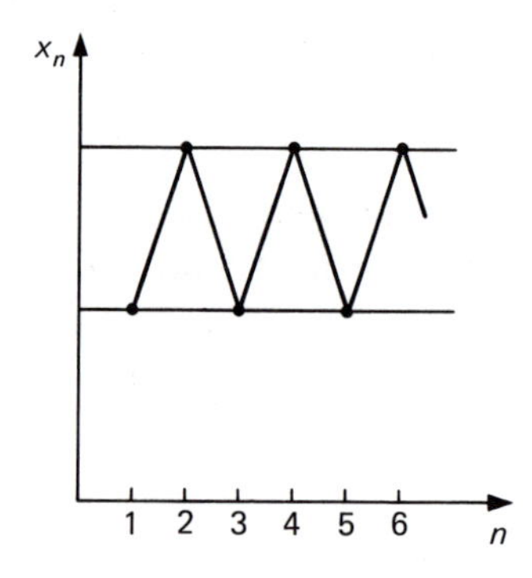

(b)

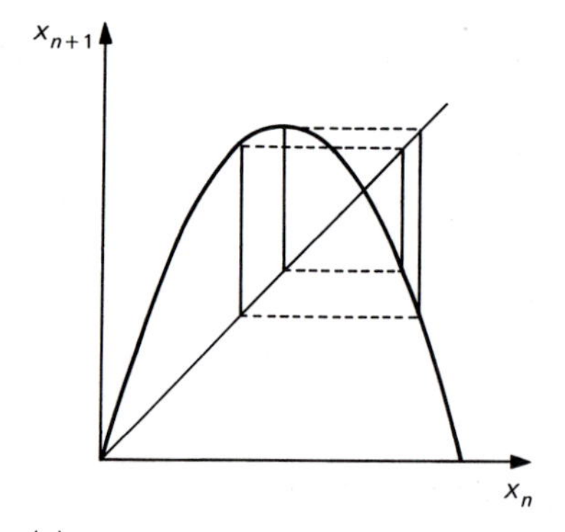

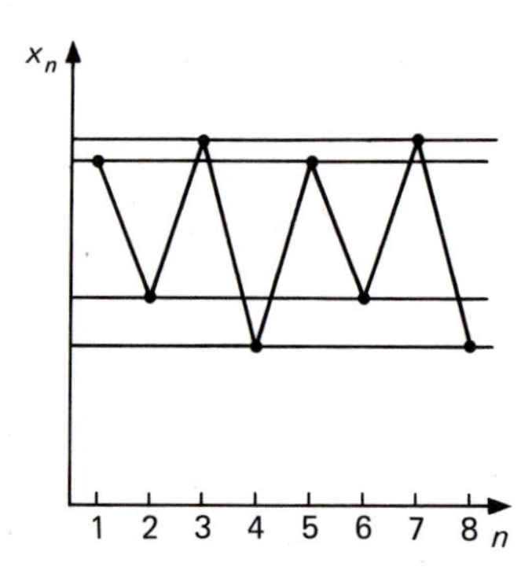

(c)

this map, $(\mu_{n+1}=\mu_n)$, corresponds to the end of the sequence of bifurcations and hence to the onset of chaos. We find $\mu_c=3.562$. Furthermore, the approach of the sequence to this value is simply $\delta\mu_{n+1}/\delta\mu_n=\dfrac{\mu_{n+1}-1}{2}$ which gives a limiting value of 1/5.12. This should be compared with the exact values given above (10.2.14). (May and Oster (1976) have given a somewhat similar but more elaborate theory.)

For $\mu>\mu_c$ the periodic behaviour described above disappears and the solution x_n is found to jump in a seemingly random fashion from one value to the next. This behaviour is not periodic and is called CHAOTIC. Furthermore, the values of x_n depend sensitively on the initial condition x_o. By this one means that two points initially close together, x_o and $x_o+\delta, \delta \ll 1$ say, will lead to values of x after n iterates which are far apart. Initially the points diverge from one another exponentially. This may be seen as follows. Let x_n and $\bar{x}_n$ be the values of x after n iterates starting from initial conditions x_o and $x_o+\delta$ respectively. Substituting $\bar{x}_n=x_n+\phi_n$ into (10.2.1) and linearizing ϕ_n we obtain

$$\phi_{n+1}=\mu(1-2x_n)\phi_n. \tag{10.2.18}$$

This equation may be solved *approximately* to give $\phi_n=\exp(\lambda n)$ where λ is an average value defined by

$$\lambda=\operatorname*{Lt}_{m\to\infty}\frac{1}{N}\sum_{n=0}^{N}\ln[\mu|1-2x_n|], \tag{10.2.19}$$

and the summation is over the orbit of x_n appropriate to the initial condition x_o. If $|\lambda|>1$, adjacent initial points, on average, diverge and this is taken as characteristic of chaos. Of course, the divergent behaviour cannot continue indefinitely as both x_n and $\bar{x}_n$ remain between the limits 0 and 1, *see* Ott (1981). (The quantity λ introduced above is an example of a Lyapunov number and will be discussed more fully later in this Chapter.)

Another characteristic of chaotic behaviour emerges by considering the correlation function

$$C(m)=\operatorname*{Lt}_{N\to\infty}\frac{1}{N}\sum_{n=0}^{N}(x_n-\bar{x})(x_{n+m}-\bar{x}), \tag{10.2.20}$$

where

$$\bar{x}=\operatorname*{Lt}_{N\to\infty}\frac{1}{N}\sum_{n=0}^{N}x_n. \tag{10.2.21}$$

For $\mu<\mu_c$, where the x_n's relax to asymptotic states which are periodic, the above summation will be dominated by these states and $C(m)$ will thus be a periodic function of m with a period equal to the number of states. Chaotic behaviour, which is non-periodic and seemingly random, is such that $C(m) \to \delta^m$ as $m \to \infty$, where δ is a constant less than unity, so that $C(m) \to 0$ as $m \to \infty$ (*see* Li and Yorke (1975)).

Before discussing chaos in more detail we consider yet another surprise concerning equation (10.2.1). It is found numerically that a small window exists in μ space, that is a small range of μ values ($3.8284 < \mu < 3.8496$) where the solution of (10.2.1) asymptotically approaches a stable state of period 3 (Shail and Pakham (1983)). This fixed point differs in a very important way from the fixed points previously discussed. In Fig. 10.11 the third iterate of the map $f^3(x)$ is plotted for the critical value $\mu = 1 + 2\sqrt{2} (= 3.828)$. It will be noted that there are three intersections with the diagonal, the intersection with the minimum counting as a double point. For $\mu < 3.828$ there is only one intersection and this fixed point is unstable and a chaotic solution appertains. For $\mu > 3.828$ the minimum crosses the diagonal giving three fixed points. One of these associated with the minimum is such that $|(f^3(x)'| < |$, and so is stable and this corresponds to the period three solution. Thus pictorially the chaotic solution changes to one of period three by $f^3(x)$ becoming tangential at a point to the 45° line. Such a phenomenon is called a 'tangent bifurcation'.

The essential difference between a tangent bifurcation and those discussed above (pitchfork bifurcation) is best illustrated by considering the map in the vicinity of the tangent bifurcation. Suppose for the moment that one has a map of the form $x_{n+1} = F(x_n, \mu)$ where μ is the control parameter

Fig. 10.11. Illustration of a tangent bifurcation. The threefold iterate $f^3(x)$ of the logistic equation is shown as a function of x. The curve is tangent to the 45° line at three points, the central of which corresponds to a stable, period three solution. From Schuster (1984).

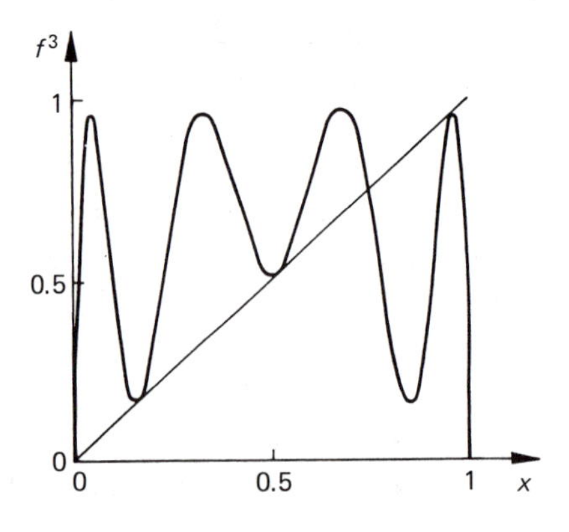

as usual. A fixed point x_s exists such that $x_s = F(x_s, \mu)$ and a linear stability analysis about this point gives $(x_n = x_s + \delta x_n)$

$$\delta x_{n+1} = F_x(x_s, \mu)\delta x_n \tag{10.2.22}$$

where the suffix denotes differentiation with respect to that variable. The stability boundary is given by $|F_x(x_s, \mu)| = 1$. This condition must be interpreted as a condition on μ. Suppose μ_s satisfies this condition. Near this new critical state one has $x_s = x + \Delta x, \mu = \mu_s + \Delta\mu$, where, from the equation defining x_s,

$$\Delta x = F_x \Delta x + \tfrac{1}{2} F_{xx} \Delta x^2 + F_\mu \Delta\mu + \ldots \tag{10.2.23}$$

Now first suppose $F_x = -1$, then

$$\Delta x \simeq \frac{F_x \Delta\mu}{2} \tag{10.2.24}$$

and

$$\delta x_{n+1} = \lambda \delta x_n = \left[-1 + \left(F_{x\mu} + \frac{F_{xx} F_\mu}{2} \right) \Delta\mu \right] \delta x_n. \tag{10.2.25}$$

Thus a change in μ, $\Delta\mu$ will take the system from a stable situation $(\lambda < 1)$ to an unstable one $(\lambda > 1)$. This is the pitchfork bifurcation.

However, now consider the condition $F_x = +1$, in which case

$$\Delta x^2 = \frac{-2F_\mu \Delta\mu}{F_{xx}} \tag{10.2.26}$$

and

$$\delta x_{n+1} = \lambda \delta x_n = \left[1 \pm F_{xx} \left(\frac{-2F_\mu \Delta\mu}{F_{xx}} \right)^{\frac{1}{2}} \right] \delta x_n. \tag{10.2.27}$$

Now for one sign of $\Delta\mu$ we have that λ is real and one solution corresponds to a stable situation $(\lambda < 1)$, namely the period three solution. However, a change in the sign of $\Delta\mu$ leads to λ being complex and a marginal stable state where $\delta x_n \simeq e^{\pm i\alpha n\sqrt{\Delta\mu}}$ with $\alpha^2 = -2F_\mu F_{xx}$. Thus there is a major difference in behaviour near a state where $F_x = -1$ from that where $F_x = +1$. The latter is by definition a tangent bifurcation. On one side (in parameter space), a tangent bifurcation shows oscillatory motion whereas on the other it is stable. On the other hand a pitchfork bifurcation simply changes a stable state to an unstable one (no oscillations).

For the logistic map there were no non-trivial solutions to both conditions for a tangent bifurcation, namely $f(x_s) = x_s$ and $f_x(x_s) = 1$. However, for the map of period three, where $F(x) = f^3(x)$, one solution exists for $\mu=1+\sqrt{2}$. A model of turbulence based on a tangent bifurcation has been suggested by Pomeau and Manneville (1980) and is discussed later. Increase of μ above the value 3.828 reveals a whole bifurcation sequence similar to that for $1<\mu<\mu_c$. In particular, the sequence has a limit and satisfies (10.1.3) with an identical value of the constant, but a new critical value of μ, above which one observes chaos once again.

In fact for $\mu>\mu_c$ a very complicated structure emerges. Windows exist in μ space for which the solution is periodic and undergoes a sequence of bifurcations similar to that discussed above for $1<\mu<\mu_c$ and which eventually becomes chaotic. The bifurcation sequence always satisfies equation (10.2.14). However, the range of μ values for which this behaviour is found is very small. The most pronounced is the one associated with period 3 where $\mu=3.8284$ is the critical value for the onset of period three behaviour and $\mu=3.8496$ the value for the onset of chaos. More generally we find that for $\mu_c<\mu<4$, there are regions of periodic asymptotic states (fixed points) intermixed with regions of chaos. For $\mu=4$ the system is chaotic and x_n can take all values between 0 and 1. In the other regions of chaos the allowed values of x_n fall within well defined limits. For $\mu>4$ any initial condition between 0 and 1 leads to values of x_n diverging as n becomes large.

The complicated structure of the solution of (10.2.1) is well revealed by looking at its 'long time' (large n) behaviour. Equation (10.2.1) is iterated numerically a sufficiently large number of times ($\simeq 100$) so that the solution has lost all memory of its initial condition and then a large number of subsequent values of x are plotted as a function of μ. If the asymptotic state is periodic, period N, then there will be N distinct values of x, whilst if the system is chaotic regions will exist where x takes a continuous range of values. Such a plot is shown in Fig. 10.12. This is an extension of Fig. 10.4 to include chaotic behaviour and also the periodic behaviour for $\mu>\mu_c$.

The period 3 behaviour is clearly seen and so are the bands of chaos. It should be remembered, though, that a closer inspection of the chaotic bands would reveal further small windows of periodic behaviour.

In view of the complexity of the solution of (10.2.1) it is natural to introduce statistical concepts as an aid to their understanding. Thus one is led to consider a function $g(x,n)$ such that $g(x,n)\mathrm{d}x$ is the probability that after n iterates the value of x lies between x and $x+\mathrm{d}x$. Then from (10.2.1) one may write down a master equation

$$g(x,n+1)=\int_0^1 g(x',n)\delta[x-\mu x'(1-x')]\mathrm{d}x' \qquad (10.2.28)$$

This merely states that the value x' is mapped to the value x according to equation (10.2.1). It is exact, though of course g may be an extremely complicated function of x. A simplification occurs if one assumes that an asymptotic state exists such that $g(x, n)$ is independent of n, $g(x)$ say. Such a state is called an invariant measure and satisfies

$$g_s(x) = \int_0^1 g_s(x')\delta[x - \mu x'(1 - x')]\mathrm{d}x'. \tag{10.2.29}$$

For $1 < \mu < 3$ we have seen that asymptotically all initial conditions give the same value of x, namely $x_s = (\mu - 1)/\mu$. In this case

$$g_s(x) = \delta(x - (\mu - 1)/\mu) \tag{10.2.30}$$

a result which is readily shown to satisfy (10.2.29). Similarly for the periodic solutions, g_s is simply a sum of delta functions, each centred on one of the periodic states.

Equation (10.2.29) may be expressed in an alternative form simply by evaluating the integral:

$$g_s(x) = \left\{ g_s\left[\frac{1 + (1 - 4x/\mu)^{\frac{1}{2}}}{2}\right] + g_s\left[\frac{1 - (1 - 4x/\mu)^{\frac{1}{2}}}{2}\right]\right\} \Big/ \mu(1 - 4x/\mu)^{\frac{1}{2}}.$$

Fig. 10.12. Complete bifurcation diagram for the logistic map, showing the bifurcation sequences and bands of chaos. From Schuster (1984).

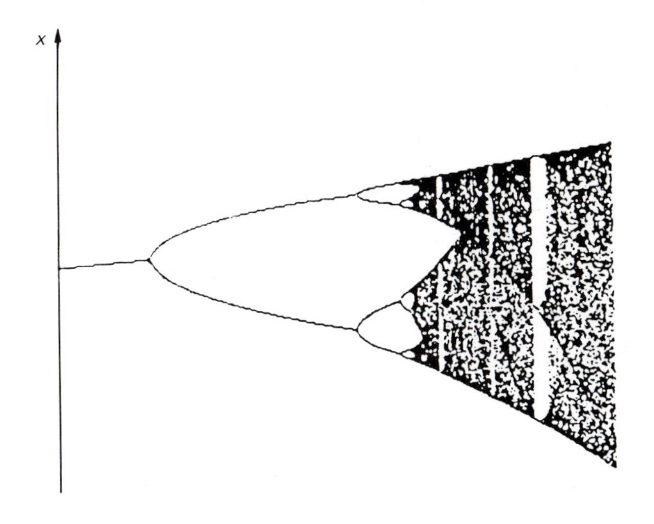

A simple change of variable $\left[y=\frac{1+(1-4x/\mu)^{\frac{1}{2}}}{2}\right]$ finally gives

$$g_s[\mu y(1-y)]=[g_s(y)+g_s(1-y)]/\mu\,|2y-1|. \tag{10.2.31}$$

Unfortunately, very little is known about such equations in general. However, for $\mu=4$ it is readily shown by substitution that

$$g_s(x)=\pi/[x(1-x)]^{\frac{1}{2}}, \tag{10.2.32}$$

where the factor π arises from a normalization. The reason one obtains a simple smooth function is that the maximum point $(x=\frac{1}{2})$ is mapped to an unstable fixed point $(x=1)$. For some other values of μ the maximum is mapped after a *finite* number of iterations to the unstable fixed point $x=1-1/\mu$. In this case the distribution function is split into a finite number of regions in each of which the function g_s is smooth but has irregularities of the form $1/(x-\bar{x})^{\frac{1}{2}}$ at the edges $(x=\bar{x})$ of the regions. The first few regions of smooth distributions have been studied numerically by Collet and Eckmann (1980). For other values of μ, the g_s's probably have a fractal structure, (Mandelbrot (1982)), and methods of studying such structures are still in their infancy.

The deceptively simple-looking logistic map (10.2.1) has revealed an astonishingly complex solution. It is totally unlike the seemingly analogous differential equation (10.2.2) whose solution it only resembles for $\mu<3$. More details of the properties of this map are given in the book by Collet and Eckmann (1980) (Note that they study the map $y_{n+1}=1-ay_y^2$, which however, is equivalent to (10.2.1) if we make the identification $a=\mu(\mu-2)/4$ and $y_n=2(1-2x_n)/(2-\mu)$). A further class of maps which received considerable attention are the so-called tent maps. These are maps which may be written in the general form of equation (10.2.13) but where $f(x)$ is defined to be a linear function of x in a series of intervals. For example take

$$\begin{aligned} f(x)&=\alpha x, 0\leqslant x\leqslant b\\ &=\frac{\alpha b(1-x)}{1-b},\quad b\leqslant x\leqslant 1. \end{aligned}$$

Though such maps do not show the richness of behaviour of the logistic map they are in some ways easier to handle analytically. In particular the case $\alpha=2, b=\frac{1}{2}$ has received considerable attention and it has been proven that such a map shows ergodic behaviour and has a constant invariant measure $g_s(x)=1$. In fact this map is identical to the logistic map for $\mu=4$ as can be seen by substituting $x_n=\sin^2\theta_n$ in (10.2.1) to give $\theta_{n+1}=2\theta_n$, modulo π. It was in fact knowledge of this information that led to (10.2.1).

By studying two discrete maps in some detail, one of which was the logistic map, we have shown how very simple one-dimensional maps can lead to very complicated behaviour such as bifurcation sequences and chaos. The question remains as to whether or not this type of behaviour is widespread or just particular to the two equations we have studied. This is a very relevant question as it is tempting to associate the properties found with experimental results such as the power spectra of turbulent systems. The universality of the results was first appreciated by Feigenbaum (1980) and it is his work that we will now consider.

We consider the general map of the form (10.2.3) but now with the control parameter considered explicitly, that is $x_{n+1}=f(x_n,\mu)$. As we have seen, fixed points exist such that $\bar{x}=f(\bar{x},\mu)$, giving $\bar{x}$ as a function of μ. The stability of this fixed point can be studied by linearization which leads to an equation of the form $\delta x_{n+1}=f'(\bar{x},\mu)\delta x_n$. In particular there is a critical value of μ such that $f'(\bar{x},\mu)=0$. Such a state is called super-stable. For maps with a single maximum, that is of the general form of a parabola, such a state always exists. For example if we consider the logistic map, $\bar{x}=1-1/\mu$ and $f(\bar{x},\mu)=\mu(1-2\bar{x})$ so $f'=0$ for $\bar{x}=\frac{1}{2}$ and $\mu=2$. Larger values of μ correspond to period doubled solutions so that the fixed point now satisfies $\bar{x}=f^2(\bar{x})$, and nearby orbits satisfy $\delta x_{n+1}=(f^2(\bar{x}))'\delta x_n$. However, by the chain rule, $(f^2(\bar{x}))'\equiv \mathrm{d}f^{(2)}(\bar{x})/\mathrm{d}\bar{x}=f'(f(\bar{x}))f'(\bar{x})$. The corresponding super-stable orbit is still given by $f'(\bar{x})=0$ or $\bar{x}=\frac{1}{2}$. For the logistic map the two fixed points are given by $\bar{x}=\{1+\mu\pm\sqrt{[(\mu-1)^2]}+4/2\mu\}$ and the critical value of μ to give a super-stable orbit is $1+\sqrt{5}$. The next bifurcation is treated in a similar manner and the condition for the super-stable orbit is now $(f^4)'=0$ which again has a solution $\bar{x}=\frac{1}{2}$. From this Feigenbaum concluded that the sequence of bifurcations could be studied by considering the properties of the maps f^{2n} about the point $\bar{x}=\frac{1}{2}$. There is always one of the fixed points of the higher bifurcations near $\bar{x}=\frac{1}{2}$. The other fixed points are enslaved to this point as they are simple iterates of it. The super-stable orbits for periods 2 and 4 are illustrated for the logistic map in Fig. 10.13. An important point to note is that the orbits shown, about $\bar{x}=\frac{1}{2}$, for both periods are very similar. To express this idea more quantitatively we consider the form of f and f^2 about $x=\frac{1}{2}$. Then, neglecting terms of order $(x-\frac{1}{2})^4$ we have with $y=x-\frac{1}{2}$

$$f(x,\mu)=\mu(\tfrac{1}{4}-y^2)$$

and

$$f^2(x,\mu)=(\mu/4)(1-\mu^2/4)(1+2\mu^2y^2/(1-\mu^2/4)).$$

The fixed point adjacent to $x=\frac{1}{2}$ is readily obtained by iteration of the map and is given from the above by $\mu/4$ and $\mu(\mu^2/4-1)/4$ respectively. These quantities give the scale of the map. Thus we see that about $x=\frac{1}{2}$ the effect of considering the $f^2(x,\mu)$ map is to reproduce the initial $f(x,\mu)$ map subject to an inversion, a rescaling of the variable $(x-\frac{1}{2})$ and a rescaling of the parameter μ. This process can now be continued to produce self-similar forms for $f^{2^n}(x)$ about $x=\frac{1}{2}$. Feigenbaum conjectured that this process of

Fig. 10.13. Super stable orbits for the logistic map, $\mu=1+\sqrt{5}$. Note the similarity between the squared boxes.

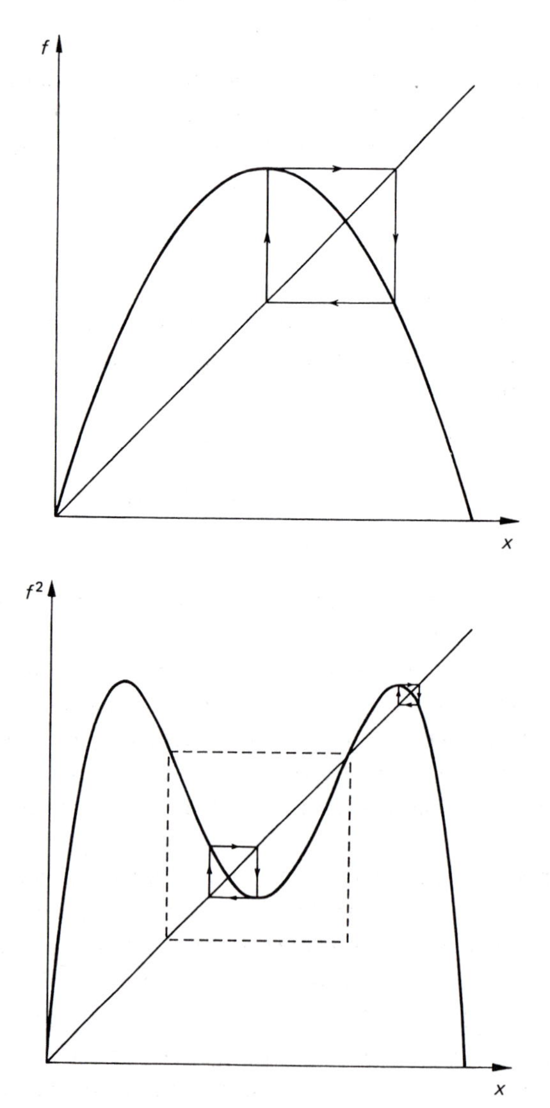

iteration of the map would eventually lead to a unique final map, this map being a fixed point in a functional space of all maps having the general form of a parabola. Moreover, this fixed point map is a stable point. Thus, no matter what the form of the initial map, subsequent iterates will always lead to this unique map. The iteration scheme thus has a universal nature which is revealed by studying the maps in the vicinity of the fixed point. We denote by T the operation of iterating a map twice, rescaling the variable $y(=x-\frac{1}{2})$ by α, and changing the control parameter μ such that we have a super stable orbit. Then

$$\mathrm{T}f(y, R+p\Delta n) = -\alpha f^2(y/\alpha n, R+\Delta_n+p\Delta_{n+1}),$$

where p is a parameter which is selected to give the super-stable orbit and $R+\Delta_n(P=1)$ corresponds to the value of μ for the onset of the next bifurcation. The basic idea of Feigenbaum is that there is a fixed point in functional space such that $p\Delta_n=\Delta_n+p\Delta_{n+1}$ and $\Delta_{n+1}=\Delta_n/\delta$ and $\mathrm{T}g=g$ so that

$$g(y,p) = -\alpha g^2(y/\alpha, p).$$

This equation can be solved numerically and has a solution only for $\alpha=2.5029\ldots$ Further one can look for solutions in the neighbourhood of this fixed point by writing

$$f(y, R+p\Delta_n) = g(y, R+\bar{p}\Delta_n) + \Delta_n(p-\bar{p})h(y,\bar{p}).$$

Substituting this into (10.2.8) and linearizing in $(p-\bar{p})$ gives the equation

$$g'(g(y))h(y) + h(g(y)) = -(\delta/\alpha)h(\alpha y).$$

Knowing $g(y)$ and α, this equation may be solved for h and more importantly for δ. One finds $\delta=4.6692016\ldots$.

The two universal constants α and δ have an immediate interpretation in terms of the shape and size of the bifurcation diagram. The scaling α is a measure of the distance between the nearest fixed points whilst δ is a measure of the change in μ between successive bifurcations. For further details see the paper by Feigenbaum contained in the collection edited by Cvitanovic (1984).

The important point of all this analysis is the universality of g, α and δ. For a wider class of maps (roughly parabolic) the higher iterates converge in the vicinity of $x=\frac{1}{2}$ to the universal function $g(y,p)$ with a scaling determined by α and δ. For large n we have $\Delta_{n+1}\simeq\Delta_n/\delta$ or $\Delta_n\simeq 1/\delta^n$. This universal scaling was first discovered by Feigenbaum who numerically iterated two distinct maps and noted that Δ_n scaled as above with the same exponent δ has also been found to arise in real experiments. Unfortunately, experimental difficulties preclude a study of more than just the first few bifurcations but they seem to confirm the theory.

It is well to note that because of the universality, the sequence of bifurcations described by the universal parameters α and δ are expected to occur in many experimental situations, but, it must also be noted that this universal nature of the structure gives little information on the structure of the basic map $x_{n+1}=f(x_n,\mu)$. This will only be found from a more detailed study of the first few bifurcations where the universal structure has not yet set in.

On the other hand, an important implication of the universality is that if a simple one-dimensional map is sufficient to describe the most important aspects of the natural system, then there is no need to consider the 'exact' one-dimensional map. Approximate ones will still show the same complicated bifurcation sequences leading to chaos. For example, perturbation theories may well yield correct qualitative features even when applied outside their strict domains of applicability.

Thus we can analyse experimental data for bifurcation behaviour and the associated universality with some confidence not because one believes it can be understood in terms of the logistic map but only that a one-dimensional map exists which is of the general form of the logistic map.

For more on the subject matter of thie Section see Kuzamoto (1984).

10.3 Flows and maps

Most time-dependent physical theories treat time as a continuous variable and furthermore they are usually assumed to be local in time. In this case they can be modelled by N coupled first order equations (evolution equations)

$$\frac{\mathrm{d}A_i}{\mathrm{d}t}=f_i(A_j), \tag{10.3.1}$$

where A_i is an N component vector and $f_i(A_j)$ are known functions, nonlinear in general, of the various A_j's. If the f_i's are not implicit functions of time then the system is called autonomous, otherwise it is non-autonomous. Newton's equations of motion for $N/6$ particles are precisely of this form, in which case the A_i's are the position and momentum components and the f_i's are simple derivatives of a Hamiltonian. For this reason, equations of the above form are said to constitute a dynamical system even when the A_i's are not related to particle position and momentum. The solution of (10.3.1) is expressed in the phraseology of real dynamical systems. One speaks for example, of the solution evolving with time in an N dimensional phase space, where the coordinates are just the A_i's as a particle trajectory or orbit. If one studies not just one initial condition but a family of them, all originally in a small compact region of space (a ball), then the time evolution constitutes a flow in the phase plane.

The Lorenz equations introduced above and defined by equation (10.1.2) are obviously of the above form. The forced Van der Pol equation

$$\frac{d^2x}{dt^2}+\varepsilon(1-x^2)\frac{dx}{dt}+\omega^2x=\alpha\cos(\eta t) \tag{10.3.2}$$

can, by defining $A_1=x, A_2=dx/dt$, be put in the form of (10.3.1) with $f_1=A_2$ and $f_2=-\omega^2A_1-\varepsilon(1-A_1^2)A_2+\alpha\cos(\eta t)$. In this form the equations are non-autonomous. However, by making the further identification $A_3=\eta t$ so that $f_2=\omega^2A_1-\varepsilon(1-A_1^2)A_2+\alpha\cos(A_3)$ and $f_3=\eta$, we obtain an autonomous set of equations, albeit at the expense of increasing the number of components.

Equations of the form of (10.3.1) appear quite naturally in the study of chemical reactions in homogeneous mixtures. Here the A_i's are the concentrations of the various chemicals and the homogeneity condition is imposed (by careful stirring) to remove possible spatial effects. (Haken (1978), Chapter 9). Similarly these equations apply in biological and ecological systems with the obvious re-interpretations of the variables (May (1976a, b); Haken (1978), Chapter 10).

Of course, many physical systems evolve spatially as well as temporally and for such systems some preliminary analysis, usually involving Fourier transformations, is needed to cast the equations into the form of (10.3.1). In such cases, N is infinite and one tacitly assumes that most of the results for finite but large N can be extrapolated to $N=\infty$. To illustrate this procedure, consider the problem of solving the nonlinear Schrödinger equation subject to periodic boundary conditions. That is solve

$$i\frac{\partial\psi}{\partial t}-a\frac{\partial^2\psi}{\partial x^2}-b|\psi^2|\psi+c\psi=0$$

subject to the condition $\psi(x)=\psi(x+L)$. The boundary condition is automatically satisfied if we write

$$\psi(x,t)=\sum_{n=-\infty}^{\infty}A_n(t)e^{i2\pi nx/L}.$$

Substitution of this expression into the nonlinear Schrödinger equation gives equations expressing the time evolution of the A's of the form of (10.3.1) with

$$f_n=i\left[c+a\left(\frac{2\pi n}{L}\right)^2\right]+ib\sum_{ml}A_m|A_l|A_{n+l-m}.$$

In this case $N=\infty$ and little progress can be made in an analytic study of such equations. However, a truncation scheme based on treating $A_0, A\pm 1$ exactly, but neglecting all other A's, gives good results, see Section 5.6 and Infeld (1981c).

For $N=2$, equations of the above type can be studied by using the powerful methods of phase plane analysis, Section 6.2. These methods are extremely useful in giving qualitative information about the solution (for example whether the motion remains in a finite region of the phase plane, that is A_1 and A_2 remain bounded in time) even when it is not possible to obtain explicit solutions. The solution can be sketched in the phase plane without knowing its explicit dependence on t.

This procedure is illustrated by application to a number of realistic problems in Section 6.2 whilst the solution of (10.3.2) with $\alpha=0$ is illustrated in Fig. 10.14.

Unfortunately, for $N>2$ we not only loose the possibility of being able to show the topological features pictorially but the solutions themselves become much more complicated. For $N=2$ the most complicated behaviour is a limit cycle (attractor) whereas even for $N=3$ we have the

Fig. 10.14. Phase plane portrait for the Van der Pol equation (10.3.2), $\alpha=0$.

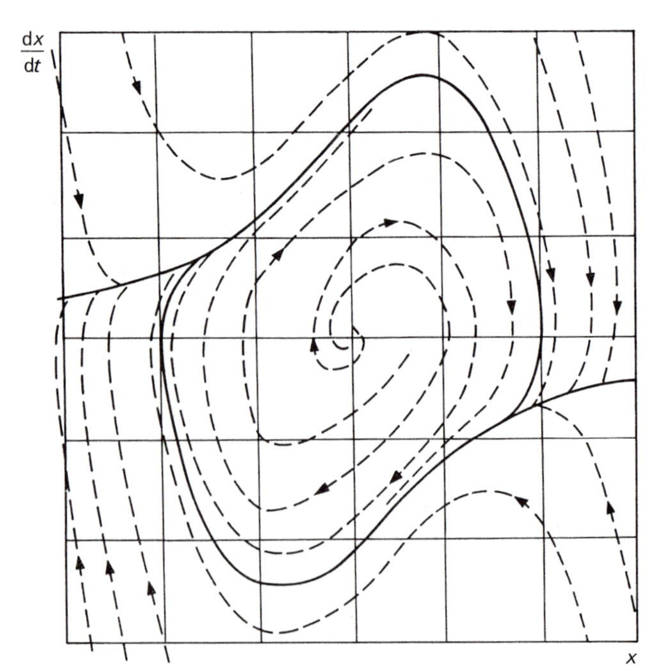

possibility of a strange attractor. These are the higher dimensional analogues of limit cycles but possess an infinitely more complex microstructure.

An extremely useful construction for these higher dimensional problems is the Poincaré map or section. For example, in the case of $N=3$, instead of studying the form of the orbit of the solution in the three-dimensional space, consider the interaction of the orbit with the $A_3=0$ plane. Each time the orbit passes through this plane it is registered by its coordinate value A_1, A_2 so that at the nth intersection of the *continuous* orbit with the $A_3=0$ plane the position in the plane is given by $A_{1,n}, A_{2,n}$. The plane $A_3=0$ is the Poincaré section and the Poincaré map is the relationship between $A_{1,n+1}, A_{2,n+1}$ and $A_{1,n}, A_{2,n}$. Though the solution $A_1(t)$ is a continuous function of t, the Poincaré map is not, and the time evolution of the solution is now represented by a discrete map of the form

$$A_{1,n+1}=F(A_{1,n}, A_{2,n})$$

and

$$A_{2,n+1}=G(A_{1,n}, A_{2,n}).$$

Here F and G are in general nonlinear functions. To obtain the form of F and G it is necessary to solve the flow equations (10.3.1). These ideas are illustrated graphically in Fig. 10.15.

The concept of a Poincaré section is easily generalized to an N component system by considering the intersection of the flow with any surface of dimension less than N (hyperplane). The corresponding relation between the points of intersections gives the Poincaré Map.

To obtain a Poincaré map it is necessary to solve the original equations and this is in general impossible. However, the concept is useful in understanding the qualitative features of the solution. For example, in studying the Lorenz equations ($N=3$) a one-dimensional map can be

Fig. 10.15. Showing the relationship between a continuous flow and a Poincaré section (map).

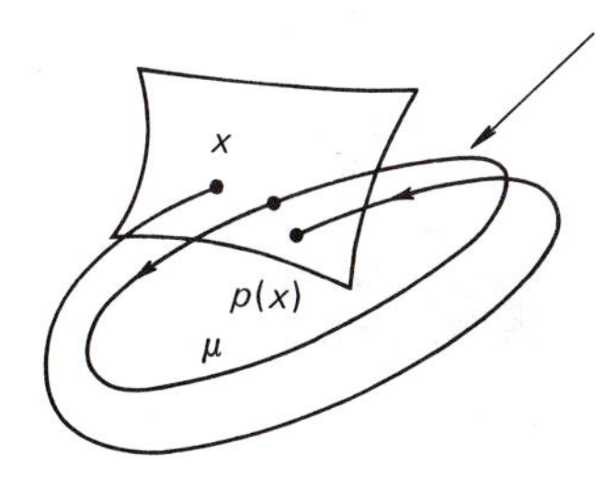

obtained, either by numerically solving the equations or by approximate analytic techniques, which can be used to study the chaotic features associated with the original differential equations. (This is discussed more fully in Section 4 of this Chapter.) Alternatively, one can postulate the form of the map and treat this as the basic model describing the physical situation (Holmes (1979)).

Finally, discrete maps arise naturally in the study of certain physical problems. Consider the motion of a ball under gravity that bounces (perfect reflection) on a table whose height oscillates with constant frequency and constant amplitude about some fixed position. The motion of the particle between successive bounces is easy to solve and it is not difficult to relate the velocity V_{n+1} and time, t_{n+1}, at which the ball hits the table for the $n+1$ time to the values of V_n, t_n. In this case the Poincaré map can be obtained analytically. Other examples, where Poincaré maps arise naturally, but where the derivation of the maps is non-trivial, are considered by Lichtenberg and Lieberman (1983).

In summary, most physical systems are described by continuous time equations (flows) but associated with them are discrete maps (Poincaré maps) where study can reveal important properties of the original system. However, it is found that properties such as chaos and the form of strange attractors are most easily studied by considering discrete maps. Though these are interesting in their own right as curious mathematical objects, it should be remembered that such maps can have a direct relation to continuous equations.

Fig. 10.16. Experimentally determined bifurcation. From Testa, *Phys. Rev. Lett.* **48**, 714 (1982).

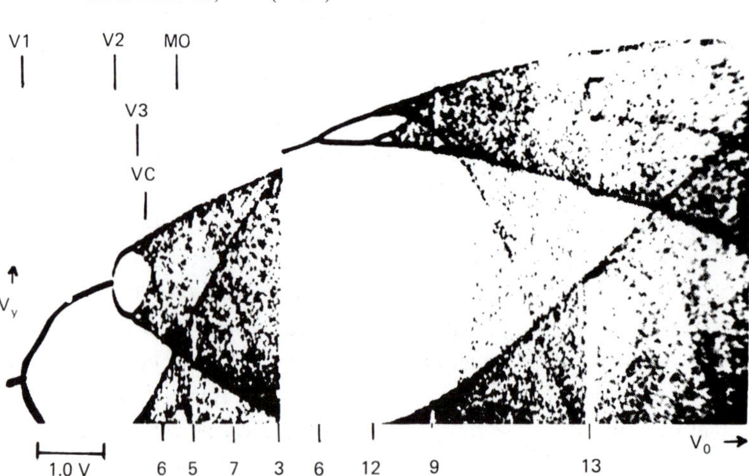

For example, the set of continuous equations describing the effect of an external time dependent force on a chemical reaction show the full range of bifurcation sequences leading to chaos. These equations have been studied numerically. The bifurcation sequence takes the form of a simple limit cycle undergoing a twist (a loop not on a plane but on a Möbius strip) which, projected onto a plane, shows a double loop. This twist bifurcation process in three-dimensional space continues until a chaotic limit is reached leading to turbulence. Various Poincaré sections can be studied and a particular one is considered in detail. It is found numerically and shown analytically by perturbation theory that a one-dimensional map exists which controls the flow. The map is of the general form of a logistic map and undergoes the universal bifurcation process.

Direct experimental evidence for a bifurcation sequence leading to chaos comes from the study of a simple, but essentially nonlinear, electric network. The experimental results in the form of the variation of the asymptotic state with a control parameter in this case is shown in Fig. 10.16. This should be compared to the same type of plot obtained from the logistic equation and shown in Fig. 10.12.

10.4 Strange attractors

A simple example of an attractor is the asymptotic value x_s (or fixed point) of the discrete map (10.2.1) with $1<\mu<3$. For any initial condition in the range $(0, 1)$ the solution evolves with n (is attracted) towards the same constant value x_s. If $3<\mu<3.449$ any initial value is attracted to the same bifurcated state (x_a, x_b). These are simple attractors. Associated with the attractor is a range of values of initial conditions, in this case $0<x<1$, which finally end up on the attractor. For $x>1$ the iterates of the map eventually become unbounded.

A second example of an attractor is a limit cycle and one such object is shown in Fig. 10.14. This arises from the solution of a continuous equation, the Van der Pol equation given by (10.3.2). Again it is important to note that the limit cycle is the long time solution and this asymptotic state is attained for a range of initial conditions.

These are examples of simple attractors, or just attractors, and their common feature is that for a range of initial conditions the solution to the equation ends up in the same asymptotic state (is attracted to this state). Physically this means that there is some dissipation in the system. By analogy one can imagine a three-dimensional attractor where the asymptotic state takes the form of a torus. This is surrounded by a region which includes all points representing initial conditions which eventually end up on the torus. This latter region is called the ball of attraction. The n-

dimensional analogue is also easily imaginable. However, the fundamental contribution of Ruelle and Takens (1971) was to show that attractors of the above form would not usually arise in practice (were non-generic). The simple toroidal shape, being prone to instabilities, evolves to a much more complicated structure, which they called a 'strange attractor'. Ruelle and Takens gave a purely mathematical example of a strange attractor. Eight years before them, Lorenz (1963) solved numerically the equations given by (10.1.2) and now known as the Lorenz equations, and obtained extremely complicated solutions having a pseudo-random nature. This is now known to be an example of a strange attractor. However, the basic properties of strange attractors are best illustrated by discussing a two-dimensional map introduced by Hénon (1976) with this reason in mind.

The Hénon map is

$$x_{n+1}=1-ax_n^2+by_n \qquad (10.4.1)$$
$$y_{n+1}=x_n,$$

where a and b are real constants. The basic properties of this map were obtained by Hénon by simply iterating the above equations. He chose the values $a=1.4$ and $b=0.3$. Starting from an initial value $x_0=0.631$,

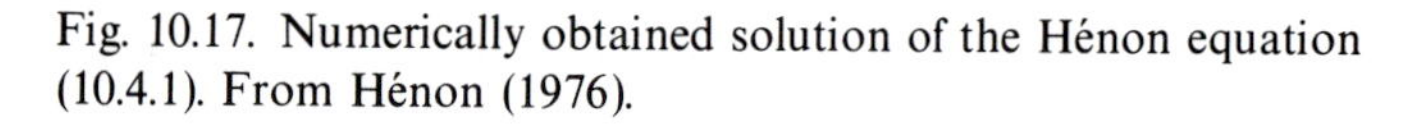

Fig. 10.17. Numerically obtained solution of the Hénon equation (10.4.1). From Hénon (1976).

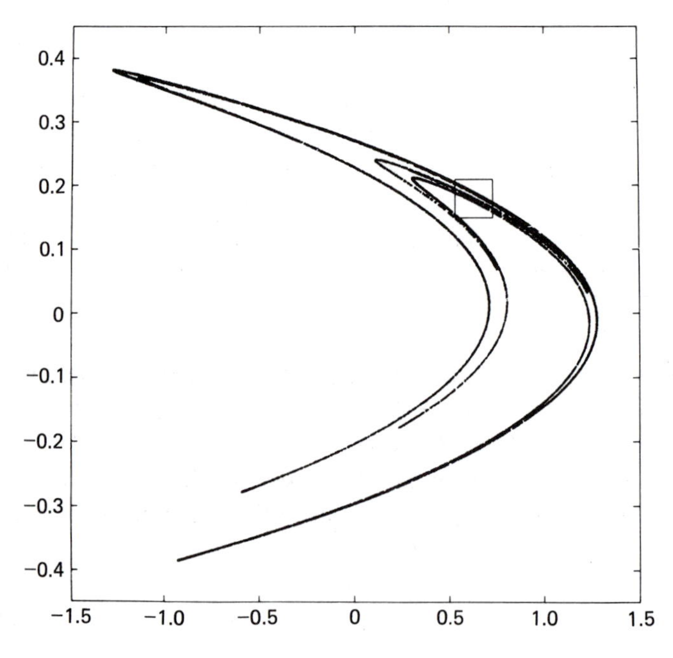

$y_0 = 0.189$ the map was iterated 10^4 times and a phase portrait obtained by plotting y_n against x_n. This is shown in Fig. 10.17. It was also found that starting from other initial conditions led to the same figure if a few low iterates ($n = 1, 2$) were ignored. Thus we can conclude that the Figure represents an attractor. Hénon also estimated the region inside which initial conditions lead to the attractor (ball of attraction).

The strangeness of the attractor is revealed by enlarging a small region of it. This is illustrated in Fig. 10.18(*a*) where the squared region in Fig. 10.17 is enlarged and the squared region in that Figure is enlarged in Fig. 10.18(*b*). This process can be continued indefinitely. The important point is that what seemed to be simple lines are revealed as being composed of a number of lines and that this sub-division continues at each stage of the enlargement process *ad infinitum*. The transverse structure of the attractor is very complicated and seemingly scale invariant, a property it has in common with a Cantor set. There is as yet no universally accepted mathematical definition of a strange attractor (see Eckmann (1981)). Here we shall loosely define a strange attractor as an attractor with complicated, scale invariant, transverse structure. When this definition is accepted, the above is a good example of such a concept.

The structure of the attractor can also be revealed by perturbation theory. Following Bridges and Rowlands (1977) we treat b in equation (10.4.1) as a small parameter and expand in powers of it. To lowest order we have

$$x_{n+1} = 1 - ax_n^2, \; y_{n+1} = x_n. \tag{10.4.2}$$

The map for x_n is simply related to the logistic map and $a = 1.4$ is equivalent to $\mu = 3.569$ (see Section 10.3). Thus the map is in the chaotic region and x_n takes values in the range $0.343 < x < 0.892$. Since to this order the above map is true for all n it must also represent the attractor, which is, by definition, the form of the solution for large n. Thus from (10.4.2) we have the lowest order approximation to the attractor, namely

$$x = 1 - ay^2, \; 0.343 < x < 0.892. \tag{10.4.3}$$

The density of points along the line of the attractor is determined by the chaotic solution of $x_{n+1} = 1 - ax_n^2$. This is not known and so little can be said about this aspect of the attractor.

The next approximation is obtained by using (10.4.2) to express y_n in terms x_n, namely $y_n = \pm(1 - x_n)/a$, and substituting into (10.4.1). The new form for the attractor is then

$$x = 1 - ay^2 \pm (b/\sqrt{a})\sqrt{(1 - y)}. \tag{10.4.4}$$

Fig. 10.18. Enlargements of segments of the Hénon map. (*a*) Enlargement of region in square in Fig. 10.17. (*b*) Enlargement of region in square in (*a*).

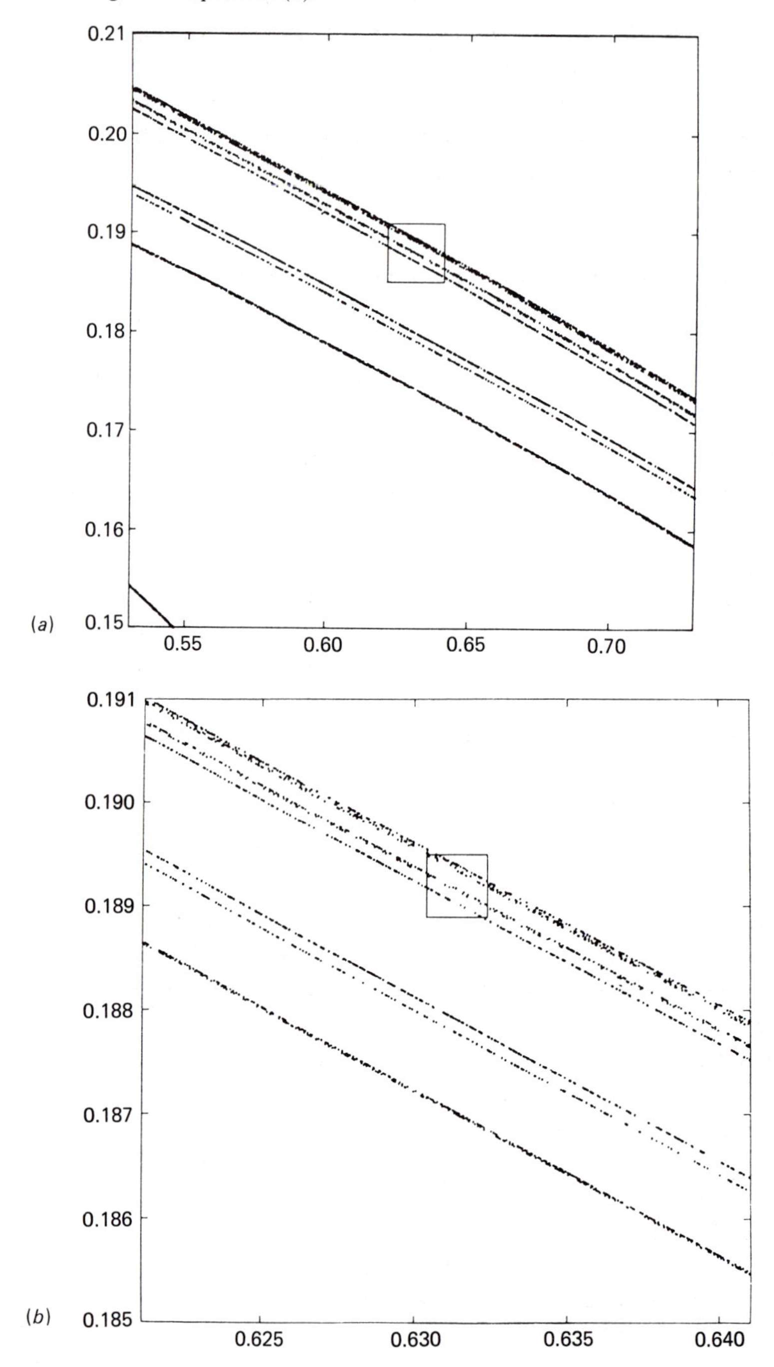

The disadvantage of the above method is that the bifurcation point found in first order at $y=1$ is present to all orders. To overcome this difficulty it is necessary to consider the perturbative scheme in more detail. What the method is doing is to replace the original two-dimensional map (10.4.1) by a pseudo one-dimensional map described by $x_{n+1}=F(x_n), y_{n+1}=x_n$ but where $F(x)$ is a multi-valued function. For example, to first order $F(x)=1-ax^2\pm(b/\sqrt{a})\sqrt{(1-x)}$. If we now assume that such a procedure can be carried out to all orders we have

$$x_{n+1}=F(x_n),$$

which must satisfy, as is seen by substitution into (10.4.1),

$$F(x)=1-ax^2+bF^{-1}(x). \tag{10.4.5}$$

Here $F^{-1}(x)$ is the inverse map. This is a functional equation for the strange attractor. In view of the known complexity of the strange attractor one would not expect to be able to solve (10.4.5) analytically. However, such equations can be solved by iteration, replacing (10.4.5) by

$$F_{m+1}(x)=1-ax^2+bF_m^{-1}(x) \tag{10.4.6}$$

and considering $F(x)$ as the limit of $F_m(x)$ as $m\to\infty$. That is, $F(x)$ is the fixed point of the above equation considered as a mapping in m, but allowing x to take a continuous range of values. The convergence of the iterative scheme is controlled by the value of b. This procedure for obtaining the form of strange attractors has been applied by Broomhead and Rowlands (1983) to other maps. It was used by Bridges and Rowlands (1977) to obtain an improvement to the first order result given by (10.4.4). They obtained

$$x=1-ay^2\pm\xi\sqrt{(1+y\pm\xi)}$$

where $\xi=b/\sqrt{a}$. This is in excellent agreement with the numerical results of Hénon, Fig. 10.19. These authors apply the same technique to elucidate the structure of the solution of a biological problem discussed numerically by Guckenheimer *et al.* (1977).

The above perturbative method or iterative method show that at each stage of the process a curve splits into two and the magnitude of the splitting is proportional to ξ^m where m is the order of the iteration. Thus we can make a one to one correspondence between the curves revealed in the transverse direction by the numerical iterations and the multi-valueness of the function $F(x)$ defined by (10.4.5).

The other celebrated example of a strange attractor, but now arising directly in a continuous system of equations, is the Lorenz model. These equations were introduced by Saltzmann and Lorenz in a study of

Fig. 10.19. Analytically determined form of the Hénon strange attractor (below) as compared to rescaled Fig. 10.17 (above).

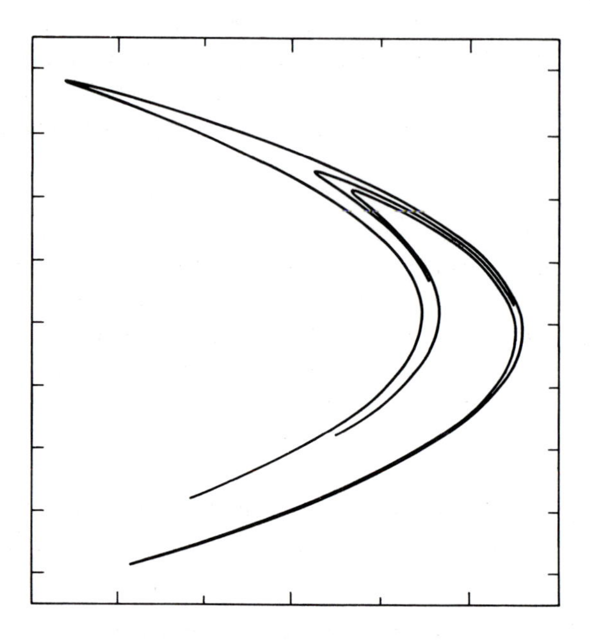

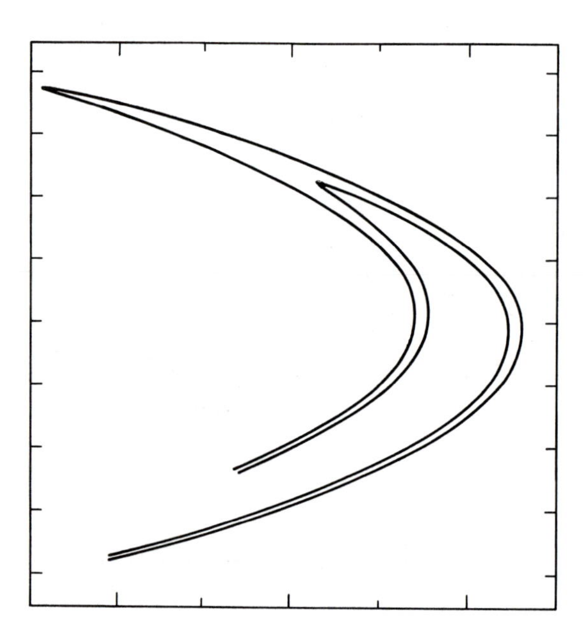

atmospheric flows and instabilities, a problem appertaining to weather forecasting. Lorenz made a detailed numerical study and obtained solutions showing complicated pseudo-random behaviour which we now associate with chaos. The sensitivity of the type of motion led Lorenz to seriously questioning the possibility of long-range weather forecasting, as small changes in the initial state (weather now) could lead to large changes at a later time (future weather).

The equations are given by (10.1.2) where x, y, and z are normalized mode amplitudes and r the normalized Reynolds number. The form of solution obtained numerically by Lorenz is illustrated in Fig. 10.20 where a projection of the flow on the x, z plane is shown. This behaviour is now associated with a strange attractor. It has the same kind of complex transverse structure as does the Hénon map. The orbit consists of a section where it spirals out from a fixed centre, then moves relatively quickly through the $x=0$ plane and then continues the spiral motion, but about a second centre.

This spiralling motion can be understood using perturbation theory, see McLaughlin and Martin (1975), Rowlands (1983). The equilibrium or

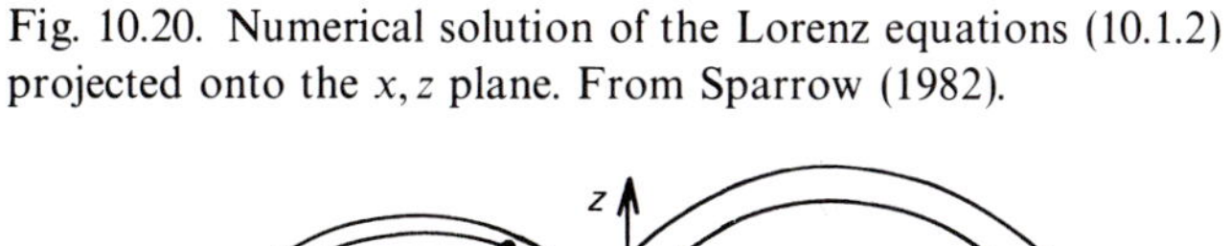

Fig. 10.20. Numerical solution of the Lorenz equations (10.1.2) projected onto the x, z plane. From Sparrow (1982).

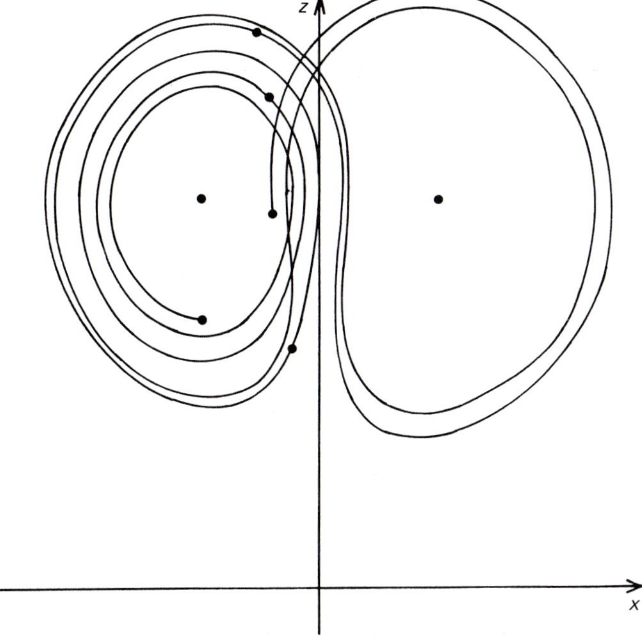

stationary points associated with (10.1.2), where the time derivatives are zero, are given by $z_o = r-1, y_o = x_o, x_o = b(r-1)$. Another stationary point exists at $x=y=z=0$ but this plays no part in the subsequent analysis. For $r > r_c \left(= \dfrac{\sigma(\sigma+b+\xi)}{\sigma-b-1} \right)$, which for the parameters used by Lorenz ($\sigma=10$, $b=8/3$) is equal to 24.7, these critical points are unstable as revealed by linear analysis. A nonlinear theory based on a multiple time perturbation theory (see for example Nayfeh (1973)) shows that the orbit is represented by

$$x = x_o + A\cos\theta + A^2(B\cos 2\theta + D\sin 2\theta + E) \qquad (10.4.7)$$

where $\theta = \lambda_o t + \chi$ and B, D and E are complicated functions, of σ and b. Similar expressions hold for y and z. Here λ_o is the linearized frequency given by $\lambda_o^2 = b(r_c + \sigma)$. The amplitude A and phase χ are slow functions of time, with A satisfying a Landau type equation

$$\frac{dA}{dt} = \alpha A + \beta A^3$$

where α is the linear growth rate proportional to $(r - r_c)$ and β a known positive quantity. Thus for $r > r_c$ the amplitude increases with time and this gives rise to the spiralling motion in three dimensions found numerically by Lorenz. The nonlinear term does not lead to saturation. The details of this spiralling motion may best be discussed by following Lorenz and calculating the maximum value of z, M say, each time the orbit trajectory completes an almost closed curve. Knowing the analytic form of the solution, as illustrated above, allows one to perform this calculation and obtain a relationship (map) between the amplitude M_n after n spirals to the value after $n+1$ spirals of the form

$$M_{n+1} = F(M_n).$$

Rowlands (1983) obtained an analytic form for this map, which for the Lorenz parameter ($r - = 28, \sigma = 10, b = 8/3$), takes the form

$$F(M) = 1.064M/(1 - 0.00172M^2)^{\frac{1}{2}}.$$

This in the strict sense is not a Poincaré map but serves to reduce the details of the flow to a discrete map. The above analytic form for F compares very well with the numerical values obtained by Lorenz (*see also* Fowler (1983)).

The above analysis does not apply as the orbit spirals out near to the $x=0$ plane. Numerically, it is found that the orbit spends a short time in this vicinity before it is 'kicked' out to spiral around the other fixed point. It then spirals about this other fixed point in an entirely analogous fashion as studied above. This is of course to be expected from the symmetry of the position of the equilibrium points and that of the equations. The full map obtained numerically by Lorenz is shown in Fig. 10.21. The above, analytic map only describes the left hand half of the cusp. The right hand half could be found following the method of Broomhead *et al.* (1981).

Strictly speaking, the map obtained numerically is not one-dimensional, but the transverse structure is not visible. This is because the parameter akin to b in the Hénon map is α/λ_1 (0.007) where $\lambda_1(=1+b+\sigma)$ is a measure of the rate of contraction to the analytic solution discussed above. The solution (10.4.7) is based on a linear theory described by two parameters A and χ, whereas the Lorenz equations have three distinct linear modes. Because of the smallness of α/λ_1, or the largeness of λ_1, it was assumed that the third mode, which decays as $e^{-\lambda_1 t}$, is identically zero.

The important point is that the pseudo-random behaviour associated with the solution of (10.1.2) is generated by iterations of this one-dimensional map. Such a map always shows chaotic behaviour. For example, since the modulus of the slope is everywhere greater than unity, the Lyapunov number (a measure of the average slope) must be greater than unity.

It is readily confirmed numerically that a whole range of initial conditions lead to attraction in the vicinity of the equilibrium points

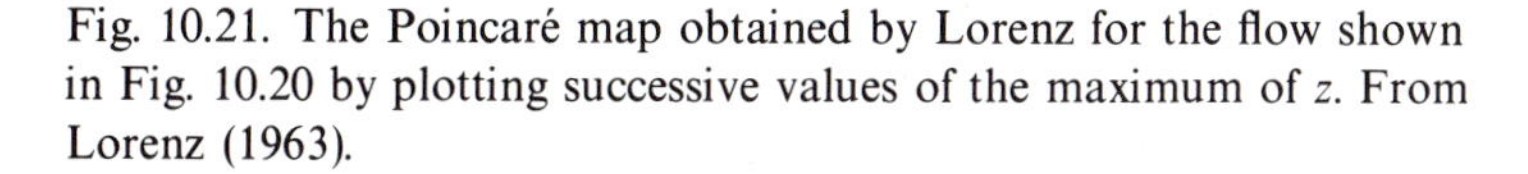
Fig. 10.21. The Poincaré map obtained by Lorenz for the flow shown in Fig. 10.20 by plotting successive values of the maximum of z. From Lorenz (1963).

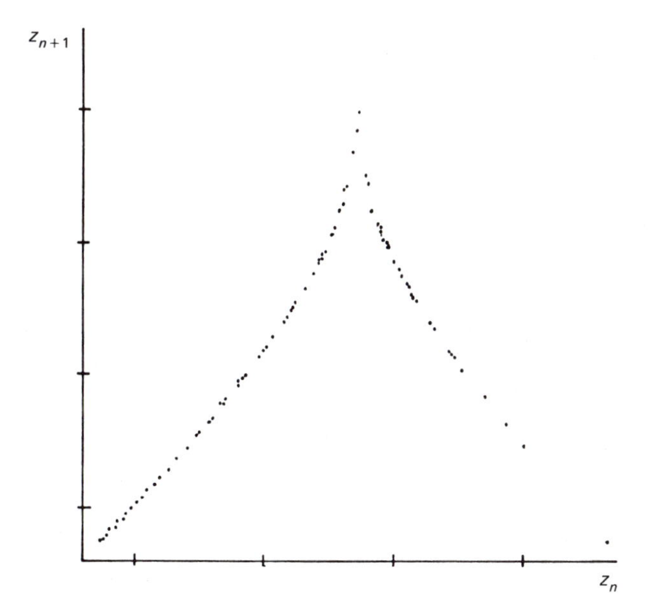

(x_0, y_0, z_0). This, together with the chaotic nature of the solution revealed by the one-dimensional map, amply confirms that the Lorenz equations constitute an example of a strange attractor.

Two general points emerge from the above discussion. The first is that just three coupled first order, ordinary, but nonlinear, differential equations, are sufficient to give rise to a strange attractor (even when there is only one nonlinear term), and secondly that the micro-structure characteristic of a strange attractor can be understood in terms of a simple one-dimensional difference equation.

Whilst perturbation theories on their own may not be good enough to capture the structure of the one-dimensional maps, such methods together with the impositions of global conditions are sufficient. Another example of this approach, as applied to the forced Brusselator problem, is discussed in Broomhead *et al.* (1981) whilst the forced Duffing model has been discussed by Holmes (1979) and Currie (1987).

The Hénon map and the Lorenz equations have been discussed above purely as examples of strange attractors. These equations do in fact show a much wider range of solution as the various parameters are changed. Sparrow (1982) has considered the Lorenz equations in detail. A natural question to emerge from the above discussion is how can one quantitatively describe the micro-structure of a strange attractor.

In view of the micro-complexity, a statistical approach is probably the best one can hope for. However, this aspect of the subject has turned out to be extremely difficult and few analytic approaches have as yet been developed.

A quantity that gives some information of the structure of these complicated solutions is the Lyapunov number. For one-dimensional maps of the form (10.2.3) the Lyapunov number λ is defined by

$$\lambda(\mu) = \underset{N\to\infty}{\mathrm{Lt}} \frac{1}{N} \sum_{j=1}^{N} \ln|f'(x_j, \mu)| \tag{10.4.8}$$

where x_i is the ith iterate of the map. This can be interpreted as an average rate of separation of neighbouring orbits. (For logistic maps this reduces to (10.2.19)). An alternate definition can be given in terms of the probability function introduced in section (10.2), namely

$$\lambda(\mu) = \underset{n\to\infty}{\mathrm{Lt}} \int_0^1 g_n(x, \mu) \ln|f'(x, \mu)|\,\mathrm{d}x$$

$$= \int_0^1 g_s(x, \mu) \ln|f'(x, \mu)|\,\mathrm{d}x.$$

Shaw (1981) has obtained the form of $\lambda(\mu)$ for the logistic map numerically whilst McCreadie and Rowlands (1982) have shown that the dominant feature can be obtained analytically. These results are shown in Fig. (10.22). It is seen from these figures that λ is a complicated function of μ. For a simple fixed point at $x=\bar{x}$, $g_s(x)$ is equal to $\delta(x-\bar{x})$ and $\lambda=\ln|f'(\bar{x})|$. If the fixed point is stable $|f'|<1$ and thus $\lambda<0$. This condition also applies to any periodic orbit. Hence λ is in effect a measure of the average stability of an orbit since $f'(x)$ is the local growth rate. We have identified $\lambda>0$ in Section 10.2 as a condition for chaos.

Shaw (1981) has introduced the idea of using information theory concepts into the study of the statistical properties and this has been partially successful in giving a new interpretation of Lyapunov numbers. The Lyapunov number is a good measure of the sensitivity to initial conditions, but does not convey much information about the global structure of the attractor.

An alternative is to introduce the concept of Hausdorff or fractal dimension, Mandelbrot (1982), Farmer *et al.* (1983). The phase space portrait of the strange attractor corresponding to the Lorenz equations, which is simply a continuous line orbit, would seem to have the dimensions of a line, namely unity. However, if the trajectory is allowed to continue for a sufficiently long time it apparently fills up a finite volume of the phase space (the degree of filling up depends on how close one looks at the orbit) and one might naturally associate it with a dimension of three. The dimension seems to depend on how close one looks at the orbits. Hausdorff introduced the concept of non-integer dimensions to give a quantitative measure to problems of the above nature. This avoids the difficulty mentioned above, namely that at low magnification the orbit seems to fill the three-dimensional space whilst in reality (high magnification) it is one-dimensional.

The Hausdorff or fractal dimension d is defined in an operational sense by

$$d=\operatorname*{Lt}_{\varepsilon\to\infty}\ \ln N(t)/\ln(1/\varepsilon) \tag{10.4.9}$$

where ε represents the size of a basic unit of volume (box) and $N(\varepsilon)$ is the minimum number of these boxes one needs to contain the strange attractor. By 'containing' we mean that the orbit passes through a box. For small ε, the above gives $N(\varepsilon)=B/\varepsilon^d$ where B is a constant which may be identified with the volume of the attractor. This relationship shows how $N(\varepsilon)$ scales with the level of coarse graining represented by ε. The fractal dimension is always smaller or equal to the dimension of the phase space (topological dimension).

Fig. 10.22. The Lyapunov number $\bar{\lambda}$ for the logistic equation as a function of μ: (*a*) analytic approximation; (*b*) numerical evaluation. From McCreadie and Rowlands, *Phys. Lett.* **91A**, 146 (1982).

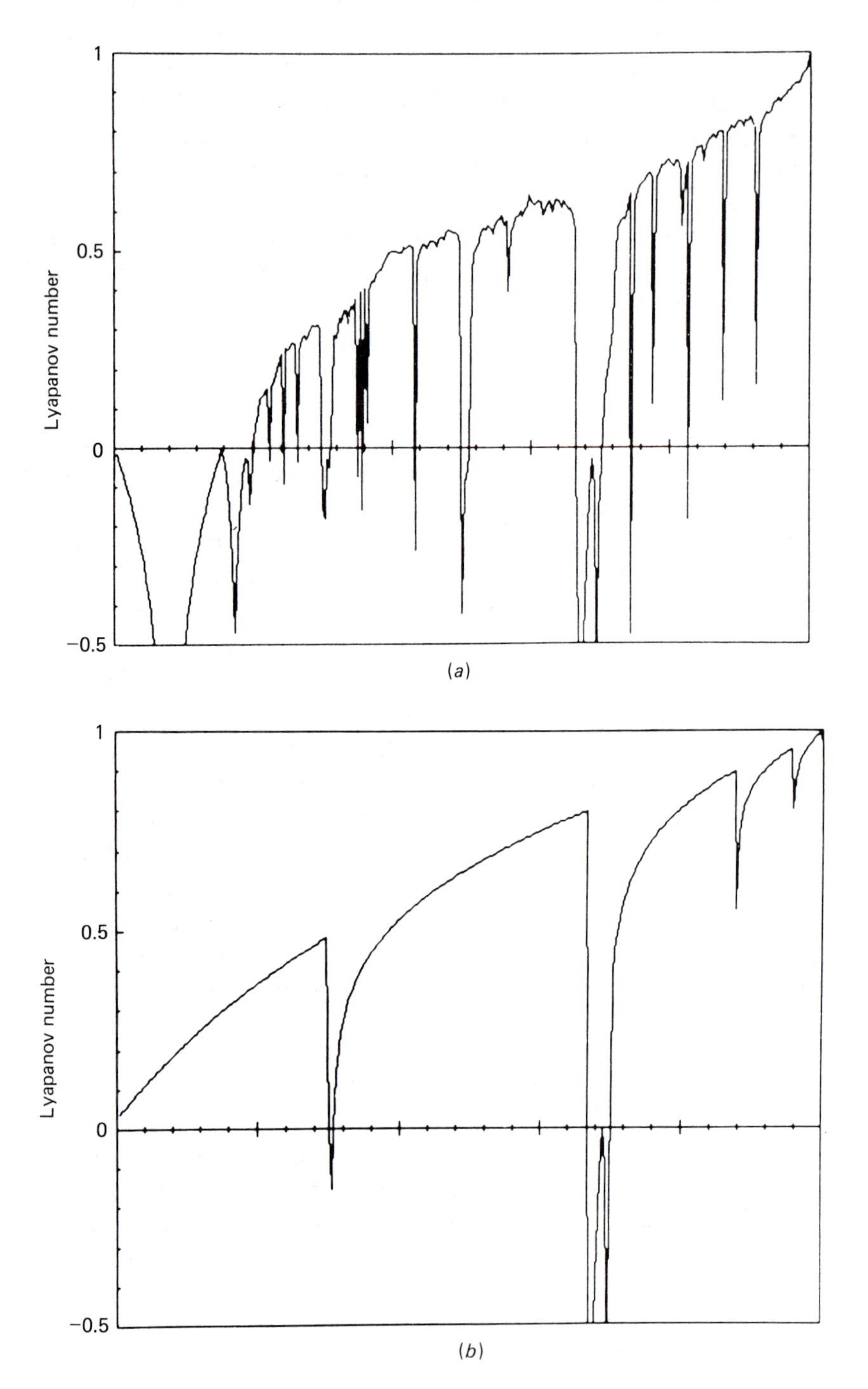

As a non-trivial example of the calculation of the fractal dimension we consider the transverse nature of the Hénon map. It was shown above that the microstructure could be revealed by the perturbation scheme based on the smallness of ξ. At each stage of the scheme a single line in the attractor is replaced by two and their distance apart scales as a power of ξ. Thus if the perturbation method has been applied n times there are 2^n lines such that the smallest distance apart is of order ξ^n (there are of course turning points where the lines come together but these represent a small fraction of the total and do not make a finite contribution to d in the limit of small ε). We identify ε with ξ^n (in this case the boxes are assumed to be square and two-dimensional) and then $N \simeq 2^n/\xi^n$ since the lines are assumed continuous and of finite length. Thus

$$d = \operatorname*{Lt}_{\xi \to 0} \frac{\ln(2^n/\xi^n)}{\ln(1/\xi^n)} = 1 + \ln 2/\ln(1/\xi).$$

This method has been used by Broomhead and Rowlands (1983) to calculate the fractal dimension for a number of two and three dimensional maps confirming conjectures of Kaplan and Yorke (1979).

In general, however, the calculation of the fractal dimension has to be done numerically. For the Hénon map, Russel *et al.* (1980) obtain the value $(a=1.4, b=0.3) d=1.26$. This is different from the value of d given above because of the structure along the 'lines' in the attractor (in obtaining the above value of d these lines were assumed continuous). These authors have also calculated the dimension of the Lorenz attractor to be $d=2.06$.

However, if $d>2$ it becomes extremely difficult to compute d by a base counting algorithm as implied by (10.4.8) and Grassberger and Procaccia (1983) have introduced another characterization of strange attractors. This is based on calculating the correlation function between points of the long time series of just one quantity or observable describing the attractor. For example, take just $x(t)$ for the Lorenz model. From this data we construct an n-dimensional vector $\mathbf{x}(t)$ with components $x(t), x(t+t_s)$ . . . $x(t+(n-1)t_s)$ where t_s is an arbitrary but fixed time. Then the correlation integral is defined by

$$C(r) = \operatorname*{Lt}_{n \to \infty} \frac{1}{N^2} \sum_{i \neq j} \theta(r - |x(t_i) - x(t_i)|),$$

where the vectors are sampled at time t_i and there are a total of N sample times. Usually the data is sampled at regular intervals τ so that $t_j = j\tau$ and then $t_s = m\tau$ with m an integer. In the above θ is the Heaviside function. A correlation exponent ν is then defined, by (assuming the limit to exist)

$$\nu = \operatorname*{Lt}_{r \to 0} \frac{\mathrm{d}(\ln C(r))}{\mathrm{d}\ln r},$$

that is $C(r) \simeq r^{\nu}$ for small r. Computationally it is easier to calculate ν rather than d if these quantities are greater than 2 but more importantly, most experimental information is given in the form of a time series of just one quantity, such as temperature, which may be identified with x in the above and used to calculate ν. This procedure will be considered in more detail at the end of this Chapter, where we will discuss the experimental evidence for the existence of strange attractors.

Grassberger and Procaccia (1983) have calculated ν for the Hénon map and the Lorenz equations and have shown that in both cases $\nu \leqslant d$. In fact ν is equal to d in both cases to within the numerical accuracy but the authors present arguments for the inequality.

In summary, it seems that the best indicator of a strange attractor is the correlation exponent as this has the advantage of being easier to compute than the fractal dimension when this exceeds 2 and also it can be calculated using just one component of the attractor.

In conclusion of this Section, we have seen how strange attractors can arise from both discrete maps and continuous flows. Their complex micro-structure can be studied both by numerical and analytical methods. The complicated micro-structure can be given a quantitative measure in terms of fractal dimensions or correlation exponents.

10.5 Effect of external noise

It will be realized from the discussion of the logistic map in Section 2 that the existence of the higher order periodic solutions demands the

Fig. 10.23. Bifurcation diagram for the noisy logistic map. From Haken (1983).

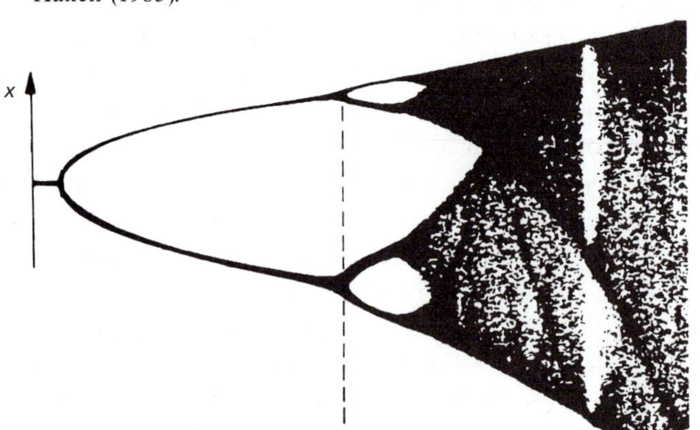

specification of μ to a very high degree of accuracy. For example, the interval for period four is just 3.449 to 3.544. To keep a control parameter of a system, for example the temperature, to within such bounds can be difficult experimentally. One would expect external fluctuations, unless very carefully controlled, to make comparable changes in the control parameter. Thus the logistic map is somewhat idealistic and one should of necessity incorporate the effects of external noise. Note that the effects of externally produced fluctuations are quite distinct from a specific change in the form of the map. This latter type of change would only alter the quantitative nature of the bifurcation sequence leading to chaos, not the basic scenario.

To illustrate the effect of external noise, consider the map studied by Mayer-Kress and Haken (1981)

$$x_{n+1} = \mu x_n(1 - x_n) + \xi_n$$

where ξ_n, which represents the external disturbance, is a random number with a given distribution. This model is not entirely satisfactory, as the external disturbance can push x_n out of its interval. Provisions must be made for this. In Fig. 10.23 the asymptotic states of the map as a function of μ for the same degree of noise is plotted. If this is compared to Fig. 10.12, the case of no external noise, it is apparent that though the lower Feigenbaum bifurcations are still present, the higher ones are smeared out and lost. As might be expected, the chaotic regions still remain. Eckmann (1981) gives some quantitative results on the significance of noise in these latter regions.

10.6 Experimental evidence for strange attractors

The experimental evidence cited in Section 10.4 gives strong support for the Ruelle–Takens scenario of turbulence based on the existence of strange attractors. Experimentally, broad band noise is found and this is identified with the chaotic nature of the attractor. However, a more detailed analysis of the experimental data is needed and this could help elucidate their basic properties.

Experimental results are usually obtained in the form of values of some quantity, x say, obtained at a series of moments of time – a time series. To illustrate the basic ideas behind the methods used to examine such data consider for the moment that x is given as a continuous function of time. Then it is possible in principle to calculate the various derivatives $\mathrm{d}x/\mathrm{d}t, \mathrm{d}^2x/\mathrm{d}t^2$ etc., and construct a phase space treating these derivatives as a set of orthogonal vectors. The time evolution of the physical system is then represented by an orbit in this phase space. The dimensions of this phase space are simply the number of derivatives constructed plus one. Suppose the time series $x(t)$ corresponds to a simple periodic disturbance, that is $x(t) = A\cos(\omega t)$. In a one-dimensional phase space $(x(t))$ motion would simply cover a continuous line between the two points $x = \pm A$. In two dimensions $(x, \mathrm{d}x/\mathrm{d}t)$ the phase portrait is a simple ellipse since $\omega^2 x^2 + (\mathrm{d}x/\mathrm{d}t)^2 = A^2\omega^2$. In three dimensions $(x, \mathrm{d}x/\mathrm{d}t, \mathrm{d}^2x/\mathrm{d}t^2)$ since $\mathrm{d}^2x/\mathrm{d}t^2 + \omega^2 x^2 = 0$ and $x^2 + ((\mathrm{d}x/\mathrm{d}t)/\omega)^2$ the orbit is the intersection of the elliptical cylinder defined by $\omega^2 x^2 + (\mathrm{d}x/\mathrm{d}t)^2 = \omega^2 A^2$, any $\mathrm{d}^2x/\mathrm{d}t^2$, and the plane defined by $\mathrm{d}^2x/\mathrm{d}t^2 + \omega^2 x^2 = 0$. This is again in general an ellipse. The natural dimension for this case is two and the thing to note is that an increase of dimension of the phase space above this value does not change the form of the orbit. Given the orbit in two dimensions we know all about the solution. Increasing the dimension of the phase space above two gives us no more information, but decreasing the dimension to unity loses it all. By analogy with this simple example we infer that for any time series there exists a natural dimension for the phase space. In this space the orbit gives a complete picture of the experimental results. A phase space of higher dimension gives the same picture, though distorted, but a phase space of lower dimension is inadequate to describe the system completely. Thus the problem is to find the minimal dimension of the associated phase space, and this is to be recognized by the fact that an increase in the dimension does not alter the nature of the orbit. This can be done for example by calculating the fractal dimension, as defined in Section 10.4 as a function of the phase space dimension N. If the value of N is less than the true fractal dimension, the attractor fills the space and the value of the fractal dimension calculated will simply be N. The value of N is increased until the calculated fractal dimension reaches a constant value. This asymptotic value is the true fractal dimension of the system. By this we mean that the experimental results $x(t)$ are equivalent to an orbit in a phase-space of N_s dimension which in principle is the solution of N_s first order nonlinear coupled equations. The fractal dimension has a value between $N_s - 1$ and N_s.

Unfortunately, experimental results are not usually smooth enough, that is measured at sufficiently small intervals, for one to obtain reasonably accurate values for the derivatives. Numerically, differentiation is notoriously difficult to carry out without a large loss in accuracy. A way out of this difficulty was suggested by Takens (1980). One defines an N component vector $\mathbf{X}(t)$ with components defined by $X(t)$, $X(t+\tau)$,. . . $X(t+(N-1)\tau)$, where τ is a fixed delay, and constructs a N-dimensional phase space from these components. Again the orbit in this phase space is a realization of the time evolution of the system. One then proceeds as above to calculate the fractal dimension as a function of N and look for saturation. In practice many experimental results are obtained by sampling at regular intervals, τ_s say. The vector $\mathbf{X}$ is then defined to have components $X(n\tau_s), X(n\tau_s+\tau) \ldots X(n\tau_s+(N-1)\tau)$, where of necessity $\tau = M\tau_s$, where M is some integer. The above ideas of phase space descriptions have been based on a firm theoretical foundation by Takens (1980).

The magnetic fluctuations in a plasma physics device have been analysed along the above lines, and the correlation exponent calculated and its convergence with N and also its dependence on the choice of τ_s discussed. Though for technical reasons the number of experimental points is limited, convergence is demonstrated giving strong evidence for the existence of a strange attractor. A value of the exponent of $\nu = 2.1$ was obtained suggesting that the system could be described in terms of just three coupled equations. Interestingly this value is close to that for the Lorenz equation (2.06) suggesting the intriguing possibility that this system can be modelled by these equations.

Similar calculations have been performed for the frequency of laser light scattered from Tokomak plasmas over a long time interval, obtaining low fractal dimensions (2–6), see Armstrong *et al.* (1987) and Barkley *et al.* (1987).

There still remain the problems of extrinsic noise and the choice of the delay time τ. All experimental data contains a noisy component and it is essential to be able to differentiate this from the pseudo-random or chaotic behaviour of a strange attractor (intrinsic noise). If there was no strange attractor present and the time series was simply noisy then it would be found that the fractal dimension as a function of the phase-space dimension would not show any convergence. Even when a strange attractor exists the expected convergence could well be masked by the noise.

Most of the calculations mentioned above use a range of values for the delay time and this seems to indicate results essentially independent of τ, but for technical reasons the range of τ values is quite small.

In an attempt to overcome these difficulties Broomhead and King (1986) have suggested an alternate method of analysis which incorporates concepts from signal processing. Though some questions still remain, the methods outlined above can be used with confidence to obtain the fractal dimension or correlation exponent of experimental data and this indicates the presence of an underlying strange attractor. To date a number of quite different physical systems have been analysed in this manner and strange attractors shown to exist. Encouragingly, the dimensions found are small (usually not exceeding five) so that one can with confidence attempt to model such systems with a small number of coupled equations. However, this is just the first step. Further experimental data or further analysis must be carried out to identify what are the variables in these equations (the quasi-modes).

10.7 Other theories of turbulence

The discussion of turbulence given in Section 10.2 and based on the concept of quasi-modes has been restricted to the three distinct scenarios linked with the names of Landau, Ruelle and Takens, and Feigenbaum. It is obvious even from this brief discussion that these scenarios are in no way complete. For example, one can easily imagine the situation where the bifurcation process takes place whilst the modes interact strongly among themselves. The necessity of introducing such complications rests with experiment and to date there is no evidence to warrant such a study.

A seemingly totally different model of turbulence has been described by Pomeau and Manneville (1980). This differs from the models mentioned above in that, though there is no clear precursor stage, once the turbulence has developed it takes a very characteristic form. Namely the time evolution of a system will, for an interval, seemingly behave as appropriate to a periodic system. This will change to typical chaotic behaviour for an interval of time and then revert to being periodic. The intervals of these two different types of behaviour appear to be random. The overall time sequence is called intermittence. There is some experimental evidence for this type of behaviour (Pomeau *et al.* (1981)).

Intermittence can be described in terms of a one-dimensional map near the parameter threshold for a tangent bifurcation. For the logistic map as μ increases beyond the critical value $\mu = 1 + 2\sqrt{2}$, chaotic motion changes to that of period 3. However, if $\mu = \mu - 0(\varepsilon)$, where $\varepsilon \ll 1$, even though the motion is chaotic, the orbit will spend a long time in the vicinity of a fixed point and its time behaviour will be characteristic of a period 3 orbit. Once it gets away from near the fixed point it will undergo chaotic motion until returning to the vicinity of the fixed point. The intervals between the visits will essentially be random as will be the time spent in the vicinity of the fixed point. This latter time scales as $|\mu - \bar{\mu}|^{\frac{1}{2}}$.

10.8 Conclusions

In recent years a number of different models have been introduced in an attempt to understand the important phenomenon of turbulence, particularly the precursor behaviour leading to fully developed turbulence. This has necessitated a study of new mathematical concepts such as chaos and strange attractors and a deeper understanding of the mathematics of dynamical systems. The important new idea is that pseudo-random, non-periodic time dependence, which physically is the important characteristic of turbulent behaviour, can arise from *very simple* deterministic equations, both discrete and continuous.

There is good experimental evidence that some of the qualitative features of these new theories are correct. Reliable measurements have been made of the dimensions of strange attractors in quite distinct types of systems and encouragingly the values of these dimensions are small ($\leqslant 5$) giving one confidence that these systems can be explained using a relatively low number of coupled equations.

However, it must be stressed that there is an enormous gap between the physically acceptable models such as the Navier–Stokes equations for ordinary liquids and the model equations such as those of Lorenz which are so good at illustrating the details of a strange attractor. The concepts of quasi-modes and enslaving may be useful in bridging this gap. The present state of the art is very reminiscent of the theory of electrons in solids. The free electron theory was and still is particularly successful, though all electron–electron interactions seemed to be ignored. It was not until the work of Bohm and Pines (1951) that the model could be justified by a theory. The present state of affairs resembles the pre-Bohm and Pines period.

Exercises on Chapter 10

Exercise 1
Show that the Landau equation

$$\frac{\mathrm{d}A}{\mathrm{d}t}=\lambda^2(\mu-\mu_c)A-\beta A^3$$

with $\beta>0$, has one real solution for $\mu<\mu_c$ and three for $\mu>\mu_c$. Use linear analysis to study the stability of these solutions. What is the time development of A for $\mu=\mu_c$?

Exercise 2
Show that equation (10.2.1) can be put in the form

$$z_{n+1}=z_n^2-Kz_n+K=H(z_n).$$

Show that, if $z_o, z_1 \ldots z_{n-1}$ are an N cycle, $z_i=H^i(z_o), H(z_{N-1})=z_o$, then

$$\prod_{i=0}^{N-1} z_i=1.$$

Hint: Take $z_o \neq k$ and write the recurrence for $n=N-1$.

Exercise 3
Generate Fig. 10.12 on a personal computer. Enlarge the window for which three values bifurcate.

Exercise 4

Generate Fig. 10.19 on a personal computer. What are the main limitations of the model?

Exercise 5

Read the article by Sander (1987) in Scientific American (Jan). Generate a 'fractal' on a personal computer as suggested on page 84.

APPENDIX 1

Parameter stretching as suggested by the linear dispersion relations

A.1.1 Ion acoustic waves in an unmagnetized plasma, $\Omega_c = 0$

From (5.1.14), dropping the subscripts, for small k, $k_{\|} \gg k_{\perp}$,

$$\omega^2 = \frac{k^2}{1+k^2} \simeq k^2(1-k^2)$$

$$k = k_{\|}(1 + \tfrac{1}{2}k_{\perp}^2/k_{\|}^2).$$

Thus

$$\omega = k_{\|} + \tfrac{1}{2}k_{\perp}^2/k_{\|} - \tfrac{1}{2}k_{\|}^3$$

and

$$e^{i(\omega t - \mathbf{kx})} = e^{i(k_{\|}t + \frac{1}{2}[k_{\perp}^2/k_{\|} - k_{\|}^3]t - k_{\|}x - k_{\perp}y)}.$$

Without loss of generality we can take $k_{\|} = \varepsilon^{\frac{1}{2}}k'_{\|}, k_{\perp} = \varepsilon^p k'_{\perp}$ to obtain

$$e^{i(-k'_{\|}\varepsilon^{\frac{1}{2}}[x-t] - k'_{\perp}\varepsilon^p y' - \frac{1}{2}[k'^3_{\|} - k'^2_{\perp}k'^{-1}_{\|}\varepsilon^{2p-2}]\varepsilon^{3/2}t)}.$$

This suggests $p = 1$ and a new choice of coordinates,

$$\xi = \varepsilon^{\frac{1}{2}}(x-t)$$
$$\sigma = \varepsilon y$$
$$\tau = \varepsilon^{3/2}t,$$

ensuring that the new frequency will depend on $k'_{\|}$ and $k'_{\perp}$. The following operator relations are implied

$$\partial_x = \varepsilon^{\frac{1}{2}}\partial_{\xi}, \partial_t = \varepsilon^{3/2}\partial_{\tau} - \varepsilon^{\frac{1}{2}}\partial_{\xi}, \partial_y = \varepsilon\partial_{\sigma}. \quad \text{(A.1.1)}$$

We expect the dependent variables to take the form

$$n = 1 + \varepsilon^{p n}n^{(1)} + \ldots$$
$$\phi = \varepsilon^{p\phi}\phi^{(1)} + \ldots$$
$$\mathbf{v}_{\|} = \varepsilon^{p_{\|}}\mathbf{v}_{\|}^{(1)} + \ldots$$
$$\mathbf{v}_{\perp} = \varepsilon^{p_{\perp}} + \mathbf{v}_{\perp}^{(1)} + \ldots$$

If we concentrate on curl free velocities, $\mathbf{v}$ becomes derivable from a potential ψ. In view of A.1.1,

$$p_{\perp}=p_{\parallel}+\tfrac{1}{2}.$$

To lowest order, meaningful equations can only be obtained if

$$p_n=p_{\phi}=p_{\parallel}, \tag{A.1.2}$$

and they will be

$$n^{(1)}=\phi^{(1)}=v_{\parallel}^{(1)}.$$

Thus $p_{\parallel}$ will generate all other ps. This, however, demands a decision on our part. We decree that in the next order, nonlinearity, $v_{\parallel}v_{\parallel x}$, be competitive with dispersion, $v_{\parallel xxx}$, giving

$$2p_{\parallel}+\tfrac{1}{2}=p_{\parallel}+3/2,$$

or $p_{\parallel}=1$. We have thus obtained the entire expansion of Section 5.2 by invoking just one nonlinear argument (the rest followed from considering long-wave, linear modes).

A.12 Magnetized plasmas, $\Omega_c>0$

The dispersion relation (5.1.13) is, for small k, without the subscripts,

$$\omega^2=\frac{k_{\parallel}^2}{1+k^2+k_{\perp}^2\Omega_c^{-2}}\simeq k_{\parallel}^2(1-k^2-k_{\perp}^2\Omega_c^{-2})$$

$$\omega\simeq k_{\parallel}-\tfrac{1}{2}(k_{\parallel}^3+k_{\parallel}k_{\perp}^2[1+\Omega_c^{-2}]).$$

The wave is now proportional to ($k_{\parallel}\simeq\varepsilon^{\frac{1}{2}}k_{\parallel}', k_{\perp}=\varepsilon^r k_{\perp}$):

$$e^{i(\omega t-\mathbf{kx})}=e^{i(-k_{\parallel}'\varepsilon^{\frac{1}{2}}[x-t]-k_{\perp}'\varepsilon^r y-\frac{1}{2}[k_{\parallel}'^3+k_{\parallel}'k_{\perp}'^2(1+\Omega_c^{-2})\varepsilon^{2r-1}]\varepsilon^{3/2}t)}$$

where z has been taken to be cyclic. This suggests $r=\frac{1}{2}$ and

$$\xi=\varepsilon^{\frac{1}{2}}(x-t)$$
$$\eta=\varepsilon^{\frac{1}{2}}y$$
$$\tau=\varepsilon^{3/2}t,$$

and

$$\partial_x=\varepsilon^{\frac{1}{2}}\partial_{\xi},\quad \partial_t=\varepsilon^{3/2}\partial_{\tau}-\varepsilon^{\frac{1}{2}}\partial_{\xi},\quad \partial_y=\varepsilon^{\frac{1}{2}}\partial_{\eta}. \tag{A.1.3}$$

For dependent variables we now take

$$n=1+\varepsilon^{r_n}n^{(1)}+\ldots$$
$$\phi=\varepsilon^{r_{\phi}}\phi^{(1)}+\ldots$$
$$v_{\parallel}=\varepsilon^{r_{\parallel}}v^{(1)}+\ldots$$
$$v_y=\varepsilon^{r_y}v_y^{(1)}+\ldots$$
$$v_z=\varepsilon^{r_z}v_z^{(1)}+\ldots$$

We again obtain A.1.2 and $r_{\parallel}$ will once more yield all other powers from the lowest order equations (5.1.7)–(5.1.9). The result is

$$r_y=r_{\parallel}+1 \text{ and } r_z=r_{\parallel}+\tfrac{1}{2}.$$

These values can also be found from drift velocity relations known to plasma physicists.

The value of r_z can be found by considering the so-called $\mathbf{E} \wedge \mathbf{B}$ drift along z, with $\mathbf{E} = -\nabla\phi$:

$$v_z = \frac{\mathbf{E} \wedge \mathbf{B}}{B^2} \cdot \mathbf{i}_z = \frac{-\nabla\phi \wedge \boldsymbol{\Omega}_c}{\Omega_c^2} \cdot \mathbf{i}_z = \partial_y \phi / \Omega_c$$

and so $r_z = r_\phi + \frac{1}{2} = r_\parallel + \frac{1}{2}$.

We now find r_y from the polarization drift which is $\dot{\mathbf{E}}\Omega_c^{-2}$ in our units and is along y

$$v_y = -\partial_{ty}^2 \phi \Omega_c^{-2}.$$

In view of A.1.3, lowest order ∂_t introduces $\varepsilon^{\frac{1}{2}}$, thus

$$r_y = r_\phi + \tfrac{1}{2} + \tfrac{1}{2} = r_\parallel + 1.$$

Again, balancing of nonlinearity and dispersion is the only nonlinear argument involved. It yields $r_\parallel = 1$.

APPENDIX 2

Relation between the trace and the inverse scattering method

A demonstration of the connection between the trace method of Section 7.9 and the inverse scattering method (Section 7.4) is of some interest. Define

$$K_o(x,x')=\sum_{n=1}^{N} f_n(x,y,t)g_n(x'),$$

where y and t dependence in K_o has been treated as parametric and suppressed in the notation. We have

$$T_r\left(\frac{\partial \mathbf{B}}{\partial x}\right)=K_o(x,x)$$

$$T_r\left(\frac{\partial \mathbf{B}}{\partial x}\cdot \mathbf{B}\right)=\int_{-\infty}^{x} K_o(x,x_1)K_o(x_1,x)\mathrm{d}x_1$$

$$T_r\left(\frac{\partial \mathbf{B}}{\partial x}\cdot \mathbf{B}^{n-1}\right)=-(-1)^{n-1}\int_{-\infty}^{x}\int_{-\infty}^{x}$$

$$\ldots\int_{-\infty}^{x} K_o(x_1,x_{n-1})K_o(x_{n-1},x_{n-2})\ldots K_o(x_1,x)\mathrm{d}x_1\mathrm{d}x_2\ldots \mathrm{d}x_{n-1}.$$

Equation (7.9.16) now takes the form

$$\phi(x,y,t)=K_o(x,x)-\int_{-\infty}^{x} K_o(x,x_1)K_o(x_1,x)\mathrm{d}x_1$$

$$+\int_{-\infty}^{x}\int_{-\infty}^{x} K_o(x_1,x_2)K_o(x_2,x_1)K_o(x_1,x)\mathrm{d}x_1\mathrm{d}x_2+\ldots,$$

and $\phi(x,y,t)$ satisfies

$$\phi(x,y,t)=-K(x,x),$$

where K in turn satisfies

$$K(x,x')=-K_o(x,x')-\int K(x,x_1)K_o(x_1,x')\mathrm{d}x_1.$$

This is just the Gelfand–Levitan–Marchenko equation that is the basis of the inverse scattering method, for which it is a general result (not limited to N soliton solutions).

APPENDIX 3

Some formulae for perturbed nonlinear ion acoustic waves and solitons

A.3.1 No magnetic field

If we linearize (8.4.1)–(8.4.3), take $\Omega_c=0$, and perform a small K expansion as in Section 8.3, we obtain the dispersion relation in the form of a quartic in Ω/K, where the coefficients depend on C, M, ψ (Infeld and Rowlands (1979a) though some expressions have been simplified here)

$$\sum_{m=0}^{4} a_m(\Omega/K)^m=0 \qquad (A.3.1)$$

$$\begin{aligned}
a_o&=\mathscr{L}_o^{-1}\cos^4\psi+(2M\mathscr{L}_1\mathscr{L}_o^{-1}-M\mathscr{L}_2\mathscr{L}_o^{-1}\langle v^{-1}\rangle+B_o)\cos^2\psi\sin^2\psi\\
&+(MA_1+A_1B_o\langle v^{-1}\rangle)\sin^4\psi, \qquad (A.3.2)\\
a_1&=-2\mathscr{L}_o^{-1}\langle v^{-1}\rangle\cos^3\psi+2[M^2(\mathscr{L}_1-\mathscr{L}_{-1}\mathscr{L}_2\mathscr{L}_o^{-1})\\
&-M-MA_1\langle v^{-1}\rangle-A_oB_o]\sin^2\psi\cos\psi,\\
a_2&=[3M(\langle v^{-1}\rangle-M\mathscr{L}_o)+\langle v^{-1}\rangle^2\mathscr{L}_o^{-1}+4M^2\mathscr{L}_{-1}\mathscr{L}_1\mathscr{L}_o^{-1}\\
&-2M\underline{\mathscr{L}_{-1}}\mathscr{L}_o^{-1}\\
&-M^2\overline{\mathscr{L}}_{-2}\mathscr{L}_2\mathscr{L}_o^{-1}]\cos^2\psi+[2MA_o+MA_1\langle v^{-1}\rangle^2+B_o(A_1A_2+A_o^2)\\
&-M\langle v^{-1}\rangle]\cdot\sin^2\psi.\\
a_3&=2[MA_o\mathscr{L}_{-1}\mathscr{L}_o^{-1}+MA_2\mathscr{L}_1\mathscr{L}_o^{-1}-M(A_1A_2+A_o^2)-MA_o\langle v^{-1}\rangle]\cos\psi\\
a_4&=M[A_1A_2+A_o^2]\langle v^{-1}\rangle-MA_2,
\end{aligned}$$

where

$$\phi_x\int_0^x v^k\phi_x^{-2}\mathrm{d}x=\mathscr{L}_k x\phi_x+\text{periodic}(x), k=-2,\ldots 2 \qquad (A.3.3)$$

$$\begin{aligned}
\overline{\mathscr{L}}_{-2}&=\mathscr{L}_{-2}+M\langle v^{-3}\rangle\\
A_o&=\langle v^{-1}\rangle-M(\mathscr{L}_o-\mathscr{L}_{-1}\mathscr{L}_1\mathscr{L}_o^{-1})\\
A_1&=M(\underline{\mathscr{L}_1^2}\mathscr{L}_o^{-1}-\mathscr{L}_2)\\
A_2&=M(\mathscr{L}_{-2}-\mathscr{L}_{-1}^2\mathscr{L}_o^{-1})\\
B_o&=\langle\phi_x^2\rangle-M\langle v\rangle.
\end{aligned}$$

Equation (A.3.3) is not suitable for finding the $\mathscr{L}_k$. Simple manipulations of (8.4.3)–(8.4.6) yield

$$\begin{pmatrix} M, & -C, & M, & O\\ -C, & M, & O, & M\\ -(v_1+v_2), & v_1v_2, & 0, & 1\\ 1, & -(v_1+v_2), & v_1v_2, & O\end{pmatrix}\begin{pmatrix}\mathscr{L}_1\\ \mathscr{L}_o\\ \mathscr{L}_{-1}\\ \mathscr{L}_2\end{pmatrix}=\begin{pmatrix}\frac{1}{2}\\ \frac{1}{2}\langle v\rangle+\langle v^{-1}\rangle\\ I_o\\ I_1\end{pmatrix}$$

where v_1 is a minimum and v_2 is a maximum of v. If we take $x=0$ such that $v(0)=v_1$, then $v(\lambda/2)=v_2$ and

$$I_k=2\lambda^{-1}\int_0^{\lambda/2}(v-v_1)(v-v_2)v^{-k}\phi_x^{-2}dx.$$

These integrals are well behaved, even though their component parts are not. Finally,

$$\overline{\mathscr{L}}_{-2}=M^{-1}(\tfrac{1}{2}\langle v^{-1}\rangle+C\mathscr{L}_{-1})-\mathscr{L}_{o}.$$

Now we just need a computer to give us $I_o(C,M)$ and $I_1(C,M)$ as well as the values of $\langle v\rangle$ and $\langle v^{-1}\rangle$. Once they are known, the problem is reduced to that of solving a quartic for Ω/K.

A.3.2 $\Omega_c>0$

For this case a_n and all pure $\cos^l\psi$ terms remain the same, whereas terms in which powers of $\sin\psi$ appear become

$$a_o^S\rightarrow(\langle\phi_x^2\rangle-M\langle\overline{\delta v_{y1}}\rangle)\cos^2\psi\sin^2\psi$$
$$a_1^S\rightarrow2A_o(M\langle\overline{\delta v_{y1}}\rangle-\langle\phi_x^2\rangle)\cos\psi\sin^2\psi$$
$$a_2^S\rightarrow(A_1A_2+A_o^2)(\langle\phi_x^2\rangle-M\langle\overline{\delta v_{y1}}\rangle)\sin^2\psi$$

where

$$\overline{\delta v_{y1}}=-iK_y^{-1}\delta v_{y1},$$

simpler than without the magnetic field! This, however, is something of a delusion, as $\langle\overline{\delta v_{y1}}\rangle$ is considerably more complicated than the other expressions. It is the average of

$$\overline{\delta v_{y1}}=D\sin\sigma(x)+E\cos\sigma(x)+v(x)-\Omega_c\int_0^x\sin[\sigma(x)-\sigma(\xi)]d\xi$$

$$\sigma(x)=\Omega_c\int_0^x d\xi v^{-1}(\xi)$$

where D and E must be chosen such that $\overline{\delta v_{y1}}$ and $\delta_x\,\overline{\delta v_{y1}}$ be λ periodic:

$$D=\Omega/2[\int_0^\lambda\cos\sigma d\xi-\cot(\sigma(\lambda)/2)\int_0^\lambda\sin\sigma d\xi]$$

$$E=\Omega/2[-\int_0^\lambda\sin\sigma d\xi-\cot(\sigma(\lambda)/2)\int_0^\lambda\sin\sigma d\xi].$$

Zeros in the denominators of D and E for $\Omega_c>0$ give the resonances of (8.4.9). The two Ω_c limits (zero and infinity) are much simpler:

$$\lim_{\Omega_c\rightarrow0}\langle\overline{\delta v_{v1}}\rangle=M(\langle v\rangle-\langle v^{-1}\rangle^{-1})$$

thus the $K\rightarrow0,\Omega_c\rightarrow0$ limit differs from the $\Omega_c\rightarrow0,K\rightarrow0$ limit and A.3.1 is not recovered except for $\psi=0$. The $\Omega_c=0$ case must be treated separately. For large Ω_c we obtain, when $\sigma(\lambda)\neq2\pi n$:

$$\langle\overline{\delta v_{y1}}\rangle=-M\Omega_c^{-2}\langle\phi_x^2v^{-1}\rangle+O(\Omega_c^{-4})$$

For $\psi=\pi/2, \Omega^2/K^2$ is always real (see Figure 8.16)

$$(\Omega^2/K^2)_{\psi=\pi/2}=\frac{(A_1A_2+A_o^2)(M\langle\overline{\delta v_{y1}}\rangle-\langle\phi_x^2\rangle)}{M(A_1A_2+A_o^2)\langle v^{-1}\rangle-MA_2}. \tag{A.3.4}$$

When this expression is negative, $\psi=\pi/2$ modes are purely growing. For large Ω_c, perpendicular perturbation growth rates scale as $1+M^2\Omega_c^{-2}\langle\phi^2v^{-1}\rangle/\langle\phi^2\rangle$. This becomes $1+\Omega_c^{-2}$ in the ZK model $(M\to 1, v\to 1)$. Laedke and Spatschek (1981) obtained a formula for the soliton limit, $C=1+M^2$, λ infinite and $\psi=\pi/2$ (see Fig. 8.17)

$$(\Omega^2/K^2)_{\psi=\pi/2,\mathrm{Sol}}=\frac{\langle\phi_x^2\rangle-M\langle(v-M)^2v^{-1}\rangle-\Omega_cJ/M}{\partial_M\langle(v-M)^2v^{-1}\rangle} \tag{A.3.5}$$

$$J=\langle n(v-M)\int_{-\infty}^{x}[v(\xi)-M]n(\xi)\sin[\sigma(\xi)-\sigma(x)]\mathrm{d}\xi\rangle$$

$$\langle f\rangle=\int_{-\infty}^{+\infty}f\mathrm{d}x.$$

To obtain this result they expanded in K_y around the *soliton* solution and then only satisfied one of the two consistency conditions so obtained. Thus (A.3.5) would seem to be suspect. However, this equation yields the proper KP and ZK limits as in Section 8.3 and, furthermore, seems to be consistent with (A.3.4) for C approaching $1+M^2$ (in a K expansion around a *nonlinear wave* there are no unsatisfied consistency conditions). Thus, all in all, (A.3.5) would appear to be correct, even if it is the product of rather cavalier mathematics. Two errors (an expansion procedure yielding perturbed quantities that misbehave at infinity; and ignoring a consistency condition) seem to have obligingly cancelled or otherwise faded out of the calculation.

APPENDIX 4

Colliding soliton theory

If a velocity potential is introduced in (9.2.1)–(9.2.3), cylindrical coordinates are used, and $\partial/\partial_\theta = 0$ assumed, we obtain from (9.2.2) once integrated:

$$\phi = -(\psi_t + \tfrac{1}{2}\psi_r^2 + \tfrac{1}{2}\psi_z^2), \mathbf{v} = \nabla\psi, \tag{A.4.1}$$

and from (9.2.3)

$$n = \mathrm{e}^{\phi} - r^{-1}(r\phi_r)_r - \phi_{zz}.$$

It is now possible to find one equation for ψ alone from (9.2.1):

$$\begin{aligned} &e^{\phi}[\phi_t + r^{-1}(r\psi_r)_r + \psi_r\phi_r + \psi_{zz} + \psi_z\phi_z] - [r^{-1}(r\phi_r)_r + \phi_{zz}]_t \\ &- r^{-1}[r\psi_r\{r^{-1}(r\phi_r)_r + \phi_{zz}\}]_r - [\psi_z\{r^{-1}(r\phi_r)_r + \phi_{zz}\}]_z = 0, \end{aligned}$$

ϕ being given by (A.4.1).

In order to derive a 'small' ψ equation, we introduce

$$\begin{aligned} &\rho = \varepsilon^{\frac{1}{2}} r \qquad && \tau = \varepsilon^{\frac{1}{2}} t \\ &\eta = \varepsilon z \qquad && \psi = \varepsilon^{\frac{1}{2}}\psi^{(1)} + \varepsilon^{3/2}\psi^{(2)} + \ldots \end{aligned}$$

We obtain the following equation up to $\varepsilon^{5/2}$

$$\begin{aligned} &\rho^{-1}(\rho\psi_\rho^{(1)})_\rho - \psi_{\tau\tau}^{(1)} + \varepsilon[\rho^{-1}(\rho\psi_\rho^{(2)})_\rho - \psi_{\tau\tau}^{(2)} \\ &+ \rho^{-1}(\rho\psi_\rho^{(1)})_{\rho\tau\tau} - 2\psi_\rho^{(1)}\psi_{\rho\tau}^{(1)} + \psi_{\eta\eta}^{(1)}] = 0. \end{aligned} \tag{A.4.2}$$

This equation spans two orders in ε and thus differs fundamentally from earlier models. In the original variables we obtain the approximate equation

$$\psi_{tt} - r^{-1}(r\psi_r)_r - r^{-1}(r\psi_r)_{rtt} + 2\psi_r\psi_{rt} - \psi_{zz} = 0$$

or, equivalently in view of lowest order (A.4.2)

$$\begin{aligned} &\mathrm{BC1}\ \psi_{tt} - \nabla_r\psi - \nabla_r^2\psi - (\psi_r^2)_t - \psi_{zz} = 0. \\ &\Delta_r \doteq r^{-1}\partial_r r\partial_r. \end{aligned} \tag{A.4.3}$$

Similarly, for cylindrical solitons moving up and down the z axis.

$$\mathrm{BC2}\ \psi_{tt} - \psi_{zz} - \psi_{zzzz} + (\psi_z^2)_t - \Delta_r\psi - r^{-2}\psi_{\theta\theta} = 0.$$

Equation (A.4.3) can be further reduced to two equations CI, and that following it to a pair of CII equations – one for each direction of propagation $\pm z$.

The spherical soliton equation is, in r, t

$$\text{BS } \psi_{tt} - \Delta_r^2 \psi + (\psi_r^2)_t = 0, \Delta_r = \Delta^{-2} \partial_r r^2 \partial_r.$$

For further considerations we will also need the Cartesian version, generalized KP, Infeld (1980a)

$$\psi_{tt} - \psi_{xx} - \psi_{xxxx} + (\psi_x^2)_t - \psi_{zz} = 0. \tag{A.4.4}$$

The Lagrangian for the original equations (9.2.1)–(9.2.3) was given in Section 5.2 by (5.2.29). It leads to the energy density

$$\mathscr{E} = L' - \dot{\psi} L_{\dot{\psi}} = \tfrac{1}{2} n (\nabla \psi)^2 + n\phi + 1 - e^{\phi} - \tfrac{1}{2} (\nabla \phi)^2.$$

We will use this expression to find the far field r dependence of the amplitude of a single imploding or expanding cylindrical or spherical soliton. Equation (A.4.4) is solved exactly by

$$\psi_x = \tfrac{3}{2} k\beta \operatorname{sech}^2 \tfrac{1}{2} (kx - \beta t + \delta) \tag{A.4.5}$$

$$k^2 - \beta^2 + k^2 \beta^2 = 0. \tag{A.4.6}$$

Head-on collisions can be described. The form of (A.4.5) suggests we look for approximate far-field solutions to BCI and BS in the form

$$\psi_r \simeq 3/2 k(r) \beta(r) \operatorname{sech}^2 \frac{\beta}{2} \left[\int^r k\beta^{-1} \mathrm{d}r - t \right]. \tag{A.4.7}$$

But what is $k(r)$? From (A.4.6) we can take $k \simeq \beta$ in the amplitude, whereas for solutions depending on r, t only, energy conservation takes the form

$$\int \mathscr{E}(r, t) r^m \mathrm{d}r = \text{constant},$$
$$(m = 0, 1, 2 \text{ for flat, cyl, sph, symmetry}) \tag{A.4.8}$$

When we calculate $\mathscr{E}$ up to terms quadratic in ψ_r, ψ_t we obtain, using (A.4.7) and (A.4.8), $k \sim r^{-m/3}$, thus the amplitude is $r^{-2m/3}$. Also amplitude x (width square)$\simeq$constant as $r \to \infty$ (see Fig. 9.7 where this is confirmed for spherical solitons).

APPENDIX 5

A model equation for spherical solitons

Assume $\mathbf{v}$ to be curl free and introduce a velocity potential

$$\mathbf{v}=\nabla\psi$$

Equations (9.2.1)–(9.2.3) become

$$\frac{\partial n}{\partial t}+\nabla(n\nabla\psi)=0 \tag{A.5.1}$$

$$\frac{\partial\psi}{\partial t}+\tfrac{1}{2}(\nabla\psi)^2+\phi=0 \tag{A.5.2}$$

$$\nabla^2\phi=\mathrm{e}^{\phi}-n. \tag{A.5.3}$$

As seen in Section 5.2, these equations are derivable from the Lagrangian

$$L=\int L'\mathrm{d}\mathbf{x}\mathrm{d}t, L'=\tfrac{1}{2}n[(\nabla\psi)^2+2\dot{\psi}+2\phi]+1-\mathrm{e}^{\phi}-\tfrac{1}{2}(\nabla\phi)^2.$$

In spherical polars r, θ, v.

$$L=\int L'r^2\sin^2\theta\mathrm{d}r\mathrm{d}\theta\mathrm{d}v\mathrm{d}t.$$

Introduce

$$\mathscr{L}=r^2\sin^2\theta L'=r^2\sin^2\theta[\tfrac{1}{2}\mathrm{n}(\psi_r^2+r^{-2}\psi_\phi^2+r^{-2}\sin^{-2}\theta\psi_v^2+2\psi_t+2\phi)$$
$$+1-\mathrm{e}^{\phi}-\tfrac{1}{2}(\phi_2^2+r^{-2}\phi_\theta^2+r^{-2}\sin^{-2}\theta\phi_v^2)].$$

Now introduce a coordinate system following a set of collapsing spheres and stretched similarly to that of Section 9.2 (CI):

$$\rho=-\varepsilon^{\frac{1}{2}}(r+t), \qquad r=-(\varepsilon^{-3/2}\tau+\varepsilon^{-\frac{1}{2}}\rho), \qquad \psi=\varepsilon^{\frac{1}{2}}\psi^{(1)}+\ldots$$
$$\tau=\varepsilon^{3/2}t, \qquad \phi=\varepsilon\phi^{(1)}+\ldots, \qquad n=1+\varepsilon n^{(1)}+\ldots$$

and θ, v unaltered. Instead of expanding (A.5.1)–(A.5.3) we expand the Lagrangian density. (This procedure was introduced in Section 5.2.) The first two orders in ε are uninteresting:

$$\mathscr{L}^{(-3)}=0, \mathscr{L}^{(-2)}=-\tau^2\sin^2\theta\psi_\rho^{(1)}$$

whereas

$$\mathscr{L}^{(-1)} = \tau^2 \sin^2\theta[-n^{(1)}\psi_\rho^{(1)} + n^{(1)}\phi^{(1)} - \tfrac{1}{2}\phi^{(1)^2} + \tfrac{1}{2}\psi_\rho^{(1)^2}]$$
$$+ (\tau^2\sin^2\theta\psi^{(1)})_\tau - (\tau^2\sin^2\theta\psi_\rho^{(2)} + 2\rho\tau\sin^2\theta\psi^{(1)})_\rho,$$

giving the following Euler–Lagrange equations

$$\begin{array}{ll} \delta n^{(1)} & \psi_\rho^{(1)} = \phi^{(1)} \\ \delta\phi^{(1)} & n^{(1)} = \phi^{(1)} \\ \delta\psi^{(1)} & n^{(1)} = \psi_\rho^{(1)}. \end{array}$$

The next order $\mathscr{L}$ in terms of ψ is

$$\mathscr{L}^{(0)} = \tau^2\sin^2\theta(\psi_\tau^{(1)}\psi_\rho^{(1)} + \tfrac{1}{3}\psi_\rho^{(1)^3} - \tfrac{1}{2}\psi_\rho^{(1)^2}) + (\tau^2\sin^2\theta\psi^{(2)} + 2\rho\tau^3\sin^2\theta\psi^{(1)})_\tau$$
$$- (\tau^2\sin^2\theta\psi_\rho^{(3)} + 2\rho\tau\sin^2\theta\psi^{(2)} + \rho^2\sin^2\theta\psi^{(1)})_\rho.$$

As perfect derivatives do not contribute to L, this is equivalent to

$$\mathscr{L}_{\text{eq}}^{(0)} = \tau^2\sin^2\theta(\psi_\tau^{(1)}\psi_\rho^{(1)} + \tfrac{1}{3}\psi_\rho^{(1)^3} - \tfrac{1}{2}\psi_{\rho\rho}^{(1)^2}) \tag{A.5.4}$$

yielding

$$\psi_{\rho\tau}^{(1)} + \frac{1}{\tau}\psi_\rho^{(1)} + \psi_\rho^{(1)}\psi_{\rho\rho}^{(1)} + \tfrac{1}{2}\psi_{\rho\rho\rho\rho}^{(1)} = 0, \tag{A.5.5}$$

the angle-differentiated terms only appearing in the next order.

A similar calculation would yield, for the two types of cylindrical equation

$$\text{CI}\ \ \mathscr{L}_{\text{eq}}^{(3/2)} = \tau(\psi_\tau^{(1)}\psi_\rho^{(1)} + \tfrac{1}{3}\psi_\rho^{(1)^3} - \tfrac{1}{2}\psi_{\rho\rho} + \tfrac{1}{2}\psi_z^{(1)^2}), \tag{A.5.6}$$

$$\text{CII}\ \ \mathscr{L}_{\text{eq}}^{(2)} = \rho\left(\psi_\tau^{(1)}\psi_z^{(1)} + \tfrac{1}{3}\psi_z^{(1)^3} - \tfrac{1}{2}\psi_{zz}^{(1)^3} + \tfrac{1}{2}\psi_\rho^{(1)^2} + \frac{1}{2\rho^2}\psi_\theta^{(1)^2}\right) \tag{A.5.7}$$

The method of Lagrangian expansion, used in Section 5.2 to derive the KP equation, is mathematically simpler than the usual expansion of all equations of motion. The price we pay is restricting the ion flow to be curl free from the start, rather than obtaining this result from the calculations.

References

Abdulloev, K.O., Bogolyubski, I.L. and Makhankov, V.G. Dynamics of Langmuir turbulence. Formation and interactions of solitons. *Phys. Lett.* **48A** 161–2 (1974).

Ablowitz, M.J. Remarks on nonlinear evolution equations and ordinary differential equations of Painlevé type 1. *Physica* **D3** 129–41 (1981).

Ablowitz, M.J., Ramani, A. and Segur, H. A connection between nonlinear evolution equations of Painlevé type. I, *J.* Math. Phys. **21**, 715–21 (1980a); II, 1006–15 (1980b).

Ablowitz, M. and Segur, H. On the evolution of packets of water waves. *J. Fluid Mech.* **92**, 691–715 (1979).

Abraham-Shrauner, B. Exact, stationary wave solutions of the nonlinear Vlasov equations. *Phys. Fluids* **11** 1162–7 (1968).

Abraham-Shrauner, B. Exact damped sinusoidal electric field of nonlinear one-dimensional Vlasov–Maxwell equations. *J. Plasma Phys.* **32** 197–205 (1984a).

Abraham-Shrauner, B. Exact, time dependent solutions of the one-dimensional Vlasov–Maxwell equations. *Phys. Fluids* **27** 197–202 (1984b).

Abraham-Shrauner, B. Lie transformation group solutions of the nonlinear one-dimensional Vlasov equation. *J. Math. Phys.* **26** 1428–35 (1985).

Abraham-Shrauner, B. and Feldman, W.C. Nonlinear Alfvén waves in high-speed solar windstreams. *J. Geophys. Res.* **82** 618–24 (1977).

Abramowitz, M. and Stegun, I. *Handbook of mathematical functions.* Dover, New York (1965).

Alberty, J.M., Koikawa, T. and Sasaki, R. Canonical structure of soliton equations I. *Physica.* **5D** 43–65 (1982).

Albritton, J. and Rowlands, G. On the relation between Lagrangian solutions and Bernstein–Greene–Kruskal modes in a cold plasma. *Nuc. Fus.* **15** 1199 (1975).

Amick, C.J. and Toland, J.F. On periodic water waves and their convergence to solitary waves in the long-wave limit. *Phil. Trans. R. Soc. London* **A303** 633 – 69 (1981).

Andersen, H.K., D'Angelo, N., Michelson, P. and Nielson, P. Investigation of Landau damping effects on shock formation. *Phys. Rev. Lett.* **19** 149–51 (1967).

Anderson, D., Bondeson, A. and Lisak, M. Transverse instability of soliton solutions to nonlinear Schrödinger equations. *J. Plasma Phys.* **21** 259–66 (1979).

Anker, D. and Freeman, N.C. On the soliton solutions of the Davey–Stewartson equation for long waves. *Proc. R. Soc. London* **A360** 529–40 (1978).

Antipov, S.V., Nezlin, M.V., Snezhkin, E.N. and Trubnikov, A.S. Rossby autosoliton and stationary model of the Jovian Great Red Spot. *Nature* **323** 238–40 (1986).

Armstrong, R.J., Greene, G.J. and Ono, M. Dimensionality in a toroidal plasma: a preliminary investigation. *International Conference on Plasma Physics, Kiev* 1987, **3** 16–19 (1987).

Armstrong, T. and Montgomery, D.T. Asymptotic state of the two stream instability, *J. Plasma Phys.* **1** 425–33 (1967).

Atiyah, M., Gibbon, J.D. and Wilson, G. (ed.). New developments in the theory and application of solitons. *Phil. Trans. R. Soc. London* **315** 333–469 (1985).

Baldwin, D.E. and Rowlands, G. Plasma oscillations perpendicular to a weak magnetic field. *Phys. Fluids* **9** 2444–53 (1966).

Baldwin, D.E. and Rowlands, G. A means of classifying convective and absolute instabilities. *Phys. Fluids* **13** 2036–38 (1970).

Barkley, H., Andreoletti, J., Gervais, F., Olivian, J., Quemeneur, A. and Truc, A. Internal structure of density fluctuations in TFR Tokomak plasmas. *International Conference on Plasma Physics*, Kiev 1987, **3** 24–7 (1987).

Barone, A., Esposito, F., Magee, C.J. and Scott, A.C. Theory and applications of the sine–Gordon equation. *Riv. Nuovo Cim.* (2) **1** 227–67 (1971).

Bartucelli, M., Carbonaro, P. and Muto, V. Kadomtsev–Petviashvili–Burgers for shallow water waves. *Lett. Nuovo Cim.* **42** 279–84 (1985).

Barut, A.O. and Raczka, R. *Theory of group representations and applications.* PWN, Warsaw (1980); World Scientific. Singapore (1987).

Bates, D.R. and Esterman, I. *Advances in atomic and molecular physics* **6**. Academic, New York (1970).

Bell, T.F. Nonlinear Alfvén waves in a Vlasov plasma. *Phys. Fluids* **8** 1829–39 (1965).

Ben-Jacob, E., Brand, H., Dee, G., Kramer, L. and Langer, J.S. Pattern propagation in nonlinear dissipative systems. *Physica* **14D** 348–64 (1985).

Benjamin, T.B. Instability of periodic wavetrains in nonlinear dispersive systems. *Proc. R. Soc. London* **A299** 59–75 (1967).

Benjamin, T.B. The stability of solitary waves. *Proc. R. Soc. London* **A328** 153–183 (1972).

Benjamin, T.B. and Feir, J.E. The disintegration of wave trains on deep water, I: Theory. *J. Fluid Mech.* **27** 417–30 (1967).

Benney, D.J. Nonlinear gravity wave interactions. *J. Fluid Mech.* **14** 557–84 (1962).

Berk, H.L. and Roberts, K.V. Numerical study of Vlasov's equation for a special class of distribution functions. *Phys. Fluids* **10** 1595–97 (1967).

Bernstein, I.B. Waves in a plasma in a magnetic field. *Phys. Rev.* **109** 10–21 (1958).

Bernstein, I.B., Greene, J.M. and Kruskal, M.D. Exact nonlinear plasma oscillations, *Phys. Rev.* **108** 546–50 (1957).

Bernstein, I.B. and Trehan, S.K. Plasma oscillations I, *Nuc. Fus.* **1** 3–41 (1960).

Bertrand, P., Feix, M.R. and Bauman, G. Electrostatic waves in periodic inhomogeneous plasma. *J. Plasma Phys.* **6** 351–66 (1971).

Bhatnagar, P.L. *Nonlinear waves in one-dimensional dispersive systems.* Oxford University Press (1979).

Bialynicki-Birula, I. Nonlinear electrodynamics: variations on a theme by Born and Infeld. *Quantum theory of particles and fields.* Jancewicz, B. and Lukierski, J. (ed.), World Scientific, Singapore (1983).

Bialynicki-Birula, I. and Mycielski, J. Wave equations with logarithmic nonlinearities. *Bull. Acad. Polon. Sci.* **III23** 461–66 (1975).

Bialynicki-Birula, I. and Mycielski, J. Nonlinear wave mechanics. *Ann. Phys.* **100** 62–93 (1976).

Bialynicki-Birula, I. and Mycielski, J. Gaussons: solitons of the logarithmic Schrödinger equation. *Phys. Scr.* **20** 539–44 (1979).

Bishop, A.R. and Schneider, T. (ed.). Solitons and condensed matter physics, **8** in *Solid State Sciences.* Springer, Berlin and New York (1981).

Bloomberg, H.W. Sideband instability in the water bag approximation. *Phys. Fluids* **17** 263–5 (1974).

Bloor, M.I.G. A note on the limiting form of shallow water waves. Chapter 3 of *Advances in nonlinear waves.* Debnath, L. (ed.), **1**. Pitman, Boston and London 59–63 (1984).

Bluman, G.W. and Cole, J.D. *Similarity methods for differential equations.* Springer, Berlin and New York (1974).

Bohm, D. and Pines, D. A collective description of electron interactions. *Phys. Rev.* I Magnetointeractions. **82** 625–34 (1951). II Collective vs individual particle aspects of the interaction. **85** 338–53 (1952). III Coulomb interactions in a degenerate electron gas. **92** 609–25 (1953). IV Electron interactions in metals. **92** 626–36 (1953), (just Pines).

Boillat, G. *La propagation des ondes.* Gauthier–Villars, Paris (1965).

Boillat, G. Nonlinear electrodynamics; Lagrangians and equations of motion. *J. Math. Phys.* **11** 941–51 (1970).

Bona, J.L., Pritchard, W.G. and Ridgway-Scott, L. Solitary wave interaction. *Phys. Fluids* **23** 438–41 (1980).

Borisov, A.B., Tankeyev, A.P., Shagalov, A.G. and Bezmaternih, G.V. Multi-vortex-like solutions of the sine–Gordon equation. *Phys. Lett.* **111A** 15–18 (1985).

Born, M. and Infeld, L. Foundations of the new field theory. *Proc. R. Soc. London* **A144** 425–51 (1934).

Boussinesq, J. Théorie de l'intumescence liquide appelée onde solitaire ou de translation se propageant dans un canal rectangulare. *Compte Rendus Acad. Sci.* **72** 755–9 (1871).

Bretherton, F.P. Resonant interaction between waves. The case of discrete oscillations. *J. Fluid Mech.* **20** 457–79 (1964).

Bridges, R. and Rowlands, G. On the analytic form of some strange attractors. *Phys. Lett.* **63A** 189–90 (1977).

Briggs, R.G. Electron stream interaction with plasmas. MIT, Cambridge Mass. (1964).

Broomhead, D.S. and King, G.P. Extracting qualitative dynamics from experimental data. *Physica* **20D** 217–36 (1986).

Broomhead, D.S., McCreadie, G. and Rowlands, G. On the analytic derivation of Poincaré maps – the forced Brusselator problem. *Phys. Lett.* **84A** 229–31 (1981).

Broomhead, D.S. and Rowlands, G. Spinodal decomposition, a phenomenological theory. *Philos. Mag.* **A44** 543–59 (1981).

Broomhead, D.S. and Rowlands, G. A simple derivation of the Mel'nikov condition for the appearance of homoclinic points. *Phys. Lett.* **89A** 63–5 (1982).

Broomhead, D.S. and Rowlands, G. On the use of perturbation theory in the calculation of the fractal dimension of strange attractors. *Physica* **D10** 340–52 (1983).

Broomhead, D.S. and Rowlands, G. On the analytic treatment of non-integrable difference equations. *J. Phys.* **A16** 9–24 (1983).
Bryant, P.J. Two-dimensional periodic permanent waves in shallow water. *J. Fluid Mech.* **115** 527–32 (1982).
Bullough, R.K. and Caudry, P.W. The multiple sine–Gordon equations in nonlinear optics and liquid ^{3}He. *Nonlinear evolution equations solvable by the spectral transform.* Calogero, F. (ed.), Pitman, Boston and London (1978).
Bullough, R.K. and Caudry, P.J. (eds). *Solitons.* Springer, Berlin and New York (1980a).
Bullough, R.K., Caudry, P.J. and Gibbs, H.M. The double sine–Gordon equations: a physically applicable system of equations. *Solitons.* Bullough, R.K. and Caudry, P.J. (ed.). Springer, Berlin and New York (1980b).
Butcher, P.N. The Gunn effect. *Rep. Prog. Phys.* **30** 97–148 (1967).
Butcher, P.N. and Rowlands, G. The stability of Gunn domains. *Phys. Lett.* **26A** 226–7 (1968).
Butler, D.S. and Gribben, R.J. Relativistic formulation for nonlinear waves in a non-uniform plasma. *J. Plasma Phys.* **2** 257–81 (1968).
Callen, J.D. Absolute and convective microinstabilities in a magnetized plasma, Ph. D. thesis, MIT Center for Space Research (1968).
Calogero, F. Nonlinear evolution equations solvable by the spectral transform. Nonlinear equations and dynamical systems, Boiti, M., Pempinelli, F. and Soliani, G. (eds), *Lecture Notes in Physics.* Springer, Berlin and New York (1978a).
Calogero, F. *Nonlinear evolution equations solvable by the spectral transform method.* Pitman, Boston and London (1978b).
Calogero, F. and Degasperis, A. Inverse spectral problem for the one-dimensional Schrödinger equation with an additional linear potential. *Lett. Nuovo Cim.* **23** 143–9 (1978a).
Calogero, F. and Degasparis, A. Solution by the spectral-transform method of a nonlinear evolution equation including as a special case the Cylindrical KdV equation. *Lett. Nuovo Cim.* **23** 150–4 (1978b).
Calogero, F. and Degasparis, A. Conservation laws for a nonlinear evolution equation that includes as a special case the Cylindrical KdV equation. *Lett. Nuovo Cim.* **23** 155–60 (1978c).
Cap, F.S. *Handbook on plasma instabilities*, 3 volumes. Academic, New York (1976, 1978, and 1982).
Case, K.M. Plasma oscillations. *Ann. Phys.* **7** 349–64 (1959).
Censor, D. Ray tracing in weakly nonlinear moving media. *J. Plasma Phys.* **16** 415–26 (1976).
Chandrasekhar, S. *Hydrodynamic and hydromagnetic stability*. Clarendon, Oxford (1961).
Chen, B. and Saffman, P.G. Steady gravity-capillary waves on deep water, *Stud. Appl. Math.* I Weakly nonlinear waves **60** 183–210 (1979). II Numerical results for finite amplitude **62** 95–111 (1980).
Chen, B. and Saffman, P.G. Three dimensional stability and bifurcations of capillary and gravity waves on deep water. *Stud. Appl. Math.* **72** 125–47 (1985).
Chen, F.F. *Introduction to plasma physics and controlled fusion* **1**. Plenum, New York (1984).
Chen, T. and Schott, L. Observation of diverging cylindrical solitons excited with a probe. *Phys. Lett.* **58A** 459–61 (1976).

Chen, T. and Schott, L. Diverging solitons and their transition to dispersive waves, *Plasma Phys.* **19** 959–67 (1977).

Clarkson, P.A., McLoed, J.B., Olver, P.J. and Ramani, A. Integrability of Klein–Gordon equations. *SIAM J. Math. Anal.* **17** 798–802 (1986).

Clemmow, P.C. and Dougherty, J.P. Electrodynamics of particles and plasmas, Addison–Wesley, Reading, Mass. (1969).

Clemmow, P.C. and Mullaly, R.F. *Report on Conference on the physics of the ionosphere.* Physical Society, London (1955).

Cohen, B.I., Watson, K.M. and West, B.J. Some properties of deep water solitons, *Phys. Fluids* **19** 345–54 (1976).

Collet, P. and Eckmann, J.P. Iterated maps on the interval as dynamical systems. *Progress in physics* **1**. Birkhaüser, Boston (1980).

Crapper, G.D. An exact solution for progressive capillary waves of arbitrary amplitude. *J. Fluid Mech.* **2** 532–40 (1957).

Cumberbatch, E. Spike soliton for radially symmetric solitary waves. *Phys. Fluids* **21** 374–6 (1978).

Currie, A. Chaotic Dynamics in Flows and Discrete Maps. Ph.D. thesis University of Warwick (1987).

Cvitanovic, P. (ed.). *Universality in chaos.* Adam Hilger, Bristol (1984).

Dashen, R.F., Hasslacher, B. and Neveu, A. Nonperturbative methods and extended hadron models in field theory II. Two-dimensional models and extended hadrons, *Phys. Rev.* **D10** 4130–8 (1974).

Davey, A. and Stewartson, K. On three-dimensional packets of surface waves. *Proc. R. Soc. London* **A338** 101–10 (1974).

Davidson, R.C. *Methods in nonlinear plasma theory.* Academic, New York (1972).

Davydov, A.S. The role of proteins in electron transport at large distances. *Phys. Stat. Sol.* b **90** 457–64 (1978).

Davydov, A.S. The role of solitons in the energy and electron transfer in a one-dimensional molecular system. *Physica* **D3** 1–22 (1981).

Davydov, A.S. *Solitons in molecular systems.* Reidel, Dorderecht and Boston (1985).

Dawson, J.M. Plasma oscillations of a large number of electron beams. *Phys. Rev.* **118** 381–9 (1960).

Debnath, L. (ed.). *Nonlinear waves.* Cambridge University Press (1983).

DeLeonardis, R.M. and Trullinger, S.E. Theory of boundary effects on sine–Gordon solitons. *J. Appl. Phys.* **51** 1211–26 (1980).

Denavit, J., Pereira, N.R. and Sudan, R.N. Two-dimensional stability of Langmuir solitons. *Phys. Rev. Lett.* **30** 1435–8 (1974).

Derfler, H. and Simonen, T.C. Landau waves: an experimental fact. *Phys. Rev. Lett.* **17** 172–5 (1966).

Dodd, R.K., Ellbeck, J.C., Gibbon, J.D. and Morris, H.C. *Solitons and nonlinear wave equations.* Academic, New York (1983).

van Dooren, R. The three-soliton solution of the two dimensional Korteweg–de Vries equation. *Advances in nonlinear waves.* Debnath, L. (ed.), **2**. Pitman, Boston 187–98 (1985).

Dougherty, J.P. Lagrangian methods in plasma dynamics; I. General theory of the method of averaged Lagrangians. *J. Plasma Phys.* **4** 761–85 (1970).

Drazin, P.G. *Solitons.* Cambridge University Press (1983).

Dryuma, V.S. On the integration of cylindrical KP equation by the method of the inverse problem of scattering theory. *Dokl Akad Nauk SSSR* **268** 15–17 (1983); *Sov. Math. Doklady* **27** 6–8 (1983).

van Dyke, M. *An album of fluid motion.* Parabolic, Stanford, California (1982).

Dysthe, K.B. Convective and absolute instability. *Fus. Nuc.* **6** 215–21 (1966).

Dysthe, K.B. Note on a modification to the nonlinear Schrödinger equation for application to deep water waves. *Proc. R. Soc. London* **A369** 105–14 (1979).

Eckmann, J.P. Roads to turbulence in dissipative dynamical system. *Rev. Mod. Phys.* **53** 643–54 (1981).

Eilbeck, J.C. Numerical studies of solitons, *Solitons and condensed matter physics.* Bishop, A.R. and Schneider, T. (eds). Springer, Berlin and New York (1981).

Eilenberger, G. Solitons, **19**, *Springer series in solid state physics.* Springer, Berlin and New York (1981).

Eleonsky, V.M., Kalagin, N.E., Novozhilova, N.S. and Silin, V.P. Methods of asymptotic expansion and of finite-dimensional models in the theory of nonlinear waves, *Advances in nonlinear waves.* Debnath, L. (ed.), **2**. Pitman, Boston and London 286–308 (1985).

Emmerson, G.S. *J.S. Russell, a biography.* John Murray, London (1971).

D'Evelyn, M. and Morales, G.J. Properties of large amplitude Langmuir solitons, *Phys. Fluids* **21** 1997–2008 (1978).

Fadeev, L.D. A Hamiltonian interpretation of the inverse scattering method. *Solitons*, Bullough, R.K. and Caudry, P.J. (eds). Springer, Berlin and New York (1980).

Fainberg, Y.B., Kurilko, V.I. and Shapiro, V.D. Instabilities in the interaction of charged particle beams with plasmas. *Zh. Tech. Phys.* **31** 633–9 (1961); *Sov. Phys. Tech. Phys.* **6** 459–63 (1961).

Falk, L. A solution of the multiple three-wave system. *International Conference on Plasma Physics* **96**. *Göteborg* 1982, 96 (1982).

Farmer, J.D., Ott, E. and Yorke, J.A. The dimension of chaotic attractors. *Physica* **7D** 158–80 (1983).

Feigenbaum, M.J. The universal metric properties of nonlinear transformations. *J. Stat. Phys.* **21** 669–706 (1978).

Feigenbaum, M.J. Universal behaviour in nonlinear systems, *Los Alamos Science* **1** 4–27 (1980).

Fenton, J. A ninth order solution for the solitary wave. *J. Fluid Mech.* **53** 257–71 (1972).

Fermi, E., Ulam, S. and Pasta, J. *Collected papers of Enrico Fermi.* Segré, E. (ed.), University of Chicago, **2** 978–88 (1965).

Fetter, A.L. Nonuniform states of an imperfect Bose gas. *Ann. Phys.* **70** 67–101 (1972).

Fornberg, B. and Whitham, G.B. A numerical and theoretical study of certain nonlinear wave phenomena. *Phil. Trans. R. Soc. London* **289** 373–404 (1978).

Fowler, A.C. Note on a paper by G. Rowlands (Chaotic trajectories of ordinary differential equations). *J. Phys.* **A16** 3139–43 (1983).

Franklin, R.N. Microinstabilities in plasmas; nonlinear effects, *Rep. Prog. Phys.* **40** 1369–1413 (1977).

Freeman, N.C. A two-dimensional distributed soliton solution of the Korteweg de Vries equation. *Proc. R. Soc. London* **A366** 185–205 (1979).

Freeman, N.C. and Davey, A. On the evolution of packets of long surface waves, *Proc. R. Soc. London* **A344** 427–33 (1975).

Friedrichs, K.O. and Hyers, D.H. The existence of solitary waves. *Comm. Pure Appl. Math.* **7** 517–50 (1954).

Frycz, P. Dynamics of ion acoustic plasma solitons. Ph.D. thesis, Institute for Nuclear Studies, Warsaw (1988).

Galdi, G.P. and Straughan, B. A nonlinear analysis of the stabilizing effect of rotation in the Bénard problem. *Proc. R. Soc. London* **A402** 257–83 (1985).

Galloway, J.J. and Kim, H. Lagrangian approach to nonlinear wave interactions in a warm plasma. *J. Plasma Phys.* **6** 53–72 (1971).

Gardner, C.S., Greene, J.M., Kruskal, M.D. and Miura, R.M. Method for solving the Korteweg de Vries equation. *Phys. Rev. Lett.* **19** 1095–7 (1967).

Gelfand, I.M. and Shilov, G.E. *Generalized functions.* Academic, New York (1964).

Gibbon, J.D., James, I.N. and Moroz, I.M. The sine–Gordon equation as a model for a rapidly rotating baroclinic fluid. *Phys. Script.* **20** 402–8 (1979).

Gibbons, J., Thornhill, S.G., Wardrop, M.J. and ter Haar, D. On the theory of Langmuir solitons. *J. Plasma Phys.* **17** 153–70 (1977).

Goldman, M.V. Theory of stability of large periodic plasma waves. *Phys. Fluids* **13** 1281–9 (1970).

Goldstein, P. and Infeld, E. The Zakharov equations: a non-Painlevé system with exact N soliton solutions. *Phys. Lett.* **103A** 8–10 (1984).

Gordon, M.J.V., Williams, G.A. and Packard, R.E. Vortex photography; a progress report. *J. Phys. Colloq.* **39** C6 172–3 (1978).

Gould, R.W., O'Neil, T.M. and Malmberg, J.H. Plasma wave echo. *Phys. Rev. Lett.* **19** 219–22 (1967).

Grassberger, P. and Procaccia, I. Characterization of strange attractors. *Phys. Rev. Lett.* **50** 346–9 (1983).

Gribben, R.J. A theory for the propagation of slowly varying nonlinear waves in a non-uniform plasma. *Nonlinear waves.* Debnath, L. (ed.), Cambridge University Press 221–44 (1983).

Grimshaw, R. Slowly varying solitary waves, I. KdV equation. *Proc. R. Soc. London* **A368** 359–75 (1979).

Guckenheimer, J., Oster, G. and Ipaktchi, A. Dynamics of density dependent population models. *J. Math. Biol.* **4** 101–47 (1977).

Haken, H. *Synergetics, a workshop.* Springer, Berlin and New York (1977).

Haken, H. *Synergetics.* Springer, Berlin and New York (1978).

Haken, H. *Advanced synergetics.* Springer, Berlin and New York (1983).

Hall, L.S. and Heckrotte, W. Instabilities: convective versus absolute. *Phys. Rev.* **166** 120–8 (1968).

Hasegawa, A. Theory and computer experiment on self-trapping instability of plasma cyclotron waves. *Phys. Fluids* **15** 870–81 (1972).

Hasegawa, A., Maclennan, C.G. and Kodama, Y. Nonlinear behaviour and turbulence spectra of drift waves and Rossby waves. *Phys. Fluids* **22** 2122–9 (1979).

Hasegawa, A. and Mima, K. Stationary spectrum of strong turbulence in magnetized nonuniform plasma. *Phys. Rev. Lett.* **39** 205–8 (1977).

Hayes, W.D. Group velocity and nonlinear dispersive wave propagation. *Proc. R. Soc. London* **A332** 199–221 (1973).

Helleman, R.H.G. Self generated chaotic behaviour in nonlinear mechanics. *Universality in chaos.* Cvitanovic, P. (ed.). Adam Hilger, Bristol (1984).

Hénon, M. A two-dimensional mapping with a strange attractor. *Comm. Math. Phys.* **50** 69–77 (1976).

Herbst, B.M., Mitchell, A.R. and Weideman, A.C. On the stability of the nonlinear Schrödinger equation. *J. Comp. Phys.* **60** 263–81 (1985).

Hershkowitz, N. and Romesser, T. Observations of ion-acoustic cylindrical solitons. *Phys. Rev. Lett.* **32** 581–3 (1974).

Hirota, R. Exact solutions of the KdV equation for multiple collisions of solitons. *Phys. Rev. Lett.* **27** 1192–4 (1971).

Hirota, R. Exact solutions to the equation describing 'cylindrical solitons'. *Phys. Lett.* **71A** 393–4 (1979a).

Hirota, R. The Bäcklund and inverse scattering transform of the KdV equation with nonuniformities. *J. Phys. Soc. Jpn.* **46** 1681–2 (1979b).

Hirota, R. and Ito, M. Resonance of solitons in one dimension. *J. Phys. Soc. Jpn.* **52** 744–8 (1983).

Hogan, S.J. The fourth-order evolution equation for deep-water gravity-capillary waves. *Proc. R. Soc. London* **A402** 359–72 (1985).

Holmes, P. A nonlinear oscillator with a strange attractor. *Phil. Trans. R. Soc. London* **292** 420–47 (1979).

Hunter, J.K. and Vanden-Broeck, J.M. Solitary and periodic gravity–capillary waves of finite amplitude. *J. Fluid Mech.* **134** 205–19 (1983a).

Hunter, J.K. and Vanden-Broek, J.M. Accurate computations for steep solitary waves. *J. Fluid Mech.* **136** 63–72 (1983b).

Hyman, J.M., McLaughlin, D.W. and Scott, A.C. On Davydov's alpha helix solitons. *Physica* **D3** 23–44 (1981).

Ichikawa, Y.H. Alternative Lax-pair operators for the sine–Gordon equation, *Advances in nonlinear waves.* Debnath, L. (ed.), **2**, Pitman, Boston and London 240–3 (1985).

Ichimaru, S. *Basic principles of plasma physics.* Benjamin, Reading, Mass. (1973).

Ikezi, H. Experiments on ion-acoustic solitary waves. *Phys. Fluids* **16**, 1668–75 (1973).

Ikezi, H., Chang, R.P.H. and Stern, R.A. Nonlinear evolution of the electron-beam-plasma instability. *Phys. Rev. Lett.* **36** 1047–51 (1976).

Ikezi, H., Schwarzenegger, K., Simons, A.L., Ohsawa, Y. and Kamimura, T. Nonlinear self-modulation of ion-acoustic waves. *Phys. Fluids* **21** 239–48 (1978).

Ince, E.L. *Ordinary differential equations.* Dover, New York (1944).

Infeld, E. A new class of equilibrium configurations of a plasma which fit into a sphere. *Phys. Fluids* **14** 2054–6 (1971).

Infeld, E. On the stability of nonlinear cold plasma waves. *J. Plasma Phys.* **8** 105–10 (1972). For sequel see Infeld and Rowlands (1973).

Infeld, E. Form and ergodic behaviour of nonlinear waves in a warm two component plasma. *Phys. Lett.* **48A** 175–6 (1974).

Infeld, E. On three-dimensional generalizations of the Boussinesq and Korteweg de Vries equations. *Q. Appl. Math.* **38** 277–87 (1980a).

Infeld, E. Self-focusing of deep water waves. *Zh. Eksp. Teor. Fiz. Lett.* **32** 97–100 (1980b); *Sov. Phys. JETP.* **32** 87–9 (1980b).

Infeld, E. Invariants of the two-dimensional Korteweg de Vries and Kadomtsev–Petviashvili equations. *Phys. Lett.* **86A** 205–7 (1981a).

Infeld, E. Three-dimensional stability of Korteweg de Vries waves and solitons, III.

Lagrangian methods, KdV with positive dispersion. *Acta Phys. Polon.* **A60** 623–43 (1981b).
Infeld, E. Quantitative theory of the Fermi–Ulam–Pasta recurrence in the nonlinear Schrödinger equation. *Phys. Rev. Lett.* **47** 717–18 (1981c).
Infeld, E. Fermi–Ulam–Pasta recurrence of nonlinear Langmuir wave envelope. *International Conference on Plasma Physics, Göteborg 1982*, 228 (1982).
Infeld, E. Fale nieliniowe (Nonlinear waves). *Postepy Fizyki* **34** 215–38 (1983).
Infeld, E. Self-focusing of nonlinear ion acoustic waves and solitons in magnetized plasmas. *J. Plasma Phys.* **33** 171–82 (1985). For sequel see Infeld and Frycz (1987).
Infeld, E. and Frycz, P. Infinite sequences of local conservation laws for the Kadomtsev–Petviashvili equation. *Acta. Phys. Polon.* **B14** 129–32 (1983).
Infeld, E. and Frycz, P. Self-focusing of nonlinear ion acoustic waves and solitons in magnetized plasmas, II. Numerical results. *J. Plasma Phys.* **37** 97–106 (1987).
Infeld, E., Frycz, P. and Czerwinska-Lenkowska, T. Dynamics of ion acoustic waves in an external magnetic field. *Lett. Nuovo Cim.* **44** 537–43 (1985).
Infeld, E. and Rowlands, G. On the stability of nonlinear cold plasma waves, II. *J. Plasma Phys.* **10** 293–300 (1973).
Infeld, E. and Rowlands, G. On the stability of nonlinear waves coexisting with plasma beams. *J. Plasma Phys.* **13** 171–87 (1975).
Infeld, E. and Rowlands, G. A sufficient condition for instability of BGK type waves in both unmagnetized and magnetized plasmas. *J. Plasma Phys.* **17** 57–68 (1977).
Infeld, E. and Rowlands, G. Stability of nonlinear ion sound waves and solitons in plasmas. *Proc. R. Soc. London* **A366** 537–54 (1979a).
Infeld, E. and Rowlands, G. Three-dimensional stability of Korteweg de Vries waves and solitons, II. *Acta Phys. Polon.* **A56** 329–32 (1979b). Part I, see Infeld, Rowlands and Hen (1978).
Infeld, E. and Rowlands, G. On the stability of electron plasma waves. *J. Phys.* **A12** 2255–62 (1979c).
Infeld, E. and Rowlands, G. Three-dimensional stability of solutions of the nonlinear Schrödinger equation. *Z. Physik.* **B37** 277–80 (1980).
Infeld, E. and Rowlands, G. Stability of weakly nonlinear ion sound waves with finite ion temperature. *J. Plasma Phys.* **25** 81–7 (1981).
Infeld, E. and Rowlands, G. Nonlinear oscillations in a warm plasma. *International Conference on Plasma Physics, Kiev 1987*, **1**, 201–4 (1987a).
Infeld, E. and Rowlands, G. Nonlinear oscillations in a warm plasma. *Phys. Rev. Lett.* **58** 2063–6 (1987b).
Infeld, E. and Rowlands, G. A model for the nonlinear stage of the Rayleigh–Taylor instability. *Phys. Rev. Lett.* **60** 2273–5 (1988).
Infeld, E., Rowlands, G. and Hen, M. Three-dimensional stability of Korteweg de Vries waves and solitons. *Acta Phys. Polon.* **A54** 131–9 (1978). Part II **A56** 329–33, see Infeld and Rowlands (1979b); Part III **A60** 623–43, see Infeld (1981b).
Infeld, E. and Skorupski, A. Convective and absolute two stream instabilities. *Nucl. Fus.* **9** 25–6 (1969).
Infeld, E. and Ziemkiewicz, J. Stability of complex solutions of the nonlinear Schrödinger equation. *Acta Phys. Polon.* **A59** 255–75 (1981a).
Infeld, E. and Ziemkiewicz, J. Two-dimensional stability of large amplitude Langmuir waves and solitons. *Phys. Lett.* **83A** 331–2 (1981b).

Infeld, E., Ziemkiewicz, J. and Rowlands, G. On the two surface dimensional dynamics of periodic and solitary wavetrains on a water surface over arbitrary depth. *Phys. Fluids.* **30** 2330–8 (1987).

Isichenko, M.B. and Marnachev, A.M. Generalized two-dimensional electron vortices and two-dimensional stability of electron flows. *International Conference on Plasma Physics, Kiev 1987*, **2**, 48–50 (1987).

Jackson, E.A. Nonlinear oscillations in a cold plasma. *Phys. Fluids* **3** 831–3 (1960).

Jackson, E.A. Perturbations of nonlinear travelling waves in a cold plasma. *Phys. Fluids* **6** 753–4 (1963).

Janssen, P.A.E.M. Modulation instability and the Fermi–Pasta–Ulam recurrence. *Phys. Fluids* **24** 23–6 (1981).

Jeffrey, A. and Kakutani, T. Weak nonlinear dispersive waves, a discussion centered around the Korteweg–de Vries equation. *SIAM Rev.* **14** 582–643 (1972).

Johnson, R.S. On the oblique interaction of a large and a small solitary wave. *J. Fluid Mech.* **120** 49–70 (1982).

Johnson, R.S. The Korteweg–de Vries equation and related problems in water wave theory. *Nonlinear waves.* Debnath, L. (ed.). Cambridge University Press 25–43 (1983).

Kadomtsev, B.B. *Phénomens collectifs dans les plasmas.* Mir, Moscow (1979).

Kadomtsev, B.B. and Petviashvili, V.I. On the stability of solitary waves in weakly dispersive media. *Dokl. Akad. Nauk. SSSR* **192** 753–6 (1970); *Sov. Phys. Dok.* **15** 539–41 (1970).

Kako, F. and Yajima, N. Interaction of ion acoustic solitons in two-dimensional space. *J. Phys. Soc. Jpn.* **49** 2063–71 (1980).

Kako, F. and Yajima, N. Interaction of ion acoustic solitons in multidimensional space, II. *J. Phys. Soc. Jpn.* **51** 311–22 (1982).

Kako, M. Nonlinear wave modulation in cold magnetized plasmas. *J. Phys. Soc. Jpn.* **33** 1678–87 (1972).

Kako, M. Nonlinear modulation of plasma waves. *Suppl. Progr. Theor. Phys.* **55** 120–37 (1974).

Kako, M. and Hasegawa, A. Stability of oblique modulation of an ion acoustic wave. *Phys. Fluids* **19** 1967–9 (1976).

Kako, M. and Rowlands, G. Two-dimensional stability of ion acoustic solitons. *Plasma Phys.* **18** 165–70 (1976).

Kakutani, T. and Kawahara, T. Weak ion acoustic shock waves. *J. Phys. Soc. Jpn.* **29** 1068–73 (1970).

Kakutani, T. and Yamasaki, N. Solitary waves on a two layer fluid. *J. Phys. Soc. Jpn.* **45** 674–9 (1978).

Kaplan, J.L. and Yorke, J.A. The onset of chaos in a fluid flow model of Lorenz, in bifurcation theory and applications in scientific disciplines, Gurel, O. and Rössler, O.E. (eds). *Ann. NY Acad. Sci.* **316** 400–7 (1979).

Karpman, V.I. Soliton evolution in the presence of a perturbation. *Phys. Scrip.* **20** 462–78 (1979).

Katyshev, Y.V. and Makhankov, V.G. Stability of some one-field solitons. *Phys. Lett.* **57A** 10–12 (1976).

Kaup, D.J. The Eastabrook–Wahlquist method with examples of application. *Physica* **1D** 391–411 (1980).

Kaup, D.J. and Newell, A.C. Solitons as particles and oscillators. *Proc. R. Soc. London* **A361** 413–46 (1978).

Kawahara, T. Nonlinear self-modulation of capillary-gravity waves on a liquid layer. *J. Phys. Soc. Jpn.* **38** 265–70 (1975a).

Kawahara, T. Derivative-expansion method for nonlinear waves on a liquid layer of slowly varying depth. *J. Phys. Soc. Jpn.* **38** 1200–6 (1975b).

Khazei, M., Bolton, J.M. and Lonngren, K.E. Resonant interaction of ion acoustic solitons in three dimensions. *International Conference on Plasma Physics, Göteborg 1982*, 216 (1982).

Kim, H.C., Stenzel, R.L. and Wong, A.Y. Developments of 'cavitons' and trapping of rf field. *Phys. Rev. Lett.* **33** 886–9 (1974).

Kittel, C. *Introduction to solid state physics.* Wiley, Chichester, 1976.

Ko, K. and Kuehl, H.H. Korteweg–de Vries soliton in a slowly varying medium. *Phys. Rev. Lett.* **40** 233–6 (1978).

Ko, K. and Kuehl, H.H. Cylindrical and spherical KdV solitary waves. *Phys. Fluids* **22** 1343–8 (1979).

Konno, K., Mitsuhashi, T. and Ichikawa, Y.H. Dynamical processes of the dressed ion acoustic solitons. *J. Phys. Soc. Jpn.* **43** 669–74 (1977).

Konyukov, M.V. Nonlinear Langmuir electron oscillations in a plasma. *Zh. Eksp. Teor. Fiz.* **37** 799–801 (1959); *Sov. Phys. JETP* **10** 570–1 (1960).

Korteweg, D.J. and de Vries, G. On the change of form of long waves advancing in a rectangular canal and on a new type of long stationary waves. *Philos. Mag.* (5) **39** 422–43 (1895).

Krall, N.A. and Trivelpiece, A.W. *Principles of plasma physics.* McGraw Hill, New York (1973).

Kruer, W.L. and Dawson, J.M. Sideband instability. *Phys. Fluids* **13** 2747–51 (1970).

Krumhansl, J.A. and Schreiffner, J.R. Dynamics and statistical mechanics of one dimensional model Hamiltonian for structural phase transitions. *Phys. Rev.* **B11** 3535–45 (1975).

Krylov, N. and Bogoliubov, N.N. *Introduction to nonlinear mechanics.* Princeton University Press, Princeton (1947).

Kumei, S. Group theoretic aspects of conservation laws of nonlinear dispersive waves: KdV type equations and nonlinear Schrödinger equations. *J. Math. Phys.* **18** 256–64 (1977).

Kunin, I.A. Teoria uprugykh sryed s mikrostrukturoy. *Moskva Izdat.* (1975).

Kuramoto, Y. (ed.). *Chaos and statistical methods.* Springer, Berlin and New York (1984).

Kursunoglu, B., Perlmutter, A. and Scott, L.F. *The significance of nonlinearity in the natural sciences.* Plenum, New York and London (1977).

Kusmartsev, F.V. Catastrophe of the Langmuir soliton. *International Conference on Plasma Physics, Kiev 1987*, **2** 59–62 (1987).

Kuznetsov, E.A., Spector, M.D. and Fal'kovich, G.E. On the stability of nonlinear waves in integrable models. *Physica* **10D** 379–86 (1984).

Kuznetsov, E.A. and Turitsyn, S.K. Two-dimensional and three-dimensional solitons in weakly dispersive media. *International Conference on Plasma Physics, Göteborg 1982*, 235 (1982).

El-Labany, S.K. and Rowlands, G. The nonlinear two stream problem – a new approach. *Plasma Phys.* **28** 1549–58 (1986).

Ladyzhenskaya, O.A. *Mathematical problems of dynamics of viscous incompressible liquid.* Fizmatgiz., Moscow (in Russian) (1961).

Laedke, E.W. and Spatschek, K.H. Limitations of two-dimensional model equations for ion acoustic waves. *Phys. Rev. Lett.* **47** 719–22 (1981).
Laedke, E.W. and Spatschek, K.H. Nonlinear ion acoustic waves in weak magnetic fields. *Phys. Fluids* **25** 985–9 (1982a).
Laedke, E.W. and Spatschek, K.H. Growth rates of bending KdV solitons. *J. Plasma Phys.* **28** 469–84 (1982b).
Laedke, E.W. and Spatschek, K.H. Stable three-dimensional envelope solitons. *Phys. Rev. Lett.* **52** 279–82 (1984).
Laird, M.J. and Knox, F.B. Exact solution for charged particle trajectories in an electromagnetic field. *Phys. Fluids* **8** 755–6 (1965).
Lakshmanan, M. and Kaliappan, P. Lie transformations, nonlinear evolution equations, and Painlevé forms. *J. Math. Phys.* **24** 795–806 (1983).
Lamb, C.L. *Elements of soliton theory.* John Wiley, New York (1980).
Lamb, H. *Hydrodynamics.* Cambridge University Press (1932).
Landau, L.D. and Lifshitz, E.M. *Fluid mechanics.* Pergamon, London (1959).
Langer, J.S. Instabilities and pattern formation in crystal growth. *Rev. Mod. Phys.* **52** 1–28 (1980).
Lavrientev, M.A. A contribution to the theory of long waves. *CR (Dokl.) Acad. Sci. SSSR* **41** 275–7 (1943); reprodiced in *Am. Math. Soc. Transl.* **102** 51–3 (1954).
Lavrientev, M.A. *Variational methods.* Noordhoft (1960).
Leslie, D.C. *Developments in the theory of turbulence.* Clarendon, Oxford (1973).
Lewis, H.R. and Symon, K.R. Exact time dependent solutions of the Vlasov–Poisson equations. *Phys. Fluids* **27** 192–6 (1984).
Li, T.Y. and Yorke, J.A. Period three implies chaos. *Am. Math. Month.* **82** 985–92 (1975).
Libchaber, A. and Maurer, J. Une experience de Rayleigh–Bénard de geometrie reduite; multiplication, accrochage, et demultiplication de frequences. *J. de Phys.* **41** *Colloq.* C3 51 (1980).
Libchaber, A. and Maurer, J. A Rayleigh–Bénard experiment: Helium in a small box. *Nonlinear phenomena at phase transitions and instabilities.* Riste, T. (ed.), Plenum, New York and London (1982).
Lichtenberg, A. and Lieberman, M.A. *Regular and stochastic motion.* Springer, Berlin and New York (1983).
Lighthill, M.J. Group velocity. *J. Inst. Maths. Applics.* **1** 1–28 (1965).
Lighthill, M.J. A discussion on nonlinear theory of wave propagation in dispersive systems. *Proc. R. Soc. London* **A299** 1–145 (1967). (Specific reference is also made to Lighthill's paper on page 28.)
Lighthill, M.J. *Waves in fluids.* Cambridge University Press (1978).
Lighthill, M.J. and Whitham, G.B. On kinematic waves. *Proc. R. Soc. London* **A229** 281–345 (1955).
Liouville, J.J. *Mathematiques pures et appliquées, Paris* **18** (1) 71–2 (1853).
Liu, A.K. Interaction of solitary waves in stratified fluids. *Advances in nonlinear waves.* Debnath, L. (ed.), **1**. Pitman, Boston and London 108–117 (1984).
Liu, C.S. Instability of a large-amplitude plasma wave due to inverted trapped particle population. *J. Plasma Phys.* **8** 169–74 (1972).
Longuet-Higgins, M.S. The instabilities of gravity waves of finite amplitude, *Proc. R. Soc. London*, I. Superharmonics, **A360** 471–88 (1978a); II. Subharmonics, **A360** 489–505 (1978b).
Longuet-Higgins, M.S. A technique for time-dependent, free-surface flows. *Proc. R. Soc. London* **A371** 441–51 (1980).

Longuet-Higgins, M.S. On the overturning of gravity waves. *Proc. R. Soc. London* **A376** 377–400 (1981).

Longuet-Higgins, M.S. Towards the analytic description of overturning waves. *Non-linear waves.* Debnath, L. (ed.), Cambridge University Press 1–24 (1983).

Longuet-Higgins, M.S. and Cockelet, E.D. The deformation of steep surface waves on water. *Proc. R. Soc. London*, I. A numerical method of computation, **A350** 1–26 (1976); II. Growth of normal mode instabilities, **A364** 1–28 (1978).

Longuet-Higgins, M.S. and Fenton, J.D. On the mass, momentum, energy and circulation of a solitary wave II, *Proc. R. Soc. London* **A340** 471–93 (1974).

Lonngren, K.E. Soliton experiments in plasmas. *Plasma Phys.* **25** 943–82 (1983).

Lonngren, K.E., Gabl, E.F., Bulson, J.M. and Khazei, M. On the resonant interaction of spherical ion acoustic solitons. *Physica* **9D** 372–8 (1983).

Lonngren, K.E., Khazei, M. and Bulson, J.M. Resonant interaction of ion acoustic solitons in three dimensions. *Phys. Fluids* **25** 759–64 (1982).

Lonngren, K. and Scott, A. *Solitons in action.* Academic, New York (1978).

Lorenz, E.N. Deterministic nonperiodic flow. *J. Atmos. Sci.* **20** 130–41 (1963).

Lugovtzov, A.A. and Lugovtzov, B.A. *Dinamika sploshnoy sryedi* **1**. Nauka, Novosibirsk (1969).

Luke, J.C. A perturbation method for nonlinear dispersive wave problems. *Proc. R. Soc. London* **A292** 403–12 (1966).

Luke, J.C. A variational principle for a fluid with a free surface. *J. Fluid Mech.* **27** 395–7 (1967).

Lutomirski, R.F. and Sudan, R.N. Exact nonlinear electromagnetic whistler modes. *Phys. Rev.* **147** 156–65 (1966).

Ma, Y.C. A note on the instability of capillary waves. *Advances in nonlinear waves*, Debnath, L. (ed.), **1**. Pitman, Boston and London 64–74 (1984).

McCreadie, G.A. and Rowlands, G. An analytic approximation to the Lyapunov number for 1D maps. *Phys. Lett.* **A91** 146–8 (1981).

McEwan, A.D., Mander, D.W. and Smith, R.N. Forced resonant second-order interaction between damped internal waves. *J. Fluid Mech.* **55** 589–608 (1972).

McLaughlin, J.B. and Martin, P.C. Transition to turbulence of statistically stressed fluid. *Phys. Rev. Lett.* **33** 1189–92 (1974).

McLaughlin, J.B. and Martin, P.C. Transition to turbulence in a statistically stressed fluid system. *Phys. Rev.* **A12**, 186–203 (1975).

McLean, J.W. Instabilities of finite amplitude water waves. *J. Fluid Mech.* **114** 315–30 (1982a).

McLean, J.W. Instabilities of finite-amplitude gravity waves on water of finite depth. *J. Fluid Mech.* **114** 331–41 (1982b).

McNamara, B. and Rowlands, G. Plasma stability and the Liapunov method. *Proc. Inst. Mech. Engrs.* **178** pt 3, 47–50 (1964).

Makhankov, V.G. Dynamics of classical solitons (in non-integrable systems). *Phys. Rep.* **35** 1–128 (1978).

Makhankov, V. Computer experiments in soliton theory. *Comp. Phys. Comm.* **21** 1–49 (1980).

Makino, M., Kamimura, T. and Taniuti, T. Two-dimensional behaviour of solitons in a low beta plasma with convective motion. *J. Phys. Soc. Jpn.* **50** 954–61 (1981a).

Makino, M., Kamimura, T. and Taniuti, T. Dynamics of two-dimensional solitary vortices in a low beta plasma with convective motion. *J. Phys. Soc. Jpn.* **50** 980–9 (1981b).

Malmberg, J.H. and Wharton, C.B. Dispersion of electron plasma waves. *Phys. Rev. Lett.* **17** 175–8 (1966).
Malmberg, J.H., Wharton, C.B., Gould, R.W. and O'Neil, T. Plasma wave echo experiment. *Phys. Rev. Lett.* **20** 95–7 (1968).
Manakov, S.V., Zakharov, V.E., Bordag, L.A. and Matveev, V.B. Two-dimensional solitons of the Kadomtsev–Petviashvili equation and their interaction. *Phys. Lett.* **63A** 205–6 (1977).
Mandelbrot, B.B. *The fractal geometry of Nature.* Freeman, San Francisco (1982).
Martin, D.U. and Yuen, H.C. Quasi-recurring energy leakage in the two-space dimensional nonlinear Schrödinger equation. *Phys. Fluids* **23** 881–3 (1980).
Martin, D.U., Yuen, H.C. and Saffman, P.G. Stability of plane wave solutions of the two-space-dimensional nonlinear Schrödinger equation. *Wave Motion* **2** 215–29 (1980).
Maurer, J. and Libchaber, A. Effect of the Prandtl number on the onset of turbulence in liquid ^{4}He. *J. Phys. Lett.* **41** L515–18 (1980).
Maxon, S. Cylindrical solitons in a warm, multi-ion plasma. *Phys. Fluids* **19** 266–71 (1976).
Maxon, S. and Viecelli, J. Cylindrical solitons. *Phys. Fluids* **17** 1614–16 (1974a).
Maxon, S. and Viecelli, J. Spherical solitons. *Phys. Rev. Lett.* **32** 4–6 (1974b).
May, R.M. *Theoretical ecology.* Blackwell, Oxford (1976a).
May, R.M. Simple mathematical models with very complicated dynamics. *Nature* **261** 159–67 (1976b).
May, R.M. and Oster, G.F. Bifurcations and Dynamic Complexity in Simple Ecological Models. *Am. Natur.* **110** 573–83 (1976).
Mayer-Kress, G. and Haken, H. The influence of noise on the logistic model. *J. Stat. Phys.* **26** 149–71 (1981).
Michell, A.G.M. The highest waves in water. *Phil. Mag.* (5) **36** 430–7 (1893).
Mikhailovskaya, L.A. and Erokhin, N.S. On stability of large gradient vortices in plasma. *International Conference on Plasma Physics, Kiev 1987*, **2** 69–72 (1987).
Miles, J.W. Korteweg–de Vries equation modified by viscosity. *Phys. Fluids* **19** 1063 (1976).
Miles, J.W. On internal solitary waves. *Tellus* **31** 456–62 (1979).
Miles, J.W. Solitary waves. *Ann. Rev. Fluid Mech.* **12** 11–43 (1980).
Miles, J.W. The Korteweg–de Vries equation, a historical essay. *J. Fluid Mech.* **106** 131–47 (1981).
Mima, K. and Hasegawa, A. Nonlinear instability of electromagnetic drift waves. *Phys. Fluids* **21** 81–86 (1978).
Miura, R.M. KdV equation and generalizations I. A remarkable explicit nonlinear transformation. *J. Math. Phys.* **9** 1202–4 (1968).
Miura, R.M. The Korteweg de Vries equation: a survey of results. *SIAM Review* **18** 412–59 (1976).
Miura, R.M., Gardner, C.S. and Kruskal, M.D. KdV equations and generalizations II. Existence of conservation laws and constants of motion. *J. Math. Phys.* **9** 1204–9 (1968).
Montgomery, D. and Joyce, G. Shock-like solutions of the electrostatic Vlasov equation. *J. Plasma Phys.* **3** 1–11 (1969).
Montgomery, D.C. and Tidman, D.A. *Plasma kinetic theory.* McGraw Hill, New York (1964).

Morse, P.M. and Feshbach, H. *Methods of theoretical physics*, **2**. McGraw Hill, New York (1953).
Morse, R.L. and Nielson, C.W. One, two, and three-dimensional numerical simulation of two beam plasmas. *Phys. Rev. Lett.* **23** 1087–90 (1969a).
Morse, R.L. and Nielson, C.W. Numerical simulation of warm two-beam plasma. *Phys. Fluids* **12** 2418–25 (1969b).
Murawski, K. and Infeld, E. On the stability of the MKdV–KdV wave. *Australian J. Phys.* **41** 1–10 (1988).
Nagasawa, T. and Nishida, Y. Virtual state of interacting ion acoustic soliton. *International Conference on Plasma Physics, Göteborg 1982*, 215 (1982).
Nagasawa, T. and Nishida, Y. Virtual states in strong interactions of plane ion-acoustic solitons. *Phys. Rev.* **A28** 3043–50 (1983).
Nagasawa, T., Shimizu, M. and Nishida, Y. Strong interaction of plane ion acoustic solitons. *Phys. Lett.* **87A** 37–40 (1981).
Nagasawa, T., Tsuruta, H. and Nishida, Y. Excitation of converging ion acoustic solitons. *Phys. Lett.* **79A** 71–3 (1980).
Nakamura, A. Bäcklund transformation of the cylindrical KdV equation. *J. Phys. Soc. Jpn.* **49** 2380–6 (1980).
Nakamura, A. The Miura transform and the existence of an infinite number of conservation laws of the cylindrical KdV equation. *Phys. Lett.* **82A** 111–12 (1981).
Nakamura, A. and Chen, H.H. Soliton solutions of the cylindrical KdV equation. *J. Phys. Soc. Jpn.* **50** 711–18 (1981).
Nakamura, Y. and Ogino, T. Numerical and laboratory experiments on spherical ion acoustic solitons. *Plasma Phys.* **24** 1295–1315 (1982).
Nakamura, Y., Tsukabayashi, I. Kako, F. and Lonngren, K.E. Oblique collision of cylindrical outgoing ion acoustic solitons. *International Conference on Plasma Physics, Göteborg*, 30 (1982).
Navet, M. and Bertrand, P. Multiple 'water bag' model and Landau damping. *Phys. Lett.* **34A** 117–18 (1971).
Nayfeh, A.H. *Perturbation methods.* John Wiley, New York (1973).
Newell, A.C. Solitons in mathematics and physics. *CBMS–NSF Regional Conference in Applied Mathematics, Philadelphia 1985* (1985).
Newell, A.C. and Redekopp, L.G. Breakdown of Zakharov–Shabat theory and soliton creation. *Phys. Rev. Lett.* **38** 377–80 (1977).
Nezlin, M.V., Snezhkin, E.N. and Trubnikov, A.S. Common mechanism of vortex structure drive in plasma, planetary atmospheres and in galaxies. *International Conference on Plasma Physics, Kiev 1987* 280–3 (1987).
Nimmo, J.J.C. and Freeman, N.C. The use of Bäcklund transformations in obtaining N soliton solutions in Wronskian form. *J. Phys.* **A17** 1415–24 (1984).
Nishida, Y. and Nagasawa, T. Oblique collision of plane ion acoustic solitons. *Phys. Rev. Lett.* **45** 1626–9 (1980).
Nishida, Y., Nagasawa, T. and Kawamata, S. Experimental verification of the characteristics of ion acoustic cylindrical solitons. *Phys. Lett.* **69A** 196–8 (1978).
Nishida, Y., Nagasawa, T. and Kawamata, S. Observation of scattering of ion acoustic cylindrical solitons. *Phys. Rev. Lett.* **42** 379–83 (1979).
Novikov, S., Manakov, S.V., Pitayevski, L.P. and Zakharov, V.E. *Theory of solitons, the inverse scattering method.* Consultants Bureau, Plenum, New York and London (1984).

Nozaki, K. Vortex solitons of drift waves and anomalous diffusion. *Phys. Rev. Lett.* **46** 184–7 (1981).

Nycander, J. Propagation of discontinuities in the Hasegawa-Mima equation. *International Conference on Plasma Physics, Kiev 1987*, **2** 84–7 (1987).

Oevel Wand Steeb, W.H. Painlevé analysis for a time-dependent Kadomstev–Petviashvili equation. *Phys. Lett.* **103A** 239–42 (1984).

Oficjalski, J. and Bialynicki-Birula, I. Collisions of gaussons. *Acta Phys. Polon.* **B9** 759–76 (1978).

Ogino, T. and Takeda, S. Computer simulation and analysis for the spherical and cylindrical ion acoustic solitons. *J. Phys. Soc. Jpn.* **41** 257–64 (1976).

Okhuma, K. and Wadati, M. The Kadomtsev–Petviashvili equation: the trace method and the soliton resonances. *J. Phys. Soc. Jpn.* **52** 749–60 (1983).

Ott, E. Strange attractors and chaotic motions of dynamical systems. *Rev. Mod. Phys.* **53** 655–71 (1981).

Panat, P.V. Improved KdV type equation for cylindrical and spherical solitary waves. *Phys. Fluids* **19** 915–16 (1976).

Pawlik, M. and Rowlands, G. The propagation of solitary waves in piezoelectric semiconductors. *J. Phys.* **C8** 1189–1204 (1975).

Payne, G.L., Nicholson, D.R. and Downie, R.M. Numerical solution of the Zakharov equations. *J. Comp. Phys.* **50** 482–9 (1983).

Penrose, O. Electrostatic instabilities of a uniform non-Maxwellian plasma. *Phys. Fluids* **3** 258–65 (1960).

Pereira, N.R. Collisions between Langmuir solitons. *Phys. Fluids* **20** 750–5 (1977).

Perring, J.K. and Skyrme, T.H.R. A model unified field equation. *Nucl. Phys.* **31** 550–5 (1962).

Petviashvili, V.I. Plasma oscillations in the presence of a periodic plasma wave. *Zh. Eksp. Teor. Fiz.* **53** 917–25 (1967); *Sov. Phys. JETP* **26** 555–9 (1968).

Petviashvili, V.I. *Solitary waves and vortexes, Plasma physics.* Kadomstev, B. (ed.), Mir, Moscow 123–43 (1981).

Pitayevski, L.P. Vortex lines in an imperfect Bose gas. *Zh. Eksp. Teor. Fiz.* **40** 646–51 (1961); *Sov. Phys. JETP* **13** 451–4 (1961).

Plebanski, J. *Lecture notes on nonlinear electrodynamics.* Nordita, Copenhagen (1970).

Polovin, R.V. Criteria for instability and gain. *Zh. Tekh. Fiz.* **31** 1220–30 (1961); *Sov. Phys. Tech. Phys.* **6** 889–95 (1962).

Pomeau, Y. and Manneville, P. Intermittent transition to turbulence in dissipative dynamical systems. *Comm. Math. Phys.* **74** 189–97 (1980).

Pomeau, Y., Roux, J.C., Rossi, A., Bachelart, S. and Vidal, C. Intermittent behaviour in the Belousov–Zhabotinski reaction. *J. Phys. Lett.* **42** L271–3 (1981).

Rasmussen, J.J. Effects of trapped particles on strongly nonlinear electron plasma waves. *Phys. Scripta* **T2** 29–40 (1982).

Rayleigh, Lord. On waves. *Phil. Mag.* (5) **1** 257–79 (1876).

Rebbi, C. and Soliani, G. *Solitons and particles.* World Scientific, Singapore (1984).

Rice, M.R. *Charge density wave system: the particle model, Solitons and condensed matter physics*, Bishop, B. and Schneider, T. (eds), **8** in Springer series in solid state physics. Springer, Berlin and New York 246–53 (1981).

Richards, P.I. Shock waves on the highway. *Oper. Res.* **4** 42–51 (1956).

Roberts, D. The general Lie group and similarity solutions for the one-dimensional Vlasov–Maxwell equations. *J. Plasma Phys.* **33** 219–36 (1985).

Rognlien, T.D. and Self, S.A. Interpretation of dispersion relations for bounded systems. *J. Plasma Phys.* **7** 13–48 (1972).

Romeiras, F. Integrability of double three-wave interaction. *International Conference on Plasma Physics, Göteborg 1982*, 186 (1982).

Rosales, R. The similarity solution of the Korteweg–de Vries equation and the related Painlevé transcendent. *Proc. R. Soc. London* **A361** 265–75 (1978).

Rose, D.J. and Clark, M. *Plasmas and Controlled Fusion.* MIT and Wiley, Boston, New York and London (1961).

Rosenbluth, M.N. and Post, R.F. High-frequency electrostatic plasma instability inherent in 'loss conc' particle distributions. *Phys. Fluids* **8** 547–50 (1965).

Rosenbluth, M.N. and Longmire, C.L. Stability of plasmas confined by magnetic fields. *Ann. Phys.* **1** 120–40 (1957).

Rowlands, G. Landau damping in the water-bag model. *Phys. Lett.* **30A** 408–9 (1969a).

Rowlands, G. Stability of nonlinear plasma waves. *J. Plasma Phys.* **3** 567–76 (1969b).

Rowlands, G. The long time behaviour of a pulse in an unstable plasma. *Plasma Phys.* **12** 89–93 (1970).

Rowlands, G. Vibrations of a quantized vortex in a weakly interacting Bose fluid. *J. Phys.* **A6** 322–8 (1973).

Rowlands, G. On the stability of solutions of the nonlinear Schrödinger equation. *J. Inst. Math. Appl.* **13** 367–77 (1974).

Rowlands, G. Time recurrent behaviour in the nonlinear Schrödinger equation. *J. Phys.* **A13** 2395–2400 (1980).

Rowlands, G. An approximate analytic solution of the Lorenz equations. *J. Phys.* **A16** 585–90 (1983).

Ruelle, D. and Takens, F. On the nature of turbulence. *Comm. Math. Phys.* **20** 167–92 (1971).

Russell, D.A., Hanson, J.D. and Ott, E. Dimensions of strange attractors. *Phys. Rev. Lett.* **45** 1175–8 (1980).

Russell, J. Scott. Report on waves. *14th meeting of the British Association Report, York*, 311–90 (1844).

Russell, J. Scott. *The wave of translation in the oceans of water, air and ether.* Trubner, London (1895).

Saffman, P.G. and Yuen, H.C. Stability of a plane soliton to infinitesimal two-dimensional perturbations. *Phys. Fluids* **21** 1450–1 (1978).

Saffman, P.G. and Yuen, H.C. Three-dimensional waves on deep water. *Advances in nonlinear waves.* Debnath, L. (ed.), **2**, Pitman, Boston and London 1–30 (1985).

Saltzman, B. Finite amplitude free convection as an initial value problem, I. *J. Atmos. Sci.* **19** 329–41 (1962).

Sander, L.M. Fractal growth. *Scientific American* **256** Jan. 82–88 (1987).

Santini, P.O. The evolution of two-dimensional packets on water waves over an uneven bottom. *Lett. Nuovo Cim.* **30** 236–40 (1981).

Satsuma, J. *N*-soliton solution of the two-dimensional Korteweg–de Vries equation. *J. Phys. Soc. Jpn.* **40** 286–90 (1976).

Sawada, K. and Kotera, T. A method for finding *N*-soliton solutions of the KdV equation and KdV-like equations. *Prog. Theor. Phys.* **51** 1355–67 (1974).

Schamel, H. Nonlinear electrostatic plasma waves. *J. Plasma Phys.* **7** 1–12 (1972).

Schmitt, J.P.M. Dispersion and cyclotron damping of pure ion Bernstein waves. *Phys. Rev. Lett.* **31** 982–6 (1973).

Schuster, H.G. *Deterministic chaos, an introduction.* Physik Verlag, Weinheim (1984).

Schwartz, L.W. Computer extension and analytic continuation of Stokes expansion for gravity waves. *J. Fluid Mech.* **62** 553–78 (1974).

Schwarzmeier, J.L., Lewis, H.R., Abraham-Shrauner, B. and Symon, K.R. Stability of Bernstein–Greene–Kruskal equilibria. *Phys. Fluids* **22** 1747–60 (1979).

Scott, A.C. *Active and nonlinear wave propagation in electronics.* Wiley-Interscience, New York (1970).

Scott, A.C., Chu, F.Y.F. and McLaughlin, D.W. The soliton: a new concept in applied sciences. *Proc. IEEE* **61** 1443–83 (1973).

Sekerka, R.F. A time dependent theory of stability of a planar interface during dilute binary alloy solidification. *Crystal Growth.* Peiser, H.S. (ed.), Pergamon, Oxford (1967).

Sen, A., Karney, C.F.F., Johnston, G.L. and Bers, A. Three-dimensional effects in the nonlinear propagation of lower-hybrid waves. *Nucl. Fus.* **18** 171–9 (1978).

Shail, R. and Pakham, B.A. On some properties of a nonlinear mapping. *Int. J. Math. Educ. Sci. Technol.* **14** 277–86 (1983).

Shaw, R.S. Strange attractors, chaotic behaviour and information flow. *Z. Natur.* **36a** 80–112 (1981).

Shimizu, K. and Ichikawa, Y.H. Automodulation of ion oscillation modes in plasmas. *J. Phys. Soc. Jpn.* **33** 789–92 (1972).

Shockley, R.C. Self trapping states in a saturable Klein–Gordon equation. *Phys. Rev.* **A35** 4729–37 (1987).

Skorupski, A.A. Linear waves and instabilities in fluid plasmas. *Modern plasma physics, IAEA, Vienna 1981*, 187–248 (1981).

Smith, A. The steady-state Vlasov equation for a monotonic electric potential. *J. Plasma Phys.* **4** 511–22 (1970a).

Smith, A. An exact electrostatic shock solution for a collisionless plasma. *J. Plasma Phys.* **4** 549–61 (1970b).

Sparrow, C. *The Lorenz equations, bifurcations, chaos, and strange attractors.* Springer, Berlin and New York (1982).

Spatschek, K.H. and Laedke, E.W. Numerical results for stability region and maximum growth rates of nonlinear waves. *International Conference on Plasma Physics, Göteborg 1982*, 174 (1982).

Stewartson, K. and Stuart, J.T. A nonlinear instability theory for a wave system in plane Poisseulle flow. *J. Fluid Mech.* **48** 529–45 (1971).

Stix, T.H. *The theory of plasma waves.* McGraw Hill, New York (1962).

Stokes, G.G. On the theory of oscillatory waves. *Camb. Trans.* **8** 441–73 (1847).

Sturrock, P.A. Kinematics of growing waves. *Phys. Rev.* **112** 1488–1503 (1958).

Sturrock, P.A. Excitation of plasma oscillations. *Phys. Rev.* **117** 1426–9 (1960).

Su, C.H. An evolution equation for a stratified flow having two characteristics coalesced. *Advances in nonlinear waves.* Debnath, L. (ed.), **1** Pitman, Boston and London 90–107 (1984).

Su, C.H. and Gardner, C.S. Korteweg–de Vries equation and generalizations, III. Derivation of the KdV equation and Burgers equation. *J. Math. Phys.* **10** 536–9 (1969).

Su, M.Y., Bergin, M., Marler, P. and Myrick, R. Experiments on nonlinear instabilities and evolution of steep gravity wave trains. *J. Fluid Mech.* **124** 45–72 (1982).

Su, M.Y. and Green, A.W. Coupled two and three-dimensional instabilities of surface gravity waves. *Phys. Fluids* **27** 2595–7 (1984).

Sudan, R.N. Classification of instabilities from their dispersion relations. *Phys. Fluids* **8** 1899–1904 (1965).

Sutton, G.W. and Sherman, A. *Engineering magnetohydrodynamics.* McGraw Hill, New York (1965).

Swinney, H.L., Festermacher, P.R. and Gallup, J.P. *Synergetics, a workshop.* Haken, H. (ed.), Springer, Berlin and New York (1977).

Sym, A. *Geometry of solitons,* **1**, Reidel, Dorderecht (in press).

Tagare, S.G. and Shukla, P.K. Nonlinear cylindrical ion-acoustic waves in a warm collisional plasma. *Phys. Fluids* **20** 868–9 (1977).

Tajima, T., Goldman, M.V., Leboeuf, J.N. and Dawson, J.M. Breakup and reconstruction of Langmuir wave packets. *Phys. Fluids* **24** 182–3 (1981).

Tajiri, M. and Kawamoto, S. Reduction of KdV and cylindrical KdV equations to Painlevé equations. *J. Phys. Soc. Jpn.* **51** 1678–81 (1982).

Tajiri, M. and Nishitani, T. Two soliton resonant interactions in one spatial dimension: solutions of Boussinesq type equations. *J. Phys. Soc. Jpn.* **51** 3720–3 (1982).

Tajiri, M., Nishitani, T. and Kawamoto, S. Similarity solutions of the Kadomtsev–Petviashvili equation. *J. Phys. Soc. Jpn.* **51** 2350–6 (1982).

Takens, F. *Preprint 7907*, Department of Mathematics, Groningen University, Holland (1980).

Taniuti, T. and Washimi, H. Self-trapping and instability of hydrodynamic waves along the magnetic field in a cold plasma. *Phys. Rev. Lett.* **21** 209–12 (1968).

Taniuti, T. and Wei, C.C. Reductive perturbation method in nonlinear wave propagation, I. *J. Phys. Soc. Jpn.* **24** 941–6 (1968).

Taylor, G. The instability of liquid surfaces when accelerated in a direction perpendicular to their planes, I. *Proc. R. Soc. London* **A201** 192–6 (1950).

Taylor, R.J., MacKenzie, K.R. and Ikezi, H. A double-plasma device for plasma beam and wave studies. *Rev. Sci. Instrum.* **43** 1675–8 (1972).

Thyagaraja, A. Recurrent motions in certain continuum dynamical systems. *Phys. Fluids* **22** 2093–6 (1979).

Thyagaraja, A. Recurrence phenomena and the number of effective degrees of freedom in nonlinear wave motions. *Nonlinear waves.* Debnath, L. (ed.), Cambridge University Press 308–25 (1983).

Torvén, S. Modified Korteweg–de Vries equation for propagating double layers in plasmas. *Phys. Rev. Lett.* **47** 1053–6 (1981).

Torvén, S. Weak double layers in a current carrying plasma. *Physica Scripta* **33** 262–5 (1986).

Tran, M.Q. Ion acoustic solitons in a plasma: review of their experimental properties and related theories. *Physica Scripta* **20** 317–27 (1979).

Tsukabayashi, I., Nakamura, Y., Kako, F. and Lonngren, K.E. Oblique collision of cylindrical ion acoustic solitons. *Phys. Fluids* **26** 790–4 (1983).

Tsukabayashi, I., Nakamura, Y. and Ogino, T. Reflection of a cylindrical ion acoustic soliton at a symmetric axis. *Phys. Lett.* **81A** 507–10 (1981).

Tsukabayashi, I. and Yagishita, T. Propagation of circular solitary wave on shallow water. *J. Phys. Soc. Jpn.* **46** 1401–2 (1979).

Turpin, F.M., Benmoussa, C. and Mei, C.C. Effects of slowly varying depth and current on the evolution of a Stokes wavepacket. *J. Fluid Mech.* **132** 1–23 (1983).

Vakhitov, N.G. and Kolokolov, A.A. Stationary solutions of the wave equation in the medium with nonlinearity saturation. *Izv. VUZ; Radiofiz.* **16** 1020–8 (1973).

Vanden-Broeck, J.M. Nonlinear gravity-capillary stern waves. *Phys. Fluids* **23** 1949–53 (1980).

Verheest, F. Ion acoustic solitons at critical densities in multicomponent plasmas with different ionic charges and temperatures. *International Conference on Plasma Physics, Kiev 1987*, **2** 115–18 (1987).

Wadati, M. and Sawada, K. New representations of the soliton solution for the Korteweg de Vries equation. *J. Phys. Soc. Jpn.* **48** 312–18 (1980a).

Wadati, M. and Sawada, K. Application of the trace method to the modified Korteweg–de Vries equation. *J. Phys. Soc. Jpn.* **48** 319–25 (1980b).

Wahlquist, H.D. and Estabrook, F.B. Bäcklund transformation for solutions of the KdV eqn. *Phys. Rev. Lett.* **31** 1386–90 (1973).

Walker, J. Edge waves from a spokelike pattern when vibrations are set up in a liquid. *Scientific American* **253** Dec. 134–8 (1984).

Walshaw, A.C. and Jobson, D.A. *Mechanics of fluids.* Longman, London (1972).

Walstead, A.E. A study of nonlinear waves described by the cubic Schrödinger equation. Ph.D. thesis, Lawrence Livermore Laboratories (1980).

Wardrop, M.J. and ter Haar, D. The stability of three-dimensional planar Langmuir solitons. *Physica Scripta* **20** 493–501 (1979).

Washimi, M. and Taniuti, T. Propagation of ion-acoustic solitary waves of small amplitude. *Phys. Rev. Lett.* **17** 996–8 (1966).

Watanabe, S. Soliton generation of tail in nonlinear dispersive media with weak dissipation. *J. Phys. Soc. Jpn.* **45** 276–82 (1978).

Weiland, J. and Wilhelmsson, H. *Coherent nonlinear interaction of waves in plasmas.* Pergamon, Oxford and New York (1977).

Weiss, J. The Painlevé property for partial differential equations, II. *J. Math. Phys.* **24** 1405–13 (1983).

Weiss, J., Tabor, M. and Carnavale, G. The Painlevé property for partial differential equations. *J. Math. Phys.* **24** 522–6 (1983).

Wesson, J.A. MHD stability theory. Chapter 9, *Plasma physics and nuclear fusion*, Gill, R.D. (ed.). Academic, New York (1981).

Wharton, C.B., Malmberg, J.H. and O'Neil, T.M. Nonlinear effects of large-amplitude plasma waves. *Phys. Fluids* **11** 1761–3 (1968).

Whipple, R.T.P. *A class of equilibrium configurations of a current carrying plasma which fit into a space.* AERE-R Report 3325, Harwell (1960).

Whitham, G.B. Nonlinear dispersive waves. *Proc. R. Soc. London* **A238** 238–61 (1965a).

Whitham, G.B. A general approach to linear and nonlinear dispersive waves using a Lagrangian. *J. Fluid Mech.* **22** 273–83 (1965b).

Whitham, G.B. *Linear and nonlinear waves.* John Wiley, New York (1974).

Wilcox, J.Z. and Wilcox, T.J. Stability of localized plasma model in two and three dimensions. *Phys. Rev. Lett.* **34** 1160–3 (1975).

Williams, J.M. Limiting gravity waves in water of finite depth. *Phil. Trans. R. Soc. London* **A302** 139–88 (1981).

Wong, A.Y. and Quon, B.H. Spatial collapse of beam-driven plasma waves. *Phys. Rev. Lett.* **34** 1499–1502 (1975).

Woodruff, D.P. *The solid–liquid interface.* Cambridge University Press (1973).
Yajima, N. and Oikawa, M. Formation and interaction of sonic-Langmuir solitons. *Prog. Theo. Phys.* **56** 1719–39 (1976).
Yajima, N., Oikawa, M. and Satsuma, J. Interaction of ion-acoustic solitons in three-dimensional space. *J. Phys. Soc. Jpn.* **44** 1711–14 (1978).
Yuen, H.C. and Ferguson, W.E. Jr. Relationship between Benjamin–Feir instability and recurrence in the nonlinear Schrödinger equation. *Phys. Fluids* **21** 1275–8 (1978a).
Yuen, H.C. and Ferguson, W.E. Jr. Fermi–Ulam–Pasta recurrence in the two space dimensional nonlinear Schrödinger equation. *Phys. Fluids* **21** 2116–18 (1978b).
Yuen, H.C. and Lake, B.M. Nonlinear deep water waves: theory and experiment. *Phys. Fluids* **18** 956–60 (1975).
Yuen, H.C., Lake, B.M. and Ferguson, W.E. Jr. *The significance of nonlinearity in natural science.* Plenum, New York (1977).
Zabusky, N.J. and Kruskal, M.D. Interaction of 'solitons' in a collisionless plasma and the recurrence of initial states. *Phys. Rev. Lett.* **15** 240–3 (1965).
Zakharov, V.E. The instability of waves in nonlinear dispersive media. *Zh. Eksp. Teor. Fiz.* **51** 1107–14 (1966). *Sov. Phys. JETP* **24**, no. 4 (1967).
Zakharov, V.E. Stability of periodic waves of finite amplitude on the surface of a deep fluid. *J. Appl. Tech. Phys.* **9** 86–94 (1968).
Zakharov, V.E. Collapse of Langmuir waves. *Zh. Eksp. Teor. Fiz.* **62** 1745–59 (1972); *Sov. Phys. JETP* **35** 908–14 (1972).
Zakharov, V.E. and Kuznetsov, E.A. On three-dimensional solitons. *Zh. Eksp. Teor. Fiz.* **66** 594–7 (1974); *Sov. Phys. JETP* **39** 285–6 (1974).
Zakharov, V.E., Manakov, S.V., Novikov, S.P. and Pitayevski, L.P. *Teoriye solitonov*, Nauka, Moscow (1980); *Theory of solitons.* Plenum, New York (1984).
Zakharov, V.E. and Shabat, A.B. Exact theory of two-dimensional self-focusing and one-dimensional self modulation of waves in nonlinear media. *Zh. Eksp. Teor. Fiz.* **61** 118–34 (1971); *Sov. Phys. JETP* **34** 62–9 (1972).
Zakharov, V.E. and Shabat, A.B. A scheme for integrating the nonlinear equations of mathematical physics by the method of the inverse scattering problem, I. *Funct. Anal. Appl.* **8** 226–35 (1974).
Zakharov, V.E. and Shulman, E.I. Degenerative dispersion laws, motion invariants and kinetic equations. *Physica* **1D** 192–202 (1980).
Ze, F., Hershkowitz, N. Chan, C. and Lonngren, K.E. Excitation of spherical ion acoustic solitons with a conducting probe. *Phys. Fluids* **22** 1554–7 (1979a).
Ze, F., Hershkowitz, N., Chan, C. and Lonngren, K.E. Inelastic collisions of spherical ion-acoustic solitons. *Phys. Rev. Lett.* **42** 1747–50 (1979b).
Ziemkiewicz, J., Infeld, E. and Rowlands, G. Stability of nonlinear hydromagnetic waves and solitons in a two component plasma: a check on the validity of the Korteweg de Vries equation. *Acta Phys. Polon.* **A60** 457–72 (1981).

Author index

Subject index